R

Modern Regression Methods

Modern Regression Methods

THOMAS P. RYAN
Department of Statistics
Case Western Reserve University

A Wiley-Interscience Publication
JOHN WILEY & SONS, INC.
New York • Chichester • Brisbane • Toronto • Singapore • Weinheim

Library of Congress Cataloging in Publication Data:
Ryan, Thomas P.
    Modern regression methods / Thomas P. Ryan.
        p. cm. – (Wiley series in probability and statistics.
    Applied probability and statistics)
        "A Wiley-Interscience publication."
        Includes bibliographical references and index.
        ISBN 0-471-52912-5 (cloth : alk. paper)
        1. Regression analysis. I. Title. II. Series.
    QA278.2.R93 1996
    519.5'36–dc20   96-6368

PREFACE

Since regression is a popular research area, regression analysis is an ever-changing collection of techniques. Many standard regression methods have been shown to produce misleading results for certain data sets. This is especially true of ordinary least squares.

Students should be taught the regression methods that are considered to be the best at a given point in time, and these methods should ideally be used by practitioners. Since the quest for better methods should be a never-ending goal, a complete presentation of what is known and not known can motivate researchers to work on developing new methods or determining the properties of existing methods.

This book was written with these objectives in mind. A concerted effort has been made to present state-of-the-art regression methodology. The vast majority of the material is suitable for students taking a first regression course, but the nature of some of the subject matter is such that it will be more easily understood by researchers and practitioners who have used regression for some time.

The book contains a considerable amount of material not previously given in regression books. This includes standard concepts that are presented in new ways, useful techniques that have been given in research papers written within the past few years, and some new methods suggested by the author.

Chapter 1 is essentially a standard introduction to regression. Chapter 2 is primarily devoted to checking model assumptions, and the corrective action that should be taken when the assumptions appear not to be met. Much of the subject matter is at a noticeably higher level than the material presented in Chapter 1. Chapter 3 is an introduction to matrix algebra and to the application of matrix algebra to regression. While Chapter 4 is an introduction to multiple regression, most of the tools of multiple regression are presented in subsequent chapters. Graphical techniques for multiple regression are presented in detail in Chapter 5, and transformations are discussed in Chapter 6. Some new techniques are presented in each of these chapters. Chapter 7 is primarily a standard treatment of variable selection techniques. The use of polynomial and trigonometric terms in regression models is discussed in Chapter 8. A detailed treatment of logistic regression is provided in Chapter 9, and this includes a discussion of both exact logistic regression and the maximum likelihood approach. Nonparametric regression is discussed in Chapter 10, and some new methods are suggested. Chapter 11 is a modern and detailed treatment of robust regression. Chapter 12 is a standard treatment of ridge regression, and Chapter 13 is an introduction to nonlinear regression. Design of experiments for regression analyses is covered in Chapter 14 and equileverage designs receive some emphasis.

It is important that students and practitioners be able to apply what they have learned after completing a statistics book. Five real data sets are analyzed extensively in Chapter 15, drawing appropriate techniques from the methods presented in the preceding 14 chapters. The reader is encouraged to "work along" in the analysis stage, and see if agreement is reached.

The following suggestions are made for college instructors. The book can be used in an undergraduate course, provided that the students have had at least one rigorous course in basic statistical concepts. The book is not mathematically rigorous (for the most part), but parts of the book are conceptually rigorous. Those sections should probably be avoided in an undergraduate course. In general, the text should be most appropriate for undergraduates in statistics, math, and the engineering and physical sciences.

A one-semester undergraduate course should include the following material. All of Chapter 1 could be covered, except perhaps Sections 1.8, 1.8.1–1.8.3, and 1.9. Chapter 2 is a very important but long chapter. Accordingly, an instructor may wish to cover lightly or skip altogether some combinations of Sections 2.1.2.1, 2.1.2.3, 2.1.3.1–2.1.3.1.2, 2.3.2.1., 2.3.3, and 2.3.4.

All of Chapter 3 should be covered unless knowledge of matrix algebra is assumed, in which case Sections 3.1 and 3.1.1 would not be covered. All of Chapter 4 should be covered, and the instructor who covers any of Chapter 5 may want to concentrate on Sections 5.1.1 and 5.5.

Chapter 6 would perhaps not be covered but all of Chapter 7 could be covered. All of the remaining chapters would probably be skipped except for Chapter 15, and the sections that are covered in this chapter will depend on what material has been covered in the preceding chapters. If the outline given here were followed, Sections 15.1–15.3 and 15.6 could be covered, except for skipping the discussion at the end of Section 15.1 on ridge regression.

The book might be used as a text for a graduate statistics course, depending on the theoretical rigor that is desired. Chapters 9 and 11, in particular, are at the appropriate level for such a course, although an instructor might wish to supplement the theoretical material in some of the earlier chapters. The book would be especially suitable as a text for graduate students in math, business, the engineering sciences, and the physical sciences.

Many people have contributed directly or indirectly to this book. In particular, Eric Ziegel carefully read the entire manuscript and provided me with a very large number of helpful suggestions and minor corrections. David Ruppert, Ken Berk, Cyrus Mehta, and Nitin Patel have made helpful comments on individual chapters, and I have also benefited from discussions with Randy Eubank, Jon Cryer, Marty Puterman, Roger Hoerl, Steve Marron, and Frank Harrell. I am indebted to Arny Stromberg for providing me with his exact LMS program, and to Doug Hawkins for providing me with his program for LTS, in addition to certain manuscripts. Several anonymous reviewers have also made suggestions that have considerably improved the book.

I also wish to acknowledge the assistance of Jim Ashton of SAS Institute, Inc., Maresha Sceats and Jennifer Row of BMDP, Inc., Carey Gersten of Statistical Sciences, Inc., Lucy Saunders of Systat, Inc., and Robert Gruen of SPSS, Inc., in addition to Minitab, Inc., and Cytel Software Corporation.

I would like to express my appreciation to Kate Roach for initiating this project while she was an editor with Wiley, and the help that I have received from my current editors, Steve Quigley and Jessica Downey, and production editor Rosalyn Farkas, has also been greatly appreciated.

The finishing touches to the book were made while I was employed by the Department of Statistics, University of Newcastle, Australia and the Department of Statistics at Case Western Reserve University. The support and assistance that I received from these two universities is gratefully acknowledged.

THOMAS P. RYAN

September 1996

CONTENTS

CHAPTER 1

Introduction

The ability to predict the future would certainly be a highly desirable skill. Soothsayers have claimed to have this ability for centuries, but history has shown them to be quite fallible. No statistical technique can be used to eliminate or explain all of the uncertainty in the world, but statistics can be used to quantify that uncertainty. Unlike the soothsayers, the user of *regression analysis* can make predictions and place error bounds on those predictions. Regression can be used to predict the value of a *dependent variable* from knowledge of the values of one or more *independent variables*. The technical details of regression are given starting in Section 1.2.

The parents of a young man who has applied to a prestigious private university would certainly welcome information on what his grade-point average (GPA) might be after four years. If his predicted average was not high enough to enable him to graduate, his parents would obviously be more willing to pay his tuition at a different school at which he would be expected to graduate.

Similarly, it would be nice to know how long a certain make of automobile would be expected to last before needing major repairs, or to know the expected yield of a manufacturing process given a certain combination of inputs, or to know what a company's sales should be for a given set of expenses.

We see predictions being made all around us. Colleges and universities *do* use regression analysis to predict what a student's four-year college GPA would be if the student were admitted, and then make a decision regarding admittance from that predicted value and from other information. There is a vast amount of literature on this subject; one of the more recent papers is Graham (1991), who considered the prediction of GPA for MBA students. Other research papers include Paolillo (1982).

Regression analysis is one of the two most widely used statistical techniques (analysis of variance is the other), and it is used in almost every field of application. The following sample of titles of research papers provides some evidence of the range of applications (italics added).

"Extending applicable ranges of *regression* equations for yarn strength forecasting" (Smith and Waters, 1985).

1

"A Cox *regression* model for the relative mortality and its application to diabetes mellitus survival data" (Andersen et al., 1985).

"The estimation of percent body fat, body density and total body fat by maximum R^2 *regression* equations" (Mukherjee and Roche, 1984).

"Comparison of the Cox model and the *regression* tree procedure in analysing a randomized clinical trial" (Schmoor et al., 1993).

"Strategies for coupling digital filtering with partial least squares *regression*: Application to the determination of glucose in plasma by Fourier transform near-infrared spectroscopy" (Small et al., 1993).

"A review of *regression* diagnostics for behavioral research" (Chatterjee and Yilmaz, 1992).

"Aggregation bias and the use of *regression* in evaluating models of human performance" (Walker and Catrambone, 1993).

Simultaneous confidence and prediction intervals for nonlinear *regression* models with application to a groundwater flow problem" (Vecchia and Cooley, 1987).

"Applications of a simple *regression* model to acid rain data" (Stein et al., 1993).

"Multiple *regression* analysis, using body size and cardiac cycle length, in predicting echocardiographic variables in dogs" (Jacobs and Mahjoob, 1988).

"Multiple *regression* approach to optimize drilling operations in the Arabian Gulf area" (Al-Betairi et al., 1988).

"Performance of three *regression*-based models for estimating monthly soil temperatures in the Atlantic region" (Dwyer et al., 1988).

"The use of *regression* models to predict spatial patterns of cattle behavior" (Senft et al., 1983).

A complete listing of all of the research papers in which regression analysis has been applied in analyzing data from subject-matter fields would require a separate book. It has even been claimed that lifespan can be predicted using regression, and we examine this issue in Section 2.4.7.

In a nonstatistical context, the word *regression* means "to return to an earlier place or state," and reversion is listed in dictionaries as a synonym. We might then wonder how regression can be used to predict the future when the word literally means to go backward in some sense.

Regression analysis can be traced to Sir Francis Galton (1822–1911) who observed that children's heights tended to "revert" to the average height of the population rather than diverting from it. That is, the future generations of offspring who are taller than average are not progressively taller than their respective parents, and parents who are shorter than average do not beget successively smaller children.

Galton originally used the word *reversion* to describe this tendency and some years later used the word *regression* instead. This early use of the word *regression* in data analysis is unrelated, however, to what has become known as regression analysis.

The user of regression analysis attempts to discern the relationship between a dependent variable and one or more independent variables. That relationship will not be a functional relationship, however, nor can a cause-and-effect relationship necessarily be inferred. (Regression can be used when there is a cause-and-effect relationship, however.) The equation $F = \frac{9}{5}C + 32$ expresses temperature in Fahrenheit as a function of temperature measured on the Celsius scale. This represents an exact functional relationship; one in which temperature in Celsius could just as easily have been expressed as a function of temperature in Fahrenheit. Thus there is no clear choice for the dependent variable.

Exact relationships do not exist in regression analysis, and in regression the variable that should be designated as the dependent variable is usually readily apparent. As a simple example, let's assume that we want to predict college GPA using high school GPA. Obviously, college GPA should be related to high school GPA and should depend on high school GPA to some extent. Thus, college GPA would logically be the dependent variable and high school GPA the independent variable.

Throughout this book the independent variables will be referred to as regressors, predictors, or regression variables, and the dependent variable will occasionally be referred to as the response variable.

1.1 SIMPLE LINEAR REGRESSION MODEL

The word *simple* means that there is a single independent variable, but the word *linear* does not have the meaning that would seem to be self-evident. Specifically, it does not mean that the relationship between the two variables can be displayed graphically as a straight line. Rather, it means that the model is linear in the parameters.

The basic model is

$$Y = \beta_0 + \beta_1 X + \epsilon \qquad (1.1)$$

in which Y and X denote the dependent and independent variable, respectively, and β_0 and β_1 are parameters that must be estimated. The symbol ϵ represents the error term. This does not mean that a mistake is being made; it is simply a symbol used to indicate the absence of an exact relationship between X and Y.

The reader will recognize that, except for ϵ, Eq. (1.1) is in the general form of the equation for a straight line. That is, β_1 is the slope and β_0 is the Y-intercept.

1.2 USES OF REGRESSION MODELS

Once β_0 and β_1 have been estimated (estimation is covered in Section 1.4), the following *prediction equation* results:

$$\hat{Y} = \hat{\beta}_0 + \hat{\beta}_1 X \tag{1.2}$$

The "hats" (as they are called) above β_0 and β_1 signify that those parameters are being estimated, but the hat above Y means that the dependent variable is being predicted. [The reader will observe that there is no error term in Eq. (1.2), as we do not estimate error terms explicitly in regression models.] The equation would be used for values of X within the range of the sample data, or perhaps only slightly outside the range.

The prediction equation implies that prediction is one of the uses of regression analysis. Simply stated, the objective is to see if past data indicate a strong enough relationship between X and Y to enable future values of Y to be well predicted. The rationale is that if past values of Y can be closely fit by the prediction equation, then future values should similarly be closely predicted. As mentioned previously, a prediction equation can be used to predict what a student's college GPA would be after four years if he or she were admitted to a particular college or university.

Regression can also be used for the related purposes of estimation and description. Specifically, once $\hat{\beta}_0$ and $\hat{\beta}_1$ have been obtained, these parameter estimates can be used to describe the relationship between Y and X. For example, $\hat{\beta}_1$ is the amount, positive or negative, by which we would predict Y to change per unit change in X, *for the range of X* in the sample that was used to determine the prediction equation. Similarly, $\hat{\beta}_0$ is the value that we would predict for Y when $X = 0$.

A seldom-mentioned but important use of regression analysis is for control. For example, Y might represent a measure of the pollutant contamination of a river, with the regressor(s) representing the (controllable) input pollutant(s) being varied (e.g., reduced) in such a way as to control Y at a tolerable level. The use of regression for control purposes is discussed by Hahn (1974) and Draper and Smith (1981, p. 413) and is discussed in Section 1.8.1.

1.3 GRAPH THE DATA!

The use of the model given in Eq. (1.1) assumes that we have already selected the model. As George E.P. Box has often stated: "All models are wrong, but some are useful." Any statistical analysis should begin with graphic displays of the data, and these displays can help us select a useful model.

The importance of graphing regression data is perhaps best illustrated by

**Table 1.1   Four Data Sets Given by Anscombe (1973)
That Illustrate the Need to Plot Regression Data**

Data Set:	1–3	1	2	3	4	4
Variable:	X	Y	Y	Y	X	Y
	10	8.04	9.14	7.46	8	6.58
	8	6.95	8.14	6.77	8	5.76
	13	7.58	8.74	12.74	8	7.71
	9	8.81	8.77	7.11	8	8.84
	11	8.33	9.26	7.81	8	8.47
	14	9.96	8.10	8.84	8	7.04
	6	7.24	6.13	6.08	8	5.25
	4	4.26	3.10	5.39	19	12.50
	12	10.84	9.13	8.15	8	5.56
	7	4.82	7.26	6.42	8	7.91
	5	5.68	4.74	5.73	8	6.89

Anscombe (1973), who showed that completely different graphs can correspond to the same regression equation and summary statistics. In particular, Anscombe showed that the graph of a data set that clearly shows the need for a quadratic term in X can correspond to the same regression equation as a graph, which clearly indicates that a linear term is sufficient. The four data sets are given in Table 1.1, and the reader is asked to compare the four (almost identical) regression equations against the corresponding scatter plots in Exercise 1.1.

Consider the data given in Table 1.2. Assume that the data have come from an industrial experiment in which temperature is varied from 375° to 420°F in 5° increments, with the temperatures run in random order and the process yield recorded for each temperature setting. (It is assumed that all other factors are being held constant.) The scatter plot for these data is given in Figure 1.1. The

Table 1.2   Process Yield Data

Y (yield)	X (°F)
26.15	400
28.45	410
25.20	395
29.30	415
24.35	390
23.10	385
27.40	405
29.60	420
22.05	380
21.30	375

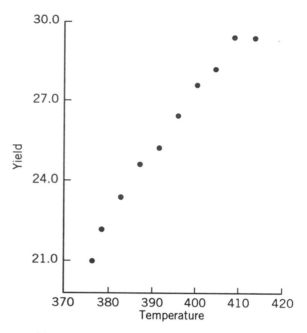

Figure 1.1   Scatter plot of the data in Table 1.2.

plot shows a strong linear relationship between X and Y, although there is some hint of curvature at the extreme values of X. Therefore, the model given by Eq. (1.1) is a good starting point, possibly to be modified later.

1.4   ESTIMATION OF β_0 AND β_1

Point estimates of β_0 and β_1 are needed to obtain the prediction equation given in Eq. (1.2). A crude approach would be to draw a line through the center of the points and then use the slope and Y-intercept of the line as the estimates of β_1 and β_0, respectively.

Before one could even attempt to do so, however, it would be necessary to define what is meant by "center." We could attempt to minimize the sum of the slant distances (with each distance measured from the point to the line), but since we will be using X to predict Y, it would make more sense to try to minimize the sum of the vertical distances. If we use the signs of those distances, we have a problem, however, because then the line is not uniquely defined (as the reader is asked to show in Exercise 1.2).

One way to eliminate this problem would be to minimize the sum of the absolute values of the distances [see, e.g., Birkes and Dodge (1993)]. This has been suggested as a way to deal with extreme X-values, and is discussed briefly in Chapter 11.

The standard approach, however, is to minimize the sum of the squares of the vertical distances, and this is accomplished by using the *method of least squares*. For the purpose of illustration we must assume that Eq. (1.1) is the correct model, although as stated previously the correct model will generally be unknown.

The starting point is to write the model as

$$\epsilon = Y - (\beta_0 + \beta_1 X) \tag{1.3}$$

Since ϵ represents the vertical distance from the observed value of Y to the line represented by $Y = \beta_0 + \beta_1 X$ that we would have if β_0 and β_1 were known, we want to minimize $\sum \epsilon^2$.

For convenience we define L as

$$L = \sum_{i=1}^{n} \epsilon_i^2 = \sum_{i=1}^{n} (Y_i - \beta_0 - \beta_1 X_i)^2 \tag{1.4}$$

where n denotes the number of data points in a sample that has been obtained. To minimize L, we take the partial derivative of L with respect to each of the two parameters that we are estimating and set the resulting expressions equal to zero. Thus,

$$\frac{\partial L}{\partial \beta_0} = 2 \sum_{i=1}^{n} (Y_i - \beta_0 - \beta_1 X_i)(-1) = 0 \tag{1.5a}$$

and

$$\frac{\partial L}{\partial \beta_1} = 2 \sum_{i=1}^{n} (Y_i - \beta_0 - \beta_1 X_i)(-X_i) = 0 \tag{1.5b}$$

Dropping the 2 and the -1 from Eqs. (1.5a) and (1.5b), the solutions for β_0 and β_1 would be obtained by solving the equations (which are generally called *normal equations*)

$$\sum_{i=1}^{n} (Y_i - \beta_0 - \beta_1 X_i) = 0$$

and

$$\sum_{i=1}^{n} (X_i Y_i - X_i \beta_0 - \beta_1 X_i^2) = 0$$

which become

$$n\beta_0 + \beta_1 \sum_{i=1}^{n} X_i = \sum_{i=1}^{n} Y_i$$

and

$$\beta_0 \sum_{i=1}^{n} X_i + \beta_1 \sum_{i=1}^{n} X_i^2 = \sum_{i=1}^{n} X_i Y_i$$

The solution of these two equations (for the estimators of β_0 and β_1) produces the least-squares estimators

$$\hat{\beta}_1 = \frac{\sum X_i Y_i - (\sum X_i)(\sum Y_i)/n}{\sum X_i^2 - (\sum X_i)^2/n} \tag{1.6}$$

and

$$\hat{\beta}_0 = \overline{Y} - \hat{\beta}_1 \overline{X}$$

(The astute reader will recognize that L in Eq. (1.4) is not automatically minimized just by setting the first derivatives equal to zero and solving the resultant equations. It can be shown, however, that the determinant of the matrix of second-order partial derivatives of L is positive, thus ensuring that a minimum, rather than a maximum, has been attained. Matrix algebra for regression is covered in Chapter 3.)

Notice that in Eq. (1.6) we have dropped the beginning and ending points of the summation and have simply used $\sum$. We shall do so frequently in this and subsequent chapters, as all summations hereinafter will be assumed to start with 1 and end with n, unless indicated otherwise. We will also drop the subscript i in most summation expressions.

If computations were performed by hand calculator, $\hat{\beta}_1$ would obviously have to be computed before $\hat{\beta}_0$, since the latter is obtained using the former.

It is advantageous (and space saving, in particular) to use shorthand notation to represent the right side of Eq. (1.6). The symbol S_{xy} can be used to represent the numerator, and S_{xx} to represent the denominator. The S can be thought of as representing "sum," and xx represents X times X, which of course equals X^2. In statistical jargon, S_{xx} represents the "corrected" sum of squares for X, and S_{xy} denotes the corrected sum of products of X and Y. That is, $S_{xx} \neq \sum X^2$ and $S_{xy} \neq \sum XY$. Rather, $S_{xx} = \sum (X - \overline{X})^2$ and $S_{xy} = \sum (X - \overline{X})(Y - \overline{Y})$, where these two expressions can be shown to be equal to the denominator and numerator, respectively, of Eq. (1.6). Thus the "correction" factors are $\overline{X}$ and $\overline{Y}$. Also, $S_{yy} = \sum (Y - \overline{Y})^2$.

When the calculations are performed using the data in Table 1.2, the following results are obtained:

$$\hat{\beta}_1 = \frac{102,523.5 - (3975)(256.9)/10}{1,582,125 - (3975)^2/10}$$

$$= \frac{405.75}{2062.5}$$

$$= 0.19673$$

and

$$\hat{\beta}_0 = 25.69 - 0.19673(397.5)$$
$$= -52.51$$

The regression equation is thus $\hat{Y} = -52.51 + 0.19673X$.

1.5   INFERENCES FROM REGRESSION EQUATIONS

What can we infer from this equation? Since the data have come from a designed experiment, we can state that, *within the range of temperatures used in the experiment*, process yield should increase by approximately 0.2 ($\times 100$) units per single degree increase in temperature. This assumes that all controllable factors that might influence process yield have been held constant during the experiment, and that the randomization of temperatures has prevented any "lurking variables" from undermining the inferences made from the data. (Lurking variables are factors that can have a systematic effect on the response variable; see Section 5.7 for further discussion of lurking variables.) See Chapter 14 for a discussion of inferences that can be made from controlled experiments versus the conclusions that can be drawn from observational studies. The latter refers to obtaining data without any intervention on the part of the regression user, such as taking a sample of data from a college registrar's records. From Table 1.2 we can see that if the temperatures are ordered from 375° to 420°F, the average change in process yield for each degree increase in temperature is 0.184, which differs only slightly from $\hat{\beta}_1$ (as we might expect).

In a study of this type the primary benefit to be derived from this experiment would be a better understanding of how temperature affects process yield and to gain a better understanding of what might seem to be the optimal temperature setting, as is done in Evolutionary Operation [see, e.g., Box and Draper (1969) or Ryan (1989)]. Notice, however, that this cannot be determined from the data in Table 1.2, as the yield is strictly increasing as temperature increases. It would appear, though, that the optimal temperature setting may be just slightly greater than 420°, as yield does level off somewhat as temperature is increased from 415° to 420°.

1.5.1   Predicting Y

To illustrate the other uses of regression analysis and the inferences that we can draw from such an analysis, let's assume that we want to use the regression equation to predict Y. That is, we are interested in predicting process yield for different temperature settings.

Let's assume that we would like to be able to predict what yield should be when the temperature is 400°. A very simple approach would be to use 26.2 as the predicted value since that was the process yield that resulted when temperature was set at 400° in the experiment. To do so, however, would be to ignore the question "Is there a relationship between X and Y?" and, if a relationship does exist, to not utilize the extent of that relationship.

Consider this. Assume that a scatter plot of Y versus X resulted in a random configuraton of points such that the line represented by the regression equation would have a slope of zero (which would be a horizontal line). It would then be foolish to use any particular value of Y in the data set as the predicted value of Y for the corresponding value of X because the regression equation would be telling us that there is no relationship between X and Y and that we might as well use the average value of Y in the data set in predicting Y. (Notice that this is what happens algebraically when $\hat{\beta}_1 = 0$, since $\hat{\beta}_0 = \overline{Y} - \hat{\beta}_1\overline{X}$ and $\hat{Y}$ then equals $\hat{\beta}_0$, which in this case equals $\overline{Y}$.)

Thus, there is no point in trying to predict Y from X unless there is evidence of some type of a relationship (linear or nonlinear) between the two, and the apparent strength of the relationship suggests that Y will be well predicted from X.

How do we discern the strength of the relationship between X and Y? Figure 1.1 shows that there is a strong linear relationship between X and Y, and when this is the case the observed values in Table 1.2 should be very close to the fitted values. In particular, when $X = 400$, $\hat{Y} = -52.51 + 0.19673(400) = 26.18$, which hardly differs from $Y = 26.15$ in Table 1.2 [We shall use the term *fitted values* to denote the $\hat{Y}$ values for the sample and will use the term *predicted value* to represent the prediction of a future (i.e., unobservable) value of Y].

We would expect the other fitted values to be close to the observed values, and Table 1.3 shows that this is true. We can see that all of the fitted values are close to the observed values, with the largest difference occurring at $X = 420$. The latter is to be expected since the difference in the Y values for $X = 415$ and

Table 1.3 Y and $\hat{Y}$ Values for Table 1.2 Data

X	Y	$\hat{Y}$	$Y - \hat{Y}$	$(Y - \hat{Y})^2$
400	26.15	26.18	−0.03	0.0009
410	28.45	28.15	0.30	0.0900
395	25.20	25.20	0.00	0.0000
415	29.30	29.13	0.17	0.0289
390	24.35	24.21	0.14	0.0196
385	23.10	23.23	−0.13	0.0169
405	27.40	27.17	0.23	0.0529
420	29.60	30.12	−0.52	0.2704
380	22.05	22.25	−0.20	0.0400
375	21.30	21.26	0.04	0.0016
			0.00	0.5212

$X = 420$ is noticeably smaller than any of the other differences for successive X values.

It should be noted that the values for $(Y - \hat{Y})^2$ differ slightly from the values that would be obtained if a computer (or more decimal places) were used, as the given values are computed directly from the two-decimal-place values for $Y - \hat{Y}$. This also means that subsequent calculations in this chapter that utilize $\sum (Y - \hat{Y})^2$ will also differ slightly from the results produced by a computer.

It can also be observed that $\sum (Y - \hat{Y}) = 0$. This is due to Eq. (1.5a), since that equation can be written, with the parameters replaced by their respective estimators, as $\sum [Y - (\hat{\beta}_0 + \hat{\beta}_1 X)] = 0 = \sum (Y - \hat{Y})$. Thus, since the second of the two equations that were solved is not involved, having $\sum (Y - \hat{Y}) = 0$ does not ensure that the correct values for $\hat{\beta}_0$ and $\hat{\beta}_1$ were obtained, as the reader is asked to investigate in Exercise 1.2.

For the correct $\hat{\beta}_0$ and $\hat{\beta}_1$ values, $\sum (Y - \hat{Y})^2$ is minimized. This is because the method of least squares was used with the initial intent being to minimize $\sum \epsilon^2 = \sum (Y - \beta_0 - \beta_1 X)^2$. We cannot observe the values for ϵ, however, since β_0 and β_1 are unknown. What we do observe are the e values, where $e = Y - \hat{Y}$ is called a *residual*, and $\sum e^2 = \sum (Y - \hat{\beta}_0 - \hat{\beta}_1 X)^2$ is minimized for a given set of data. (Obviously, e is defined only when $\hat{Y}$ denotes a fitted value rather than a predicted value yet to be observed.)

1.5.2   Worth of the Regression Equation

For this example it is clear that the regression equation has value for prediction (or for control or descriptive purposes).

A scatter plot will not always show as obvious a linear relationship as was seen in Figure 1.1, however, and the size of the $Y - \hat{Y}$ values will generally depend on the magnitude of Y. Consequently, it would be helpful to have a numerical measure that expresses the strength of the linear relationship between X and Y.

A measure of the variability in Y is $\sum (Y - \overline{Y})^2$. Since

$$\sum (Y - \overline{Y})^2 = \sum (Y - \hat{Y})^2 + \sum (\hat{Y} - \overline{Y})^2 \qquad (1.7)$$

as is shown in the chapter appendix, it would be logical to use either $\sum (Y - \hat{Y})^2$ or $\sum (\hat{Y} - \overline{Y})^2$ as a measure of the worth of the prediction equation, and divide the one that is used by $\sum (Y - \overline{Y})^2$ so as to produce a unit-free number.

It is a question of whether we want the measure to be large or small. If we define

$$R^2 = \frac{\sum (\hat{Y} - \overline{Y})^2}{\sum (Y - \overline{Y})^2} \qquad (1.8)$$

then R^2 represents the percentage of the variability in Y [as represented by $\sum (Y - \overline{Y})^2$] that is explained by using X to predict Y. Notice that if $\hat{\beta}_1 = 0$, then $\hat{Y} = \overline{Y}$ and $R^2 = 0$. At the other extreme, R^2 would equal 1 if $Y = \hat{Y}$ for each value of Y in the data set. Thus, $0 \leq R^2 \leq 1$ and we would want R^2 to be as close to 1 as possible.

From the data in Table 1.3,

$$R^2 = \frac{\sum (\hat{Y} - \overline{Y})^2}{\sum (Y - \overline{Y})^2}$$

$$= 1 - \frac{\sum (Y - \hat{Y})^2}{S_{yy}} \quad \text{[from Eq. (1.7)]} \qquad (1.9)$$

$$= 1 - \frac{0.5212}{80.3390}$$

$$= 0.9935$$

Thus, R^2 is very close to 1 in this example.

How large must R^2 be for the regression equation to be useful? That depends upon the area of application. If we could develop a regression equation to predict the stock market (which unfortunately we cannot), we would be ecstatic if $R^2 = 0.50$. On the other hand, if we were predicting college GPA, we would want the prediction equation to have strong predictive ability, since the consequences of a poor prediction could be quite serious.

Although R^2 is a well-accepted measure, a few criticisms can be made both for one X and for more than one X. For example, with a single X the slope of the plotted points will affect R^2, as is discussed by Barrett (1974). The argument is as follows. Consider the expression for R^2 given in Eq. (1.9). If we let the closeness of $\hat{Y}_i$ to Y_i be fixed so that $\sum (Y_i - \hat{Y}_i)^2$ remains constant, but then rotate the configuration of points so that the slope of the regression line is increased, R^2 must increase because $\sum (Y - \overline{Y})^2$ will increase. Under the assumption that the values of X are preselected, Ranney and Thigpen (1981) show that the expected value of R^2 can also be increased by increasing the range of X. The value of R^2 may also be artificially large if the sample size is small relative to the number of parameters. Also, as pointed out by Draper and Smith (1981, p. 91), when there are repeated X values, it is the number of *distinct* X values relative to the number of parameters that is important, not the sample size. Another disturbing feature of R^2 is that we can expect it to increase for each variable that is added to a known (true) model. (This result is discussed in the chapter appendix.)

Despite these shortcomings, R^2 has value as a rough indicator of the worth of a regression model. Another form of R^2, R^2_{adjusted}, is sometimes used. Since

$$R^2_{\text{adjusted}} = 1 - \frac{\text{SSE}_p/(n - p)}{\text{SST}/(n - 1)}$$

with p denoting the number of parameters in the model, R^2_{adjusted} can decrease as p is increased if $n - p$ declines at a faster rate than SSE_p. The latter represents $\sum e^2$ and $\text{SST} = \sum (Y - \overline{Y})^2$. Thus, R^2_{adjusted} might be used to determine the "point of diminishing returns." R^2_{adjusted} is discussed further in Section 7.4.1.

1.5.3 Regression Assumptions

What has been presented to this point in the chapter has hardly involved statistics. Rather, calculus was used to minimize a function, and the estimators that resulted from this minimization, $\hat{\beta}_0$ and $\hat{\beta}_1$, were used to form the prediction equation. Therefore, there are very few assumptions that need to be made at this point.

One assumption that obviously must be made is that the model given in Eq. (1.1) is a suitable proxy for the "correct" (but unknown) model. We also need to assume that the variance of ϵ_i [i.e., $\text{Var}(\epsilon_i)$] is the same for each value of i ($i = 1, 2, \ldots, n$). If this requirement is not met, then *weighted least squares* (see Section 2.1.3.1) should be used.

A typical analysis of regression data entails much more than the development of a prediction equation, however, and additional inferences require additional assumptions. These assumptions are stated in terms of the error term ϵ in the regression model. One very important assumption is that the error terms are uncorrelated. Specifically, any pair of errors (ϵ_i, ϵ_j) should be uncorrelated, and the errors must also be uncorrelated with X. This means that if we knew the value of ϵ_i, that value would not tell us anything about the value of ϵ_j. Similarly, the value of ϵ_i should not depend upon the value of X_i.

The assumption of uncorrelated errors is frequently violated when data are collected over time, and the consequences can be serious (see Chapter 2). Another important assumption is that ϵ should have approximately a normal distribution. The method of least squares should not be used when the distribution of ϵ is markedly nonnormal. For example, positive and negative residuals with equal absolute values should clearly not be weighted the same if the distribution of the errors is not symmetric. As discussed by Rousseeuw and Leroy (1987, p. 2) and others, Gauss introduced the normal distribution as the optimal error distribution for least squares, so using least squares when the error distribution is known (or believed) to be nonnormal is inappropriate. Although some theoretical statisticians would argue that asymptotic results support the use of least squares for nonnormal error distributions when certain conditions are met, such a stance will often lead to very bad results. It is important to realize that the methods of Chapter 11 (for nonnormal error distributions) will often be needed. With any statistical analysis (using regression or some other technique) it is a good idea to analyze the data first assuming that the assumptions are met and then analyze the data not assuming that the assumptions are met. If the results differ considerably, then the results of the second analysis will generally be the more reliable.

This last point cannot be overemphasized, as the confidence intervals, prediction interval, and hypothesis tests that are presented in subsequent sections are sensitive (i.e., not robust) to more than a slight departure from normality [see, e.g., Rousseeuw and Leroy (1987, p. 41) or Hamilton (1992, p. 113)]. Nonnormality, bad data, and good data that are far removed from the rest of the data can create serious problems. Consequently, the regression user and serious student of regression are urged to study Chapter 11 carefully. Methods for checking the assumptions of normality, independence, and a constant error variance are given in Chapter 2.

Another assumption on ϵ is that the mean of ϵ is zero. This assumption is never checked; it simply states that the "true" regression line goes through the center of a set of data. These assumptions on ϵ can be represented by $\epsilon \sim$ NID$(0, \sigma^2)$, with the additional assumption of normality meaning that the ϵ_i, ϵ_j are not only uncorrelated but are also independent [i.e., normal and independent (NID)].

Another assumption that is necessary for the theory that immediately follows, but is not necessary for regression analysis in general, is for the values of X to be selected by the experimenter rather than being allowed to occur at random. (The use of regression analysis when X is random is discussed in Section 1.9.)

When X is fixed, the assumptions on ϵ translate into similar assumptions on Y. This is because $\beta_0 + \beta_1 X$ is then an (unknown) constant, and adding a constant to ϵ causes $Y = \beta_0 + \beta_1 X + \epsilon$ to have the same distribution and variance as ϵ. Only the mean is different. Specifically, for a given value of $X, Y \sim N(\beta_0 + \beta_1 X, \sigma^2)$, and the assumption of uncorrelated errors means that the Y_i, Y_j are also uncorrelated. As with the errors, the assumption of normality allows us to state further that the Y_i, Y_j are also *independent*. These assumptions of normality and independence for Y are necessary for the inferential procedures that are presented in subsequent sections. [In subsequent sections and chapters we will write Var(Y) to represent the conditional variance, Var($Y|X$), as the variance of Y is assumed to be the same for each value of X.]

1.5.4 Inferences on β_1

Confidence intervals and hypothesis tests are covered in introductory statistics courses, and the relationship between them is often discussed. Although the information provided by a hypothesis test is also provided by a confidence interval, the confidence interval provides additional information. Obviously, it would be helpful to have a confidence interval on the true rate of change of Y per unit change in X, acting as if Eq. (1.1) is the true model.

To obtain a confidence interval for β_1, we need an estimate of the standard deviation of $\hat{\beta}_1$, after first deriving the expression for the standard deviation. The latter can be obtained as follows. We first write $\hat{\beta}_1$ as a linear combination of the Y values. Specifically,

$$\hat{\beta}_1 = \frac{\sum (X_i - \overline{X})(Y_i - \overline{Y})}{\sum (X_i - \overline{X})^2}$$

$$= \frac{\sum (X_i - \overline{X})Y_i}{S_{xx}} \tag{1.10}$$

with the second line resulting from the fact that $\sum (X_i - \overline{X})\overline{Y} = 0$. [Notice that the expression for $\hat{\beta}_1$ in Eq. (1.10) differs from the expression in Eq. (1.6). The latter is used for hand calculation; the reader is asked to show the equivalence of the two expressions in Exercise 1.3.]

Using the expression in the second line, we may write $\hat{\beta}_1 = \sum k_i Y_i$, with $k_i = (X_i - \overline{X})/S_{xx}$. Recall that Y is assumed to have a normal distribution, so the fact that $\hat{\beta}_1$ can be written as a linear combination of normally distributed random variables means that $\hat{\beta}_1$ also has a normal distribution. Furthermore, writing $\hat{\beta}_1$ in this manner makes it easy to obtain its standard deviation.

We may also use this expression for $\hat{\beta}_1$ to show that it is an unbiased estimator of β_1, as is shown in the chapter appendix. In general, an *unbiased estimator* is one for which the expected value of the estimator is equal to the parameter, the unknown value of which the value of the estimator serves to estimate. It will be seen later in this section that $\hat{\beta}_1$ is in the center of the confidence interval for β_1. This would be illogical if $\hat{\beta}_1$ were a biased estimator, and the amount of the bias were known. It should be noted that the unbiasedness property of $\hat{\beta}_1$ is based upon the assumption that the fitted model is the true model. This is a rather strong assumption, which will generally be false. But since the true model is unknown, we do not know the extent to which $\hat{\beta}_1$ is biased. Therefore, the usual textbook approach is to regard the $\hat{\beta}_i$ as being unbiased when the method of least squares is used.

We initially obtain the variance of $\hat{\beta}_1$ as

$$
\begin{aligned}
\text{Var}(\hat{\beta}_1) &= \text{Var}\left(\sum k_i Y_i \right) \\
&= \sum \text{Var}(k_i Y_i) \\
&= \sum k_i^2 \text{Var}(Y_i) \\
&= \sigma^2 \sum k_i^2 \\
&= \frac{\sigma^2}{S_{xx}}
\end{aligned}
$$

Thus, the standard deviation of $\hat{\beta}_1$ is $\sigma/\sqrt{S_{xx}}$, so the estimated standard deviation is $\hat{\sigma}/\sqrt{S_{xx}}$. Therefore, we need to estimate σ. (The estimated standard deviation of an estimator is frequently called the *standard error* of the estimator.)

Before doing so, however, we need to clearly understand what σ and σ^2 represent. To this point in the chapter (and in Section 1.5.3, in particular) σ^2 has been used to represent $\text{Var}(\epsilon_i)$ and $\text{Var}(Y_i|X_i)$. Can we state that σ^2 is also the variance of Y? Not quite.

Our interest is in the variance of Y given X, not the variance of Y ignoring X. Whereas we could estimate σ_y^2 as $s_y^2 = \sum (Y - \overline{Y})^2/(n - 1)$, this would ignore the fact that X is being used to predict Y. Consequently, the estimate of σ_ϵ^2 should be less than s_y^2, and the variance of Y that we should speak of is, in statistical parlance, the conditional variance of Y given X, as was stated previously, remembering that the variance is assumed to be the same for each value of X. We need not be concerned about the latter, however, since it is equal to σ_ϵ^2 when the postulated model is the true model. As stated previously, we cannot expect to know the true model, but we act as if these two variances are the same in the absence of information that would suggest that the model given by Eq. (1.1) is not an appropriate model.

In estimating σ_ϵ^2 we can think about how the variance of a random variable is estimated in an introductory statistics course. For a sample of observations on some random variable, W, σ_w^2 is estimated by $s_w^2 = \sum (W - \overline{W})^2/(n - 1)$, with the divisor making s_w^2 an unbiased estimator of σ_w^2.

If we proceed to estimate σ_ϵ^2 in an analogous manner, we would compute $s_e^2 = \sum (e - \overline{e})^2/(n - 2)$. Note that e is substituted for ϵ since ϵ is not observable.

Notice also that the divisor is $n-2$ instead of $n-1$. A divisor of $n-2$ is needed to make s_e^2 an unbiased estimator of σ_ϵ^2, as is shown in the chapter appendix. (Another reason why the divisor must be $n-2$ is given later in this section.)

Since $e = Y - \hat{Y}$ and $\bar{e} = 0$, s_e^2 can be written as $s_e^2 = \sum (Y - \hat{Y})^2/(n-2)$. Putting these results together, we obtain

$$s_{\hat{\beta}_1} = \frac{s_e}{\sqrt{S_{xx}}}$$

$$= \sqrt{\frac{\sum (Y - \hat{Y})^2}{(n-2)S_{xx}}}$$

If we write

$$t = \frac{\hat{\beta}_1 - \beta_1}{s_{\hat{\beta}_1}}$$

it can be shown that t has a Student's-t distribution with $n-2$ degrees of freedom (see the chapter appendix).

It follows that

$$P\left(-t_{\alpha/2, n-2} \leq \frac{\hat{\beta}_1 - \beta_1}{s_{\hat{\beta}_1}} \leq t_{\alpha/2, n-2}\right) = 1 - \alpha \qquad (1.11)$$

provides the starting point for obtaining a $(1-\alpha)\%$ confidence interval for β_1, with the value of $t_{\alpha/2, n-2}$ obtained from a t table with a tail area of $\alpha/2$ and $n-2$ degrees of freedom. Rearranging Eq. (1.11) so as to give the end points of the interval produces

$$P(\hat{\beta}_1 - t_{\alpha/2, n-2}s_{\hat{\beta}_1} \leq \beta_1 \leq \hat{\beta}_1 + t_{\alpha/2, n-2}s_{\hat{\beta}_1}) = 1 - \alpha$$

The lower limit (LL) is thus $\hat{\beta}_1 - t_{\alpha/2, n-2}s_{\hat{\beta}_1}$ and the upper limit (UL) is $\hat{\beta}_1 + t_{\alpha/2, n-2}s_{\hat{\beta}_1}$.

For the data in Table 1.2, the values of $\sum (Y - \hat{Y})^2$ and S_{xx} were given previously. We thus have

$$s_{\hat{\beta}_1} = \sqrt{\frac{\sum (Y - \hat{Y})^2}{(n-2)S_{xx}}}$$

$$= \sqrt{\frac{0.5212}{(8)2062.5}}$$

$$= 0.0056$$

A 95% confidence interval for β_1 would then have LL = 0.19673 − 2.306(0.0056) = 0.1838, and UL = 0.19673 + 2.306(0.0056) = 0.2096.

We would certainly want our confidence interval to not include zero because if $\beta_1 = 0$ there would be no regression equation. Since R^2 was quite large (0.9935), we would expect the interval not to include zero. The converse is not true, however. That is, if the interval does not include zero, R^2 could be much less than one, and could even be less than 0.5. This will be discussed further in the context of hypothesis testing.

Statistics books that cover regression analysis generally present a hypothesis test of $\beta_1 = 0$ in which the null and alternative hypotheses are $H_0: \beta_1 = 0$ and $H_a: \beta_1 \neq 0$, respectively. The rejection of H_0, using a (typical) significance level of $\alpha = .05$ or $\alpha = .01$, will not ensure that the regression equation will have much value, however.

This can be demonstrated as follows, continuing with the current example. For testing $H_0: \beta_1 = 0$,

$$t = \frac{\hat{\beta}_1 - 0}{s_{\hat{\beta}_1}}$$

$$= \frac{0.19673}{0.0056}$$

$$= 35.13$$

Since $t_{8,.025} = 2.305$ and $t_{8,.005} = 3.355$, we would reject $H_0: \beta_1 = 0$ at either significance level. As inferred earlier, the hypothesis test and confidence interval for β_1 are *not* robust (i.e., not insensitive) to nonnormality or to extreme data points that can result from nonnormality. Therefore, the hypothesis test and confidence interval should be used with caution, with methods such as those given in Chapter 2 used for detecting a clear departure from normality.

With some algebra we can show (see chapter appendix) that

$$R^2 = \frac{t^2}{n - 2 + t^2} \tag{1.12}$$

so $R^2 < 0.50$ if $t^2 < n - 2$. If the calculated value of t had equaled $t_{8,0.25}$, then R^2 would have equaled 0.3993. Using a significance level of $\alpha = .01$ would not help much, because if $t = t_{8,.005}$ then $R^2 = 0.5845$.

It should thus be clear that for the regression equation to be of value (for not only prediction but also for the other uses of regression), the calculated value of t should exceed some multiple of the tabular value. If we adapt to a t-test the recommendation of Draper and Smith (1981, p. 133), which is based upon the work of Wetz (1964), we obtain the result that the calculated t-value should be *at least* twice the tabular value.

For the current example, if $t \geq 2t_{8,.025}$ then $R^2 \geq 0.7267$, and if $t \geq 2t_{8,.005}$ then $R^2 \geq 0.8491$. Obviously these numbers look better than the pair of R^2 values given previously. As stated earlier, there is no clear dividing line between high and low R^2 values, but we can see that the "double t" rule could result in some rather small R^2 values being declared acceptable when n is much larger than 10. For example, for $n = 20, t \geq 2t_{18,.025}$ implies $R^2 \geq 0.4952$, whereas $t \geq 3t_{18,.025}$ implies $R^2 \geq 0.6882$. Therefore, 3 might be a more suitable multiplier if n is much larger than 10.

How large should n be? This decision is also somewhat arbitrary, just as are the choices for α and the multiplier of t. Draper and Smith (1981, p. 417) suggest that n should be at least equal to 10 times the number of regressors. Thus, we might consider having $n \geq 10$ in simple regression. This should be considered as only a very rough rule-of-thumb, however, as data are expensive in certain fields of application, such as the physical sciences. As stated by Frank and Friedman (1992), "in chemometrics applications the number of predictor variables often (greatly) exceeds the number of observations."

Because of the relationship between confidence intervals and hypothesis tests for β_1, the results that were stated in terms of hypothesis tests also apply to confidence intervals. Specifically, if $t = t_{8,.025}$ the lower limit of the 95% confidence interval would be zero, and if $t > t_{8,.025}$ the lower limit would exceed zero (and similarly the upper limit would be less than zero if $t < -t_{8,.025}$). Thus, a 95% (or 99%) confidence interval for β_1 that does not cover zero does not guarantee a reasonable value for R^2.

1.5.5 Inferences on β_0

We are usually not interested in constructing a confidence interval for β_0, and rarely would we want to test the hypothesis that $\beta_0 = 0$. There are, however, a few situations in which a confidence interval for β_0 would be useful, and the form of the confidence interval can be obtained as follows.

Analogous to the form of the confidence interval for β_1, the confidence interval for β_0 is given by

$$\hat{\beta}_0 \pm t_{\alpha/2, n-2} s_{\hat{\beta}_0}$$

It is shown in the chapter appendix that $\text{Cov}(\overline{Y}, \hat{\beta}_1) = 0$, with Cov denoting covariance. It then follows that $\text{Var}(\hat{\beta}_0) = \text{Var}(\overline{Y} - \hat{\beta}_1 \overline{X}) = \text{Var}(\overline{Y}) + \overline{X}^2 \text{Var}(\hat{\beta}_1) = \sigma^2/n + \overline{X}^2(\sigma^2/S_{xx})$, which can be written more concisely as $\sigma^2(\sum X^2/nS_{xx})$. It then follows that

$$s_{\hat{\beta}_0} = \hat{\sigma} \left(\frac{\sum X^2}{n S_{xx}} \right)^{1/2}$$

The use of a hypothesis test of $H_0: \beta_0 = 0$ against $H_a: \beta_0 \neq 0$ is another matter, however, as this relates to choosing between the models $Y = \beta_0 + \beta_1 X + \epsilon$ and $Y = \beta_1 X + e'$. A choice between the two models should be made before any data are collected, and rarely would we use the no-intercept model. We would generally use the latter when (1) we know that $Y = 0$ when $X = 0$, (2) we are using a data set (for parameter estimation) where the points are at least very close to the point $(0,0)$, and preferably cover that point, and (3) we are interested in using the regression equation (for prediction, say) when X is close to zero.

Certainly there are many potential applications of regression analysis for which the first condition is met. One such application is described by Casella (1983), in which the independent variable is the weight of a car and the dependent variable is gallons per mile. The author contends that a no-intercept model (whith is also termed *regression through the origin*) is appropriate in view of the physical considerations. Obviously, no car weighs close to zero pounds, however, so the second and third conditions are not met, and to say that the first condition is met in this example is to state the obvious: If we do not drive a car, then we will not realize any mileage from a car, so no gas will be used. Thus, since the first condition is satisifed only trivially, one might argue that a hypothesis test of $H_0: \beta_0 = 0$ is needed to determine if a no-intercept model should be used, and in this example Casella (1983) found that H_0 was not rejected. It was also shown, however, that the intercept model had a slightly higher R^2 value, so one could argue that the intercept model should be used. (Note: Different forms of R^2 have been recommended for different types of regression models; this is discussed in Section 1.6 and in subsequent chapters.)

Recall a potential application of regression for control that was mentioned briefly in Section 1.2 in which the objective is to control the level of a certain type of river pollutant (and, ideally, to eliminate the pollutant). If the dependent variable were a measure of pollution that could have a nonzero value only if the pollutant were nonzero, then regression through the origin would be appropriate because we would probably be interested in predicting Y when X is close to zero.

Even if we know that Y must equal zero when $X = 0$, that is not a sufficient reason for using the no-intercept model, especially when this condition is trivially satisfied. (The latter will certainly be true in many regression applications because if X and Y are related and X is "absent" in the sense that $X = 0$ is an implausible value, then Y is also likely to be absent). For the "absent–absent" scenario we are not likely to have data that are close to the point $(0,0)$, so if we force the regression line to go through the origin, the line may not fit the data as well as a regression line with an intercept. An example of this is given by Draper and Smith (1981, p. 185), who indicate that the intercept model provides a better fit to the data even though it was known that $Y = 0$ when $X = 0$. (Y was the height of soap suds in a dishpan, and X was grams of the detergent.)

Thus, there are limited conditions for which linear regression through the origin is appropriate. Accordingly, the topic is treated only briefly in Section 1.6.

1.5.6 Inferences for Y

Fitted values of Y have been illustrated previously, but we would generally like to have an interval about the predicted values. Such an interval is termed a *prediction interval* rather than a confidence interval because the latter is constructed only for a parameter, and Y is not a parameter.

1.5.6.1 Prediction Interval for Y

Confidence intervals for a parameter θ often have the general form $\hat{\theta} \pm ts_{\hat{\theta}}$. The confidence interval presented in the next section will thus be of this form, but the prediction interval for Y, which is considerably more useful, is not of this form. Specifically, the prediction interval is given by

$$\hat{Y} \pm t_{\alpha/2, n-2} \sqrt{\hat{\sigma}_{\hat{y}}^2 + \hat{\sigma}_{\epsilon}^2} \qquad (1.13)$$

Notice that we would have the general form of a confidence interval (for the parameter that $\hat{Y}$ estimates) if it were not for the second term under the radical in Eq. (1.13). The presence of that term can be explained as follows. Since the mean of ϵ is zero, it follows that the conditional mean of Y given $X, \mu_{y|x}$, is equal to $\beta_0 + \beta_1 X$. Because we are predicting an individual Y value, we must account for the variability of Y about its mean. Since $\epsilon = Y - (\beta_0 + \beta_1 X) = Y - \mu_{y|x}$, the additional variance component must be σ_{ϵ}^2. Thus, since $Y = \mu_{y|x} + \epsilon$ and $\hat{Y} = \hat{\mu}_{y|x}$, we must add $\sigma_{\hat{y}}^2$ to σ_{ϵ}^2 and then use the appropriate estimators for each of the two variance components. [Note that $\text{Var}(\hat{Y} + \epsilon)$ is equal to the sum of the individual variances because ϵ is assumed to be independent of $\hat{Y}$, which follows from the assumption stated earlier that ϵ must be uncorrelated, and hence independent under normality, of X.]

It is shown in the chapter appendix that

$$\text{var}(\hat{Y}) = \sigma_{\epsilon}^2 \left[\frac{1}{n} + \frac{(x - \bar{x})^2}{S_{xx}} \right] \qquad (1.14)$$

Thus, $\hat{\sigma}_{\epsilon}^2 + \hat{\sigma}_{\hat{y}}^2 = \hat{\sigma}_{\epsilon}^2 [1 + 1/n + (x - \bar{x})^2/S_{xx}]$, and the square root of this last expression would be used in obtaining the prediction interval given in Eq. (1.13).

It was stated previously that σ_{ϵ}^2 could be estimated as $\hat{\sigma}_{\epsilon}^2 = \sum (Y - \hat{Y})^2/(n-2)$.

This is not the best form of $\hat{\sigma}_\epsilon^2$ for computational purposes, however, as each of the n $\hat{Y}$ values would have to be computed. It is shown in the chapter appendix that $\sum(Y - \hat{Y})^2 = S_{yy} - \hat{\beta}_1 S_{xy} = S_{yy} - \hat{\beta}_1^2 S_{xx}$. Thus, $\hat{\sigma}_\epsilon^2 = (S_{yy} - \hat{\beta}_1 S_{xy})/(n - 2)$, which is frequently denoted as simply s^2.

Since $\hat{\sigma}_\epsilon^2$ and $\hat{\sigma}_{\hat{\beta}_1}^2$ both contain s^2, Eq. (1.13) may be written in the form

$$\hat{Y}_0 \pm t_{\alpha/2, n-2} \, s \sqrt{1 + \frac{1}{n} + \frac{(x_0 - \bar{x})^2}{S_{xx}}} \tag{1.15}$$

where $\hat{Y}_0$ is the predicted value of Y using a particular value of X denoted by x_0. The latter may be one of the values used in developing the prediction equation, or it may be some other value as long as that value is within the range of X values in the data set used in constructing the interval (or at least very close to that range). (Notice that here we are using a lowercase letter to denote a specific value of a random variable, as is customary.)

To obtain predicted values and prediction intervals for values of X outside that range would be to engage in *extrapolation*, as we would be extrapolating from a region where we can approximate the relationship between Y and X to a region where we have no information regarding that relationship.

For the data in Table 1.2, a 95% prediction interval for Y given that $x_0 = 380$ can be obtained as follows. For $x_0 = 380$, $\hat{y}_0 = 22.05$, as was given in Table 1.3. Using $s^2 = (S_{yy} - \hat{\beta}_1 S_{xy})/(n - 2)$ we obtain

$$s^2 = \frac{80.339 - 0.19673(405.75)}{8}$$

$$= 0.0645$$

so that $s = 0.2540$. The prediction interval would then be obtained as

$$\hat{Y}_0 \pm t_{\alpha/2, n-2} \, s \sqrt{1 + \frac{1}{n} + \frac{(x_0 - \bar{x})^2}{S_{xx}}}$$

$$= 22.05 \pm 2.306(0.2540) \sqrt{1 + \frac{1}{10} + \frac{(380 - 397.5)^2}{2062.5}}$$

$$= 22.05 \pm 0.65 \tag{1.16}$$

Thus, the prediction interval is $(21.4, 22.7)$ when $x = 380$. Prediction intervals are interpreted analogous to the way that confidence intervals are interpreted. That is, if the experiment described in Section 1.2 were repeated 100 times, $x = 380$ was one of the temperature readings used, the normality assumption was met, and the true model was given by Eq. (1.1), the expected number of times that the interval $(21.4, 22.7)$ covers the observed value of Y when $x = 380$ is 95.

One might ask why we are interested in obtaining a prediction interval for Y for $x = 380$ when we have an observed value of Y corresponding to that value of X. That value of Y, 22.05 in this example, is of course a random variable, and we would much prefer to have an interval estimate for a future value of Y than a point estimate obtained by using an observed sample value. Therefore, even if we wish to estimate future process yield for an x_0 that is one of the sample values, it is desirable to have an interval estimate in addition to the predicted value.

It should be noted that the width of the prediction interval given by Eq. (1.16) will depend on the distance that x_0 is from $\bar{x}$. This should be intuitively apparent, as we should certainly be able to do a better job of predicting Y when x_0 is in the middle of a data set than when it is at either extreme.

There are many areas of application in which prediction intervals are extremely important. Consider again the scenario where Y = college GPA, and consider two possible prediction intervals given by (1.8, 2.6) and (2.1, 2.3). The value of $\hat{Y}$ is 2.2 in each case (since $\hat{Y}$ lies in the center of the interval), but an admissions officer would undoubtedly look more favorably upon the second interval than the first interval. In other areas of application, such as fisheries management, having a well-estimated prediction interval is thought to be probably as important as having good estimates of the regression parameters (Ruppert and Aldershof, 1989).

1.5.6.2 *Confidence Interval for* $\mu_{Y|X}$

A confidence interval for $\mu_{Y|X}$ is similar to the prediction interval in terms of the computational form, but quite different in terms of interpretation. In particular, a prediction interval is for the future, whereas a confidence interval is for the present. The only difference computationally is that the "1" under the radical in Eq. (1.16) is not used for the confidence interval. That is, the latter is of the form $\hat{Y}_0 \pm t_{\alpha/2, n-2} s_{\hat{Y}}$. The confidence interval would be an interval on the average value of Y in a population when $X = x_0$. This will generally be less useful than the prediction interval, however. For example, using Y = college GPA and X = high school GPA, of what value would it be to have a confidence interval for the average GPA of all students who have entered a certain college with a particular value of X and subsequently graduated, when we are trying to reach a decision regarding the admittance of a particular student.

1.5.7 ANOVA Tables

An *analysis of variance (ANOVA)* table is usually constructed in analyzing regression data. Loosely speaking, the information provided by such tables is essentially the square of the information used in conjunction with the t-distribution presented in Section 1.5.4. The hypothesis test and confidence and prediction intervals presented in the preceding sections have utilized the standard deviation of the appropriate estimators, whereas ANOVA tables utilize variance-type information in showing variation due to different sources.

An ANOVA table for the example used in this chapter will have components that correspond to Eq. (1.7). Those three components are all *sums of squares*. Since $\sum(Y-\overline{Y})^2$ represents the total variability in Y, disregarding X, it would be logical to call it the *total sum of squares* (SS_{total}). The calculation of $\sum(Y-\hat{Y})^2$ produces the *residual sum of squares* ($SS_{residual}$); $\sum(\hat{Y}-\overline{Y})^2$ is the reduction in SS_{total} that results from using X to predict Y, and this is labeled $SS_{regression}$. Thus, $SS_{total} = SS_{regression} + SS_{residual}$.

Mean squares in ANOVA tables are always obtained by dividing each sum of squares by the corresponding *degrees of freedom (df)*. When SS_{total} is defined as $\sum(Y-\overline{Y})^2$, which is the usual way, "total" will always have $n-1$ df. This can be explained as follows. First, there are $n-1$ of the n components of $\sum(Y-\overline{Y})$ that are free to vary since $\sum(Y-\overline{Y}) = 0$. We then obtain $n-1$ as the number of degrees of freedom by applying the rule that states that degrees of freedom are given by the number of observations, n, minus the number of linear restrictions. Another way to view df (total) is to recognize that a degree of freedom is used, in general, whenever a parameter is estimated, and it can be shown that $SS(\hat{\beta}_0) = (\sum Y)^2/n$, which is part of SS_{total} since $\sum(Y-\overline{Y})^2 = \sum Y^2 - (\sum Y)^2/n$. Thus, this "corrected" sum of squares that corrects for the mean also incorporates the sum of squares for one of the parameters that has been estimated. In order to compute $\sum(Y-\hat{Y})^2$, both parameters must have been estimated, so this sum of squares has $n-2$ df. Since degrees of freedom are generally additive, $\sum(\hat{Y}-\overline{Y})^2$ must have one df, and this one df corresponds to the estimation of β_1.

Using the data in Table 1.2, we obtain the ANOVA table given in Table 1.4. The number in the column labeled F is obtained by computing the ratio of the two mean squares.

In introductory courses the relationship between the t and F distributions is usually discussed. Specifically, $t^2_{\nu_1, \alpha/2} = F_{1, \nu_1, \alpha}$ where ν_1 is the degrees of freedom of the t-statistic, and 1 and ν_1 are the degrees of freedom for the numerator and denominator, respectively, of the two components whose ratio comprises the F-statistic. (The upper tail areas for the t and F distributions are here denoted by $\alpha/2$ and α, respectively.)

It is easy to show that $t^2 = F$ where $t = \hat{\beta}_1/s_{\hat{\beta}_1}$. We proceed by writing

$$t = \frac{\hat{\beta}_1}{s_e/\sqrt{S_{xx}}}$$

Table 1.4   ANOVA Table for Data in Table 1.2

Source	df	Sum of Squares	Mean Square	F
Regression	1	79.822	79.822	1235.38
Residual	8	0.517	0.065	
Total	9	80.339		

so that $t^2 = \hat{\beta}_1^2 S_{xx}/s_e^2$. It was mentioned in Section 1.5.6.1 that $\sum (Y - \hat{Y})^2 = S_{yy} - \hat{\beta}_1^2 S_{xx}$, where $\sum (Y - \hat{Y})^2 = SS_{residual}$ and $S_{yy} = SS_{total}$. It follows that $\hat{\beta}_1^2 S_{xx} = SS_{regression}$. Since $s_e^2 = MS_{residual}$, $t^2 = SS_{regression}/MS_{residual} = MS_{regression}/MS_{residual} = F$. (It is shown in the chapter appendix that the ratio of these two mean squares has an F-distribution, starting from the assumption of a normal distribution for the error term.)

Whether we look at $t = 35.13$ or $F = 1235.38$, the magnitude of each of these two numbers coupled with the fact that $R^2 = 0.9935$ indicates that there is a strong relationship between yield and temperature for the range of temperatures covered in the experiment.

1.5.8 Lack of Fit

Frequently a regression model can be improved by using nonlinear terms in addition to the linear terms. In simple regression the need for nonlinear terms is generally indicated by the scatter plot of the data, but with more than one regressor the need for nonlinear terms will be more difficult to detect.

A *lack-of-fit test* can be performed to see if there is a need for one or more nonlinear terms. We should note that we are still concerned here with linear regression models, which are linear in the parameters; we are simply trying to determine the possible need for terms such as polynomial terms or trigonometric terms that would enter the model in a linear manner.

The general idea is to separate $SS_{residual}$ into two components: $SS_{pure\,error}$ and $SS_{lack\,of\,fit}$. Simply stated, $SS_{pure\,error}$ is the portion of $SS_{residual}$ that cannot be reduced by improving the model. This can be seen from the formula for $SS_{pure\,error}$ which is

$$SS_{pure\,error} = \sum_{j=1}^{n_i} \sum_{i=1}^{k} (Y_{ij} - \overline{Y}_i)^2 \tag{1.17}$$

where $\overline{Y}_i$ is the average of the n_i Y values corresponding to X_i, k is the number of different X values, and Y_{ij} is the jth value of Y corresponding to the ith distinct value of X.

Consider the data in Table 1.5 for which $k = 6$ and, for example, $Y_{42} = 33$. Thus, $SS_{pure\,error} = (15-15.5)^2 + (16-15.5)^2 + (20-21)^2 + (22-21)^2 + (31-32)^2 + (33-32)^2 + (46-47.5)^2 + (49-47.5)^2 = 9.0$. For a given value of X, different values of Y will plot vertically on a scatter plot, and there is no way that a regression model can be modified to accommodate a vertical line segment, as the slope would be undefined.

Using the formulas given in the preceding sections, $SS_{residual} = 261.85$. Thus, $SS_{lack\,of\,fit} = 261.85 - 9.00 = 252.85$, so almost all of the residual is due to lack of fit, thereby indicating that the wrong model is being used. When this is the

Table 1.5 Data for Illustrating Lack of Fit

Y	X
15	10
16	10
14	15
20	20
22	20
31	25
33	25
46	30
49	30
60	35

case, σ_ϵ^2 should *not* be estimated by $MS_{residual}$ as the latter contains more than just experimental error. Some authors use *error* instead of *residual* in ANOVA tables such as Table 1.4. The two terms are conceptually different as the former is the *experimental error* (in Y) that results when an experiment is repeated, whereas the latter represents the collection of factors that result in the model not providing an exact fit to the data. The latter might consist of not having the right functional form of the regressor(s) or not using all relevant regressors, in addition to experimental error (i.e., variation) resulting from factors that cannot be identified and/or controlled during the experiment.

By extracting the experimental error component from the residual, we can see whether we should attempt to improve the model. A formal lack-of-fit test is performed by computing $F = MS_{lack\,of\,fit}/MS_{pure\,error}$, and rejecting the hypothesis of no lack of fit if F exceeds the value from the F-table.

Each mean square is computed in the usual manner by dividing the sum of squares by the corresponding degrees of freedom. The df for pure error is best viewed as the sum of the degrees of freedom for each X that is repeated, where each such df is $n - 1$. When $SS_{pure\,error}$ is computed using Eq. (1.17) and then divided by the df, this is equivalent to taking a weighted average of the s_i^2 values, where s_i^2 is the sample variance of the Y values corresponding to X_i, with the weights being $n_i - 1$. Thus, $\sigma_{pure\,error}^2$ is estimated in a logical manner. The df for lack of fit is obtained from df (residual) − df (pure error).

For the Table 1.5 data

$$F = \frac{252.85/4}{9.00/4}$$
$$= 28.09$$

Since $F_{4,4,.05} = 6.39$ and $F_{4,4,.01} = 15.98$, we would conclude that there is evidence of lack of fit.

Since the degree of lack of fit is so extreme, we would expect that the non-

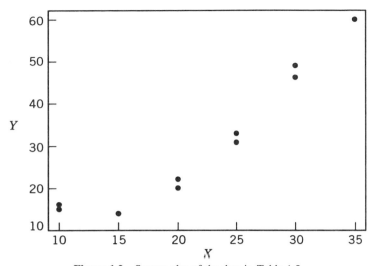

Figure 1.2   Scatter plot of the data in Table 1.5.

linearity would be evident in the scatter plot. The latter is shown in Figure 1.2, and we can see the obvious curvature.

It is important to recognize that this lack-of-fit test does not indicate how the model should be modified, it simply shows that some modification is necessary. Other methods must be employed to determine how the model should be modified. Some of these methods are discussed in detail in Chapter 5.

When the regressor values are not preselected, we may not have any repeated values. When this happens there are other methods that can be used. Daniel and Wood (1980, p. 133) discuss a *method of nearest neighbors* that is based on forming pseudo-replicates from regressor values that are close together.

1.6 REGRESSION THROUGH THE ORIGIN

This topic was mentioned briefly in Section 1.5.5 in which it was stated that linear regression through the origin will rarely be applicable. The model is

$$Y = \beta X + \epsilon \tag{1.18}$$

and to derive the least squares estimator of β we could use the same approach as was used in Section 1.4. Doing so would produce the estimator

$$\hat{\beta} = \frac{\sum XY}{\sum X^2}$$

Inferences for the model given by Eq. (1.18) are similar to the inferences that were developed in Section 1.5 for the model given by Eq. (1.1). There are, however, a few notable differences. In particular, the residuals will not sum to zero. For the intercept model, the solution of Eq. (1.5a) is what causes the residuals to sum to zero. There is no counterpart to that equation for the no-intercept model, however, so the residuals are not forced to sum to zero.

Other differences include the fact that the estimate of σ^2 is based on $n - 1$ df instead of $n - 2$ since there is only one parameter in the model. A major difference is the way in which R^2 is defined. As discussed in Section 1.5.2, the definition of R^2 for the intercept model is quite intuitive in that it measures the (scaled) improvement in the prediction of Y that results for $\hat{\beta}_1 \neq 0$ over $\hat{\beta}_1 = 0$. For the intercept model, $\hat{Y} = \overline{Y}$ if $\hat{\beta}_1 = 0$, but for the no-intercept model $\hat{Y} = 0$ if $\hat{\beta} = 0$. Therefore, an analogous definition of R^2 for the no-intercept model would be one that measures improvement over $\hat{Y} = 0$. This suggests that R^2 *might* be defined as $R^2 = \sum \hat{Y}^2 / \sum (Y - \hat{Y})^2$.

Some reflection should indicate that this is not logical, however, because if we knew that there was no relationship between X and Y, the most reasonable choice for $\hat{Y}$ is still $\hat{Y} = \overline{Y}$. Also, it should be apparent that R^2 would exceed one if we had a perfect fit (i.e., $\hat{Y}_i = Y_i, i = 1, 2, \ldots, n$). Therefore, something must be subtracted from the numerator.

The use of $\hat{Y} = \overline{Y}$ when there is no regression relationship might suggest that we also use R^2 as given in Eq. (1.8) for the no-intercept model, but whereas we know from Eq. (1.7) that R^2 when defined in this manner has an upper bound of one, we have no such assurance for the no-intercept model. Kvålseth (1985) gives an example in which R^2 exceeds one when defined by Eq. (1.8), and recommends that R^2 be defined as in Eq. (1.9) for both the intercept and no-intercept models. Although Eq. (1.9) is equivalent to Eq. (1.8) for an intercept model, they are not equivalent for a no-intercept model. Simple algebra reveals that using Eq. (1.9) for a no-intercept model produces

$$R_0^2 = \frac{\sum \hat{Y}^2 - (\sum Y)^2 / n}{\sum (Y - \overline{Y})^2} \tag{1.19}$$

Using Eq. (1.9) avoids the issue of how $\hat{Y}$ would be computed if a regression relationship did not exist, but the use of either (1.9) or (1.19) is not totally devoid of deficiencies, however, as R_0^2 could be negative. This would simply indicate the total inappropriateness of the model given in Eq. (1.18). We also note that Eq. (1.19) is equivalent to the form of R^2 recommended by Gordon (1981).

It is worthwhile to reiterate a point made in Section 1.5.5 regarding a no-intercept model. In discussing such models, Kvålseth (1985) states: "Occasionally, an analyst may force such a model to be fitted by empirical data by relying purely on theoretical reasoning, whereas a careful analysis of the data may reveal that an intercept model is preferable."

1.7   CORRELATION

If X is a *random* variable, we may properly speak of the *correlation* between X and Y. This refers to the extent that the two random variables are related, with the strength of the relationship measured by the (sample) correlation coefficient, r_{xy}. The latter is computed as

$$r_{xy} = \frac{S_{xy}}{\sqrt{S_{xx}S_{yy}}} \tag{1.20}$$

with the components of Eq. (1.20) as previously defined. This is sometimes called the Pearson correlation coefficient to distinguish it from the other correlation coefficients that are used. The relationship between X and Y is assumed to be linear when Eq. (1.20) is used, and X and Y are assumed to have a bivariate normal distribution, although the second assumption is generally not checked. Formally, the population correlation coefficient, ρ_{xy}, is defined as the covariance between X and Y divided by the standard deviation of X times the standard deviation of Y. When those parameters are estimated by the sample statistics, the result reduces to Eq. (1.20).

The possible values of r_{xy} range from -1 to $+1$. The former represents perfect negative correlation, as would happen if all of the points fell on a line with a negative slope, and the latter represents perfect positive correlation. In regression there is nothing really "negative" about negative correlation, as a strong negative correlation is just as valuable as a strong positive correlation as far as estimation and prediction are concerned. A zero correlation between X and Y would signify that it would not be meaningful to construct a regression equation using that regressor, but when there is more than one regressor we would prefer that the regressors be "uncorrelated" (orthogonal) with each other. This is discussed and illustrated in Chapter 14.

When there is only a single regressor, $r_{xy}^2 = R^2$. Thus, for the one-regressor random-X case, R^2 is the square of the correlation between X and Y. When X is fixed, R^2 must then be viewed (and labeled) somewhat differently, and it is customary to refer to R^2 as the *coefficient of determination* (and the coefficient of multiple determination when there is more than one regressor).

Another connection between the square of a correlation coefficient and R^2 is $r_{y\hat{y}}^2 = R^2$, as the reader is asked to show in Exercise 1.10. (Notice that this last result holds whether X is random or not, since both Y and $\hat{Y}$ are random variables.) Clearly, Y and $\hat{Y}$ must be highly correlated for the regression model to have value.

1.8   MISCELLANEOUS USES OF REGRESSION

Three uses of simple linear regression are discussed in the next three sections.

1.8.1. Regression for Control

As mentioned in Section 1.2, an important but infrequently discussed application of regression is to attempt to control Y at a desired level through the manipulation of X.

Doing so may be difficult, however, for a number of reasons. First, since a cause-and-effect relationship is being inferred, the prediction equation must have been produced from preselected X values, and there must not be any other independent variables that are related to Y. The latter is not likely to be true when only a single X is being used, however, and the consequences when it is not true may be great.

Box (1966) gives a good example of this in regard to a hypothetical chemical process in which Y = process yield, X_1 = pressure, and X_2 = an unsuspected impurity. The scenario is that undesirable frothing can be reduced by increasing pressure, but a high value of X_2 is what actually produces frothing and also lowers yield, with the latter unrelated to pressure. Assume that the regression equation $\hat{Y} = \hat{\beta}_0 + \hat{\beta}_1 X_1$ is developed with the intention of increasing Y through manipulation of X_1. Even if an experiment consisted of systematically varying X_1 and recording the corresponding Y value, the regression equation would have no value for control because Y depends solely upon X_2. Therefore, an experimenter who attempted to rely upon the value of $\hat{\beta}_1$ to effect a desired change in Y through manipulation of X_1 would not be successful.

The equation could still have value for prediction, however, because if R^2 is large, X_1 is essentially serving as a suitable proxy for X_2. So even though the (high) correlation between Y and X_1 is a spurious correlation, X_1 can still be used to predict Y. Thus, a spurious correlation can produce suitable predictions, but $\hat{\beta}_1$ has no meaning by itself because the correct model does not include X_2. If the true relationship between X_1 and X_2 were known, then that relationship could be incorporated into the prediction equation, but the appropriate coefficient would be different from $\hat{\beta}_1$.

Following Box (1966), let's assume that $X_2 = \beta_0^* + \beta_1^* X_1$ and the true model in terms of Y is $Y = \beta_0' + \beta_2 X_2$. Then with $\hat{X}_2 = \hat{\beta}_0^* + \hat{\beta}_1^* X_1$, we have

$$\hat{Y} = \hat{\beta}_0' + \hat{\beta}_2(\hat{\beta}_0^* + \hat{\beta}_1^* X_1)$$

$$= (\hat{\beta}_0' + \hat{\beta}_0^*\hat{\beta}_2^*) + \hat{\beta}_2\hat{\beta}_1^* X_1$$

Thus, the experimenter who attempts to manipulate Y through X_1 will do poorly if $\hat{\beta}_2\hat{\beta}_1^*$ differs from $\hat{\beta}_1$ by more than a small amount, and there should generally be no reason to expect them to be close.

Additional information on the use of regression for control can be found in Hahn (1974) and Draper and Smith (1981, p. 413).

1.8.2   Inverse Regression

Two specialized uses of simple linear regression are discussed in this section and in the following section. The first of these, *inverse regression*, can be used effectively in calibration work, although its applications are not limited to calibration.

Assume that we have two measuring instruments; one is quite accurate but is both expensive to use and slow, and the other is fast and less expensive to use, but is also less accurate. If the measurements obtained from the two devices are highly correlated, then the measurement that would have been made using the expensive measuring device could be predicted fairly well from the measurement that is actually obtained using the less expensive device.

In particular, if the less expensive device has an almost constant bias and we define X = measurement from the accurate instrument and Y = measurement from the inaccurate device, and then regress X on Y to obtain $\hat{X}^* = \hat{\beta}_0^* + \hat{\beta}_1^* Y$, we would expect $\hat{\beta}_1$ to be close to 1.0 and $\hat{\beta}_0$ to be approximately equal to the bias.

This is termed *inverse regression* because X is being regressed on Y instead of Y regressed on X.

Since X and Y might seem to be just arbitrary labels, why not reverse them so that we would then have just simple linear regression? Recall that Y must be a random variable and classical regression theory holds that X is fixed. A measurement from an accurate device should, theoretically, have a zero variance, and the redefined X would be a random variable.

But we have exactly the same problem if we regress X on Y, with X and Y as originally defined. The fact that the independent variable Y is a random variable is not really a problem, as regression can still be used when the independent variable is random, as is discussed in Section 1.8.

If the *dependent* variable is not truly a random variable, however, then classical regression is being "bent" considerably, and this is one reason why inverse regression has been somewhat controversial. Krutchkoff (1967) reintroduced inverse regression, and it was subsequently criticized by other writers. [See Montgomery and Peck (1992, pp. 406–407) for a summary.]

There is an alternative to inverse regression that avoids these problems, however, and which should frequently produce almost identical results. In the *classical theory of calibration* Y is regressed against X and X is then solved for in terms of Y for the purpose of predicting X for a given value of Y. Specifically,

$$\hat{Y} = \hat{\beta}_0 + \hat{\beta}_1 X$$

so that

$$X = (\hat{Y} - \hat{\beta}_0)/\hat{\beta}_1$$

For a given value of Y, say Y_c, X is then predicted as

$$\hat{X}_c = (Y_c - \hat{\beta}_0)/\hat{\beta}_1$$

As in Section 1.7, let r_{xy} denote the correlation between X and Y. If $r_{xy} \doteq 1$, then $\hat{X}_c$ and $\hat{X}_c^*$ (from inverse regression) will be almost identical. Recall that for inverse regression to be effective the two measurements must be highly correlated, so under conditions for which we would want to use either inverse regression or the classical method of calibration, the two approaches will give very similar results. Therefore, we need not be concerned about the controversy surrounding inverse regression.

The real data given in Table 1.6 will be used to illustrate the two approaches. Regressing Y on X produces the equation

$$\hat{Y} = 0.76 + 0.9873X$$

so for a given value of Y, X would be estimated as

Table 1.6   Calibration Data

Measured Amount of Molybdenum (Y)	Known Amount of Molybdenum (X)
1.8	1
1.6	1
3.1	2
2.6	2
3.6	3
3.4	3
4.9	4
4.2	4
6.0	5
5.9	5
6.8	6
6.9	6
8.2	7
7.3	7
8.8	8
8.5	8
9.5	9
9.5	9
10.6	10
10.6	10

Source: The data were originally provided by G.E.P. Box and are given (without X and Y designated) as Table 1 in Hunter and Lamboy (1981).

$$\hat{X} = \frac{Y - 0.76}{0.9873}$$

$$= -0.7698 + 1.0129Y$$

Regressing X on Y produces the inverse regression estimator

$$\hat{X}^* = -0.7199 + 1.0048Y$$

Thus, the equations differ only slightly, and the $\hat{X}$ and $\hat{X}^*$ values are given for comparison in Table 1.7. We can see that the estimated values produced by the two procedures differ only slightly, and this is because X and Y are very highly correlated. (The value of the correlation coefficient is .996.)

It is also important to note that $\hat{X}$ and $\hat{X}^*$ are closer to X than is Y in all but two cases, and that the differences in most cases is a considerable amount.

For this example the average squared error for $\hat{X}^*$ is 0.0657 compared to 0.0662 for $\hat{X}$, but this small difference is hardly justification for using $\hat{X}^*$. Furthermore, such a comparison would be considered inappropriate by some peo-

Table 1.7   Predicted Values for Table 1.6 Data Using Inverse Regression ($\hat{X}^*$) and Calibration ($\hat{X}$)

X	$\hat{X}^*$	$\hat{X}$
1	1.08882	1.05341
1	0.88786	0.85083
2	2.39509	2.37017
2	1.89268	1.86373
3	2.89751	2.87662
3	2.69654	2.67404
4	4.20378	4.19338
4	3.50040	3.48436
5	5.30908	5.30757
5	5.20860	5.20628
6	6.11294	6.11788
6	6.21342	6.21917
7	7.51970	7.53593
7	6.61535	6.62433
8	8.12259	8.14367
8	7.82114	7.83980
9	8.82597	8.85269
9	8.82597	8.85269
10	9.93127	9.96688
10	9.93127	9.96688

ple since the classical estimator has an infinite mean (expected) square error, whereas the inverse regression estimator has a finite mean square error. (The mean square of an estimator is defined as the variance plus the square of the bias. This is discussed further in the chapter appendix.)

Various methods have been proposed for obtaining a confidence interval for X_0; see Montgomery and Peck (1992, p. 406) for details.

The reader is referred to Chow and Shao (1990) for a related discussion on a comparison of the classical and inverse methods, and an application for which the two approaches were not equally useful. See also Brown (1993), Graybill and Iyer (1994), and Hunter and Lamboy (1981).

1.8.3 Regression Control Chart

A regression (quality) control chart was presented by Mandel (1969). The objective is similar to the use of regression for control, although with a regression control chart the intention is to see whether Y is "in control" rather than trying to produce a particular value of Y.

For example, let Y be some measure of tool wear, with X = time. Since Y can be expected to change as X changes, it would be reasonable to determine a prediction interval for Y for a given X, and this is what is done. Specifically, Eq. (1.15) is used with $t_{\alpha/2, n-2}$ replaced by 2.

A regression control chart has the same potential weakness as the use of regression for control, however, in that Y could be affected by other variables in addition to X.

1.9 FIXED VERSUS RANDOM REGRESSORS

It is important to realize at the outset that even though classical regression theory is based on the assumption that the values of X are selected, regression can still be used when X is a random variable. This is important because regression data frequently come from observational studies in which the values of X occur randomly, and we could probably go further and state that most regression data sets have random regressors.

The issue of random regressors has been discussed by many writers, and there is some disagreement. The problem is compounded somewhat by the fact that in the literature it sometimes isn't clear whether regressors "with error" are considered to have random error, or measurement error, or both. Here we consider only random error; measurement error is discussed in Section 2.6.

One popular position is that we may proceed as if X were fixed, provided that the conditional distribution of Y_i given X_i is normal and has a constant variance for each i, and the X_i are independent (as are the Y_i), and the distribution of X_i does not depend on β_0, β_1, or σ^2 (see Neter et al., 1989, p. 86). Other writers speak of random regressors, while simultaneously assuming true, unobservable values for each X_i (see Seber, 1977, pp. 210–211). If X is a random variable that is devoid of measurement error, there is clearly no such thing as a true value, so the assumption of a true value that is different from the observed value implies that there is measurement error. In compiling and elaborating on results given in earlier papers, Sampson (1974) concludes that in the random regressor case we may proceed the same as in the fixed regressor case. (A detailed discussion of that work is beyond the scope of this chapter.) A somewhat different view can be found in Mandel (1984), however. Properties of the regression coefficients when X is random are discussed by Shaffer (1991).

1.10   SOFTWARE

The use of statistical software is essential if regression data are to be thoroughly analyzed. Discussion of the regression capabilities of popular mainframe and personal computer (PC) statistical software begins in Section 2.7.

SUMMARY

The material presented in this chapter constitutes the first part of an introduction to the fundamentals of simple linear regression; the introduction is completed in the next chapter in which diagnostic procedures and additional plots are discussed.

   When there is only a single regressor, a scatter plot of the data can be very informative and can frequently provide as much information as the combined information from all of the inferential procedures for determining model adequacy that were discussed in this chapter. The situation is quite different when there is more than one regressor, however, so it is desirable to master the techniques used in simple regression so as to have a foundation for understanding multiple regression. (We will discover in subsequent chapters, however, that additional techniques are needed for multiple regression.)

   We may use regression when X is either fixed or random, with X quite frequently random because regression data are often obtained from observational studies.

   Unthinking application of regression can produce poor results, so it is important for the reader to understand the array of available methods and when they should be used. This is emphasized in Chapter 15, in which the regression methods presented in this and subsequent chapters are reviewed in the context of a strategy for analyzing regression data. The reader should also study Watts (1981) in which the knowledge and skills necessary to analyze regression data are discussed and a step-by-step approach to the analysis of linear and nonlinear regression data is given. A detailed discussion of the considerations that should be made in analyzing regression data is given in Section 15.6, and a discussion of the steps to follow in analyzing regression data can also be found in Ryan (1990).

APPENDIX

1.A   Analysis of Variance Identity

We wish to show that $\sum (Y - \overline{Y})^2 \equiv \sum (Y - \hat{Y})^2 + \sum (\hat{Y} - \overline{Y})^2$. We proceed by writing

$$\sum (Y - \overline{Y})^2 \equiv \sum (Y - \hat{Y} + \hat{Y} - \overline{Y})^2$$
$$\equiv \sum [(Y - \hat{Y}) + (\hat{Y} - \overline{Y})]^2$$
$$= \sum (Y - \hat{Y})^2 + \sum (\hat{Y} - \overline{Y})^2 + 2 \sum (Y - \hat{Y})(\hat{Y} - \overline{Y})$$

The cross-product term can be shown to vanish as follows:

$$\sum (Y - \hat{Y})(\hat{Y} - \overline{Y}) = \sum (Y - \hat{Y})\hat{Y} - \sum (Y - \hat{Y})\overline{Y}$$
$$= \sum (Y - \hat{Y})\hat{Y}$$

since $\overline{Y} \sum (Y - \hat{Y}) = 0$. Thus, we need only show that $\sum (Y - \hat{Y})\hat{Y} = 0$. Notice that this can be written as $\sum e\hat{Y} = 0$, so as a by-product we will obtain the important result that the residuals are orthogonal to $\hat{Y}$. We obtain

$$\sum Y\hat{Y} = \sum Y[\overline{Y} + \hat{\beta}_1(X - \overline{X})]$$
$$= n\overline{Y}^2 + \hat{\beta}_1 \sum (X - \overline{X})Y$$
$$= n\overline{Y}^2 + \hat{\beta}_1 S_{xy}$$

and

$$\sum \hat{Y}^2 = \sum [\overline{Y} + \hat{\beta}_1(X - \overline{X})]^2$$
$$= \sum \overline{Y}^2 + \hat{\beta}_1^2 \sum (X - \overline{X})^2 + 2\overline{Y}\hat{\beta}_1 \sum (X - \overline{X})$$
$$= n\overline{Y}^2 + \hat{\beta}_1^2 S_{xx}$$

Thus, $\sum Y\hat{Y} = \sum \hat{Y}^2$ since $\hat{\beta}_1 S_{xy} = (S_{xy}/S_{xx})S_{xy} = S_{xy}^2/S_{xx} = (S_{xy}/S_{xx})^2 S_{xx} = \hat{\beta}_1^2 S_{xx}$.

1.B   Unbiased Estimator of β_1

It was stated in Section 1.5.4 that $\hat{\beta}_i$ is an unbiased estimator of β_i, under the assumption that the fitted model is the true model. We show this for β_1 as follows. Writing $\hat{\beta}_1 = \sum (X_i - \overline{X}) Y_i / S_{xx}$ and assuming that X is fixed, we obtain $E(\hat{\beta}_1) = (1/S_{xx}) \sum (X_i - \overline{X}) E(Y_i) = (1/S_{xx}) \sum (X_i - \overline{X})(\beta_0 + \beta_1 X_i) = (1/S_{xx})(0 + \beta_1 \sum (X_i - \overline{X}) X_i) = (1/S_{xx})(\beta_1 S_{xx}) = \beta_1$.

1.C Unbiased Estimator of σ_ϵ^2

The fact that s_e^2 is an unbiased estimator of σ_ϵ^2 can be shown as follows. Since Y is assumed to have a normal distribution with variance σ^2, it follows that $(Y - \mu_y)^2 / \sigma^2$ will have a chi-square distribution with one degree of freedom, and $\sum (Y - \mu_y)^2 / \sigma^2$ will have a chi-square distribution with n degrees of freedom. But μ_y is unknown, and two degrees of freedom are lost (corresponding to β_0 and β_1) when $\hat{Y}$ is used in place of μ_y, so $\chi_{n-2}^2 = \sum (Y - \hat{Y})^2 / \sigma^2$ must be chi-square with $n - 2$ degrees of freedom. Since $E(\chi_{n-2}^2) = n - 2$, it follows that $E(s_e^2) = \sigma^2$ if s_e^2 is defined as $s_e^2 = \sum (Y - \hat{Y})^2 / (n - 2)$.

This proof depends upon $E(\hat{Y}) = \mu_y$, but it can also be shown (Myers, 1990) that s^2 is unbiased when the fitted model contains the true model in addition to extraneous variables.

1.D Mean Square Error of an Estimator

The term *mean square error* was used in Section 1.8.2. When an estimator is unbiased, the mean square error of an estimator is equal to the variance of that estimator. For an arbitrary estimator $\hat{\theta}_i$, the mean square error, $\text{MSE}(\hat{\theta}_i)$, is defined as $\text{MSE}(\hat{\theta}_i) = \text{Var}(\hat{\theta}_i) + [E(\hat{\theta}_i) - \theta]^2$. If $\hat{\theta}_i$ is an estimator of a regression parameter, we must know the true model in order to determine $E(\hat{\theta}_i)$. Thus, whereas we may speak conceptually of the mean square error of an estimator, we will rarely be able to derive it.

1.E How Extraneous Variables Can Affect R^2

Since a degree of freedom is lost whenever a variable is added to a regression model, the fact that s^2 is an unbiased estimator of σ^2 for an overfitted model implies that the expected value of SS_{error} must decrease, and so the expected value of $\text{SS}_{\text{regression}}$, and hence R^2, must increase.

1.F The Distribution of $(\hat{\beta}_1 - \beta_1)/s_{\hat{\beta}_1}$

We will use the fact that a t random variable with ν degrees of freedom results from a standard normal random variable divided by the square root of a chi-square random variable that is first divided by its degrees of freedom, which is also ν.

Let

$$T = \frac{(\hat{\beta}_1 - \beta_1)/\sigma_{\hat{\beta}_1}}{s_{\hat{\beta}_1}/\sigma_{\hat{\beta}_1}}$$

The numerator of T is normal$(0,1)$, and the denominator reduces to s_e^2/σ^2. From the result given in Section 1.B, we know that $(n-2)s_e^2/\sigma^2$ has a chi-square distribution with $n-2$ degrees of freedom, so $s_{\hat{\beta}_1}/\sigma_{\hat{\beta}_1}$ is a chi-square random variable divided by its degrees of freedom, and hence T has a t-distribution with $n-2$ degrees of freedom.

1.G   Relationship Between R^2 and t^2

If we expand the numerator of R^2 in Eq. (1.8) we obtain

$$\sum (\hat{Y} - \overline{Y})^2 = \sum \hat{Y}^2 - 2\overline{Y} \sum \hat{Y} + \sum \overline{Y}^2$$
$$= \sum \hat{Y}^2 - n\overline{Y}^2$$

with the second line resulting from the fact that $\sum Y = \sum \hat{Y}$ since $\sum (Y - \hat{Y}) = 0$. From the derivations in Section 1.A we have $\sum \hat{Y}^2 - n\overline{Y}^2 = \hat{\beta}_1^2 S_{xx}$. Therefore, we may write R^2 as

$$R^2 = \frac{\hat{\beta}_1^2 S_{xx}}{S_{yy}}$$

with $S_{yy} = \sum (Y - \overline{Y})^2$. If we write t as

$$t = \frac{\hat{\beta}_1 \sqrt{S_{xx}}}{s_e}$$

so that

$$t^2 = \frac{(n-2)\hat{\beta}_1^2 S_{xx}}{S_{yy} - \hat{\beta}_1^2 S_{xx}}$$

simple algebra then shows the result given in Eq. (1.12).

1.H   Variance of $\hat{Y}$

We first write $\hat{Y}$ as $\hat{Y} = \overline{Y} + \hat{\beta}_1(X - \overline{X})$. Then

$$\text{Var}(\hat{Y}) = \text{Var}[\overline{Y} + \hat{\beta}_1(X - \overline{X})]$$
$$= \text{Var}(\overline{Y}) + \text{Var}[\hat{\beta}_1(X - \overline{X})] + 2\text{Cov}(\overline{Y}, \hat{\beta}_1(X - \overline{X}))$$
$$= \frac{\sigma^2}{n} + (X - \overline{X})^2 \sigma_{\hat{\beta}_1}^2$$

since Cov $(\overline{Y}, \hat{\beta}_1(X - \overline{X})) = 0$, where Cov represents covariance. The covariance result may be established as follows. Since X is assumed to be fixed (i.e., not a random variable), it follows that $\text{Cov}(\overline{Y}, \hat{\beta}_1(X - \overline{X})) = (X - \overline{X})\text{Cov}(\overline{Y}, \hat{\beta}_1)$, and

$$
\begin{aligned}
\text{Cov}\,(\overline{Y}, \hat{\beta}_1) &= \text{Cov}\left(\overline{Y}, \frac{\sum (X - \overline{X})Y}{S_{xx}}\right) \\
&= \frac{1}{S_{xx}}\left\{\text{Cov}\left(\frac{Y_1 + Y_2 \cdots + Y_n}{n}, X_1Y_1 + X_2Y_2 + \cdots + X_nY_n \right.\right. \\
&\quad \left.\left. - \overline{X}(Y_1 + Y_2 + \cdots + Y_n)\right)\right\} \\
&= \frac{1}{S_{xx}}\left\{\sigma^2\left(\frac{\sum X_i}{n}\right) - \overline{X}\left(\frac{n\sigma^2}{n}\right)\right\} \\
&= 0
\end{aligned}
$$

Since $\text{Var}(\hat{\beta}_1) = \sigma^2/S_{xx}$, we may write the final result for $\text{Var}(\hat{Y})$ as in Eq. (1.14).

1.I Distribution of MS_regression/MS_residual

When two independent chi-square random variables are each divided by their respective degrees of freedom, and a ratio of the two quotients is formed, that ratio will be a random variable that has an F-distribution. Therefore, we need to show that SS_regression and SS_residual are independent chi-square random variables.

From the derivation in Section 1.B we know that $\text{SS}_{\text{residual}}/\sigma^2$ has a chi-square distribution with $n - 2$ degrees of freedom. Applying the same approach to the distribution of $\sum (Y - \overline{Y})^2/\sigma^2$, it is apparent that this random variable has a chi-square distribution with $n - 1$ degrees of freedom (one df is lost since $\overline{Y}$ is used to estimate μ_y. It then follows that $\sum (\hat{Y} - \overline{Y})^2/\sigma^2$ must have a chi-square distribution with 1 df because the difference of two chi-square random variables is also a chi-square random variable with degrees of freedom equal to the difference between the two degrees of freedom. Independence of the two chi-square random variables could be established by applying Cochran's (1934) theorem on the decomposition of squared functions of normal random variables.

REFERENCES

Al-Betairi, E. A., M. M. Moussa, and S. Al-Otaibi (1988). Multiple regression approach to optimize drilling operations in the Arabian Gulf area. *SPE Drilling Engineering*, **3**, 83–88.

Andersen, P. K., K. Borch-Johnsen, and T. Deckert (1985). A Cox regression model for the relative mortality and its application to diabetes mellitus survival data. *Biometrics*, **41**, 921–932.

Anscombe, F. J. (1973). Graphs in statistical analysis. *The American Statistician*, **27**, 17–21.

Barrett, J. P. (1974). The coefficient of determination—some limitations. *The American Statistician*, **28**, 19–20.

Birkes, D. and Y. Dodge (1993). *Alternative Methods of Regression*. New York: Wiley.

Box, G. E. P. (1966). Use and abuse of regression. *Technometrics*, **8**, 625–629.

Box, G. E. P. and N. R. Draper (1969). *Evolutionary Operation*. New York: Wiley.

Brown, P. J. (1993). *Measurement, Regression, and Calibration*. Oxford: Clarendon Press.

Casella, G. (1983). Leverage and regression through the origin. *The American Statistician*, **37**, 147–152.

Chatterjee, S. and M. Yilmaz (1992). A review of regression diagnostics for behavioral research. *Applied Psychological Measurement*, **16**, 209–227.

Chow, S.-C. and J. Shao (1990). On the difference between the classical and inverse methods of calibration. *Applied Statistics*, **39**, 219–228.

Cochran, W. G. (1934). The distribution of quadratic forms in a normal system with applications to the analysis of variance. *Proc. Cambridge Philos. Soc.*, **30**, 178–191.

Daniel, C. and F. S. Wood (1980). *Fitting Equations to Data*, 2nd edition. New York: Wiley.

Draper, N. R. and H. Smith (1981). *Applied Regression Analysis*, 2nd edition. New York: Wiley.

Dwyer, L. M., A. Bootsma, and H. N. Hayhoe (1988). Performance of three regression-based models for estimating monthly soil temperatures in the Atlantic region. *Canadian Journal of Soil Science*, **68**, 323–335.

Frank, I. E. and J. H. Friedman (1992). A statistical view of some chemometrics regression tools. *Technometrics*, **35**, 109–135 (discussion: 136–148).

Gordon, H. A. (1981). Errors in Computer Packages. Least squares regression through the origin. *The Statistician*, **30**, 23–29.

Graham, L. D. (1991). Predicting academic success of students in a master of business administration program. *Educational and Psychological Measurement*, **51**, 721–727.

Graybill, F. A. and H. K. Iyer (1994). *Regression Analysis: Concepts and Applications*. Belmont, CA: Duxbury.

Hahn, G. J. (1974). Regression for prediction versus regression for control. *Chemtech*, Sept., 574–576.

Hamilton, L. C. (1992). *Regression with Graphics*. Pacific Grove, CA: Brooks/Cole Publishing Company.

Hunter, W. G. and W. F. Lamboy (1981). A Bayesian analysis of the linear calibration problem. *Technometrics*, **23**, 323–328 (discussion: 329–350).

Jacobs, G. and K. Mahjoob (1988). Multiple regression analysis, using body size and cardiac cycle length, in predicting echocardiographic variables in dogs. *American Journal of Veterinary Research*, **49**, 1290–1294.

Kvålseth, T. O. (1985). Cautionary note about R^2. *The American Statistician*, **39**, 279–285.

Krutchkoff, R. G. (1967). Classical and inverse regression methods of calibration. *Technometrics*, **9**, 425–439.

Mandel, B. J. (1969). The regression control chart. *Journal of Quality Technology*, **1**, 1–9.

Mandel, J. (1984). Fitting straight lines when both variables are subject to error. *Journal of Quality Technology*, **16**, 1–14.

Montgomery, D. C. and E. A. Peck (1992). *Introduction to Linear Regression Analysis*, 2nd edition. New York: Wiley.

Mukherjee, D. and A. F. Roche (1984). The estimation of percent body fat, body density and total body fat by maximum R^2 regression equations. *Human Biology*, **56,** 79–109.

Myers, R. H. (1990). *Classical and Modern Regression with Applications*, 2nd edition. North Scituate, MA: PWS-Kent.

Neter, J., W. Wasserman, and M. H. Kutner (1989). *Applied Linear Regression Models*, 2nd edition. Homewood, IL: Irwin.

Paolillo, J. (1982). The predictive validity of selected admissions variables relative to grade point average earned in a master of business administration program. *Educational and Psychological Measurement*, **42,** 1163–1167.

Ranney, G. B. and C. C. Thigpen (1981). The sample coefficient of determination in simple linear regression. *The American Statistician*, **35,** 152–153.

Rousseeuw, P. J. and A. Leroy (1987). *Robust Regression and Outlier Detection*. New York: Wiley.

Ruppert, D. and B. Aldershof (1989). Transformations to symmetry and homoscedasticity. *Journal of the American Statistical Association*, **84,** 437–446.

Ryan, T. P. (1989). *Statistical Methods for Quality Improvement*. New York: Wiley.

Ryan, T. P. (1990). Linear Regression, in *Handbook of Statistical Methods for Engineers and Scientists*, H. M. Wadsworth, ed. New York: McGraw-Hill, Chapter 13.

Sampson, A. R. (1974). A tale of two regressions. *Journal of the American Statistical Association*, **69,** 682–689.

Schmoor, C., K. Ulm, and M. Schumacher (1993). Comparison of the Cox model and the *regression* tree procedure in analysing a randomized clinical trial. *Statistics in Medicine*, **12,** 2351–2366.

Seber, G. A. F. (1977). *Linear Regression Analysis*. New York: Wiley.

Senft, R. L., L. R. Rittenhouse, and R. G. Woodmansee (1983). The use of regression models to predict spatial patterns of cattle behavior. *Journal of Range Management*, **36,** 553–557.

Shaffer, J. P. (1991). The Gauss–Markov theorem and random regressors. *The American Statistician*, **45,** 269–273.

Small, G. W., M. A. Arnold, and L. A. Marquardt (1993). Strategies for coupling digital filtering with partial least squares regression: Application to the determination of glucose in plasma by Fourier transform near-infrared spectroscopy. *Analytical Chemistry*, **65,** 3279–3289.

Smith, B. and B. Waters (1985). Extending applicable ranges of regression equations for yarn strength forecasting. *Textile Research Journal*, **55,** 713–717.

Stein, M. L., X. Shen, and P. E. Styer, (1993). Applications of a simple regression model to acid rain data. *Canadian Journal of Statistics*, **21,** 331–346.

Vecchia, A. V. and R. L. Cooley (1987). Simultaneous confidence and prediction intervals for nonlinear regression models with an application to a groundwater flow model. *Water Resources Research*, **23,** 1237–1250.

Walker, N. and R. Catrambone (1993). Aggregation bias and the use of regression in evaluating models of human performance. *Human Factors*, **35,** 397–411.

Watts, D. G. (1981). A task-analysis approach to designing a regression course. *The American Statistician*, **35**, 77–84.

Wetz, J. (1964). Criteria for Judging Adequacy of Estimation by an Approximating Response Function. Ph.D. thesis, University of Wisconsin.

EXERCISES

1.1. Plot each of the four data sets in Table 1.1 and then determine the prediction equation $\hat{Y} = \hat{\beta}_0 + \hat{\beta}_1 X$. Do your results suggest that regression data should be plotted before any computations are performed?

1.2. Given the following data:

X	1.6	3.2	3.8	4.2	4.4	5.8	6.0	6.7	7.1	7.8
Y	5.6	7.9	8.0	8.2	8.1	9.2	9.5	9.4	9.6	9.9

Compute $\hat{\beta}_0$ and $\hat{\beta}_1$ using either computer software or a hand calculator. Now make up five different values for $\hat{\beta}_{1(\text{wrong})}$ that are of the same order of magnitude as $\hat{\beta}_1$, and for each of these compute $\hat{\beta}_{0(\text{wrong})} = \bar{Y} - \hat{\beta}_{1(\text{wrong})}$.

Then compute $\sum (Y - \hat{Y})$ and $\sum (Y - \hat{Y})^2$ for each of these six solutions, using four or five decimal places in the calculations. Explain your results relative to what was discussed in Sections 1.4 and 1.5.1.

1.3. Show that Eq. (1.10) is equivalent to Eq. (1.6).

1.4. Express R^2 as a function of F, the test statistic used in the ANOVA table.

1.5. Graph the data in Exercise 1.2, and then compute R^2 and construct the ANOVA table. Could the magnitude of the values of R^2 and F have been anticipated from your graph?

1.6. Assume that $\hat{\beta}_1 = 2.54$. What does this number mean relative to X and Y?

1.7. What does $1 - R^2$ represent in words?

1.8. Given the following numbers:

Y	3	5	6	7	2	4	8
$\hat{Y}$	2.53	___	5.30	6.68	3.22	5.30	8.06

Fill in the blank. Could the prediction equation be determined from what is given? Explain.

1.9. Construct an example in which the correlation coefficient, r_{xy}, is equal to (+) 1.

1.10. Prove that $r^2_{Y\hat{Y}} = R^2$.

1.11. Compute the value of the correlation coefficient for the data in Exercise 1.2, and show that the sign of the correlation coefficient must be the same as the sign of $\hat{\beta}_1$.

1.12. Using the data in Exercise 1.2, construct a 95% prediction interval for Y when $X = 5.7$.

1.13. Consider the following data.

X	1	2	3	4	5	6	7
Y	5	6	7	8	7	6	5

First obtain the prediction equation and compute R^2. *Then*, graph the data. Looking at your results, what does this suggest about which step should come first?

1.14. Explain why it would be inappropriate to use the regression equation obtained in Exercise 1.2 when $X = 9.5$.

1.15. Perform a lack-of-fit test for the following data (use $\alpha = .05$).

X	3.2	2.1	2.6	3.2	3.0	2.1	3.2	2.9	3.0	2.6	2.4
Y	3.6	3.7	3.0	3.7	3.3	3.8	3.4	2.7	2.8	3.1	3.4

Now construct a scatter plot of the data. Does the plot support the result of the lack-of-fit test?

1.16. Explain the conditions under which an experimenter should consider using regression through the origin. Consider the following data.

X	0.6	0.7	0.4	1.1	1.3	0.9	1.6	1.2
Y	0.7	0.8	0.6	1.4	1.6	0.7	1.3	1.5

Assume that the (other) conditions are met for the proper use of regression through the origin, and obtain the prediction equation.

CHAPTER 2

Diagnostics and Remedial Measures

There is much more to regression analysis than developing a regression equation and performing the type of analyses that were given in Chapter 1.

A data analyst must also consider the following:

1. Can the model be improved by transforming Y and/or X, or by using nonlinear terms in X?
2. Do the assumptions appear to be met (approximately), and if not, what corrective action should be taken?
3. Are they any outliers?
4. Do any of the n observations exert considerably more influence than the other observations in determining the regression coefficients?

Each of these questions will be addressed separately in this chapter and again in Chapter 5 in the context of multiple regression (in which there is more than one regressor). We note at the outset that the use of the word *diagnostics* in this chapter is more inclusive than its use by some other regression authors. Specifically, we include the detection of influential observations as a diagnostic procedure.

We also note that the first and third questions may be answered by examining a scatter plot of the data. Therefore, the diagnostics that are presented in this chapter are given while recognizing that a scatter plot may sometimes tell us as much as the diagnostics, although the identification of influential observations, in particular, cannot always be done visually. Identification of outliers and influential observations is not easily accomplished when there is more than one regressor, however, so it is meaningful to first become acquainted with diagnostics in the context of simple regression, where we can usually *see* from the scatter plot why we are receiving a flag from a diagnostic, so that we gain an understanding and appreciation of the diagnostics before we apply them to multiple regression.

2.1 ASSUMPTIONS

As stated in Section 1.5.3, we assume that the model being used is an appropriate one, and that $\epsilon_i \sim \text{NID}(0, \sigma_\epsilon^2)$. The appropriateness of the model in simple regression can generally be assessed from the X–Y scatter plot. Therefore, we shall consider the assumptions of $\epsilon_i \sim \text{NID}(0, \sigma_\epsilon^2)$ first. Since the errors are not observed, the residuals are used in testing these assumptions. That is, the e_i are used in place of the ϵ_i, and this creates some problems as we shall see.

2.1.1 Independence

In certain ways the e_i are not good substitutes for the ϵ_i. For example, even if the ϵ_i are independent, the e_i won't be because they must sum to zero. This follows from Eq. (1.5a), and since Eq. (1.5b) can be written as $\sum_{i=1}^{n} X_i e_i = 0$, it follows that only $n - 2$ of the residuals could be independent. To see this, assume that we want to solve those two equations for the e_i. Since we have two equations to solve, we need two unknowns (e_i); the other $n - 2 e_i$ are free to assume any values and will thus be independent if the ϵ_i are independent.

This built-in interdependence of the e_i is not a problem when n is large, but could give a false impression of the ϵ_i when n is small. Therefore, when n is small, the use of uncorrelated residuals is desirable. [See Cook and Weisberg (1982, p. 34) for a discussion of uncorrelated residuals.]

When regression data are collected over time and the time intervals are equal, the e_i should be plotted against time. In the hypothetical example described in Section 1.3, it was stated that temperature was changed in a random manner, so temperature varied sequentially over time. Therefore, the residuals (or the standardized residuals) should be plotted against time.

The random variable

$$e_i' = \frac{e_i}{s\sqrt{1 - h_i}} \tag{2.1}$$

with

$$h_i = \frac{1}{n} + \frac{(x_i - \bar{x})^2}{\sum (x_i - \bar{x})^2} \tag{2.2}$$

will be called a *standardized residual* in this chapter and in subsequent chapters. Some authors have used the term *studentized residual* for what we are calling a standardized residual, with a standardized residual defined as e_i/s. Why the deliberate inconsistency? Whenever we subtract from a random variable its mean and then divide by its standard deviation we are standardizing that random variable. Therefore, since the (estimated) standard deviation of e_i

is given by the denominator in Eq. (2.1), as can be inferred from the derivation of Var(e_i) in the chapter appendix, e_i' is a standardized residual. In decrying the improper use of these terms in the literature, Welsch (1983) states: "It seems hopeless to overcome years of misuse of these terms." Perhaps not.

A time-sequence plot of the standardized residuals resulting from the use of the Table 1.2 data in Section 1.3 is given in Figure 2.1. The residuals do not exhibit a clear pattern and thus provide no indication of a violation of the assumption of uncorrelated errors.

Although graphical displays of residuals are an important part of regression analysis, a time series plot of residuals should generally be supplemented by a test for randomness, particularly when n is large, as it might then be difficult to detect a pattern of nonrandomness.

A simple test of nonrandomness is a *runs test* in which the number of runs above and below zero is determined. Since the third value of $e = Y - \hat{Y}$ in Table 1.3 would have been greater than zero if more decimal places had been used, the number of runs of the e_i is 6. The probability of having 6 or more runs in a sample of size 10 with 6 plus signs and 4 minus signs is .595 (see Draper and Smith, 1981, p. 160). Thus, 6 is not an extreme value, and so the runs test does not lead to rejection of the hypothesis of independent errors.

It should be noted, however, that because of the simplicity of the test, it will not be able to detect certain types of nonrandomness. For example, if the plot of the standardized residuals over time resembled a sine curve with a low

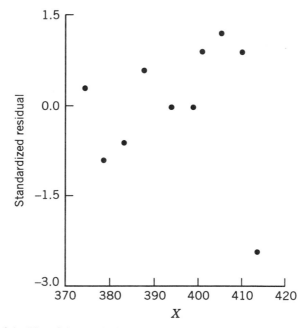

Figure 2.1 Plot of the standardized residuals against X for the data in Table 1.2.

periodicity, as in Figure 2.2, a runs test would not detect this nonrandomness. With six plus signs and eight minus signs, the probability of at most five runs is .086, so with $\alpha = .05$ we would not reject the hypothesis of randomness, even though we clearly should do so.

This type of nonrandomness can be detected with the *Durbin–Watson* test, however, which is designed to detect correlation between consecutive errors. We compute

$$D = \frac{\sum_{i=2}^{n} (e_i - e_{i-1})^2}{\sum_{i=1}^{n} e_i^2}$$

and compare D with the appropriate tabular values. Note that D will be small when consecutive e_i have a strong positive correlation and will be large when there is a strong negative correlation.

For the Table 1.2 data, $D = 2.23$, but since Durbin and Watson did not give tabular values for $n < 15$, we must extrapolate from the values for $n = 15$. Roughly, we would not reject the hypothesis of no correlation between con-

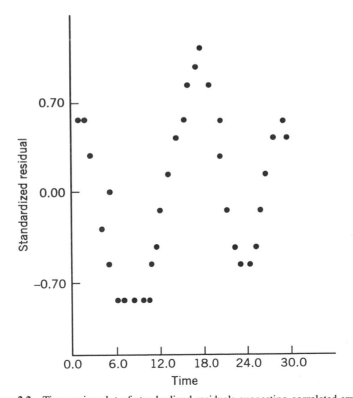

Figure 2.2 Time series plot of standardized residuals suggesting correlated errors.

secutive errors if $D > 1.2$ and $4 - D > 1.2$. This provides a two-sided test of $H_0: \rho = 0$ against $H_a: \rho \neq 0$, with ρ representing the correlation between consecutive errors, and 1.2 results from the use of $\alpha = .05$. Here we obviously conclude that there is no evidence of such correlation.

If H_0 had been rejected, this would mean that a standard least squares regression analysis cannot be used, and a time series–regression approach would be necessary. An example of the type of corrective action that may be needed is given later in this section.

A more general approach to testing for correlated errors would be to test for more than just the lag-one autocorrelation. This can be accomplished by using the *Box–Ljung–Pierce statistic* (also called the modified Box–Pierce statistic), which incorporates autocorrelations up to lag k. If n is sufficiently large, we could try to fit a time series model to the residuals and then check the appropriateness of the model using this statistic. See, for example, Cryer (1986, p. 153) or Ljung and Box (1978) for a discussion of the statistic.

2.1.1.1 Correlated Errors

We will now examine the consequences of using simple linear regression when the errors *are* correlated. We will assume that the errors are correlated over time, as will frequently occur when regression data are collected over time, and that the experimenter has not thought to use any of the diagnostic checks that have been discussed in this section.

A classic example of the deleterious effects of such misuse can be found in Box and Newbold (1971), who commented on a paper by Coen et al. (1969). The latter seemingly showed that car sales seven quarters earlier could be used to predict stock prices, as $\hat{\beta}_1$ was 14 times its standard deviation. Unfortunately, they failed to examine the residuals, and a residuals plot would have provided strong evidence that the errors were correlated. After fitting an appropriate model, Box and Newbold showed that there was no significant relationship between the two variables. [See also the discussion in Box et al. (1978, p. 496).]

In the remainder of this section we show how correlated errors can create such problems. We will assume that the errors follow a *first-order autoregressive process* [AR(1)], as this is a commonly occurring process. Specifically, we will make the usual assumption that the mean of the errors is zero, so the model is

$$\epsilon_t = \phi \epsilon_{t-1} + a_t \tag{2.3}$$

where $|\phi| < 1$, and a_t has the error term properties [NID$(0, \sigma^2)$] that the experimenter has erroneously assumed are possessed by the ϵ_t, where $Y_t = \beta_0 + \beta_1 X_t + \epsilon_t$ is the assumed regression model.

The correct model might seem to be

$$Y_t = \beta_0 + \beta_1 X_t + \phi \epsilon_{t-1} + a_t \tag{2.4}$$

where ϵ_{t-1} might be regarded as an additional regressor (with a known coefficient if ϕ is known).

Unfortunately, however, the Y_t will be correlated since the ϵ_{t-1} are obviously correlated, so Eq. (2.4) is not a model for which ordinary least squares can be applied. In particular, it can be shown (see the chapter appendix) that $\rho_{Y_t, Y_{t+1}} = \phi$, where $\rho_{Y_t, Y_{t+1}}$ is the correlation between Y_t and Y_{t+1}. It can also be shown that the correlation between Y_t and Y_{t+k} is ϕ^k.

Since this is the correlation structure of an AR(1), the appropriate corrective action is to use the model for that process on the Y_t. Specifically, if $Y_t = \alpha + \phi Y_{t-1} + b_t$, then the $Y_t - \phi Y_{t-1} - \alpha = b_t$ will be uncorrelated, just as the $\epsilon_t - \phi \epsilon_{t-1} = a_t$ are uncorrelated in Eq. (2.3).

It can be shown, as the reader is asked to do in Exercise 2.1, that letting $Y'_t = Y_t - \phi Y_{t-1}$ produces

$$Y'_t = \beta'_0 + \beta_1 X'_t + a_t \tag{2.5}$$

where $\beta'_0 = \beta_0(1 - \phi), X'_t = X_t - \phi X_{t-1}$, and $a_t = \epsilon_t - \phi \epsilon_{t-1}$.

Thus, we should regress Y'_t on X'_t. To do so, however, requires that we know ϕ. In general, ϕ will be unknown, as will be the appropriate model for ϵ_t. Thus, ϕ will have to be estimated from the data, after first determining that an AR(1) is an appropriate model.

A logical first step would be to use the autocorrelation function (and partial autocorrelation function) of the residuals in determining (1) whether the errors appear to be correlated and (2) if correlated, what an appropriate model might be. [The *autocorrelation function* gives the correlation between the residuals (or whatever is being examined) at time lags $1, 2, \ldots$, whereas the *partial autocorrelation function* gives functions of the autocorrelations.]

2.1.1.1.1 An Example

Values for X and Y are given in Table 2.1, with the Y values generated as $Y_t = 10 + 2X_t + \epsilon_t$ where $\epsilon_t = 0.8\epsilon_{t-1} + a_t$ and $a_t \sim \text{NID}(0, 1)$. Before the Y_t were generated, 77 values of a_t were randomly generated and the ϵ_t subsequently calculated with $\epsilon_0 = a_0$. Only the last 50 were used, however, and this was done so as to essentially remove the effect of setting $\epsilon_0 = a_0$. The regressor values were randomly generated from the discrete uniform distribution defined on the interval (10,20). (The discrete uniform distribution is a probability distribution such that the random variable has a finite number of possible values, and the probability of the random variable assuming any one of those values is the same, regardless of the value.)

Table 2.1 Data with Correlated Errors

Y	X	Y	X
46.8310	17	36.3575	12
50.9132	20	46.9015	18
46.4617	18		
30.5711	10		
39.8404	15		
31.1080	10		
35.4954	12		
36.1882	13		
34.0138	13		
47.2470	20		
40.5826	17		
37.1003	15		
37.6943	15		
44.0767	18		
44.0843	18		
31.4254	11		
31.3305	11		
49.1607	20		
41.8043	17		
34.8469	13		
42.4542	17		
36.3825	14		
29.1990	11		
29.6211	12		
36.2948	15		
38.2859	15		
39.9781	16		
35.0621	13		
39.8755	15		
39.8197	15		
48.9931	20		
38.7739	14		
45.1977	17		
41.3773	15		
47.9279	19		
32.4987	12		
44.2598	17		
49.7332	19		
43.4409	16		
45.0817	17		
36.0555	13		
34.5373	12		
37.7932	14		
32.2893	11		
39.3821	14		
33.4188	11		
38.5081	13		
44.8539	17		

The regression equation is $\hat{Y} = 10.2 + 1.96X$, with $R^2 = .916$ and $t = 22.95$ is the value of the t-statistic for testing the hypothesis that $\beta_1 = 0$. A time sequence plot of the residuals is given in Figure 2.3. This shows that the residuals are highly correlated, as they should be from the manner in which the errors were generated.

When the appropriate corrective action is taken by using $X_t' = X_t - 0.8X_{t-1}$ as the new regressor and $Y_t' = Y_t - 0.8Y_{t-1}$ as the new dependent variable, the regression equation is

$$\hat{Y}_t' = 2.06 + 1.94X_t'$$

with the t-statistic equal to 48.09 and $R^2 = .980$. [The change in the intercept estimate is expected since $\beta_0' = \beta_0(1 - \phi)$.] Thus, both the t-statistic and the value of R^2 indicate that the fit has been appreciably improved, and a time sequence plot of the residuals would suggest that the correlation in the errors has been virtually removed. (Note: R^2 values cannot be directly compared when a transformation of Y has been made. This is not a major problem here, however. See Section 6.4 for more information on computing R^2 after Y has been transformed.)

Although this result is not in line with the Box and Newbold scenario in that the correct model shows an even stronger relationship between the variables than the incorrect model, it does illustrate another important result. The ordinary least squares estimators will not have minimum variance among all unbiased estimators when the errors are correlated (and in general when any of the assumptions are violated). Here the estimated variance of $\hat{\beta}_1$ has decreased from 0.0073 to 0.0016 as a result of the transformation, and this explains the increase in the value of the t-statistic.

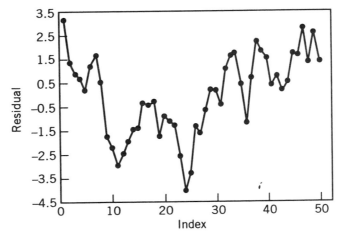

Figure 2.3   Time sequence plot of the residuals for Table 2.1 data.

Other corrective action that can be taken for correlated errors is discussed extensively by Neter et al. (1989, Chapter 13), and the reader is referred to that source for additional details.

See Section 2.5 for a detailed discussion of transformations that are used for other purposes.

It should be noted that the errors are also assumed to be independent of X, and residual plots are frequently used in checking this assumption. These are discussed in detail in Chapter 5.

In checking for possible nonindependence of the errors by using residual plots it is somewhat better to use the standardized residuals than the ordinary ("raw") residuals, and this preference should especially be exercised when residual plots are used for other purposes, such as checking for nonnormality and a possible nonconstant error variance. Certainly if we are looking for a nonrandom pattern, we want the variances of the random variables that we are plotting to be the same; this is true only if we use the standardized residuals.

2.1.2 Normality

A *normal probability plot* of the standardized residuals is frequently used for checking the normality assumption of ϵ. This is a plot of the standardized residuals against their expected values, which are computed under the assumption of normality. When a normal probability plot is used, we want the random variables that correspond to the plotted points to have the same distribution, but this will not occur if we use the raw residuals, which have different standard deviations, and hence have different distributions. Therefore, we must use the standardized residuals.

The usefulness of such plots is questionable when n is small, however [see Cook and Weisberg (1982, p. 56) or Weisberg (1980)], and there are other problems.

Checking the normality assumption on ϵ can be difficult since the distribution of e need not closely resemble the distribution of ϵ. Assume that ϵ has a normal distribution and the model is correct. As shown by Weisberg (1980), we may write e_i in terms of the ϵ_i as

$$e_i = (1 - h_i)\epsilon_i - \sum_{\substack{j=1 \\ j \neq i}}^{n} h_{ij}\epsilon_j \tag{2.6}$$

where the h_i are as defined in Eq. (2.2), and the h_{ij} are given by

$$h_{ij} = \frac{1}{n} + \frac{(x_i - \bar{x})(x_j - \bar{x})}{\sum_{k=1}^{n} (x_k - \bar{x})^2} \tag{2.7}$$

The NID assumption on the ϵ_i implies that the e_i must be normally distributed from the linear combination of the ϵ_i in Eq. (2.6). More specifically, e will have a singular normal distribution (singular because the residuals are not independent) with $\mathrm{Var}(e_i) = \sigma^2(1 - h_i)$. (This variance is derived in the chapter appendix.) Thus, unless the data have come from a G-optimum design (see Chapter 14), the variances of the residuals will differ, and will also differ from σ^2.

2.1.2.1 *Supernormality Property of Residuals*

A more serious problem occurs when the errors are not normally distributed. First, if the h_i differ, the e_i will then have different distributions. When h_i is large the corresponding e_i will be approximately normally distributed because of the central limit theorem, but when h_i is small, e_i will then have approximately the same distribution as ϵ_i, but will be slightly closer to a normal distribution. Thus, we have the disturbing result that the residuals will have a distribution that is closer to normal than will be the distribution of the errors when the errors are not normally distributed. This property has been referred to as *supernormality* by many authors. See, for example, Quesenberry and Quesenberry (1982), Boos (1987), and Atkinson (1985).

To remedy this problem, Atkinson (1981) proposed a "simulation envelope" to be used in conjunction with a half-normal plot of the e_i^*, with

$$e_i^* = \frac{Y_i - \hat{Y}_{(i)}}{s_{(Y - \hat{Y}_{(i)})}} \tag{2.8}$$

where $\hat{Y}_{(i)}$ is the predicted value of Y_i that is obtained without using the ith data point in computing $\hat{\beta}_0$ and $\hat{\beta}_1$, and the denominator in Eq. (2.8) is the estimated standard deviation of the numerator. (The half-normal plot is a plot of $|e_i^*|$ against the expected value of $|e_i^*|$, assuming normality.)

After using some algebra we are able to write

$$e_i^* = \frac{e_i}{s_{(i)}\sqrt{(1 - h_i)}} \tag{2.9}$$

where $s_{(i)}$ is the value of s that results when the ith observation is not used in the computations, and h_i is as defined in Eq. (2.2).

2.1.2.2 *Standardized Deletion Residuals*

We will call the e_i^* the *standardized deletion residuals* to reflect the fact that we have a type of standardized residual [Eq. (2.9) is of the same general form as Eq. (2.1)], and to indicate that a point is being deleted in the computations. There is, unfortunately no agreement in the literature concerning the label for e_i^*. Atkinson (1985) calls it the *deletion residual*; Cook and Weisberg (1982,

p. 20) call it an *externally studentized residual*; Montgomery and Peck (1992, p. 174) call it both the latter term and *R*-student; and Belsley et al. (1980, p. 20) call it RSTUDENT. The term that we are using here seems to be the most appropriate, however, and we shall continue to use it.

Why are the statistics in Eq. (2.8) computed without using the *i*th observation? The rationale is that if the *i*th observation is a bad data point, then it should not be allowed to inflate the estimate of σ, and thus help camouflage itself. That is certainly logical, but here we are considering methods for detecting nonnormal errors rather than trying to identify bad data points. Even though each e_i^* has a *t*-distribution with $n - p - 1$ degrees of freedom, the e_i^* are well removed from ordinary residuals, and in testing for nonnormality using residuals it seems desirable to use the standardized version of some statistic that well represents the errors. Therefore, the use of standardized residuals seems preferable. We discuss this further in the next section.

2.1.2.3 Methods of Constructing Simulation Envelopes

The procedure proposed by Atkinson (1981) is as follows. Nineteen sets (samples) of *Y* values are simulated where $Y \sim N(0, 1)$, and the values of *X* in the original sample are used in conjunction with these 19 sets of *Y*-values to produce 19 sets of e_i^*. (Thus, *X* is taken as fixed.) The envelope boundaries are then obtained as the two sets of values that contain the 19 maxima and 19 minima, respectively, with one maximum and one minimum coming from each set. (The reason for simulating 19 sets of *Y*-values is that this provides estimates of the 5th and 95th percentiles of the distribution of the *i*th-ordered standardized deletion residual.)

These simulated residuals (the e_i) are normally distributed since *Y* has a normal distribution. Thus by using the extrema of the e_i^*, we would be constructing boundaries that essentially provide a limit on how nonnormal the residuals can *appear* to be when the errors have a normal distribution.

There are some problems with this approach, however. As discussed by Atkinson (1985, p. 36), the probability that the envelope contains the sample residuals, assuming normality, is not easily determined. This is due to the fact that the envelope is constructed so as to provide an approximate probability of a particular ordered residual being contained by the envelope, rather than *all* of the residuals being inside the envelope. Another problem is that two data analysts could reach a different conclusion for a data set since the boundaries are determined by simulation. The envelopes will also be quite sensitive to the simulation of outliers in *Y* since such outliers would become part of the envelope. To remedy this last deficiency, Flack and Flores (1989) present some outlier-resistant modifications to Atkinson's method by adapting the methods of Hoaglin et al. (1986) and Hoaglin and Iglewicz (1987) to regression residuals.

One of the suggested modifications is as follows. The lower (L_i) and upper (U_i) envelope boundaries for the *i*th-ordered standardized deletion residual would be given by

$$L_i = F_i^L - k(F_i^U - F_i^L)$$
$$U_i = F_i^U + k(F_i^U - F_i^L)$$

where F_i^U, F_i^L, and k could be defined in one of several possible ways, such as letting $k = 1.5$ and letting the first two denote the 75th and 25th percentiles, respectively, of the simulated errors corresponding to the ith residual.

How does the Flack and Flores approach compare with the Atkinson method if the errors are normally distributed? For a normal distribution, the third and first quartiles are each a distance of approximately 0.67σ from μ, so it follows from a table of the standard normal distribution that the probability of a normally distributed random variable being classified as an outlier is approximately .007. As indicated by Flack and Flores (1989), the ith simulated residual will similarly fall outside the envelope less than 1% of the time. Thus, we might view the simulated envelopes as providing boundaries that approximate this probability for a single residual, but the probability that the entire residual vector is contained by the envelope cannot be calculated directly since the residuals are not independent, even if unordered, and the ordering of them also creates a dependency.

It is clear, however, that the inclusion probability for the entire set of residuals will decrease as n increases, and that probability could be unacceptably small if n is large, as can also be surmised from Table 2 of Flack and Flores (1989). This can especially be a problem with the Atkinson approach since the probability is only approximately .9 that the ith standardized deletion residual will be contained by the envelope.

If we use the Flack and Flores approach, we can attempt to correct for this problem by increasing k, but the price paid for that is less sensitivity in detecting nonnormality.

Regardless of which procedure is used, a simulated envelope should be viewed as a device that helps provide some insight into possible nonnormality of the error term. It is inappropriate to view them as providing a formal hyothesis test for normally distributed errors.

Another problem with the envelopes is as follows. The placing of bounds on each standardized deletion residual puts the emphasis on detecting extreme values, not curvature in values that are not extreme. If the errors have a highly skewed distribution, a plot of the standardized deletion residuals can exhibit (and should exhibit) considerable curvature, without any of the points necessarily being extreme. But considerable curvature could exist and all of the points still fall within the envelope, as can be seen by examining Figures 1 and 2 in Flack and Flores (1989). Perhaps in recognition of this, these authors indicate that abnormal skewness in the plot can be determined more easily by superimposing a plot with errors from a skewed distribution over the plotted points.

The envelope approach should also be viewed in light of the following. The methods of Atkinson (1981) and Flack and Flores (1989) assume the use of standardized deletion residuals, but some thought should be given to the type of residual that is to be used. Although this type of residual is useful in detecting extreme observations that do not adhere to the regression relationship formed by the "good" data, it is not necessarily the best choice of residual for testing normality. In particular, the deviation of an observation from the line formed by the other $n - 1$ points can be magnified when the fit to the remaining points is exceptionally good, as $\hat{\sigma}_{(i)}$ will then be quite small. In such an instance, the standardized deletion residual will be quite large and much larger than the corresponding standardized residual, even though a scatter plot of the data may indicate that there are not any extreme points. (The reader will see this in Exercise 2.9.) A very large standardized deletion residual will probably fall outside the simulated envelope, regardless of which approach is used. Since the envelope is to be used to roughly assess normality, we should be concerned that points can fall outside the envelope simply because of the choice of residual statistic. Obviously, we do not want nonnormality and the presence of "extreme" data points to be confounded in the plot, so under certain conditions it would be safer to use standardized residuals.

There is still the problem, however, of not having available numerical results to indicate the coverage probability for a given X matrix, so the user of simulated envelopes will probably have some difficulty in interpreting the results.

To this point we have addressed the fact that the residuals can be closer to being normally distributed than the errors when the errors have a nonnormal distribution. But what if the residuals are such that it is "clear" that the errors could *not* be normally distributed, so the hypothesis of normality is rejected? If we simulate some regression data in such a way that the errors have a normal distribution, a normal probability plot of the residuals almost certainly will not form a straight line. That is, the correlation between the residuals and their expected values will not be one. Therefore, the *sample* residuals will not be normally distributed even when the errors have a normal distribution. Daniel and Wood (1980, pp. 34–43) give many examples of plots of normal random variables, several of which are not even close to a straight line, especially when n is small, so properly interpreting a normal probability plot can be difficult. Another approach would be to use a test statistic that is a function of the residuals. One possibility is to compute the correlation between the residuals and their expected values under normality, which is essentially equivalent to the Shapiro–Wilk test for normality. Critical values for this test are given by Looney and Gulledge (1985); see also Minitab, Inc. (1989). (Obviously this test will also be undermined by supernormality, however.)

To illustrate these points, we consider the data given in Table 2.2 and the corresponding scatter plot in Figure 2.4. The data were randomly generated with an error term that has an exponential distribution with a mean of 0.5—a highly skewed distribution.

The Atkinson (1981) envelope approach and the Flack and Flores (1989) modification are illustrated in Figures 2.5(*a*) and 2.5(*b*), except that a full normal plot is used for each so that they may be compared. Although Atkinson prefers a half-normal plot, Flack and Flores give an example in which the half-normal plot is inferior to the full plot in detecting extreme values. Since the plots can be easily obtained, a safe approach would be to use both, as is also suggested by Flack and Flores.

**Table 2.2   Data with Nonnormal Errors
for Illustrating Simulation Envelopes**

Y	X
16.7801	5
10.1034	2
12.1897	3
14.1901	4
10.3819	2
22.1771	8
20.2360	7
16.3304	5
14.1002	3
18.2106	6
20.2689	7
12.5605	3
14.0517	4
16.1840	5
10.2889	2
12.8639	3
15.3753	4
22.5566	8
24.0377	9
24.4948	9
11.5980	2
14.6324	4
18.0385	6
20.2252	7
14.2256	4

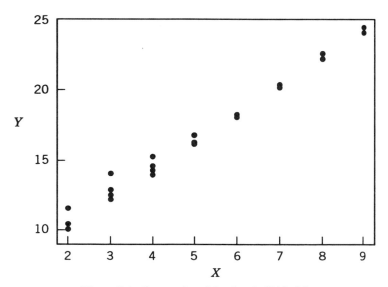

Figure 2.4   Scatter plot of the data in Table 2.2.

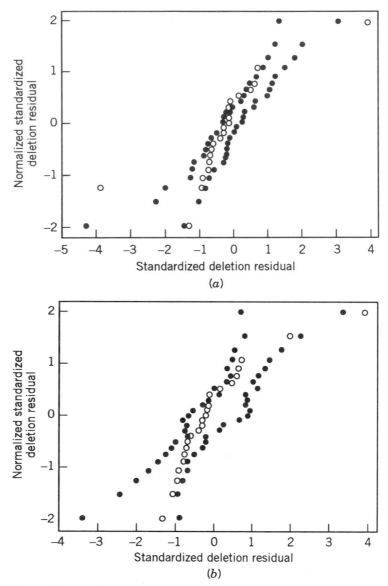

Figure 2.5 (*a*) Atkinson simulation envelope for Table 2.2 data. (*b*) Flack and Flores simulation envelope for Table 2.2 data.

Each envelope provides clear evidence of nonnormality, especially the Atkinson envelope where several standardized deletion residuals fall outside the envelope, and several others are essentially on the boundary. We note that for this example there is a considerable difference between the largest standardized deletion residual (3.92) and the largest standardized residual (3.07) (which, of course, occurs at the same point), so the evidence of nonnormality is likely exaggerated somewhat at that point.

How should the graphs appear when the errors are normally distributed? Using the same X values in Table 2.2 and the same functional relationship between X and Y, except that the errors are generated from a normal distribution, an Atkinson envelope is given for the standardized deletion residuals in Figure 2.6(a), and a Flack–Flores envelope is given in Figure 2.6(b). The latter, in particular, is what we would expect to observe as almost all of the residuals are well inside the envelope, with only a few values at or close to

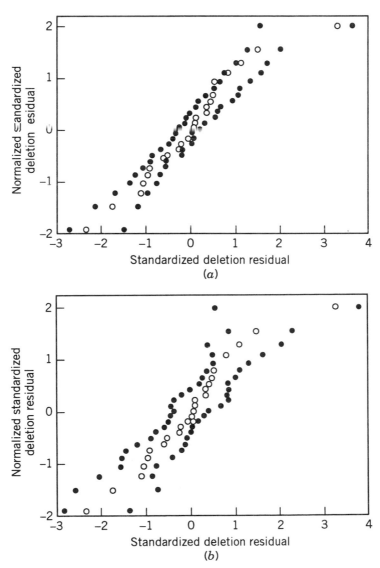

Figure 2.6 (a) Atkinson simulation envelope for example with normal errors. (b) Flack and Flores simulation envelope for example with normal errors.

either boundary. (Obviously, with a large number of points we will have one or more points outside the boundaries due to chance alone, just as some significant results from hypothesis tests will appear if enough tests are conducted. Therefore, it is a matter of determining what constitutes an extreme configuration of residuals relative to the boundaries, and that determination has to be somewhat of an art. Here we have two extremes where the message is clear, however.)

Statistical software that can be used in constructing simulated envelopes is discussed in Section 2.7 and in the chapter appendix.

2.1.3 Constant Variance

In addition to being independent and having a normal distribution, the errors must also have a constant variance. In particular, the variance should be constant over the different values of X.

Unfortunately, a nonconstant error variance occurs quite frequently in practice. Box and Hill (1974) discuss how this can happen with chemical data, and a major focus of Carroll and Ruppert (1988) is discussion and illustration of the problem as it occurs in various areas of application.

If the constant error variance assumption is not valid, then ordinary least squares cannot be used. This can be explained as follows. Recall that the method of least squares minimizes the sum of the squared distances from the points to the regression line. But if σ_ϵ^2 is not constant over X so that we must write $\sigma_{\epsilon \cdot x}^2$ to indicate that the variance depends on X, then obviously those vertical distances, the magnitude of which will depend on $\sigma_{\epsilon \cdot x}^2$, should not be weighted equally. Otherwise, the equation for the regression line will be strongly influenced by points at values for X for which $\sigma_{\epsilon \cdot x}^2$ is large. Consequently, points for which $\sigma_{\epsilon \cdot x}^2$ is comparatively large should be downweighted, and in general the weights should be a function of $\sigma_{\epsilon \cdot x}^2$. This is accomplished when *weighted least squares* (WLS) is used.

2.1.3.1 *Weighted Least Squares*

Specifically, instead of minimizing $L = \sum_{i=1}^{n} (Y_i - \beta_0 - \beta_1 X_i)^2$ as given in Eq. (1.4), the expression to be minimized is $W = \sum_{i=1}^{n} w_i(Y_i - \beta_0 - \beta_1 X_i)^2 = \sum_{i=1}^{n} w_i \epsilon_i^2$, where the w_i are the reciprocals of the $\sigma_{\epsilon_i}^2$. Thus, large errors receive small weights, as should be the case. As with the minimization of L, the minimization of W leads to two equations that when solved simultaneously produce

$$\hat{\beta}_{1(w)} = \frac{\sum w_i x_i y_i - \dfrac{\left(\sum w_i x_i \right)\left(\sum w_i y_i \right)}{\sum w_i}}{\sum w_i x_i^2 - \dfrac{\left(\sum w_i x_i \right)^2}{\sum w_i}} \qquad (2.10a)$$

and

$$\hat{\beta}_{0(w)} = \frac{\sum w_i y_i - \hat{\beta}_{1(w)} \sum w_i x_i}{\sum w_i} \qquad (2.10b)$$

Notice that if each $w_i = 1/n$, which would be equivalent to using ordinary least squares, $\hat{\beta}_{1(w)}$ and $\hat{\beta}_{0(w)}$ reduce to the ordinary least squares (OLS) estimators $\hat{\beta}_1$ and $\hat{\beta}_0$, respectively.

Although, strictly speaking, weighted least squares should be used whenever it is believed that σ_ϵ^2 is not constant, the question arises as to how "nonconstant" σ_ϵ^2 must be before the use of WLS would result in an appreciable improvement. Carroll and Ruppert (1988, p. 16) suggest that if the $s_{i(y)}$ do not differ by more than a factor of $1.5:1$, then weighting would not be necessary, whereas a ratio of at least $3:1$ would be a clear signal that weighting should be used. [Here $s_{i(y)}$ denotes the standard deviation of Y at x_i, which is equivalent of course to $s_{i(\epsilon)}$.] To apply any such rule-of-thumb we would obviously need enough data to compute each $s_{i(y)}$, and the number of replicates of each X_i would also have to be large enough to provide a good estimate of the $\sigma_{i(y)}$.

Unless prior information exists regarding the variability of Y, it may be difficult to determine the need for WLS before a regression analysis is performed. Consequently, it will frequently be necessary to perform a regular regression analysis and then examine the residuals to see if there is a need to use weighted least squares.

What are the consequences of using OLS when WLS should be used? The estimators $\hat{\beta}_0$ and $\hat{\beta}_1$ are unbiased, but the variance of each estimator is inflated because σ_ϵ^2 is not constant. [These results are proven in the chapter appendix. See also Carroll and Ruppert (1988, p. 11) for further discussion.]

To see this numerically we can randomly generate four values of ϵ for each of five values of σ_ϵ: 1.0, 1.5, 2.0, 2.5, and 3.0. (Since the ratio of largest to smallest is $3:1$, there is a clear need to use weighted least squares.) We use $x = 2, 3, 4, 5$, and 6 corresponding to the five values of σ_ϵ. Then forming the values of Y as $Y_i = X_i + \epsilon_i$ produces the data given in Table 2.3. The scatter plot of these data is given in Figure 2.7, which shows the increasing spread of Y values as X increases. If OLS is erroneously used, the prediction equation is $\hat{Y} = -0.8995 + 1.198X$, as compared with the "expected" prediction equation of $\hat{Y} = X$. It can be shown, using the expressions for the variances of $\hat{\beta}_1$ and $\hat{\beta}_0$ given in Sections 1.5.4 and 1.5.5, that $\text{Var}(\hat{\beta}_1)$ is estimated by 0.1035, and $\text{Var}(\hat{\beta}_0)$ is estimated by 1.8626.

For this example the $\sigma_{i(y)}^2$ are known (and for the remainder of this section these will be denoted as σ_i^2). We first illustrate how to use weighted least squares using these known values and will then proceed as if the weights were unknown (the usual case), with the data in Table 2.3 used to estimate the weights.

One way to view weighted least squares is to recognize that it is equivalent to using OLS on a transformed model. The model $Y_i = \beta_0 + \beta_1 X_i + \epsilon_i$ is inappropriate because $\sigma_{\epsilon_i}^2$ is not constant over i. Therefore, it is necessary to transform the model in such a way so as to make $\sigma_{\epsilon_i}^2$ constant. If we multiply

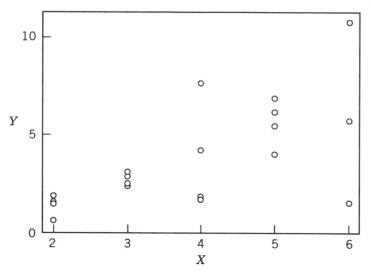

Figure 2.7   Scatter plot of the data in Table 2.3.

Table 2.3   Data for Illustrating Weighted Least Squares

Y	X
1.58	2
1.47	2
1.88	2
0.63	2
2.93	3
3.02	3
2.39	3
2.53	3
7.64	4
1.58	4
1.79	4
4.20	4
4.00	5
6.85	5
6.19	5
5.47	5
1.50	6
10.78	6
5.70	6
5.72	6

the model by $\sqrt{w_i}$ we obtain

$$Y_i^* = \beta_0\sqrt{w_i} + \beta_1 X_i^* + \epsilon_i^* \tag{2.11}$$

which is a regression model with two regressors ($\sqrt{w_i}$ and X_i^*), and with no intercept. Thus, we could, if desired, obtain the weighted least squares regression equation by applying ordinary least squares.

Our objective is to have $\text{Var}(\epsilon_i^*) = \sigma^2$, and since $\epsilon_i^* = \sqrt{w_i}\epsilon_i$ it follows that we must define $w_i = \sigma^2/\sigma_i^2$. When weighted least squares is implemented, however, we can simply use $w_i = 1/\sigma_i^2$ since the constant σ^2 in the numerator of σ^2/σ_i^2 would drop out when Eqs. (2.10) are used. Using these equations with w_i so defined produces $\hat{\beta}_{0(w)} = -1.0708$ and $\hat{\beta}_{1(w)} = 1.2434$. Notice that these values are close to the OLS estimates, $\hat{\beta}_0 = -0.8995$ and $\hat{\beta}_1 = 1.198$. This should not be surprising since both sets of estimators are unbiased.

It can also be shown, however, that the estimates of the variances of the WLS estimators are considerably less than the estimates of the variances of the OLS estimators, as will generally be the case.

In the WLS analysis, σ^2 could be estimated as $\hat{\sigma}_{(w)}^2 = \sum w_i(Y_i - \hat{Y}_i)^2/(n-2)$. This is obviously equal to $\sum (Y_i^* - \hat{Y}_i^*)^2/(n-2)$, so $\hat{\sigma}_{(w)}^2$ could also be obtained from an OLS analysis of the transformed model. It can be shown that $\hat{\sigma}_{(w)}^2 = 0.71$, whereas $\hat{\sigma}^2 = 4.14$. These are close to what we would expect since the average of the σ_i^2 is 4.0 and $\text{Var}(\sqrt{w_i}\,\epsilon_i) = 1.0$ for all i.

It can be shown that $\text{Var}(\hat{\beta}_{1(w)})$ is estimated by 0.0613, and the estimate of $\text{Var}(\hat{\beta}_{0(w)})$ is 0.6239. (The variance estimates are obtained in a manner analogous to the way that the other estimates are obtained. This is most easily handled with matrix algebra, however, so the derivation of the variances of the estimators is given in the appendix to Chapter 3.) Notice that these variance estimates are much smaller than the variance estimates of the (inappropriate) OLS estimators, which were given previously as 0.1035 and 1.8626, respectively.

Thus, in addition to satisfying the assumption that the variance of the error term be constant, we have also obtained estimators with smaller variance estimates than those of the OLS estimators. [This will always be true for large samples; see Carroll and Ruppert (1982, p. 11).]

2.1.3.1.1 *Unknown Weights*

In this hypothetical example the weights were known because the variances were known. This is not the usual case, however, and the use of weighted least squares is much more difficult when the weights are unknown. The weights have the form $w_i = 1/\hat{\sigma}_i^2$, but estimating the σ_i^2 can be quite difficult with a small amount of data.

The simplest (and most naive) approach would be to compute s_y^2 at each X_i, and then let $\hat{\sigma}_i^2 = s_y^2$. If the X values occur at random, however, we would expect to have very few, if any, repeats, and if the X values are preselected we would need to have each X_i, replicated more than just a few times before we could expect to obtain reasonable estimates of the σ_i^2.

Research papers that have addressed this issue include Carroll and Cline (1988) and Deaton et al. (1983). The former claim that each X_i should be replicated at least 10 times, whereas the latter favor at least 9 replicates. If X is random, we would not likely find clusters of 9 or 10 similar X values; and if the X values are to be selected, an experimenter would probably not want to use so many replicates for each X.

Consequently, this method of estimating the σ_i^2 should not be given serious consideration. In particular, it is inferior to the other methods to be discussed. [See Carroll and Ruppert (1988, p. 111) for a related discussion.]

A better approach would be to use what Carroll and Cline (1988) term the *average squared error*. Specifically, one would compute $\sum (Y - \hat{Y})^2/m_i$ at each X_i, where m_i is the number of replicates of X_i. [They assume that the m_i are equal; what to do when the m_i differ is discussed by Davidian (1990).] We would expect this approach to be superior to using s_i^2 because the latter is equivalent to using $\hat{Y} = \bar{Y}$ (i.e., not taking advantage of the postulated relationship between Y and X). Carroll and Cline (1988) do not suggest a necessary minimum value of m_i, but indicate that more than just a few replicates will be necessary.

Neither of these approaches can be expected to produce satisfactory results for a small-to-moderate number of replicates. For the Table 2.3 data we have $m_i = m = 4$, and the known weights are compared with the weights estimated using each approach in Table 2.4. We can see that both methods produce poor results (especially for $x = 2$ and $x = 3$), as should be expected for $m = 4$.

Carroll and Cline (1988) indicate that the estimators (of β_i) obtained using these two approaches have the same limiting distribution as the estimators obtained using known weights, but that would be of interest only to experimenters with large m.

For whichever approach is used in estimating the σ_i^2, it is likely that *iteratively reweighted least squares (IRLS)* will be necessary. As indicated previ-

Table 2.4 Comparison of Methods for Obtaining the Weights for Weighted Least Squares Applied to the Data in Table 2.3

X	$W_1 = 1/\sigma_i^2$	$W_2^a = 1/s_{yi}^2$	$W_3 =$ (average squared error)$^{-1}$
2	1.00	4.65	4.42
3	0.44	14.33	14.23
4	0.25	0.17	0.17
5	0.16	0.89	0.71
6	0.11	0.09	0.09

[a]The divisor for the sample variance in W_2 is n.

ously, we might obtain the initial weights using OLS, as has been mentioned by Carroll and Ruppert (1988, p. 13), Box and Draper (1986, p. 85), and Draper and Smith (1981, p. 110). We would then obtain new weights, if necessary, from the first WLS analysis, and repeat this process as long as is necessary.

How long will that be? This will clearly depend on how good the initial estimates are. If the initial estimates are as bad as the ones in Table 2.4, we cannot expect the estimates to converge very quickly to what the estimates would be using known weights. Carroll and Ruppert (1988, pp. 14–18) recommend that at least two iterations be used if the initial weights are obtained from using OLS, but considerably more than two iterations will be necessary if we start out with poor estimates. They also indicate (p. 14) that IRLS estimators are not guaranteed to converge, but their experience suggests that convergence is typical.

Remembering that our objective is to make $\text{Var}(\sqrt{w_i}\epsilon_i) = \sigma^2$ at each iteration, we could plot the standardized residuals against X and see if the spread of the former is approximately constant over the different values of the latter.

Table 2.5 contains the results for the Table 2.3 data with 10 iterations, and we can see that the estimates converge to $\hat{\beta}_{0(w)} = -1.32927$ and $\hat{\beta}_{1(w)} = 1.35180$ after 5 iterations beyond the initial estimates that are obtained using OLS. These estimates clearly differ considerably from the known parameter values, and this relates to the fact that the σ_i^2 are also estimated poorly. Specifically, $\sigma_i^2 = 1, 2.5, 4, 6.25,$ and 9 are estimated by 0.22, 0.07, 6.04, 1.16, and 11.5447, respectively. We note that in this instance these estimates are very close to the s_i^2, and that is because R^2 is only .435 for the OLS fit. Thus, the Carroll and Cline method cannot benefit very much from the utilization of the relationship between Y and X, as that relationship is not very strong. When this is the case the (naive) approach using the s_i^2 is apt to do about as well as the approach using $\hat{Y}$.

Either approach does have the desired effect of creating a constant error variance, however, as can be seen from Figure 2.8. In particular, the "correction" can be seen by comparing Figure 2.9, using OLS, with Figure 2.8.

Table 2.5 Iteratively Reweighted Least Squares for the Data in Table 2.3 Using the Carroll and Cline Approach

Iteration No.	$\hat{\beta}_{0(w)}$	$\hat{\beta}_{1(w)}$
1 (OLS)	−0.8995	1.1980
2	−1.30421	1.34240
3	−1.32803	1.35134
4	−1.32921	1.35178
5	−1.32926	1.35180
6–10	−1.32927	1.35180

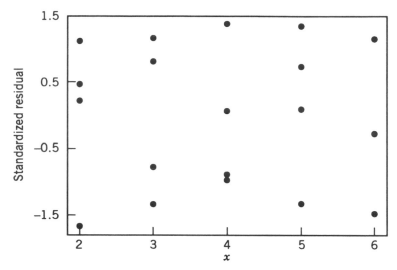

Figure 2.8 Plot of the standardized residuals for the iteratively reweighted least squares fit for the data in Table 2.3.

2.1.3.1.2 Modeling the Variance

Another approach that has been suggested in the literature is to obtain the weights by modeling s^2. Carroll and Ruppert (1988, p. 12) take the position that by modeling the variance we can avoid using many replicates, and indeed need not have every X replicated. A regression model would be developed using

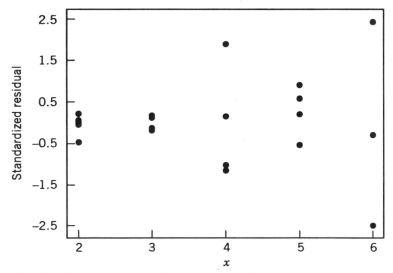

Figure 2.9 Plot of the standardized residuals from the ordinary least squares fit for the data in Table 2.3.

some function of s^2 (such as s) as the dependent variable, and the weights would then be determined from the appropriate function of the reciprocals of the predicted values of the dependent variable.

Draper and Smith (1981, p. 112) give an illustrative example in which they use this type of approach. For a sample of $n = 35$ observations, there were few repeats, but similar X-values were clustered so as to form five groups. The pure error mean square (s_e^2) was then computed for each group and regressed against $\overline{X}$ and $\overline{X}^2$. (A plot of s_e^2 against $\overline{X}$ suggested the use of a quadratic term.) The weights were then formed as the reciprocals of the predicted values.

In using such approaches to obtaining the weights, it is obvious that a model must be selected, analogous to the selection of a model for the dependent variable Y. Certainly there are many possibilities, and the reader is referred to Chapters 2 and 3 of Carroll and Ruppert (1988) for a detailed discussion.

A simple, commonly used model is a *power-of-the-mean* model in which σ_i is modeled as

$$\sigma_i = \sigma(\mu_{y_i})^\theta \tag{2.12}$$

in which σ is the constant of proportionality. With this model the σ_i (and hence the σ_i^2) are thought to vary as the mean varies. Notice that if we take the logarithm of each side of Eq. (2.12) we then have a simple linear regression model. Hence, after substituting appropriate estimators for σ_i and μ_{y_i}, which would be obtained from the OLS fit using Y as the dependent variable, the weights would be obtained from the appropriate function of the predicted values from the logarithmic model.

For the illustrative model that is used in this section,

$$\sigma_i = \tfrac{1}{2} X_i \tag{2.13}$$

and with the way the data were generated $\mu_{y_i} = X_i$. Therefore, Eq. (2.13) is in the general form of Eq. (2.12). Thus, if we were able to identify that the appropriate model seems to be a constant times X, we could use regression through the origin to estimate the multiplier, which in this example is known to be 0.5. For the moment we will use s_i as the dependent variable (other forms of the dependent variable are discussed in Section 2.4). It can be shown that the regression equation is $\hat{s}_i = 0.468X$. Thus, the estimated multiplier is very close to the actual multiplier. The values for $\hat{s}_i^2$ are 0.87789, 1.97527, 3.51161, 5.48697, and 7.90077. Notice that the reciprocals of these values differ only slightly from the reciprocals of the actual variances. Using $1/\hat{s}_i^2$ as the weights produces $\hat{Y} = -1.07 + 1.24X$. This is the same regression equation (to two decimal places) that was obtained using the known variances as weights. Although a plot of the standardized residuals shows that the spread of the residuals has not been equalized as well as when the Carroll and Cline approach was used, there is not strong evidence that the variances are unequal, considering the small number of repeats of the regressor values.

For the Table 2.3 data it was known that Eq. (2.13) gave the proper functional form for the σ_i. In general, the functional form will, of course, be unknown. It was suggested previously that one might postulate a model that has the general form given by Eq. (2.12), and that by taking the logarithm of each side of the equation we obtain a simple linear regression model. Specifically, we obtain

$$\log(\sigma_i) = \log(\sigma) + \theta \log(\mu_i)$$

Since σ_i and μ_i are unknown, we need a substitute for each. As indicated by Carroll and Ruppert (1988, p. 30), we can regard $\log(|e_i|)$ as a substitute for $\log(\sigma_i)$, and $\hat{Y}_i$ as a substitute for μ_i. [This is quite reasonable since we might estimate σ_i as $\hat{\sigma}_i = \sqrt{\sum_{j=1}^{m} e_{ij}^2 / m}$, which is the square root of the average squared error that was mentioned earlier (with e_{ij} denoting the jth residual for X_i), and $\hat{Y}_i$ is the point estimator of $\mu_i = \mu_{y_i} = \beta_0 + \beta_1 X_i$.]

Thus, we would regress $\log(|e_i|)$ against $\log(\hat{Y}_i)$ and see if a strong straight-line relationship is indicated. If so, we would then have a good estimate of θ. (A linear relationship is certainly suggested by Figure 2.10.)

For the Table 2.3 data we obtain $\log(|e_i|) = -1.540 + 0.8982 \log(\hat{Y}_i)$. Thus, θ, which we know to be equal to 1.0, is estimated by 0.8982. Working backward we obtain $\hat{\sigma}_i = 0.2144 \hat{\mu}_i^{0.8982}$, where σ_i is substituted for $|e_i|$, and antilog$(-1.540) = 0.2144$.

Although this regression approach does a reasonably good job of estimating

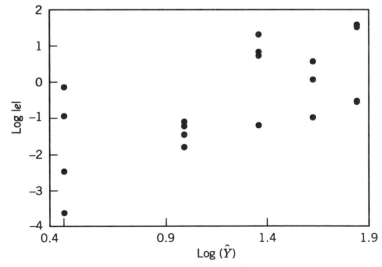

Figure 2.10 Scatter plot of log $|e|$ against $\log(\hat{Y})$ for the data in Table 2.3.

θ, it does not fare as well in estimating σ, as $\sigma = 0.5$ is estimated by $\hat{\sigma} = 0.214$. (This should not be surprising, however, as we would expect the intercept estimate to be off by more than the slope estimate when the slope estimate exceeds 1.) The comparatively poor estimation of σ in this case will cause us to underestimate the σ_i. Specifically, the estimates of $\sigma_i = 1.0, 1.5, 2.0, 2.5$, and 3.0 are 0.3080, 0.5222, 0.7261, 0.9248, and 1.1181, respectively. This is not necessarily a problem, however. We can see from Eqs. (2.10a) and (2.10b) that if we misestimate each σ_i by the same multiplicative constant, the constant will cancel from the numerator and denominator of $\hat{\beta}_{1(w)}$ and $\hat{\beta}_{0(w)}$.

Thus, if $\hat{\sigma}_i^2/\sigma_i$ is approximately constant over i, the weighted least-squares estimators should be meaningful. The ratios are 0.0948, 0.1212, 0.1320, 0.1368, and 0.1389.

Therefore, the use of weighted least squares with the indicated weights should produce good results. The process could be repeated to improve the estimates of the σ_i^2, and since we are starting with weights obtained in an efficient manner, only one more iteration may be necessary.

To accomplish the iteration we would simply regress $\log(|e_i'|)$ against $\log(\hat{Y}_i')$ where $e_i' = \sqrt{w_i}e_i$ and $\hat{Y}_i' = \sqrt{w_i}\hat{Y}_i$, and then estimate the new weights, say w_i', in the same way that the w_i were estimated. (The computations would, of course, generally be performed by computer. See Section 2.7 for information regarding the use of commercially available software to perform weighted least squares and obtain the other statistics discussed in this chapter. See also the chapter appendix for information regarding the Minitab macros that are included.)

Before looking at the computations for this iteration, as well as the estimates produced by the first weighting, there is an important point that needs to be made. When an absolute residual is close to zero, as some will inevitably be, the logarithm will be a large negative number, so the process of taking logs might inadvertently create a few influential data points. For example, $\log(.01) = -4.605$ and $\log(.02) = -3.912$, so the difference between them is approximately 0.70, whereas the difference between $\log(.51)$ and $\log(.52)$ is approximately 0.02.

Recognizing this potential problem, Carroll and Ruppert (1988, p. 35) imply that it would be wise to delete the "smallest few percent of the absolute residuals before taking logarithms." That approach should certainly work well for a large data set where the slope and intercept could be expected to be well determined from the sizable number of "noninfluential" data points, but may not work for a small-to-moderate size data set. We can see from Figure 2.9 that if we were to delete the smallest value of $\log(|e_i|)$, which is -3.6306 corresponding to -0.0265 in Table 2.3, we would expect the regression line to flatten out somewhat, and if we were to delete the other large negative value, corresponding to the point above it, the slope would then be close to zero.

Specifically, deleting only the first point causes the slope to change to 0.40, and although this produces an improved estimate of the intercept, that is irrelevant because, as we have seen, the estimate of θ is the only one that matters. When θ is (poorly) estimated by 0.40, the ratios $\hat{\sigma}_i/\sigma_i$ and $\hat{\sigma}_i^2/\sigma_i^2$ will differ considerably, and the first-stage estimates may be inadequate. For $\hat{\theta} = 0.40$, the ratios of $\hat{\sigma}_i/\sigma_i$ range from 0.31 to 0.61, and the ratios of $\hat{\sigma}_i^2/\sigma_i^2$ range from 0.10 and 0.31. (And, as indicated, also deleting the other point would make matters even worse.)

Using the "good" estimate $\hat{\theta} = 0.8982$ obtained using all of the data, the (initial) weighted least squares estimates are $\hat{\beta}_{0(w)} = -1.0838$ and $\hat{\beta}_{1(w)} = 1.2469$. These differ only very slightly from the weighted least squares estimators of -1.0708 and 1.2434, respectively, which result from using the known weights. Thus, in this instance the weights are virtually optimal without iterating.

It is worth noting that this iterative method is not based on any distributional assumption for Y (or ϵ), and as such will be inferior to a maximum likelihood approach when Y is normally distributed. Such a contrast is beyond the intended level of this book, however. For details the reader is referred to Carroll and Ruppert (1988, p. 21) and Stirling (1985).

We have seen that modeling σ has produced the best results for this example, and we would expect this to hold true in general. For additional information regarding the modeling of σ or σ^2, the reader is referred to Carroll and Ruppert (1988) and Davidian and Carroll (1987).

2.2 RESIDUAL PLOTS

One question that needs to be addressed early is the type of plot(s) that should be used in trying to detect unequal variances, and also for determining how to model the variance (or standard deviation) when the use of weighted least squares seems appropriate.

The simplest (and most naive) plot for detecting *heteroscedasticity* (i.e., unequal variances) is to plot the residuals against $\hat{Y}$ or against X. This plot should not be used to check the assumption of a constant σ^2 because the residuals do not have a constant variance even when σ^2 is constant.

Specifically, $\text{Var}(e_i) = \sigma^2(1 - h_i)$, as was previously given. Since h_i reflects the distance that x_i is from $\bar{x}$, the $\text{Var}(e_i)$ may differ considerably if there are any extreme X values. Consequently, a plot of the (raw) residuals against X could exhibit nonconstant variability of the e_i for this reason alone, or the degree of nonconstancy could perhaps be exacerbated. [See, e.g., Cook and Weisberg (1982, p. 38).]

This is not a serious problem for the data in Table 2.3, however, as there are no extreme X values. It is also worth noting that no h_i could be very large when each X_i is replicated several times because $1/n \leq h_i \leq 1/c$, where c denotes the number of replicates. With $n = 20$ and $c = 4$ for the Table 2.3 data, the h_i could not differ greatly. [We may note at this point that the upper bound on h_i shows the need for multiple deletion diagnostics (discussed briefly in Section 2.5.6), because an extreme X_i that would otherwise have a large value for h_i can have a greatly reduced value simply because X_i is replicated. This type of problem also occurs when there are near replicates.]

In general, a standardized residual, as defined by Eq. (2.1), is preferred over an ordinary (raw) residual. The plot of the standardized residuals against X for the Table 2.3 data was given in Figure 2.9. If the raw residuals had been used instead, the plot would have been similar since the h_i differ only slightly, ranging from 0.05 to 0.15. (This is not to suggest that the h_i should be examined to see if the ordinary residuals can be used; it is preferable to routinely use alternatives to the ordinary residuals.)

It has been noted previously that other writers have labeled $r_i = e_i/s$ the standardized residuals. But these r_i do not have a constant variance, so in that respect they have the same shortcoming as the e_i. (The only advantage that e_i/s has over e_i is that with e_i/s the residuals are scaled so that their values should be independent of the magnitude of Y and of its unit of measurement.)

As previously implied, the "preferred" residuals are the standardized residual given by Eq. (2.1) and the standardized deletion residual given by Eq. (2.9). [This assumes that n is sufficiently large so that a transformation to uncorrelated residuals is unnecessary. See Cook and Weisberg (1982, p. 34) for a discussion of uncorrelated residuals.]

Standardized residuals are frequently plotted against $\hat{Y}$, but when the regression model has a single regressor, the plot will have the same configuration as the plot against X when $\hat{\beta}_1$ is positive, and will be the mirror image of that plot when $\hat{\beta}_1$ is negative. The reason for this is that $\hat{Y}$ and X have a perfect positive correlation (of $+1$) when $\hat{\beta}_1$ is positive, and a perfect negative correlation of -1 when $\hat{\beta}_1$ is negative. Thus, nothing would be gained by constructing both plots.

2.3   TRANSFORMATIONS

Transformations are an extremely important part of regression analysis, but the use of transformations can be somewhat tricky. As Draper and Smith (1981, p. 221) state: "The choice of what, if any, transformation to make is often difficult to decide."

Transformations are made for essentially four purposes:

1. To transform a nonlinear model to a linear model
2. To transform X and/or Y in such a way that the strength of the linear relationship between the new variables is better than the linear relationship between X and Y
3. To obtain a relatively constant error variance
4. To obtain approximate normality for the distribution of the error term

These uses will be covered in the order given.

2.3.1 Transforming the Model

An experimenter will often have a strong belief regarding the true equation that relates two variables, simply from physical considerations of those variables. That equation could be nonlinear, but if it is nonlinear in such a way that it can be transformed to a linear equation (model), then linear regression can be applied to the transformed model.

A physical model that can be so transformed has been termed an *intrinsically linear model* by Draper and Smith (1981, p. 222) and a *transformably linear model* by other writers. A model that cannot be transformed into a linear model has been called an *intrinsically nonlinear model*.

Examples of each can be given as follows. If

$$Y = \exp{(\beta_0 + \beta_1 X)}\epsilon \tag{2.14}$$

then

$$\log{(Y)} = \beta_0 + \beta_1 X + \log{(\epsilon)}$$

which can be written in the form $Y' = \beta_0 + \beta_1 X + e'$; so a linear regression analysis would be performed on the variables Y' and X.

If the error term in Eq. (2.14) had been additive rather than multiplicative, the model would be intrinsically nonlinear, and *nonlinear regression* would have to be used (see Chapter 13).

There are various other transformably linear models in which the error term is additive rather than multiplicative. For example, $Y = 1/(\beta_0 + \beta_1 X + \epsilon)$ can easily be transformed into a linear model simply by taking the reciprocal of each side of the equation.

2.3.2 Transforming the Regressors to Improve the Fit

An X–Y scatter plot may sometimes suggest a transformation of X such that a linear model for the transformed variable will be appropriate. Consider Figure 2.11. Here a "necessary" transformation may be "in the eyes of the beholder." There are many recorded instances of two or more experienced statisticians drawing different conclusions from a graph, and Figure 2.11 might similarly evoke differences of opinion. In particular, should we try to use the model $Y = \beta_0 + \beta_1 X + \epsilon$, or does the graph suggest that we should perhaps try $Y = \beta_0 + \beta_1 \ln(X) + \epsilon$?

There are several considerations that need to be made regarding the transformation of X. It is more than just a matter of trying to determine which transformation of X to try. We must also determine whether a transformation is really necessary or desirable. In particular, will the model with the transformed X be a significant improvement over the model using X itself.

These issues are discussed in detail by Mosteller and Tukey (1977, pp.

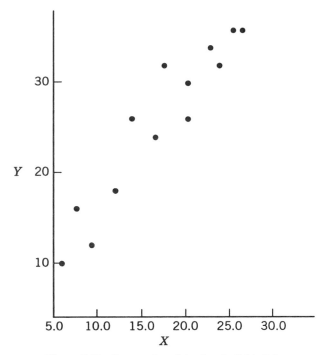

Figure 2.11 Scatter plot of the data in Table 2.6.

449–466). A transformation is not likely to provide much improvement unless the values of the transformed regressor differ more than slightly from the values of X.

One scenario where we would definitely not want to use a transformation would be when R^2 using X is very close to 1, and the transformed X values are highly correlated with the original values. A (possibly) very small increase in R^2 is certainly not going to be enough to offset the fact that the unit of measurement is lost when X is transformed, thus introducing unnecessary complexity.

Obviously, the potential for improvement is much greater when R^2 is not close to 1. Mosteller and Tukey (1977, p. 466) conclude that transforming X to $\ln(X)$ or to $\sqrt{X}$ or to $-X^{-1}$ is likely to be beneficial unless $1 - (R^2)'$ is only a small fraction of $1 - R^2$, where $(R^2)'$ is the square of the correlation between X' and X, with X' denoting the transformed regressor. Certainly it is the correlation between the transformed and untransformed regressor that is of most importance in indicating whether a transformation might be helpful; the model R^2 value using the untransformed regressor would simply indicate whether there is much room for improvement.

The range of the X values could obviate a transformation from X to $\log(X)$, as the correlation between the two will generally be at least 0.99 for practically any value of n when the ratio of largest X to smallest X does not exceed 3 (see Mosteller and Tukey, 1977, p. 458).

2.3.2.1 *Box–Tidwell Transformation*

Box and Tidwell (1962) provided a method for determining the appropriate transformation of X, assuming that a transformation is necessary. The data for the points that were graphed in Figure 2.11 are given in Table 2.6, and we shall use this example to illustrate the Box–Tidwell approach, and view the result relative to the Mosteller and Tukey recommendations.

With the Box–Tidwell procedure one assumes that

$$E(Y) = \beta_0 + \beta_1 X'$$

where X' is the ("correctly") transformed X, with

$$X' = \begin{cases} \ln(X) & \alpha = 0 \\ X^\alpha & \alpha \neq 0 \end{cases}$$

We could let X be our initial guess for X'. (This would be logical considering Figure 2.11, and is also the usual starting point for the procedure.)

Using this guess and expanding $E(Y)$ in a Taylor series produces

$$E(Y) = \beta_0 + \beta_1 X + (\alpha - 1)\left\{ \frac{\partial[E(Y)]}{\partial \alpha} \right\}_{\substack{x'=x \\ \alpha = \alpha_0}}$$

Employing the chain rule for derivatives, we can write the expression for the

Table 2.6 Data for Illustrating Box–Tidwell Transformation

Y	X
10	5
11	10
16	8
18	13
25	15
24	18
26	22
29	21
31	19
31	26
34	25
35	29
36	28

derivative as

$$\left\{ \frac{\partial[E(Y)]}{\partial\alpha} \right\}_{\substack{x'=x \\ \alpha=\alpha_0}} = \left\{ \frac{\partial(E(Y))}{\partial x'} \right\}_{x'=x} \left\{ \frac{\partial x'}{\partial\alpha} \right\}_{\alpha=\alpha_0}$$

Since $X' = X^\alpha, \partial X'/\partial\alpha = X^\alpha \ln(X) = X \ln(X)$ with $\alpha = \alpha_0 = 1$, and $\{\partial(E(Y))/\partial X'\}_{x'=x} = \beta_1$, the model becomes

$$E(Y) = \beta_0 + \beta_1 X + (\alpha - 1)\beta_1 [X \ln(X)] \qquad (2.15)$$

If we let $\gamma = (\alpha - 1)\beta_1$, we then have

$$E(Y) = \beta_0 + \beta_1 X + \gamma[X \ln(X)]$$

[Note that this is a *multiple linear regression* model. Such models are discussed in detail starting in Chapter 4.]

Using least squares in the usual fashion produces

$$\hat{Y} = \hat{\beta}_0 + \hat{\beta}_1 X + \hat{\gamma}[X \ln(X)] \qquad (2.16)$$

We would then solve for $\hat{\alpha}$ and that would give the revised estimate of α.

Specifically, since $\gamma = (\alpha - 1)\beta_1$, we would estimate α as $\alpha = \hat{\gamma}/\hat{\beta}_1^* + 1$, where $\hat{\beta}_1^*$ is obtained from $\hat{Y} = \hat{\beta}_0^* + \hat{\beta}_1^* X$. Thus, the estimate of β_1 is obtained from the regression of Y on X without considering a transformation. Even though β_1 appears twice in Eq. (2.15) and $\hat{\beta}_1$ is in Eq. (2.16), the estimate of β_1 could not be obtained from Eq. (2.16). This is because we are interested in determining the extent to which $\hat{\gamma}$ differs from zero (and thus the extent to which α differs from one) when X is in the model, so the estimate of α obviously cannot be a function of the coefficient of X when both X and $X[\ln(X)]$ are in the model.

For the data in Table 2.6 we obtain

$$\hat{Y} = \hat{\beta}_0^* + \hat{\beta}_1^* X$$
$$= 5.183 + 1.082X$$

and

$$\hat{Y} = \hat{\beta}_0 + \hat{\beta}_1 X + \hat{\gamma}X \ln(X)$$
$$= -0.04 + 2.45X - 0.361X \ln(X)$$

Therefore, the revised estimate of α (say, α_1) is $\alpha_1 = \hat{\gamma}/\hat{\beta}_1^* + 1 = -0.361/1.082 + 1 = 0.666$.

Box and Tidwell (1962) give four examples in which the correct transformations are known to be $\sqrt{X}$, $\ln(X)$, X^{-1}, and X^{-2}; and for each example their procedure essentially converges with the second iteration. In those examples the model had no error, however (i.e., there was an exact fit), and the ability of the procedure to closely approximate the appropriate transformation, if it were known, will generally depend upon the amount of error in the model.

Performing a second iteration for the current example, we define a new variable $X' = X^{.666}$ and repeat the process using X' in place of X. This produces $\alpha_2 = 0.713$, and further iterations produce $\alpha_3 = 0.7082$, $\alpha_4 = 0.7087$, and α_5 through α_{10} are also 0.7087. Thus, there is convergence, and for simplicity we will use $\alpha = 0.7$. Note that X' for each iteration is *not* obtained as X^α, but rather X' for the $(j + 1)$st iteration is obtained as $(X^{\alpha_{j-1}})^{a_j}$, where $a_j = (\hat{\gamma}_j/\hat{\beta}^*_{1,j} + 1)$. Similarly, $\alpha_j = a_j\alpha_{j-1}$.

With $\alpha = 1.0$ $R^2 = .900$, and with $\alpha = 0.7$, $R^2 = .905$. Thus, there is only slight improvement, and this is because the need for a transformation was not strongly indicated. (In Exercise 2.4 the reader is asked to obtain the Box–Tidwell transformation when a transformation *is* strongly indicated.) Frequently, we will want to use both the linear and a nonlinear form of a regressor, but not so for this example since the correlation between X and X^{-7} is .998. Thus, the two forms are almost linearly related, so including both would create the type of problem that is discussed in Section 4.3. The Box–Tidwell approach is geared toward detecting the need for a nonlinear term *instead* of the corresponding linear term. If both are needed, this could always be checked by using both in the equation, but sometimes problems will ensue, as is discussed in Section 6.5.

Cook and Weisberg (1982, p. 81) point out that the standard error of $\hat{\beta}_1$ needs to be small for the Box–Tidwell approach to be effective. This condition will hold when the linear term is a good proxy for the "true" term, as was the case in this example.

2.3.3 Transform Y to Obtain a Better Fit?

Sometimes it is preferable to try to improve the fit by transforming Y instead of X. The objective would thus be to determine a transformed variable Y' such that $Y' = \beta_0 + \beta_1X + \epsilon$ is a good model.

The following point is frequently made by writers in discussing the transformation of Y. Assume that we have fit the model $Y = \beta_0 + \beta_1X + \epsilon$ and the R^2 value is low, but the assumptions of approximate normality and constant error variance appear to be met. It we transform Y we will risk losing approximate normality and constant error variance. In particular, if we make a nonlinear transformation of Y in such a way that Y' is not very highly correlated with Y, then Y' will not be approximately normally distributed. (We should keep in mind, however, that it is the distribution of Y for each X_i that is assumed to be normally distributed.)

The (approximately) constant variance could also be disturbed. Assume that $10 \leq Y \leq 14$ when $X = 1$, $12 \leq Y \leq 16$ when $X = 2,\ldots$, and $20 \leq Y \leq 24$ when

$X = 6$, and the transformation is $Y' = 1/Y$. When Y is regressed on X the plot of the standardized residuals against X will have the same spread for each value of X. But when Y' is used the spread will decrease over X, with the spread being much greater for the smaller values of X. (This transformation would not, of course, be appropriate for this set of data; the intent here is simply to indicate that a transformation could cause the error variance to be highly unequal for different values of X, after being constant for the untransformed Y.)

Conversely, we might use a transformation to *remove* the problem of a non-constant error variance, which would happen for this example if we started with what we are calling Y' and transformed back to Y. We might also improve the fit, but assessing the fit when Y has been transformed is not as straightforward as it might seem, as the proper form of R^2 must be used. Specifically, it is necessary to convert the fitted values back to what they would be on the original scale before computing R^2. This is discussed in detail in Section 6.4.

Transforming Y in an attempt to obtain a better fit must thus be done cautiously, as the transformation might not appear to be beneficial when the proper form of R^2 is used. Box and Cox (1964) provided methods for estimating the transformation parameter for the family of transformations that they considered, and their approach is discussed in Section 2.3.4 and illustrated in Section 6.4.

There are some important points to keep in mind when using a transformation of Y and/ or X to improve the fit of a regression model. First, there might be one or more extreme points that could be indicating the need for a transformation, and these points could be bad data points. Also, extreme data points on the original scale could be highly influential data points on the transformed scale. Influential data points are discussed in Section 2.4.

2.3.4 Transforming to Correct Heteroscedasticity and Nonnormality

It is often possible to eliminate nonnormality and a nonconstant error variance with the same transformation. Therefore, both problems will be discussed in the same section.

Let's assume that either the basic simple linear regression model is adequate or that use of the Box–Tidwell transformation approach (or some other approach) has resulted in a transformation such that the regression of Y on the transformed X produces a good fit, but that there is a problem with nonnormality and/or heteroscedasticity. If we are to use ordinary least squares, we must either transform Y to eliminate the problem(s) or else use weighted least squares if the problem is heteroscedasticity.

A (corrective) transformation of Y might then result in an unacceptable fit when the transformed Y is regressed on X, however. To prevent this from happening, the transformation that is selected for Y would also be used for X, thus preserving the original fit between Y and X.

This is the philosophy behind the "transform both sides" (TBS) technique discussed in Carroll and Ruppert (1988). See also Carroll and Ruppert (1984). They consider the family of Box–Cox (1964) transformations given by

$$Y^{(\lambda)} = \left\{ \begin{array}{ll} (Y^{\lambda} - 1)/\lambda & \text{if } \lambda \neq 0 \\ \log(Y) & \text{if } \lambda = 0 \end{array} \right\}$$

For a given transformation $Y^{(\lambda)}$, use of the TBS approach produces the model

$$Y^{(\lambda)} = (\beta_0 + \beta_1 X)^{(\lambda)} + \epsilon$$

Nonlinear least squares would have to be used, however, since the model is nonlinear in β_0 and β_1. For all practical purposes we may consider the transformation Y^{λ} in place of $(Y^{\lambda} - 1)/\lambda$, since we will assume that the model has an intercept.

It would be fortuitous if a transformation of Y could correct for both nonnormality and heteroscedasticity. This issue has been addressed by Ruppert and Aldershof (1989), who assume that an experimenter already has a model that provides a good fit to the data, so it is then a matter of removing nonnormality and/or heteroscedasticity that may be present. They provide three estimators of λ: one that is designed to create a symmetric distribution for ϵ, one that is designed to remove heteroscedasticity, and one that will do both. A discussion of those results would necessarily be at a somewhat higher level than the intended level of this book (and especially of this chapter), however, so the interested reader is referred to their paper for details.

There are some simple guidelines that can be given, however. Information given by van Zwet (1964) and described by Carroll and Ruppert (1988, p. 122) is useful in trying to determine how to transform Y so as to reduce skewness. If the distribution of Y has right skewness (i.e., the "tail" of the distribution is on the right), we can reduce that skewness by making a *concave* transformation of Y. (Note: A concave transformation is one for which the first derivative is strictly *decreasing*, whereas a *convex* transformation is one where the first derivative is strictly *increasing*.) Thus, to reduce right skewness we could transform Y using Y^{λ} with $\lambda < 1$, or possibly use $\log(Y)$. Conversely, if the distribution of Y was left skewed, which is thought to occur much less frequently, we could use Y^{λ} with $\lambda > 1$. If we already had a symmetric distribution, but we had nonnormality caused by *kurtosis* ("peakedness"), the kurtosis would be reduced if we used $\lambda > 1$.

A two-stage procedure for selecting λ in the context of multiple regression so as to eliminate both nonnormality and heteroscedasticity is discussed and illustrated in Section 6.6.

One thing that should be kept in mind regarding transformations is that the apparent need for a particular transformation may be due to a few extreme observations, so that diagnostics are needed to detect this when it occurs. Preferably, the entire data set should signal the need for a transformation, not one or two extreme points that might be bad data points. Conversely, even if there were no influential data points before a transformation is made, there may be influential values among the transformed data values. Consequently, it is also important to use diagnostics *after* a transformation has been made.

2.3.5 Which R^2?

Whenever Y is transformed, there is a question of how R^2 should be computed. This issue is discussed in detail in Section 6.4. Since we may view weighted least squares as resulting from a transformation of both X and Y, as indicated in Eq. (2.11), some attention must be given to the selection of the form of R^2, and similarly when a Box–Cox transformation is used.

Kvålseth (1985) suggests using the form that was given in Eq. (1.9), with $\hat{Y}$ replaced by $\hat{Y}_{raw}$ (the predicted value of Y converted back to the original scale). Willett and Singer (1988) also indicate a preference for an R^2 measure that is in the original metric; their version of R^2 differs from Eq. (1.9) only in that $\hat{\beta}_{0(w)}$ and $\hat{\beta}_{1(w)}$ are (appropriately) used in computing $\hat{Y}$.

2.4 INFLUENTIAL OBSERVATIONS

In computing various sample statistics such as the mean, variance, and standard deviation, each of the n observations has the same weight in determining the value of those sample statistics, although bad data points can render useless any such sample statistic.

In regression analysis the situation is somewhat similar, but more complicated due in part to the fact that there is more than one dimension. For example, a point that is far removed from the others in a scatter plot may not have any effect on the regression equation, whereas an extreme value will have a considerable effect on many one-dimensional statistics. For example, an (X, Y) point could be far removed from the other points, but have (virtually) no effect on the regression equation if it fits a strong pattern formed by the other points. But certainly $\overline{X}$ and $\overline{Y}$ (i.e., one-dimensional statistics) will be greatly affected by that point.

Conversely, an extreme point could completely determine the regression line, and thus be more influential than in the one-dimensional case. (Consider a balanced configuration of $n - 1$ points that form a circle. The placement of the remaining point will completely determine the regression line. We would not, of course, attempt to fit a linear regression line to such a configuration of points, but we might encounter real data that approach this extreme scenario.)

Consider what happens when the sample variance is computed in the presence of an extreme (bad) data point. The sample mean moves in the direction of the bad point, with the result that the good points are, as a group, further from the mean than they would be if the bad point had been deleted. The same thing can happen in regression with the result that the good points are further from the regression line than they should be, and the regression relationship may appear to be weaker than it really is.

When will the deletion of a point *not* affect the parameter estimates? It can be easily shown that a point that lies on the regression line will not, by itself, have any influence on the parameter estimates (except in the trivial case where there is only two points). This can be shown as follows. Using

$$\hat{\beta}_1 = \frac{\sum (X - \overline{X})(Y - \overline{Y})}{\sum (X - \overline{X})^2} \tag{2.17}$$

[which can be shown to be equivalent to Eq. (1.6)], and $\hat{Y}_0 = \overline{Y} + \hat{\beta}_1(X_0 - \overline{X}) = Y_0$ when the point (x_0, y_0) lies on the line, the deletion of (x_0, y_0) means that the numerator of Eq. (2.17) is reduced by $(X_0 - \overline{X})(Y_0 - \overline{Y}) = (X_0 - \overline{X})[\overline{Y} + \hat{\beta}_1(X_0 - \overline{X}) - \overline{Y}] = \hat{\beta}_1(X_0 - \overline{X})^2$. The denominator is reduced by $(X_0 - \overline{X})^2$, so the new value of $\hat{\beta}_1$ must be the same as the original value because the numerator and denominator of the fraction are reduced by components whose ratio is $\hat{\beta}_1$.

Similarly, it can be shown that $\hat{\beta}_0$ remains unchanged by again using $(X_0, \overline{Y} + \hat{\beta}_1(X_0 - \overline{X}))$ as the coordinates of the deleted point and performing the necessary algebra.

Thus, a point that lies on a simple regression line has no influence on the line. Does this mean that points that are very close to a regression line must have very little influence? We will see with the following example that this is not true.

2.4.1 An Example

To illustrate this last point and influence in general, we consider the data in Table 2.7 and the corresponding scatter plot in Figure 2.12. The regression equation is $\hat{Y} = 1.5 + X$, with the line splitting the distance between each pair of points that has a common X value. For this example we might expect the

Table 2.7 Data for Illustrating Influence

Y	X
2	1
3	1
3	2
4	2
4	3
5	3
5	4
6	4
6	5
7	5

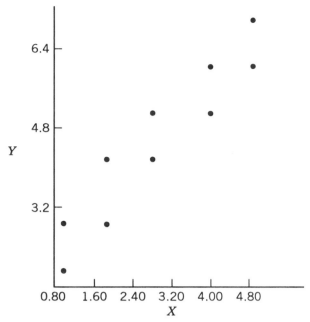

Figure 2.12   Scatter plot of the data in Table 2.7.

deletion of any one point to have very little effect on $\hat{\beta}_0$ and $\hat{\beta}_1$. And obviously if we deleted any pair of points with the same X value, $\hat{\beta}_0$ and $\hat{\beta}_1$ would retain their respective numerical values because there would then be the same general configuration with two fewer points.

If, however, we delete one point in one pair, the remaining point is then (slightly) influential in that it will deviate from the general configuration of the remaining points. Thus, we would expect the regression line to be drawn to that point with a slight change in $\hat{\beta}_1$ and a compensating change in $\hat{\beta}_0$.

Let c_i represent the change in $\hat{\beta}_1$ that occurs when the point (i, Y_S) is deleted, and let c_i^* be similarly defined when the point (i, Y_L) is deleted, with Y_L denoting the larger and Y_S the smaller of the two Y values when $X = i, i = 1, 2, \ldots, 5$. We would expect the c_i to differ, but we would also expect that $c_i^* = -c_i$. That is, $\hat{\beta}_1 + c_i$ would be the new value of the regression coefficient if (i, Y_S) is deleted, and $\hat{\beta}_1 - c_i$ is the new value if (i, Y_L) is deleted. If we define d_i similarly relative to $\hat{\beta}_0$, the new coefficients would be $\hat{\beta}_0 \pm d_i$, and the d_i would also differ.

We would also expect the magnitude of c_i (and d_i) to depend on the difference between i and 3 (the center of the data). Specifically, we would expect to observe $c_5 > c_4$, but $c_5 = c_1$. The reason for this is that at $X = 5$ there is no point to the right of it that falls in line with the configuration of the other four pairs, which would diminish the influence of the remaining point at $X = 5$. At $X = 3$, however, the points on each side form their own configuration, which is the same as the general configuration. Thus, $\hat{\beta}_1$ should not change; the line will simply move up or down to accommodate the remaining point.

These changes in $\hat{\beta}_0$ and $\hat{\beta}_1$ are shown in Table 2.8, along with the values of four influence statistics that are explained in Section 2.5.4.

We can gain further insight into the concept of influence if we add an eleventh point that is some distance from the original 10 points. We know from past work that if the point has coordinates $(X, 1.5 + X)$ it will have no influence, regardless of the value of X. If X is much larger than 5, however, the point would likely be classified as an *outlier*, a point that is a considerable distance from the center of the other points. (More specifically, an outlier is not easily defined in terms of a specific distance from the rest of the data, but a commonly used rule is that a point that is at least three or four standard deviations from the center of the data is considered an outlier. Outliers are discussed further in Section 2.6.) For example, an (X, Y) pair such as (3,4) might have been erroneously recorded as (30,40), and a scatter plot would reveal the outlying observation.

Assume that the eleventh point is (30,40). The point is 8.5 units above the (extended) regression line obtained from the 10 points. To what extent will that point draw the regression line to it? The regression equation using that point is $\hat{Y} = 0.6059 + 1.3056X$. When $X = 30$, $\hat{Y} = 39.77$, so the extra point is almost on the line. Notice that the coefficients have changed considerably (especially $\hat{\beta}_0$), so the fact that a point is very close to a regression line does not necessarily mean that the point is not influential.

Conversely, a point could be far removed from a regression line and still be highly influential; it would simply not be influential enough to draw the line close to it because of its relationship to the other points and the general configuration of those points.

Table 2.8 Influence Statistics for Table 2.7 Data and Corresponding Changes in Parameter Estimates for Single-Point Deletions

Point Deleted (X, Y)	$\hat{\beta}_0$	$\hat{\beta}_1$	Cook's D	DFFITS	C_i	DFBETAS $(\hat{\beta}_0)$	$(\hat{\beta}_1)$
(1, 2)	1.79	0.93	0.245	−0.707	1.414	−0.696	0.577
(1, 3)	1.21	1.07	0.245	0.707	1.414	0.696	−0.577
(2, 3)	1.65	0.97	0.083	−0.406	0.812	−0.353	0.234
(2, 4)	1.35	1.03	0.083	0.406	0.812	0.353	−0.234
(3, 4)	1.56	1.00	0.049	−0.312	0.624	−0.133	0
(3, 5)	1.44	1.00	0.049	0.312	0.624	0.133	0
(4, 5)	1.47	1.03	0.083	−0.406	0.812	0.071	−0.234
(4, 6)	1.53	0.97	0.083	0.406	0.812	−0.071	0.234
(5, 6)	1.36	1.07	0.245	−0.707	1.414	0.348	−0.577
(5, 7)	1.64	0.93	0.245	0.707	1.414	−0.348	0.577

Therefore, we cannot easily determine from a scatter plot whether a point will be influential unless we use a more sophisticated scatter plot. Figure 2.13 shows a scatter plot of the data in Table 2.7. The size of the symbol that denotes each plotted point indicates the size of the effect that the point has on the correlation coefficient, with an open circle signifying a positive effect and a filled-in circle indicating a negative effect.

Consider the two leftmost points. We know from Table 2.8 and the earlier discussion that these two points have an equal influence on the slope. Notice, however, that they have an unequal influence on r_{xy}, and thus have an unequal influence on R^2. Why is that? Recall the discussion in Section 1.5.2 that the value of R^2 is influenced by the slope of the regression line. Here we see an example of that because $\sum (Y - \hat{Y})^2 = 2.143$ if either of these points is deleted, but the slope is obviously increased by the presence of the point (1,2), whereas the point (1,3) causes the slope to decrease, as can be seen from Table 2.8. [If X and Y were in correlation form (see Section 3.2.5) when the plot is produced, the plot would then show the direct influence on the slope, since the slope is r_{xy} when the data are in correlation form.]

Figure 2.14 illustrates another point made in Section 1.5.2; namely, that R^2 is influenced by the spread of X. This is the scatter plot of the Table 2.7 data plus the point (30,40). Notice that the presence of this point causes a noticeable increase in the correlation, even though the point lies well off the line that is determined by the other points, as has been noted previously. The reason for the increase is that the spread of X has been considerably increased.

Simple examples such as the one used in this section not only help us understand influence but least squares as well.

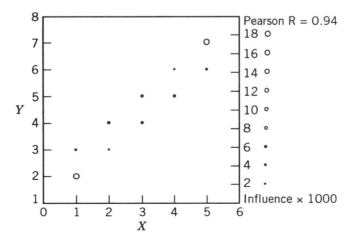

Figure 2.13 Scatter-influence plot of Table 2.7 data.

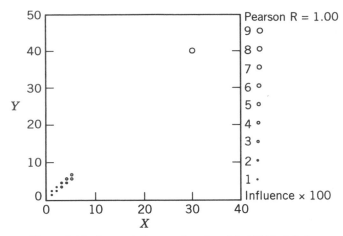

Figure 2.14   Scatter-influence plot of modified Table 2.7 data.

2.4.2   Influence Statistics

There are also various influence statistics that are routinely used, and we could view the numerical values of these statistics and/or construct scatter plots in which the sizes of the symbols reflect the relative magnitude of the influence statistic at each point.

In selecting one or more influence statistics to use, we must consider what type of influence we want to identify—influence on predicted values, coefficient estimates, or something else. (Coefficient estimates can change greatly without having much effect on predicted values, as is explained in Chapter 12.)

One obvious way to measure influence on coefficient estimates would be to compute each estimate with and without each point, as in Table 2.8. But the magnitude of the change will depend on the magnitude of the original estimate, so the change must clearly be standardized in some way. This leads to the consideration of the statistic DFBETAS that was proposed by Belsley et al. (1980, p. 13).

$DFBETAS_{ij}$ is the (standardized) effect that the ith observation has on the jth regression coefficient. For simple linear regression we can let $j = 0$ corresponding to $\hat{\beta}_0$, and $j = 1$ corresponding to $\hat{\beta}_1$. We then have

$$DFBETAS_{i0} = \frac{\hat{\beta}_0 - \hat{\beta}_0(i)}{s(i)\sqrt{\sum X^2/nS_{xx}}}$$

$$DFBETAS_{i1} = \frac{\hat{\beta}_1 - \hat{\beta}_1(i)}{s(i)\sqrt{1/S_{xx}}}$$

where $\hat{\beta}_j(i)$ and $s(i)$ signify that the ith observation is not used in computing the parameter estimates. [If s rather than $s(i)$ had been used, the denominator of each statistic would have been the estimated standard deviation of $\hat{\beta}_i$.]

If we prefer to focus attention on prediction rather than estimation, it would be logical to use a standardized measure of the difference in the predicted values, with and without the ith observation. Such a measure is $DFFITS_i$, which is defined as

$$DFFITS_i = \frac{\hat{Y}_i - \hat{Y}_{(i)}}{s(i)\sqrt{h_i}}$$

with the denominator being the estimate of the variance of $\hat{Y}_i$, and $\hat{Y}_{(i)}$ denotes the ith predicted value obtained without using the ith observation. Since we may also write $DFFITS_i$ as

$$DFFITS_i = \frac{e_i}{s(i)\sqrt{1 - h_i}}$$

we see that $DFFITS_i$ differs from the standardized residual, e'_i, that was given in Eq. (2.1) only in that $s(i)$ is used instead of s.

Another influence measure is $DFTSTAT_{ij}$, which measures the influence of the ith data point on the t-statistic of the jth regression coefficient. [See Belsley, et al. (1980, p. 14) for details.]

Unfortunately, there is not an obvious threshold value for either DFBETAS or DFFITS. The cutoff values that have been suggested by Belsley et al. are $2/\sqrt{n}$ and $2\sqrt{(p + 1)/n}$, respectively, with p equal to the number of parameters. Not everyone agrees with the cutoff value for DFFITS, however. In particular, Staudte and Sheather (1990, p. 213) indicate that they often find questionable points with a DFFITS value less than the suggested cutoff value and suggest using 1.5 as the multiplier instead of 2. The reader will see the value of this suggestion in Exercise 2.6.

A frequently used influence statistic that is somewhat related to DFFITS is Cook's $-D$ statistic, which is due to Cook (1977). Cook's statistic can be written as

$$D_i = \frac{1}{ps^2} \sum_{j=1}^{n} (\hat{Y}_{j(i)} - \hat{Y}_j)^2 \tag{2.18}$$

where $\hat{Y}_{j(i)}$ denotes the jth predicted value when observation i is not used in the computations, and p is the number of parameters. Thus, $p = 2$ for simple regression. There is also not an obvious cutoff value for this statistic. One suggested value is the 50th percentile of the F-distribution with p degrees of freedom for the numerator and $n - p$ for the denominator. This is not entirely satisfactory, however, because when the statistic is written in the form of Eq. (2.18) it is apparent that D_i should increase with n, whereas the 50*th* percentile of the F-distribution *decreases* (slightly) as n increases. It is well known that D_i does not have an F-distribution, so the suggested cutoff should not be viewed rigidly. A frequently used cutoff is 1.0.

2.4.3 Different Schools of Thought Regarding Influence

Notice that Cook's-D uses the ith observation in estimating σ^2, whereas the ith observation was not used in estimating σ in DFFITS and DFBETAS. An argument against using the ith observation in these computations is that extreme (influential) points can inflate the estimate of σ, and thus deflate the value of the influence statistic, with the consequence that an influential data point might not be detected. While indicating a preference for a single estimate of σ, so that the influential observations can be unambiguously ordered, Cook and Weisberg (1982, p. 124) do suggest the possibility of using a robust estimate of σ so as to overcome the possible deleterious effects of an inflated estimate of σ. (This and related topics are discussed in some detail in Chapter 11.)

2.4.4 Modification of Standard Influence Measures

Although DFFITS, DFBETAS, and Cook's-D are most commonly used, another well-known influence measure was given by Atkinson (1981, 1985). This is a modification of Cook's-D and can be written as

$$C_i = \left\{ \frac{n - p}{p} \cdot \frac{h_i}{1 - h_i} \right\}^{1/2} |e_i^*| \tag{2.19}$$

where h_i, p, and e_i^* are as previously defined.

The relationship between D_i and C_i may be seen more clearly if we write D_i equivalently as

$$D_i = \left\{ \frac{1}{p} \cdot \frac{h_i}{1 - h_i} \cdot \frac{s_{(i)}^2}{s^2} \right\} (e_i^*)^2$$

Although D_i and C_i are still not easily comparable since the former is a squared quantity and the latter is not, it is clear that an extreme value that causes s^2 to underestimate σ^2 (and accordingly causes $s_{(i)}^2/s^2$ to be much less than 1), will tend to deflate D_i instead of inflating it as we would prefer.

Atkinson (1981) proposed that C_i be used graphically, with C_i plotted against the index number of the observation. It is possible to select values of the regressors in such a way that $h_i = p/n, i = 1, 2, \ldots, n$. (Doing so produces a G-optimum experimental design, as discussed in Chapter 14.) This results in $C_i = |e_i^*|$, which is the motivation for the constant $(n - p)/p$ in Eq. (2.19).

2.4.5 Application of Influence Measures to Table 2.7 Data

The values of each of these four influence statistics were given in Table 2.8 for the Table 2.7 data. We may observe that the values for DFFITS, DFBETAS, and Cook's-D are less than the suggested threshold values, and this is because there are no extreme data points in Figure 2.11.

The question arises as to what constitutes a large C_i value, and as with D_i, the question is not easily answered. If the data have come from a G-optimum design, then 2 might be used as a rough cutoff since e_i^* has a t-distribution.

When a G-optimum design is not used (such as when X is a random variable), the determination of a threshold value for C is more difficult. For this case $C_i > |e_i^*|$ when an X-value has a high leverage value and is thus potentially influential. [Some authors have referred to the "potential" of the set of X-values; see, e.g., Cook and Weisberg (1982, p. 115).] A point can be potentially influential if its X-value is far removed from the others, but whether it is influential will depend on the corresponding Y-value.

Since $|e_i^*|$ can be small when a point is influential, and the value of the bracketed expression in Eq. (2.19) does not have an upper bound for X-values that are not repeated, it is difficult to determine a threshold value for C_i from consideration of those two components. Some authors have stated that the two components should be examined separately, but as Atkinson (1985, p. 26) indicates, a high leverage point is not necessarily influential. In particular, a point could be a high leverage point and yet have $C_i = 0$, which would happen if the line went through the point. When $h_i = 1$, e_i^* must equal zero, and hence C_i must equal zero, although $h_i = 1$ is not likely to occur in practice. [See Cook and Weisberg (1982, p. 14).] Having $C_i = 0$ is appropriate in this case since, as shown in Section 2.5, when a point that falls on a simple regression line is deleted, the regression equation does not change.

Putting all of this together, a point could be influential if the leverage is high or low, or if $|e_i^*|$ is large or small. A point will not likely be highly influential if both are small, however, so this suggests that a benchmark value for C_i might be determined not from the product of threshold values for the two components but rather from the maximum of those values.

A frequently used threshold value for h_i is $3p/n$, and the use of this value produces $\{3(n - 2)/(n - 6)\}^{1/2}$ for the bracketed expression. The numerical value of this expression will be close to 2 for most values of n. As indicated previously, 2 is also a reasonable cutoff for $|e_i^*|$, so if an experimental design has not been used, one possible rule-of-thumb would be to investigate the ith point if the smaller of the bracketed expression and $|e_i^*|$ exceeds 2. With this rule-of-thumb we would identify data points with an extreme x-coordinate at which the fit is poor, but we would fail to identify influential points with extreme

x-coordinates at which the fit is good. Therefore, it seems preferable to spotlight those points for which the *maximum* of the two expressions exceeds 2 (i.e., *either* $|e_i^*|$ or the bracketed expression is extreme). Although this is equivalent to doing what Atkinson advises against, there is no obvious way to determine a reasonable threshold value for C_i since a large leverage value could produce a large value for C_i without the point being influential.

All of the C_i values in Table 2.8 are clearly less than 2, and it can also be shown that each component is less than 2, so no influential points are identified.

2.4.6   Multiple Unusual Observations

It is important to recognize that whereas the diagnostics given in Sections 2.4.2 and 2.4.4 will work well under conditions such as the existence of only one extreme value, they can fail when there are multiple extreme values. For example, assume that we have a cluster of four bad data points, and we apply the previously discussed diagnostics in an effort to identify them. All of the diagnostics are likely to fail because deleting any one of the points will not have very much, if any, effect on the regression equation. This phenomeon is called *masking*, as the bad data points are masked by their number. The detection of multiple unusual observations is addressed by Belsley et al. (1980, pp. 31–39), who discuss several possible sequential procedures, including a multiple-point generalization of DFFITS. A multiple-point generalization of Cook's-D statistic is given by Takeuchi (1991), and methods for detecting multiple outliers are discussed further in Chapter 11.

2.4.7   Predicting Lifespan: An Influential Data Problem

A paper in the *Journal of the Royal Society of Medicine* (Newrick et al., 1990) claimed that lifespan could be predicted by the length of one's lifeline, with the left hand presumably a better indicator than the right hand. This conclusion was carried by the wire services and subsequently appeared in newspapers around the world. For example, *The Philadelphia Inquirer* had a lengthy article on it in its July 27, 1990, edition. The statement run on July 24, 1990 by UPI was as follows: "A study of the hands of 100 corpses revealed a "highly significant association" between the age of the deceased and the lifelines traversing their palms, the Royal Society of Medicine said today."

Certainly this would be a very practical and useful application of regression analysis, but dire consequences could obviously result if the conclusion were incorrect. Imagine someone feeling poorly but refraining from seeing a doctor because his predicted lifespan exceeded his current age by 18.6 years.

Interest in predicting lifespan from lifeline length is not new, it has simply been revived. Draper and Smith (1981, p. 67) provide a chapter exercise that contains lifespan data gathered by two researchers and discussed in a Letter to the Editor of the *Journal of the American Medical Association* (1974, Vol.

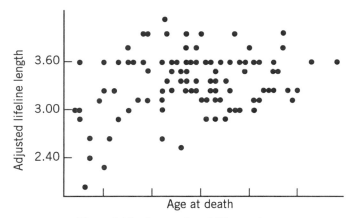

Figure 2.15   Scatter plot of lifespan data.

229, No. 11, 1421–1422). The analysis of Draper and Smith (p. 552) shows that there is no relationship between lifespan and lifeline length. (This data set is analyzed in Chapter 15.)

Could this suggest that residents in the United States differ from those in the United Kingdom in regard to the ability to predict lifespan? Not likely. The original data were not given in the paper by Newrick, Affie, and Corrall; the paper simply contained a scatter plot of the data in the sample, and that scatter plot is (approximately) reproduced in Figure 2.15, with the dependent variable defined in such a way as to compensate for differences in hand size. It should be apparent that a line with a zero slope will fit the vast majority of the data, but when lifespan is regressed against the adjusted lifeline length, the t-statistic for the slope has a p-value of .002. Thus, the relationship between the two variables is "highly significant" even though it is visually apparent that there is no relationship between them! Obviously, there are some influential data points in the lower left quadrant, and the sample size is quite large. Thus, the "significance" is not surprising.

The data were not obtainable, but a "reasonable facsimile" of the data are analyzed in Section 10.3.1, and the influential data points are spotlighted in that section.

2.5   OUTLIERS

Outlying observations are those that are at least 3 or 4 standard deviations from the center of the data. For any (x, y) point, we should not have an X-outlier if a designed experiment is being used, but could have such an outlier if the X-values are not preselected but rather occur at random. A Y-outlier could occur with either scenario, as could a residual outlier. (See Chapter 11 for additional information regarding the different types of outliers.)

The detection of outliers is tied in to the detection of influential observations, as outliers will frequently be influential points. X-outliers can frequently be detected by the leverage values, the h_i. One rule-of-thumb that has been advanced is that a point is an X-outlier if $h_i > 2p/n$, which equals $4/n$ in the one-regressor case. (The other suggested cutoff value for h_i is $3p/n$, as was mentioned in Section 2.4.5.) Either of these rules should be used with caution, however, as several observations can easily be flagged when p is small, even if they are not very outlying. Conversely, assume that we use the first of these two rules, and that the data are such that $x_1 - x_9$ occur once, x_{10} occurs twice, and x_{11} occurs three times. As indicated in Section 2.2, the upper bound on h_i when X_i is replicated is $1/c$, with c denoting the number of replicates. Here $4/n = \frac{4}{12} = \frac{1}{3} = 1/c$, so the value of x_{11} could not be labeled an outlier regardless of how extreme that value might be relative to the other 10 values. (And using $3p/n$ as the cutoff value, which is a better choice when p is small, obviously wouldn't help.)

Such a combination of X values is unlikely to occur when X is a continuous random variable, but might occur if X is an integer-valued random variable. This type of problem can also occur when the values of X are preselected. Assume that an experimenter suspects heteroscedasticity, so he replicates each of four X values five times so as to attempt to estimate the σ_i^2. No point could be labeled an X-outlier since $2p/n = 1/c$. There should not be any X-outliers when an experimental design is used, but if X-outliers did exist it could be difficult to detect them from the h_i. Assume that each design point is replicated c times. It follows that $2p/n \geq 1/c$ when the number of distinct X values is at most 4 (and 6 if $3p/n$ is used). Experimental designs for regression usually have more than four or six distinct values, however (see Chapter 14).

Data points whose y-coordinate (only) makes the point an outlier will generally be flagged as having a large standardized residual. Points that have extreme values for both coordinates may represent recording errors. They may be flagged by h_i or by the standardized residual if the point is not in line with the other points.

Remember that in simple regression we should be able to rely primarily on the X–Y scatter plot in identifying outliers, but it is helpful to also have some formal criteria to invoke.

It is important to detect outliers as they can not only significantly affect the parameter estimates, but they can also have a major effect on the other statistics that are computed. For example, from Section 2.5 we know that if an outlier falls on the regression line it does not have any effect on $\hat{\beta}_0$ or $\hat{\beta}_1$, but it might cause a sizable increase in S_{xx} and thus cause the regression relationship to appear to be stronger than it is.

Consequently, it is important to detect outliers, but it is equally important to realize that outliers should be discarded only if it can be determined that they are not valid data points. Outliers that *are* valid data points can often indicate that the postulated is incorrect, and that it should be modified to accommodate those points. If it is believed that the model is correct, then an outlier that is also an influential data point might be downweighted to accommodate the model. These issues are discussed extensively in Chapter 11.

2.6 MEASUREMENT ERROR

Classical regression theory holds that the regressors are measured without error. The words "measured with error" should mean that there is measurement error, but many writers have used this expression to also mean that the regressors are random variables. There is a considerable conceptual difference between *measurement error* and *random variation*, and it is necessary to consider them separately. (Random variation in the regressors was discussed in Section 1.9). We first consider measurement error in Y before addressing the issue of measurement error in X.

2.6.1 Measurement Error in Y

The type of problem that measurement error in Y presents depends on the nature of the error. Assume that

$$Y_i = \beta_0 + \beta_1 X_i + \epsilon_i + \delta_i$$

where δ_i is the measurement error in Y_i with $\delta_i \sim \text{NID}(0, \sigma_\delta^2)$. If ϵ_i and δ_i are independent, then we may write the model as

$$Y_i = \beta_0 + \beta_1 X_i + \epsilon_i'$$

where $\epsilon_i' \sim \text{NID}(0, \sigma_\epsilon^2 + \sigma_\delta^2)$, with $\epsilon_i' = \epsilon_i + \delta_i$. The measurement error thus adds to the model error, thereby reducing the percentage fit (R^2) of the regression model, and leaving more variability in Y unexplained by the model. The least squares estimators are still unbiased, however (Chatterjee and Hadi, 1988, p. 247).

If δ_i is markedly nonnormal and, in particular, δ_i is not small relative to ϵ_i, then an alternative to least squares such as nonparametric regression or robust regression may need to be used (see Chapters 10 and 11, respectively).

If the δ_i are correlated, we have a problem that is very similar to the problem of having the ϵ_i correlated, with the magnitude of the problem depending on the magnitude of the δ_i.

We would also have a major problem if the mean of each δ_i is not zero.

2.6.2 Measurement Error in X

The effect caused by measurement error in X will depend on the nature of the error, and will also depend on whether X is a random variable.

As shown by Draper and Smith (1981, p. 123), $\hat{\beta}_0$ and $\hat{\beta}_1$ will be unbiased if the measurement error, δ_i, is NID(0, σ_δ^2) and X is fixed. The least squares estimators will be biased when X is random, however, and the magnitude of the problem will depend on the spread of the true (i.e., correctly measured) X values relative to σ_δ^2. Consequently, it would be fortuitous to have considerable variability in X, relative to σ_δ^2, when measurement error is apt to be a problem. See also the derivation and explanation in Gunst and Mason (1980, p. 182). Fuller (1987) gives a full treatment of measurement error in statistical models.

2.7 SOFTWARE

The well-known statistical packages provide the user with many of the influence statistics that were discussed in Section 2.1, as well as some that were not discussed.

For example, Minitab gives Cook's-D, DFFITS, and h_i (leverage values) as subcommands that can be used with the REGRESS command. (Note that Minitab uses DFITS as the label for DFFITS.) (See the chapter appendix for information regarding how to obtain DFBETAS as a by-product of a robust regression Minitab macro, RWLS1DFB, that is used in Chapter 11.)

BMDP Statistical Software program 2R is BMDP's general-purpose (multiple) linear regression program, and it provides virtually all of the statistics that were discussed in this chapter, as well as some that were not covered. In particular, the following can be obtained: Cook's-D and Atkinson's modification of Cook's-D, DFFITS, leverage values and modified leverage values, standardized residuals, standardized deletion residuals (which are called *deleted standardized residuals*), and deletion residuals (called *deleted residuals*, which when divided by their standard error produces the deleted standardized residuals). Simulation envelopes for the residuals (the Atkinson type) can also be obtained with this program. Furthermore, a Box–Cox transformation can be produced with program 1R and program 2R. A 95% confidence interval on the power in the power transformation is given in each program. See Dixon (1990).

PROC REG is the main program for regression in the SAS Software System. The user can obtain Cook's-D, DFFITS, leverages, standardized residuals (termed studentized residuals), and standardized deletion residuals (termed RSTUDENT). Optimal transformations of essentially any type (relative to a squared-error loss function), can be obtained using PROC TRANSREG.

The use of SAS Software for regression is described in detail by Freund and Littell (1991). See also SAS Institute, Inc. (1989).

The primary regression program in SPSS is REGRESSION, and many diagnostics are available, although the terms are not defined mathematically in the SPSS manual (but are given on p. 599) (SPSS, 1990). In the terminology that is used therein, the user can obtain standardized residuals, studentized residuals, studentized deletion residuals, Cook's-D, (centered) leverage values, DFBETA, DFFIT, and other diagnostics.

STATA provides the capability for, using their terminology, Cook's distance, DFBETA, DFITS, standardized and studentized residuals, and other diagnostics. See Stata Corporation (1995).

The capabilities of SYSTAT include leverage values, Cook's D, and studentized residuals. For more details see Wilkinson (1990).

[Minitab is the registered trademark of Minitab, Inc. BMDP is the registered trademark of BMDP, Inc. SAS is the registered trademark of SAS Institute, Inc. SPSS is the registered trademark of SPSS, Inc. SYSTAT is the registered trademark of Systat, Inc. STATA is a registered trademark of Stata Corporation.]

SUMMARY

This chapter completes the introduction to simple linear regression that was begun with Chapter 1. Unfortunately, many regression users essentially stop with the material in that chapter, not recognizing that outliers, influential data, and assumptions that are not met can undermine their analyses. A complete analysis is important, and the need to look at the diagnostics that were discussed in this chapter is also emphasized in the user's manuals for the software packages discussed in Section 2.7.

The assumption of normality can be checked by constructing a simulation envelope, the homoscedasticity assumption and the assumption of uncorrelated errors can be checked by plotting the standardized residuals against X, and a simple runs test or a more sophisticated test such as the Durbin–Watson test can be used to assess possible correlation between the error terms.

Regressor transformation to produce a better fitting model and/or to meet model assumptions can be accomplished through use of the Box–Tidwell transformation approach, and Y would most often be transformed to remove heteroscedasticity and/or nonnormality.

There are various statistics that are available for detecting outliers and influential data points.

The diagnostic procedures that were given in this chapter are also applicable to mulitple linear regression in which additional diagnostics (see Chapter 5) are also used.

APPENDIX

2.A   Derivation of $\rho_{Y_t, Y_{t+1}} = \phi$

From Eq. (2.3) we have $\epsilon_t = \phi\epsilon_{t-1} + a_t$, so that $\sigma_{\epsilon_t}^2 = \phi^2\sigma_{\epsilon_t}^2 + \sigma_{a_t}^2$ since $\sigma_{\epsilon_t}^2 = \sigma_{\epsilon_{t-1}}^2$ and a_t is independent of ϵ_{t-1}. Thus, $\sigma_{\epsilon_t}^2 = \sigma_{a_t}^2/(1 - \phi^2)$.

From this result used in conjunction with Eq. (2.4) to find $\mathrm{Var}(Y_t)$, we obtain, assuming X_t to be fixed,

$$\mathrm{Var}(Y_t) = \phi^2 \mathrm{Var}(\epsilon_{t-1}) + \mathrm{Var}(a_t)$$

$$= \phi^2 \left(\frac{\sigma_a^2}{1 - \phi^2} \right) + \sigma_a^2$$

$$= \sigma_a^2 \left(1 + \frac{\phi^2}{1 - \phi^2} \right) = \sigma_a^2 \left(\frac{1}{1 - \phi^2} \right)$$

so that $\mathrm{Var}(Y_t) = \mathrm{Var}(\epsilon_t)$.

We may obtain $\mathrm{Cov}(Y_t, Y_{t+1})$ by combining Eqs. (2.3) and (2.4) so as to obtain

$$Y_{t+1} = \beta_0 + \beta_1 X_{t+1} + \phi^2 \epsilon_{t-1} + \phi a_t + a_{t+1}$$

Since $Y_t = \beta_0 + \beta_1 X_t + \phi \epsilon_{t-1} + a_t$, it follows that

$$\mathrm{Cov}(Y_t, Y_{t+1}) = \phi^3 \sigma_\epsilon^2 + \phi \sigma_a^2$$

$$= \sigma_a^2 \phi \left(\frac{1}{1 - \phi^2} \right)$$

after making the appropriate substitution for σ_ϵ^2. It then follows using the definition of ρ (given in Section 1.7) that $\rho_{Y_t, Y_{t+1}} = \phi$.

2.B   Derivation of Var(e_i)

For convenience we will write Eq. (2.6) in the equivalent form

$$e_i = \epsilon_i - \sum_{j=1}^{n} h_{ij} \epsilon_j \tag{A.1}$$

Equation (A.1) can be easily seen to be equivalent to Eq. (2.6) since the summation in the former contains the term $h_{ii}\epsilon_i (= h_i\epsilon_i)$, whereas the summation in Eq. (2.6) does not.)

If follows that

$$\mathrm{Var}(e_i) = \sigma^2 + \sigma^2 \left(\sum_{j=1}^{n} h_{ij}^2 \right) - 2\sigma^2 h_i$$

Using the expression for h_{ij} given in Eq. (2.7) and performing the necessary algebra, we find that $\sum_{j=1}^{n} h_{ij}^2$ is equal to the expression for h_i given in Eq. (2.2). Therefore, $\mathrm{Var}(e_i) = \sigma^2 + \sigma^2 h_i - 2\sigma^2 h_i = \sigma^2(1 - h_i)$.

2.C   Minitab Macros

2.C.1   For Obtaining Simulated Envelopes

There are two Minitab macros that can be used in working Exercise 2.8: ATKINSON and FLACFLRS. These implement the Atkinson and Flack and Flores approaches, respectively, to constructing simulated envelopes that were discussed in Section 2.1.2.3. The former generates a half-normal probability plot using 19 sets of simulated Y-values, and the latter produces both a full-normal plot and a half-normal plot, also using 19 simulated sets of Y-values. It should be noted that the macro FLACFLRS deviates slightly from what was suggested in the Flack and Flores paper in that the (approximate) values at the first and third quartiles are found by rounding the ordered residual numbers to the nearest integer rather than interpolating.

The macros MATKINSO and MFLACFLR can be used in working Exercise 2.9. These are slightly modified versions of ATKINSON and FLACFLRS in which the standardized residuals are plotted rather than the standardized deletion residuals.

2.C.2   For Obtaining DFBETAS

DFBETAS is not provided explicitly by Minitab, but it can be easily obtained as a by-product of RWLS1DFB, a macro for robust regression that is discussed in the appendix to Chapter 11. A remark in RWLS1DFB identifies the line of code in which DFBETAS is computed.

2.C.3   For the Box–Tidwell Transformation

BOXTID can be used for computing the Box–Tidwell transformation for simple linear regression and in particular for working exercise 2.4.

REFERENCES

Atkinson, A. C. (1981). Two graphical displays for outlying and influential observations in regression. *Biometrika*, **68,** 13–20.

Atkinson, A. C. (1985). *Plots, Transformations, and Regression.* New York: Oxford University Press.

Belsley, D. A., E. Kuh, and R. E. Welsch (1980). *Regression Diagnostics: Identifying Influential Data and Sources of Collinearity.* New York: Wiley.

Boos, D. (1987). Detecting skewed errors from regression residuals. *Technometrics*, **29,** 83–90.

Box, G. E. P. and D. R. Cox (1964). An analysis of transformations (with discussion). *Journal of the Royal Statistical Society, Series B*, **26,** 211–246.

Box, G. E. P. and N. R. Draper (1986). *Empirical Model-Building and Response Surfaces.* New York: Wiley.

Box, G. E. P. and W. J. Hill (1974). Correcting inhomogeneity of variance with power transformation weighting. *Technometrics*, **16,** 385–389.

Box, G. E. P. and P. Newbold (1971). Some comments on a paper by Coen, Gomme, and Kendall. *Journal of the Royal Statistical Society, Series A*, **134**, 229–240.

Box, G. E. P. and P. W. Tidwell (1962). Transformations of the independent variables. *Technometrics*, **4**, 531–550.

Box, G. E. P., W. G. Hunter, and J. S. Hunter (1978). *Statistics for Experimenters*. New York: Wiley.

Carroll, R. J. and D. B. H. Cline (1988). An asymptotic theory for weighted least squares with weights estimated by replication. *Biometrika*, **75**, 35–43.

Carroll, R. J. and D. Ruppert (1982). Robust estimation in heteroscedastic linear models. *Annals of Statistics*, **10**, 429–441.

Carroll, R. J. and D. Ruppert (1984). Power transformations when fitting theoretical models to data. *Journal of the American Statistical Association*, **79**, 321–328.

Carroll, R. J. and D. Ruppert (1988). *Transformation and Weighting in Regression*. New York: Chapman and Hall.

Chatterjee, S. and A. S. Hadi (1988). *Sensitivity Analysis in Linear Regression*. New York: Wiley.

Coen, P. J., E. Gomme, and M. G. Kendall (1969). Lagged relationships in economic forecasting. *Journal of the Royal Statistical Society, Series A*, **132**, 133–152.

Cook, R. D. (1977). Detection of influential observations in linear regression. *Technometrics*, **19**, 15–18.

Cook, R. D. and S. Weisberg (1982). *Residuals and Influence in Regression*. New York: Chapman and Hall.

Cryer, J. D. (1986). *Time Series Analysis*. Boston: Duxbury.

Daniel, C. and F. S. Wood (1980). *Fitting Equations to Data*, 2nd edition. New York: Wiley.

Davidian, M. (1990). Estimation of variance components in assays with possibly unequal replication and nonnormal data. *Biometrika*, **77**, 43–54.

Davidian, M. and R. J. Carroll (1987). Variance function estimation. *Journal of the American Statistical Association*, **82**, 1079–1091.

Deaton, M. L., M. R. Reynolds, Jr., and R. H. Myers (1983). Estimation and hypothesis testing in regression in the presence of nonhomogeneous error variances. *Communications in Statistics, Simulation and Computation*, **12**, 45–66.

Dixon, W. J., chief editor (1990). BMDP Statistical Software Manual, Volumes 1 and 2. Los Angeles: BMDP Statistical Software, Inc.

Draper, N. R. and H. Smith (1981). *Applied Regression Analysis*, 2nd edition. New York: Wiley.

Flack, V. F. and R. A. Flores (1989). Using simulated envelopes in the evaluation of normal probability plots of regression residuals. *Technometrics*, **31**, 219–225.

Freund, R. J. and R. C. Littell (1991). *SAS System for Regression*, 2nd edition. Cary, NC: SAS Institute, Inc.

Fuller, W. A. (1987). *Measurement Error Models*. New York: Wiley.

Gunst, R. F. and R. L. Mason (1980). *Regression Analysis and Its Applications*. New York: Marcel Dekker.

Hoaglin, D. and B. Iglewicz (1987). Fine-tuning some resistant rules for outliers labeling. *Journal of the American Statistical Association*, **82**, 1147–1149.

Hoaglin, D., B. Iglewicz, and J. W. Tukey (1986). Performance of some resistant rules for outlier labeling. *Journal of the American Statistical Association*, **81**, 991–999.

Kválseth, T. O. (1985). Cautionary note about R^2. *The American Statistician*, **39**, 279–285.

Looney, S. W. and T. R. Gulledge, Jr. (1985). Use of the correlation coefficient with normal probability plots. *The American Statistician*, **39**, 75–79.

Ljung, G. M. and G. E. P. Box (1978). On a measure of lack of fit in time series models. *Biometrika*, **65**, 297–303.

Minitab, Inc. (1989). *Minitab Reference Manual*. State College, PA: Minitab, Inc.

Montgomery, D. C. and E. A. Peck (1992). *Introduction to Linear Regression Analysis*, 2nd ed. New York: Wiley.

Mosteller, F. and J. W. Tukey (1977). *Data Analysis and Regression*. Reading, MA: Addison-Wesley.

Neter, J., W. Wasserman, and M. H. Kutner (1989). *Applied Linear Regression Models*, 2nd edition. Homewood, IL: Irwin.

Newrick, P. G., E. Affie, and R. J. M. Corrall (1990). Relationship between longevity and lifeline: a manual study of 100 patients. *Journal of the Royal Society of Medicine*, **83**, 498–501.

Quesenberry, C. and C. Quesenberry, Jr. (1982). On the distribution of residuals from fitted parametric models. *Journal of Statistical Computation and Simulation*, **15**, 129–140.

Ruppert, D., and B. Aldershof (1989). Transformations to symmetry and homoscedasticity. *Journal of the American Statistical Association*, **84**, 437–446.

SAS Institute, Inc. (1989). *SAS/STAT® User's Guide, 4th edition*. Cary, NC: SAS Institute, Inc.

SPSS, Inc. (1990). *SPSS Reference Guide*. Chicago: SPSS, Inc.

Stata Corporation (1995). *Stata Statistical Software: Release 4.0, Reference Manual*, Volume Two. College Station, TX: Stata Press.

Staudte, R. G. and S. J. Sheather (1990). *Robust Estimation and Testing*. New York: Wiley.

Stirling, W. D. (1985). Heteroscedastic models and an application to block designs. *Applied Statistics*, **34**, 33–41.

Takeuchi, H. (1991). Detecting influential observations by using a new expression of Cook's distance. *Communications in Statistics, Theory and Methods*, **20**, 261–274.

van Zwet, W. R. (1964). *Convex Transformations of Random Variables*. Amsterdam: Mathematisch Centrum.

Weisberg, S. (1980). Comment on "Some large-sample tests for nonnormality in the linear regression model" by H. White and G. M. MacDonald. *Journal of the American Statistical Association*, **75**, 28–31.

Welsch, R. E. (1983). Discussion of "Developments in Linear Regression Methodology: 1959–1982" by R. R. Hocking. *Technometrics*, **25**, 245–246.

Wilkinson, L. (1990). *SYSTAT: The System for Statistics*. Evanston, IL: SYSTAT, Inc.

Willett, J. B. and J. D. Singer (1988). Another cautionary note about R^2: its use in weighted least-squares regression analysis. *The American Statistician*, **42**, 236–238.

EXERCISES

2.1. Derive Eq. (2.8).

2.2. Prove that a simple regression line must pass through the point $(\overline{X}, \overline{Y})$. How will the regression equation be affected if a point with these coordinates in a data set is not used in the computations?

2.3. Assume that an appropriate (nonlinear) model is $Y = \beta_0 \exp(\beta_1 X)\epsilon$. Is this a transformably linear model? If so, what would be the transformed model?

2.4. Consider the data in Table 2.6. Use the X values from that table, but let Y be defined as $Y = 4 + 3X^{1.7}$. Then plot Y against X and notice the nonlinearity. What should be the value of the residual sum of squares if use of the Box–Tidwell procedure permits identification of the correct model? Use five iterations of the Minitab macro BOXTID described in the chapter appendix to determine the transformation that is suggested by application of the Box–Tidwell procedure. What is the (approximate) value of α to which the procedure appears to be converging? What is the value of the residual sum of squares when that value of α is used? Will use of the Box–Tidwell procedure identify the need for a nonlinear term in addition to the linear term, or will it identify the nonlinear term that would be used without a linear term? How could the procedure be modified to detect the need for a nonlinear term in addition to the linear term?

    Compute the correlation between X and $X^{1.7}$. Then consider the value of R^2 and comment on the Mosteller–Tukey rough rule-of-thumb regarding the apparent need for a transformation of these data.

2.5. Assume that X is random and the leverage (h_i) values differ considerably. It is known from analysis of past data that $\text{Var}(\epsilon_i)$ is essentially constant over the range of the X_i in the sample, yet a plot of the (raw) residuals against X shows considerable evidence of heteroscedasticity. Explain how this can happen. How would you advise an experimenter who uses this form of a residual plot?

2.6. Consider the following data, with the last three points constituting a cluster of bad data points.

X	1	2	5	3	6	2	4	7	8	9	5	3	4	5	6	3.4	3.7	3.5
Y	5	4	6	2	7	3	5	8	9	9	5	5	7	8	7	15.6	16.2	16.1

    Compute the values of DFFITS, DFBETAS, and Cook's-D for the 18 data points, and use 2 as the multiplier in the cutoff value for DFFITS. (If MINITAB is used for this, DFFITS and Cook's-D can be obtained directly, and DFBETAS can be obtained as a by-product from the macro RWLS1DFB. See the discussion in the chapter appendix). Then construct a scatter plot of the data, and use 1.5 as the DFFITS multiplier, as suggested by Staudte and Sheather. What does this suggest about the choice for the multiplier, and more generally, what does it suggest about using single-point diagnostics for detecting multiple outliers? Do the outliers have standardized residuals greater than 2? How many standardized residuals in a sample of $n = 18$ should exceed 2 if the errors have a normal distribution and there are no bad data points? What does this suggest (combined with what was discussed in the chapter) about routinely using standardized residuals to detect bad data points?

Also compute R^2 with and without the bad data points. Does this suggest that we should first identify bad data points before (hastily) concluding that there is no linear regression relationship.

2.7. The following data were generated using $\sigma_\epsilon = 3X_i$. First, compute the ordinary least squares estimates. Then use weighted least squares with the known weights, and finally, act as if the function for the variance is unknown and proceed to model the variance, using the reciprocals of the predicted values as the weights. (In modeling the variance it may be necessary to not use certain values which do not fit well with the others.)

Y	X	Y	X
24.1932	1	56.9511	5
17.6791	1	45.5276	5
24.3537	1	53.3301	6
17.3957	2	46.3803	6
7.8591	2	54.9366	6
64.6113	3	47.9322	7
8.1764	3	20.4456	7
36.5301	3	39.0552	8
16.0788	4	71.9443	8
		78.7415	8

2.8. The following data were generated using $Y = 4 + 3X + \epsilon$, where ϵ has a chi-square distribution with two degrees of freedom (a highly skewed distribution). Use the Atkinson simulated-envelope approach to check for nonnormality of the errors, and then use the Flack and Flores modified approach. Do this three times for each approach (remember that the envelopes will almost certainly differ because simulation is being used). Then construct a normal probability plot of the standardized residuals. Knowing that the errors do have a highly skewed distribution, which seems to be the more informative, the normal plot with or without the envelopes? Notice that you are asked to construct a normal probability plot of the standardized residuals, whereas the envelopes are constructed using standardized deletion residuals. Should it make any difference which is used, especially for this data set? Note that the Minitab macros ATKINSON and FLACFLRS can be used to generate the envelopes. See the chapter appendix for details.

Y	X	Y	X
39.0195	11	37.3032	11
49.0387	14	40.5741	12
54.9488	15	53.2339	16
49.2608	13	32.2264	9
23.8919	6	34.2078	10
48.6478	14	59.0119	16
35.1707	10	16.1829	3
32.0008	8	25.0679	7
34.1163	9	21.1483	5
23.9719	6	29.9091	8

2.9. Consider the data given in Table 1.1 and the graph of the data set that was given in Figure 1.1. Would it be reasonable to construct a normal probability plot of the standardized residuals in testing for normality of the error term? Construct a simulated envelope for the residuals using both the Minitab macros ATKINSON and FLACFLRS. Do these envelopes suggest evidence of nonnormality? Does the one extreme standardized deletion residual suggest possible nonnormality, or something else. Now use the Minitab macros MATKINSO and MFLACFLR that utilize standardized residuals, and compare the results.

2.10 Assume that X could be transformed in such a way so as to dramatically improve the fit of a regression model, and a transformation of Y is discovered that produces the same improvement. Would it make any difference which transformation was used? Explain.

2.11 Assume that a scatter plot shows one point to be far removed from the others. Is that point apt to be more influential when the R^2 value for the other points is large or small?

2.12. Derive Eq. (2.9).

CHAPTER 3

Regression with Matrix Algebra

Although simple linear regression can be presented with the algebraic expressions used in Chapters 1 and 2, multiple regression (i.e., more than one X) can best be presented using matrix algebra.

3.1 INTRODUCTION TO MATRIX ALGEBRA

Matrix algebra encompasses arithmetic operations on vectors and matrices. A *matrix* is a rectangular array of numbers arranged in rows and columns and is represented by a capital letter. (And in this and subsequent chapters, letters representing matrices and vectors will be in boldface type.)

For example,

$$\mathbf{A} = \begin{bmatrix} 25 & 15 \\ 15 & 20 \end{bmatrix}$$

is a square matrix with 2 rows and 2 columns. It is customary to refer to the i,jth element of a matrix; this is the element in the ith row and the jth column. Thus, the (2,1) element of $\mathbf{A}$ is 15. Notice that the (1,2) element is also 15. When the i,jth element equals the j,ith element for all combinations of i and j ($i \neq j$), the matrix is a *symmetric* matrix.

A *vector* can be viewed as a matrix that has either one row or one column. The former is termed a *row vector*, and the latter is a *column vector*. For example,

$$\mathbf{b} = \begin{bmatrix} 3 \\ 2 \\ 1 \\ 4 \end{bmatrix}$$

is a column vector and $\mathbf{b}' = [3214]$ is a row vector that is obtained by transforming a column vector, with the prime being used to denote transpose.

When a matrix is transposed, the rows and columns are simply interchanged. If this were done with matrix $\mathbf{A}$, there would be no change. That is, $\mathbf{A}' = \mathbf{A}$ because $\mathbf{A}$ is symmetric. All of the square matrices that we will encounter in subsequent chapters will be symmetric.

A common operation performed on vectors and matrices is multiplication. Not all vectors and matrices will be conformable for multiplication, however. A matrix product can only be obtained when the number of columns in the first vector or matrix is equal to the number of rows in the second vector or matrix.

We also note that a string of vectors and/or matrices may be multiplied together (as the reader is asked to do in Exercise 3.3), provided that the rule given in the preceding paragraph is satisfied for each adjacent pair.

The dimensions of a vector or matrix are generally written as the number of rows by the number of columns. For example, $\mathbf{A}$ is 2×2 since it has 2 rows and 2 columns. Similarly, $\mathbf{b}$ is 4×1 and $\mathbf{b}'$ is 1×4. Thus, $\mathbf{b}'\mathbf{b}$ can be computed as can $\mathbf{bb}'$, but neither $\mathbf{b}$ nor $\mathbf{b}'$ can be multiplied times $\mathbf{A}$.

Multiplication is row times column, with the products of corresponding numbers summed to produce each element in the product of the vectors or matrices. For example,

$$
\mathbf{b}'\mathbf{b} = \begin{bmatrix} 3 & 2 & 1 & 4 \end{bmatrix} \begin{bmatrix} 3 \\ 2 \\ 1 \\ 4 \end{bmatrix}
$$
$$
= (3 \times 3) + (2 \times 2) + (1 \times 1) + (4 \times 4)
$$
$$
= 30
$$

In this example $\mathbf{b}'$ has one row and four columns, and $\mathbf{b}$ has four rows and one column. Thus, we are multiplying a 1×4 vector times a 4×1 vector, and the dimension of the product will be given by the two outside numbers when the dimensions of the components of the product are written side-by-side. That is, a $(1 \times 4) \times (4 \times 1)$ is a 1×1, and that single number is often called a *scalar*. Notice that the two inside numbers are the same, as they must be for the product to be defined.

In regular algebra multiplication is commutative, but this is not true of matrix algebra. For this example $\mathbf{bb}'$ is a 4×4 matrix with elements of

$$
\mathbf{bb}' = \begin{bmatrix} 9 & 6 & 3 & 12 \\ 6 & 4 & 2 & 8 \\ 3 & 2 & 1 & 4 \\ 12 & 8 & 4 & 16 \end{bmatrix}
$$

The operations of addition and subtraction are also defined for vectors and matrices, provided that each of two or more matrices or vectors to be added or subtracted have the same number of rows and columns. The operations are then performed on corresponding elements. For example, if

$$\mathbf{C} = \begin{bmatrix} 10 & 13 \\ 13 & 18 \end{bmatrix}$$

then $\mathbf{A} - \mathbf{C}$ equals

$$\begin{bmatrix} 25 & 15 \\ 15 & 20 \end{bmatrix} - \begin{bmatrix} 10 & 13 \\ 13 & 18 \end{bmatrix} = \begin{bmatrix} 15 & 2 \\ 2 & 2 \end{bmatrix}$$

Division of matrices or vectors is not defined (except when each is divided by a scalar), but the *inverse* of a matrix is defined. For example, the (unique) inverse of matrix $\mathbf{A}$ is a matrix such that $\mathbf{A}^{-1}\mathbf{A} = \mathbf{A}\mathbf{A}^{-1} = \mathbf{I}$, where $\mathbf{I}$ is called an *identity matrix*. This is a matrix that has a specific form: there are 1's on the main diagonal from upper left to lower right and 0's elsewhere. Furthermore, $\mathbf{I}$ will always be a square matrix of the same dimensions as the matrix whose inverse is being obtained.

When a matrix is of *order* 2 (i.e., two rows and two columns), the inverse can be very easily obtained by hand calculation. For example, for

$$\mathbf{A} = \begin{bmatrix} 25 & 15 \\ 15 & 20 \end{bmatrix}$$

$$\mathbf{A}^{-1} = \frac{1}{275} \begin{bmatrix} 20 & -15 \\ -15 & 25 \end{bmatrix}$$

where 275, which equals $(25 \times 20) - (15 \times 15)$, is the *determinant* of the matrix. [The determinant of a matrix cannot be briefly defined. See Searle (1966, p. 61) for a formal definition.] Thus, the elements of the main diagonal of $\mathbf{A}$ are switched, a minus sign is placed in front of the off-diagonal elements, and the resulting matrix is multiplied by the reciprocal of the determinant. (The reader may readily verify that $\mathbf{A}^{-1}\mathbf{A} = \mathbf{A}\mathbf{A}^{-1} = \mathbf{I}$.) We note here that there exist various types of pseudo-inverses that can be obtained when the determinant of the matrix to be inverted is zero. We will not have the occasion to use any of those in this book, however. The interested reader is referred to books on matrix algebra such as Searle (1982) or Graybill (1983).

3.1.1 Eigenvalues and Eigenvectors

Some of the material in subsequent chapters will involve *eigenvalues* and *eigenvectors*. Eigenvalues of a square matrix $\mathbf{A}$ are obtained by solving the determinantal equation

$$|\mathbf{A} - \lambda\mathbf{I}| = 0$$

where $\mathbf{I}$ is an identity matrix of the same order as $\mathbf{A}$, λ represents an eigenvalue (the number of eigenvalues will equal the order of $\mathbf{A}$), and $|\cdot|$ designates the determinant of the difference of the matrices.

Consider the matrix $\mathbf{A}$ used in Section 3.1. Combining that matrix with the fact that here

$$\mathbf{I} = \begin{bmatrix} 1 & 0 \\ 0 & 1 \end{bmatrix}$$

we obtain

$$\mathbf{A} - \lambda\mathbf{I} = \begin{bmatrix} 25 - \lambda & 15 \\ 15 & 20 - \lambda \end{bmatrix}$$

Using the rule for obtaining the determinant of a 2×2 matrix given at the end of Section 3.1, we obtain

$$|\mathbf{A} - \lambda\mathbf{I}| = (25 - \lambda)(20 - \lambda) - (15)^2 = 0$$

This equation can be written as

$$\lambda^2 - 45\lambda + 275 = 0 \tag{3.1}$$

The solution of Eq. (3.1) produces the two eigenvalues $\lambda_1 = 37.707$ and $\lambda_2 = 7.293$.

One of the eigenvalues would have been close to zero if the determinant had been close to zero. In general, matrices whose determinant is close to zero are often encountered in regression analyses. Such matrices will have at least one eigenvalue that is close to zero. When this occurs, something other than ordinary least squares is frequently used (see, e.g., Chapter 12).

The corresponding eigenvectors are obtained by solving the equations

$$\lambda_i\mathbf{u_i} = \mathbf{A}\mathbf{u_i} \qquad i = 1, 2, \ldots, n \tag{3.2}$$

where n is the order of the matrix $\mathbf{A}$ and $\mathbf{u_i}$ is the eigenvector that corresponds to λ_i.

If we denote the elements of $\mathbf{u_i}$ as

$$\mathbf{u_i} = \begin{bmatrix} a_i \\ b_i \end{bmatrix}$$

we have from Eq. (3.2)

$$(\mathbf{A} - \lambda_i)\mathbf{u_i} = 0$$

so that

$$\begin{bmatrix} 25 - \lambda_i & 15 \\ 15 & 20 - \lambda_i \end{bmatrix} \begin{bmatrix} a_i \\ b_i \end{bmatrix} = 0$$

Using $\lambda_1 = 37.707$ produces

$$\begin{bmatrix} -12.707 & 15 \\ 15 & -17.707 \end{bmatrix} \begin{bmatrix} a_1 \\ b_1 \end{bmatrix} = 0$$

so that

$$-12.707a_1 + 15b_1 = 0$$
$$15a_1 - 17.707b_1 = 0$$

Since zero is on the right side of each equation, there is no way these two equations could be solved to produce nonzero solutions for a_1 and b_1. Consequently, it is necessary to pick an arbitrary value for one of the two and then solve for the other one. (This implies, of course, that eigenvectors are not unique). If we let $a_1 = 1$, we obtain $b_1 = 0.847$.

For $\lambda_2 = 7.293$ we obtain the equations

$$17.707a_2 + 15b_2 = 0$$
$$15a_2 + 12.707b_2 = 0$$

If we let $a_1 = 1$ we obtain $b_2 = -1.180$. Thus, the eigenvectors are

$$\mathbf{u_1} = \begin{bmatrix} 1.000 \\ 0.847 \end{bmatrix} \quad \text{and} \quad \mathbf{u_2} = \begin{bmatrix} 1.000 \\ -1.180 \end{bmatrix}$$

Note: Although the elements of $\mathbf{u_1}$ and $\mathbf{u_2}$ are not unique, the *direction* of each vector, which is determined by the relationship of the two elements, *is* unique.

For example, if we had arbitrarily picked $a_1 = 2$ and $a_2 = 2$, we would have had $b_1 = 1.694$ and $b_2 = -2.36$, so that the vector direction would be the same (i.e., the new values for b_1 and b_2 are twice the original values).

Eigenvalues and eigenvectors are also called *latent roots* and *latent vectors* and *characteristic roots* and *characteristic vectors*, respectively. A function of the largest and smallest eigenvalues that is sometimes used in describing a matrix is the *condition number*, which is defined as $\sqrt{\lambda_{max}/\lambda_{min}}$, where λ_{max} and λ_{min} denote the largest and smallest eigenvalues respectively, of a matrix.

We will now give some results that will be useful in later chapters. If we write Eq. (3.2) in matrix form so as to incorporate all of the eigenvalues and eigenvectors, we obtain

$$\mathbf{VE} = \mathbf{AV} \tag{3.3}$$

with $\mathbf{E}$ a diagonal matrix that has the eigenvalues as diagonal elements, and $\mathbf{V}$ is the matrix of eigenvectors, with the eigenvectors given by the columns of the matrix. Since by definition $\mathbf{V}$ is an orthonormal matrix so that $\mathbf{V'V} = \mathbf{VV'} = \mathbf{I}$, it follows from multiplying both sides of Eq. (3.3) by $\mathbf{V'}$ that $\mathbf{E} = \mathbf{V'AV}$.

An *orthonormal matrix* is one whose rows and columns are orthogonal and have unit length, with the *length* of a row or column defined as the square root of the sum of squares of the elements.

3.2   MATRIX ALGEBRA APPLIED TO REGRESSION

In this section we illustrate how matrix algebra can be applied to regression, regardless of the number of regressors, when ordinary least squares is applicable. Weighted least squares can also be handled with matrix algebra; the results are given in the chapter appendix.

In general, the regression model can be written in the form $\mathbf{Y} = \mathbf{X}\boldsymbol{\beta} + \boldsymbol{\epsilon}$, so for simple regression

$$\mathbf{Y} = \begin{bmatrix} Y_1 \\ Y_2 \\ \vdots \\ Y_n \end{bmatrix} \quad \mathbf{X} = \begin{bmatrix} 1 & X_1 \\ 1 & X_2 \\ \vdots & \vdots \\ 1 & X_n \end{bmatrix} \quad \boldsymbol{\beta} = \begin{bmatrix} \beta_0 \\ \beta_1 \end{bmatrix} \quad \boldsymbol{\epsilon} = \begin{bmatrix} \epsilon_1 \\ \epsilon_2 \\ \vdots \\ \epsilon_n \end{bmatrix}$$

The reader will observe that when the appropriate multiplication and addition operations are performed in computing $\mathbf{X}\boldsymbol{\beta} + \boldsymbol{\epsilon}$, one result is $Y_1 = \beta_0 + \beta_1 X_1 + \epsilon_1$. Thus, the column of ones in the $\mathbf{X}$ matrix is there for the purpose of producing the β_0 term when the multiplication is performed.

The two normal equations in Section 1.4 can be represented in matrix form by

$$(\mathbf{X}'\mathbf{X})\boldsymbol{\beta} = \mathbf{X}'\mathbf{Y} \qquad (3.4)$$

To solve the two equations represented by Eq. (3.4) requires a matrix operation that is analogous to the operation that is performed in ordinary algebra. Specifically, to solve the equation $3X = 9$ we could think of multiplying each side of the equation by the *multiplicative inverse* of the coefficient of X. Similarly, to solve Eq. (3.4) for β we multiply each side of the equation by $(\mathbf{X}'\mathbf{X})^{-1}$, being careful to multiply on the left instead of the right since the latter would not be defined.

Thus,

$$(\mathbf{X}'\mathbf{X})^{-1}\mathbf{X}'\mathbf{X}\boldsymbol{\beta} = (\mathbf{X}'\mathbf{X})^{-1}\mathbf{X}'\mathbf{Y}$$

so that

$$\hat{\boldsymbol{\beta}} = (\mathbf{X}'\mathbf{X})^{-1}\mathbf{X}'\mathbf{Y} \qquad (3.5)$$

It should be noted that this result also applies to multiple linear regression, as do the other general results given in this chapter. That is, regardless of the number of regressors, the least squares estimates of the parameters are obtained using Eq. (3.5).

We will use the data that were given in Table 1.1 to illustrate the computation of the least-squares estimates using Eq. (3.5). The matrix $\mathbf{X}$ and vector $\mathbf{Y}$ are given by

$$\mathbf{X} = \begin{bmatrix} 1 & 400 \\ 1 & 410 \\ 1 & 395 \\ 1 & 415 \\ 1 & 390 \\ 1 & 385 \\ 1 & 405 \\ 1 & 420 \\ 1 & 380 \\ 1 & 375 \end{bmatrix} \qquad \mathbf{Y} = \begin{bmatrix} 26.15 \\ 28.45 \\ 25.20 \\ 29.30 \\ 24.35 \\ 23.10 \\ 27.40 \\ 29.60 \\ 22.05 \\ 21.30 \end{bmatrix}$$

The matrix $\mathbf{X}'\mathbf{X}$ will have the general form

$$\mathbf{X}'\mathbf{X} = \begin{bmatrix} n & \sum X \\ \sum X & \sum X^2 \end{bmatrix} \qquad (3.6)$$

and $\mathbf{X'Y}$ will have components given by

$$\mathbf{X'Y} = \begin{bmatrix} \sum Y \\ \sum XY \end{bmatrix}$$

For this example we thus have

$$\mathbf{X'X} = \begin{bmatrix} 10 & 3{,}975 \\ 3{,}975 & 1{,}582{,}125 \end{bmatrix} \qquad \mathbf{X'Y} = \begin{bmatrix} 256.9 \\ 102{,}523.5 \end{bmatrix}$$

Inverting $\mathbf{X'X}$ produces

$$(\mathbf{X'X})^{-1} = \frac{1}{20{,}625} \begin{bmatrix} 1{,}582{,}125 & -3{,}975 \\ -3{,}975 & 10 \end{bmatrix}$$

Multiplying this times $\mathbf{X'Y}$ produces

$$\hat{\boldsymbol{\beta}} = (\mathbf{X'X})^{-1}\mathbf{X'Y} = \begin{bmatrix} \hat{\beta}_0 \\ \hat{\beta}_1 \end{bmatrix} = \begin{bmatrix} -52.51 \\ 0.19673 \end{bmatrix}$$

Thus, we obtain the same values of $\hat{\beta}_0$ and $\hat{\beta}_1$ that were given in Section 1.4. [The reader is asked to show in Exercise 3.1 that obtaining the elements of $(\mathbf{X'X})^{-1}\mathbf{X'Y}$ in symbols produces the same expressions for $\hat{\beta}_0$ and $\hat{\beta}_1$ that were given in Section 1.4.]

Although regression computations are generally performed with a computer, especially in multiple regression, it is worth noting that roundoff error can easily occur in regression, especially with hand computation. Consequently, division should usually be performed last, so for this example division by 20,625 was performed last in computing $\hat{\beta}_0$ and $\hat{\beta}_1$.

3.2.1 Predicted Y and R^2

Since the model is $\mathbf{Y} = \mathbf{X}\boldsymbol{\beta} + \boldsymbol{\epsilon}$, it follows that the predicted values for $\mathbf{Y}$ would be obtained as $\hat{\mathbf{Y}} = \mathbf{X}\hat{\boldsymbol{\beta}}$, and the vector of residuals would be given by $\mathbf{e} = \hat{\mathbf{Y}} - \mathbf{X}\hat{\boldsymbol{\beta}}$. It follows that one could compute R^2 as

$$R^2 = 1 - \frac{\mathbf{e'e}}{\mathbf{Y'Y} - (\sum Y)^2/n}$$

where $\mathbf{e'e} = \sum(Y - \hat{Y})^2$, with the latter given in Section 1.5.2, and $\mathbf{Y'Y} - (\sum Y)^2/n = \sum(Y - \bar{Y})^2$.

Since it is easier to compute R^2 without having to first compute the vector of residuals, it is preferable to compute R^2 as

$$R^2 = \frac{\hat{\boldsymbol{\beta}}'\mathbf{X}'\mathbf{Y} - (\sum Y)^2/n}{\mathbf{Y}'\mathbf{Y} - (\sum Y)^2/n} \tag{3.7}$$

where the numerator of Eq. (3.7) equals $\sum(\hat{Y} - \overline{Y})^2$, as the reader is asked to show in Exercise 3.2.

3.2.2 Estimation of σ_ϵ^2

It follows from Eq. (3.7) and from the discussion in Sections 1.5.4 and 1.5.7 that $\mathbf{SS}_{\text{residual}} = \mathbf{Y}'\mathbf{Y} - (\sum Y)^2/n - [\hat{\boldsymbol{\beta}}'\mathbf{X}'\mathbf{Y} - (\sum Y)^2/n] = \mathbf{Y}'\mathbf{Y} - \hat{\boldsymbol{\beta}}'\mathbf{X}'\mathbf{Y}$, so that

$$\hat{\sigma}_\epsilon^2 = \frac{\mathbf{Y}'\mathbf{Y} - \hat{\boldsymbol{\beta}}'\mathbf{X}'\mathbf{Y}}{n - 2}$$

3.2.3 Variance of Y and $\hat{\mathbf{Y}}$

It was stated in Section 1.5.3 that $\text{Var}(Y|X) = \text{Var}(\epsilon)$, assuming that the Y's are independent and that the model is correct. As in that section, we will write $\text{Var}(Y)$ to represent $\text{Var}(Y|X)$. Therefore, for a data set of n (X, Y) values, we may write $\text{Var}(Y)$ as $\text{Var}(Y) = \sigma_\epsilon^2\mathbf{I}$, where $\mathbf{I}$ is an $n \times n$ identity matrix, and the equal diagonal elements of $\sigma_\epsilon^2\mathbf{I}$ represent $\text{Var}(Y_1), \text{Var}(Y_2), \ldots, \text{Var}(Y_n)$.

To obtain $\text{Var}(\hat{\mathbf{Y}})$ we would write $\hat{\mathbf{Y}} = \mathbf{X}\hat{\boldsymbol{\beta}}$, so that

$$\begin{aligned}
\text{Var}(\hat{\mathbf{Y}}) &= \text{Var}(\mathbf{X}\hat{\boldsymbol{\beta}}) \\
&= \mathbf{X}[\text{Var}(\hat{\boldsymbol{\beta}})]\mathbf{X}' \\
&= \mathbf{X}(\text{Var}[(\mathbf{X}'\mathbf{X})^{-1}\mathbf{X}'\mathbf{Y}])\mathbf{X}' \\
&= \mathbf{X}\{(\mathbf{X}'\mathbf{X})^{-1}\mathbf{X}'(\sigma_\epsilon^2\mathbf{I})\mathbf{X}(\mathbf{X}'\mathbf{X})^{-1}\}\mathbf{X}' \\
&= \mathbf{X}\{\sigma_\epsilon^2(\mathbf{X}'\mathbf{X})^{-1}\}\mathbf{X}' \\
&= \sigma_\epsilon^2\mathbf{X}(\mathbf{X}'\mathbf{X})^{-1}\mathbf{X}' \tag{3.8}
\end{aligned}$$

The expressions given in lines two and four of this derivation can be explained as follows. When one obtains $\text{Var}(aW)$ where a is a scalar and W is a random variable, the result is $a^2\text{Var}(W)$. But when a matrix is multiplied times a vector of random variables, as in $\mathbf{X}\hat{\boldsymbol{\beta}}$, the matrix $\mathbf{X}$ is not squared because a matrix product cannot be obtained by multiplying a matrix times itself unless the matrix is square, and $\mathbf{X}$ will generally not be a square matrix. (The exception to this is when the *direct product* of two matrices is obtained, as then the two matrices simply have to have the same number of rows and columns. We will not have the need for direct products in this book.)

Let's focus attention upon $\text{Var}(\hat{Y}_1)$, where $\hat{Y}_1$ is the first element of $\hat{\mathbf{Y}}$. Since $\hat{\mathbf{Y}} = \mathbf{X}\hat{\boldsymbol{\beta}}$, it follows that $\hat{Y}_1 = \mathbf{x}_1'\hat{\boldsymbol{\beta}}$ where $\mathbf{x}_1'$ is the first row of $\mathbf{X}$. We know that the first element of $\mathbf{x}_1$ is a 1, and assume that the second value is a 2. Thus, $\mathbf{x}_1' = [1 \ 2]$. Then $\text{Var}(\mathbf{x}'\hat{\boldsymbol{\beta}}) = \text{Var}(1\hat{\beta}_0 + 2\hat{\beta}_1) = \text{Var}(\hat{\beta}_0) + 4\,\text{Var}(\hat{\beta}_1) + 4\,\text{Cov}(\hat{\beta}_0, \hat{\beta}_1)$.

Notice that this is the result that we obtain if we define $\text{Var}(\mathbf{x}'\hat{\boldsymbol{\beta}})$ as $\mathbf{x}'\text{Var}(\hat{\boldsymbol{\beta}})\mathbf{x}$ where

$$\text{Var}(\hat{\boldsymbol{\beta}}) = \begin{bmatrix} \text{Var}(\hat{\beta}_0) & \text{Cov}(\hat{\beta}_0, \hat{\beta}_1) \\ \text{Cov}(\hat{\beta}_0, \hat{\beta}_1) & \text{Var}(\hat{\beta}_1) \end{bmatrix}$$

since $\mathbf{x}'\text{Var}(\hat{\boldsymbol{\beta}})\mathbf{x}$ equals

$$[1 \ \ 2] \begin{bmatrix} \text{Var}(\hat{\beta}_0) & \text{Cov}(\hat{\beta}_0, \hat{\beta}_1) \\ \text{Cov}(\hat{\beta}_0, \hat{\beta}_1) & \text{Var}(\hat{\beta}_1) \end{bmatrix} \begin{bmatrix} 1 \\ 2 \end{bmatrix}$$

$$= [\text{Var}(\hat{\beta}_0) + 2\,\text{Cov}(\hat{\beta}_0, \hat{\beta}_1), \text{Cov}(\hat{\beta}_0, \hat{\beta}_1) + 2\,\text{Var}(\hat{\beta}_1)] \begin{bmatrix} 1 \\ 2 \end{bmatrix}$$

$$= \text{Var}(\hat{\beta}_0) + 4\,\text{Cov}(\hat{\beta}_0, \hat{\beta}_1) + 4\,\text{Var}(\hat{\beta}_1)$$

Therefore, by defining $\text{Var}(\mathbf{x}'\hat{\boldsymbol{\beta}}) = \mathbf{x}'\text{Var}(\hat{\boldsymbol{\beta}})\mathbf{x}$, we obtain the same result as we would obtain without using this definitional result. Since this result must obviously hold for each row of $\mathbf{X}$, it follows that $\text{Var}(\mathbf{X}\hat{\boldsymbol{\beta}}) = \mathbf{X}\,\text{Var}(\hat{\boldsymbol{\beta}})\mathbf{X}'$.

Notice that the same general form of this result is used in obtaining $\text{Var}(\hat{\boldsymbol{\beta}})$ (in the fourth line of the derivation) since $\hat{\boldsymbol{\beta}} = (\mathbf{X}'\mathbf{X})^{-1}\mathbf{X}'\mathbf{Y}$. Let $\mathbf{C} = (\mathbf{X}'\mathbf{X})^{-1}\mathbf{X}'$ so that $\hat{\boldsymbol{\beta}} = \mathbf{C}\mathbf{Y}$. Then $\text{Var}(\hat{\boldsymbol{\beta}}) = \text{Var}(\mathbf{C}\mathbf{Y}) = \mathbf{C}\,\text{Var}(\mathbf{Y})\mathbf{C}'$ where $\mathbf{C}' = [(\mathbf{X}'\mathbf{X})^{-1}\mathbf{X}']' = \mathbf{X}(\mathbf{X}'\mathbf{X})^{-1}$ since the transpose of a product is the product of the transposes (of the components of the product) in reverse order, and the transpose of $(\mathbf{X}'\mathbf{X})^{-1}$ is $(\mathbf{X}'\mathbf{X})^{-1}$ since the letter is symmetric.

It can then be seen from the derivation of $\text{Var}(\hat{\mathbf{Y}})$ that $\text{Var}(\hat{\boldsymbol{\beta}}) = \sigma_{\epsilon}^2(\mathbf{X}'\mathbf{X})^{-1}$. By applying the rule for inverting a 2×2 matrix to the form of $\mathbf{X}'\mathbf{X}$ given in Eq. (3.5), it can be shown that

$$(\mathbf{X}'\mathbf{X})^{-1} = \frac{1}{n\sum X^2 - (\sum X)^2} \begin{bmatrix} \sum X^2 & -\sum X \\ -\sum X & n \end{bmatrix}$$

It then follows that $\text{Var}(\hat{\beta}_1) = \sigma^2/S_{xx}$, which agrees with the result given in Section 1.5.4, and $\text{Var}(\hat{\beta}_0) = \sigma^2(\sum X^2/nS_{xx})$, as was given in Section 1.5.5.

We may also note that $\text{Cov}(\hat{\beta}_0, \hat{\beta}_1) = \sigma^2(-\sum X/nS_{xx})$, so the covariance will be zero only if $\sum X = 0$. This should be intuitive since $\hat{\beta}_0$ will then not be a function of $\hat{\beta}_1$, as $\hat{\beta}_0$ will then equal $\overline{Y}$, and it was shown in the appendix to Chapter 1 that $\text{Cov}(\overline{Y}, \hat{\beta}_1) = 0$.

It should be noted that we can always force this covariance to be zero simply by subtracting $\overline{X}$ from each X value. This has the simple effect of producing a different, but equivalent, form of the regression equation that is given by

$$\hat{Y} = \overline{Y} + \hat{\beta}_1 X' \tag{3.9}$$

where $X' = X - \overline{X}$. This has the effect of essentially eliminating β_0, so that the column of 1's could subsequently be removed from the $\mathbf{X}$ matrix. This can be advantageous for numerical purposes since numerical problems could result in producing the regression equation if the values of X were all very close to 1. This is discussed further in Section 4.2.1.

Regardless of how the regression equation is written, it follows from Eq. (3.8) that the variance of a single value of $\hat{Y}$ is given by

$$\text{Var}(\hat{Y}) = \sigma_\epsilon^2 \mathbf{x}'(\mathbf{X}'\mathbf{X})^{-1}\mathbf{x} \tag{3.10}$$

where $\mathbf{x}'$ may be a row of the $\mathbf{X}$ matrix, or it may contain a new value of X at which it is desired to predict Y.

3.2.4   Centered Data

Centering X and/or Y refers to the operation of creating a new variable by subtracting the respective mean from either a regressor or from Y, as was illustrated briefly in the preceding section. In Chapter 4 there is extensive use of the centered-data form of the regression model, and *correlation form* (a related form that is discussed in Section 3.2.5) is used extensively in software packages.

$$Y - \overline{Y} = \beta_1(X - \overline{X}) + \epsilon \tag{3.11}$$

Notice that the constant term β_0 is embedded in the model since $\beta_0 = \overline{Y} - \beta_1 \overline{X}$.

The prediction equation is given by

$$\hat{Y} - \overline{Y} = \hat{\beta}_1(X - \overline{X}) \tag{3.12}$$

[Notice that it might seem more appropriate to put the "hat" over $(Y - \overline{Y})$, but that would be equivalent to the left side of Eq. (3.12).]

In matrix notation, we will let $\mathbf{X}^+$ denote the $n \times 1$ matrix that contains the values of $X_i - \overline{X}$. Similarly, $\mathbf{Y}^+$ contains the values of $Y_i - \overline{Y}$, and $\boldsymbol{\beta}$ contains just β_1. Then, $(\mathbf{X}^+)'\mathbf{Y}^+ = \sum(X_i - \overline{X})(Y_i - \overline{Y})$ and $(\mathbf{X}^+)'\mathbf{X}^+ = \sum(X_i - \overline{X})^2$, and

$$\hat{\boldsymbol{\beta}} = [(\mathbf{X}^+)'\mathbf{X}^+]^{-1}(\mathbf{X}^+)'\mathbf{Y}^+$$

$$= \left[\sum (X_i - \overline{X})^2\right]^{-1} \sum (X_i - \overline{X})(Y_i - \overline{Y})$$

$$= \frac{S_{xy}}{S_{xx}}$$

as was given for $\hat{\beta}_1$ in Section 1.4.

The expression for R^2 is given by

$$R^2 = \frac{\hat{\boldsymbol{\beta}}'(\mathbf{X}^+)'\mathbf{Y}^+}{(\mathbf{Y}^+)'\mathbf{Y}^+} \tag{3.13}$$

Notice that $(\sum Y)^2/n$ is subtracted in the numerator and denominator of Eq. (3.7) but not in Eq. (3.13). This is because $(\sum Y)^2/n$ is the *sum of squares due to β_0*, and it is embedded in Eq. (3.13), just as $\hat{\beta}_0$ is embedded in Eq. (3.12).

The numerator of Eq. (3.13) clearly equals S_{xy}^2/S_{xx}, and the reader is asked to show in Exercise 3.6 that this is equivalent to the numerator of Eq. (3.7).

It then follows that $\hat{\sigma}_\epsilon^2$ is obtained as $\hat{\sigma}_\epsilon^2 = [(\mathbf{Y}^+)'\mathbf{Y}^+ - \hat{\boldsymbol{\beta}}'(\mathbf{X}^+)'\mathbf{Y}^+]/(n-2)$, which is the same expression that was given in Section 3.2.2.

We will "uncenter" Y before obtaining $\text{Var}(\hat{\mathbf{Y}})$ so as to avoid having to evaluate $\text{Var}(\hat{\mathbf{Y}} - \overline{\mathbf{Y}})$. Thus, the regression equation will have the form

$$\hat{\mathbf{Y}} = \overline{\mathbf{Y}} + (\mathbf{X}^+)\hat{\boldsymbol{\beta}}$$

Since $\hat{\boldsymbol{\beta}} = \hat{\beta}_1$ and using the previously mentioned result that $\text{Cov}(\overline{Y}, \hat{\beta}_1) = 0$, we have

$$\text{Var}(\hat{\mathbf{Y}}) = \text{Var}(\overline{\mathbf{Y}}) + \text{Var}[(\mathbf{X}^+)\hat{\boldsymbol{\beta}}]$$

$$= \sigma_\epsilon^2 \left[\frac{1}{n}\right]\mathbf{J} + \sigma_\epsilon^2 \mathbf{X}^+[(\mathbf{X}^+)'\mathbf{X}^+]^{-1}(\mathbf{X}^+)'$$

$$= \sigma_\epsilon^2 \left\{\left(\frac{1}{n}\right)\mathbf{J} + \mathbf{X}^+[(\mathbf{X}^+)'\mathbf{X}^+]^{-1}(\mathbf{X}^+)'\right\}$$

where $\mathbf{J}$ is an $n \times n$ matrix whose elements are all equal to 1, and the dimensions of the two matrices in the bracketed expression are also of course $n \times n$.

It then follows that the variance of a single $\hat{Y}$ is given by

$$\mathrm{Var}(\hat{Y}) = \sigma_\epsilon^2 \left\{ \frac{1}{n} + (\mathbf{x}^+)'[(\mathbf{X}^+)'\mathbf{x}^+]^{-1}\mathbf{x}^+ \right\} \qquad (3.14)$$

which is the counterpart to Eq. (3.10) for an uncentered regressor.

3.2.5 Correlation Form

When X and Y are centered and then divided by $\sqrt{S_{xx}}$ and $\sqrt{S_{yy}}$, respectively, the data are then in "correlation form," and the regression coefficients are termed *standardized regression coefficients*. With one regressor, the matrix that contains the values of $(X_i - \overline{X})/\sqrt{S_{xx}}$ is $n \times 1$ and will be denoted by $\mathbf{X}^*$. Similarly, $\mathbf{Y}^*$ will denote the standardized Y values. It follows that $(\mathbf{X}^*)'\mathbf{X}^* = 1$ and $(\mathbf{X}^*)'\mathbf{Y}^* = \sum (X_i - \overline{X})(Y_i - \overline{Y})/(\sqrt{S_{xx}} \sqrt{S_{yy}}) = S_{xy}/\sqrt{S_{xx}S_{yy}}$, which is r_{xy}, the correlation between X and Y, as was given in Section 1.7. Thus,

$$\begin{aligned}
\hat{\boldsymbol{\beta}} &= [(\mathbf{X}^*)'\mathbf{X}^*]^{-1}(\mathbf{X}^*)'\mathbf{Y}^* \\
&= (1)^{-1}r_{xy} \\
&= r_{xy}(-\hat{\beta}_1)
\end{aligned}$$

The words *correlation form* do not result from the fact that $\hat{\boldsymbol{\beta}} = r_{xy}$, however. Rather, the term refers to the fact that when there is more than one regressor, $(\mathbf{X}^*)'\mathbf{X}^*$ will be a correlation matrix that contains the correlations between the regressors, and $(\mathbf{X}^*)'\mathbf{Y}^*$ will have as elements the correlations between each regressor and Y.

Since we know that $\hat{\beta}_1$ in raw form equals S_{xy}/S_{xx}, it follows that to return to raw form from correlation form we would need to multiply the correlation-form $\hat{\beta}_1$ (say $\hat{\beta}_1^*$) by $\sqrt{S_{yy}/S_{xx}}$. That is,

$$\hat{\beta}_1 = (S_{yy}/S_{xx})^{1/2}\hat{\beta}_1^*$$

Since with correlation form we lose the intercept just as we do with the centered-data form, the intercept would have to be "recreated" as $\hat{\beta}_0 = \overline{Y} - \hat{\beta}_1\overline{X}$.

Defining R^2 for correlation form analogous to the definition when the data are (only) centered, we have (using the aforementioned $\mathbf{X}^*$ and $\mathbf{Y}^*$)

$$\begin{aligned}
R^2 &= \frac{\hat{\boldsymbol{\beta}}^*(\mathbf{X}^*)'\mathbf{Y}^*}{(\mathbf{Y}^*)'\mathbf{Y}^*} \\
&= \frac{(r_{xy})(r_{xy})}{1} \\
&= r_{xy}^2
\end{aligned}$$

Thus, when correlation form is used it is clear that $R^2 = r_{xy}^2$ (which is also true, of course, when raw form is used).

3.2.6 Influence Statistics in Matrix Form

The influence statistics that were described in Sections 2.4.2 and 2.4.4 were presented without using matrices. It is of interest to consider the appropriate expressions for these statistics when matrices are used. For example, h_i given in Eq. (2.2) is obtainable from the matrix $\mathbf{H} = \mathbf{X(X'X)}^{-1}\mathbf{X'}$ which was called the "hat matrix" by John W. Tukey (see Hoaglin and Welsch, 1978) since its multiplication times $\mathbf{Y}$ produces $\hat{\mathbf{Y}}$). The leverages are the diagonal values of the hat matrix and could thus also be denoted by h_{ii}.

Analogous to the expressions for DFBETAS$_{i0}$ and DFBETAS$_{i1}$ that were given for simple linear regression in Section 2.4.2, using matrix notation we may write DFBETAS$_{ij}$ as

$$\text{DFBETAS}_{ij} = \frac{\hat{\beta}_j - \hat{\beta}_j(i)}{s(i)\sqrt{(\mathbf{X'X})_{jj}^{-1}}}$$

where as in Section 2.4.2, i denotes the observation, j denotes the parameter, and (i) means that the ith observation is not used in computing the indicated statistic. Also, $(\mathbf{X'X})_{jj}^{-1}$ denotes the jth diagonal element of $(\mathbf{X'X})^{-1}$.

We also consider the various ways that Cook's-D statistic can be written, both with and without matrices, in addition to the expression that was given in Section 2.4.2. One such expression is

$$D_i = \frac{(\hat{\mathbf{Y}}_{(i)} - \hat{\mathbf{Y}})'(\hat{\mathbf{Y}}_{(i)} - \hat{\mathbf{Y}})}{ps^2}$$

Another equivalent expression for D_i results from writing the numerator as $(\hat{\boldsymbol{\beta}}_{(i)} - \hat{\boldsymbol{\beta}})'\mathbf{X'X}(\hat{\boldsymbol{\beta}}_{(i)} - \hat{\boldsymbol{\beta}})$. A more useful alternative expression is to write D_i as

$$D_i = \left[\frac{e_i}{s(1 - h_i)^{1/2}}\right]^2 \left(\frac{h_i}{1 - h_i}\right)\left(\frac{1}{p}\right) \tag{3.15}$$

where the first fraction in Eq. (3.15) is the standardized residual, and the second fraction will be large when the leverage is large. Thus, a large value of D_i can be caused by a large standardized residual or a large leverage value, or both.

SUMMARY

The use of matrix algebra allows formulas and derivations in regression to be presented more compactly and efficiently than when ordinary algebra is used.

Although regression computations can be easily performed without the use of matrix algebra when there is only one regressor, it is not practical to use ordinary algebra when there is more than one regressor.

APPENDIX

3.A Weighted Least Squares Formulas

Assume that the error terms have different variances but are uncorrelated. We would then not want to minimize the sum of the squares of the errors but some function of that quantity. As was discussed in Section 2.1.3.1, errors that have a large variance should receive a smaller weight than those that have a smaller variance, with the objective being to use a weighted function of the errors such that the weighted errors all have the same variance, so that minimizing the sum of the squares of the weighted errors would be practical.

Consider the expression for W given in Section 2.1.3.1, and let $\text{Var}(\epsilon) = \sigma^2 \mathbf{V}$ where $\mathbf{V}$ is a diagonal matrix whose diagonal elements [which give $\text{Var}(\epsilon_i)$] are not all equal. Let ϵ^+ denote the "transformed error" such that $\text{Var}(\epsilon^+) = \sigma^2 \mathbf{I}$, and ordinary least squares can then be applied. It should be apparent that we must let $\epsilon^+ = \mathbf{V}^{-1/2}\epsilon$, with each element of the diagonal matrix $\mathbf{V}^{-1/2}$ being the square root of the reciprocal of the corresponding element in $\mathbf{V}$. Then $\text{Var}(\epsilon^+) = (\mathbf{V}^{-1/2})^2 \sigma^2 \mathbf{V} = \sigma^2 \mathbf{I}$.

Thus, the function of the errors to be minimized is $(\epsilon^+)'\epsilon^+ = \epsilon'\mathbf{V}^{-1}\epsilon$. The form of the least squares estimators would be obtained by taking the derivative of this expression with respect to $\boldsymbol{\beta}$, and then solving the resultant matrix equation. Specifically,

$$\epsilon'\mathbf{V}^{-1}\epsilon = (\mathbf{Y} - \mathbf{X}\boldsymbol{\beta})'\mathbf{V}^{-1}(\mathbf{Y} - \mathbf{X}\boldsymbol{\beta})$$
$$= \mathbf{Y}'\mathbf{V}^{-1}\mathbf{Y} - 2\mathbf{Y}'\mathbf{V}^{-1}\mathbf{X}\boldsymbol{\beta} + \boldsymbol{\beta}'\mathbf{X}'\mathbf{V}^{-1}\mathbf{X}\boldsymbol{\beta}$$

Let L represent this last expression. Then

$$\frac{\partial L}{\partial \beta} = -2\mathbf{X}'\mathbf{V}^{-1}\mathbf{Y} + 2\mathbf{X}'\mathbf{V}^{-1}\mathbf{X}\boldsymbol{\beta}$$

(Notice that the first term in the derivative is the transpose of what we might expect. This is because the derivative with respect to a column vector is written as a column vector when the result is a vector.) Setting this derivative equal to zero and solving for $\boldsymbol{\beta}$ produces

$$\hat{\boldsymbol{\beta}} = (\mathbf{X}'\mathbf{V}^{-1}\mathbf{X})^{-1}\mathbf{X}'\mathbf{V}^{-1}\mathbf{Y}$$

Since $\text{Var}(\epsilon) = \sigma^2\mathbf{V}$ implies that $\text{Var}(\mathbf{Y}) = \sigma^2\mathbf{V}$, it follows that

$$\mathrm{Var}(\hat{\boldsymbol{\beta}}) = (\mathbf{X}'\mathbf{V}^{-1}\mathbf{X})^{-1}\mathbf{X}'\mathbf{V}^{-1}(\sigma^2\mathbf{V})(\mathbf{V}^{-1})'\mathbf{X}(\mathbf{X}'\mathbf{V}^{-1}\mathbf{X})^{-1}$$
$$= \sigma^2(\mathbf{X}'\mathbf{V}^{-1}\mathbf{X})^{-1}$$

with the two variances given by the diagonal elements and the covariances given by the off-diagonal elements.

REFERENCES

Graybill, F. A. (1983). *Matrices with Applications in Statistics*, 2nd edition. Belmont, CA: Wadsworth.

Hoaglin, D. C. and R. Welsch (1978). The hat matrix in regression and ANOVA. *The American Statistician*, **33**, 108–115.

Searle, S. (1982). *Matrix Algebra Useful for Statistics*. New York: Wiley.

Searle, S. (1966). *Matrix Algebra for the Biological Sciences*. New York: Wiley.

EXERCISES

3.1. Using the vector $\mathbf{x}_1'$ given in Section 3.2.3 and the matrix $\mathbf{A}$ given in Section 3.1, compute $\mathbf{x}_1'\mathbf{A}\mathbf{x}_1$. What must be the general form of the result whenever a string of vectors and matrices are multiplied together where the first component of the product is a row vector and the last component is a column vector?

3.2. Compute the eigenvalues and determine a set of corresponding eigenvectors for the matrix $\mathbf{C}$ given in Section 3.1.

3.3. Explain the difference between raw form, centered form, and correlation form.

3.4. Show that the numerator of Eq. (3.7) is equivalent to the numerator of Eq. (3.13).

3.5. Consider matrix $\mathbf{A}$ given in Section 3.1 and matrix $\mathbf{X}$ given in Section 3.2. Which of the following matrix products is defined, $\mathbf{AX}$ or $\mathbf{XA}$? Compute the product that is defined.

3.6. Using matrices $\mathbf{A}$ and $\mathbf{C}$ given in Section 3.1, demonstrate the fact that matrix multiplication does not have the commutative property of ordinary algebra by first computing $\mathbf{AC}$ and then computing $\mathbf{CA}$, and compare the results.

3.7. Compute the inverse of matrix $\mathbf{C}$ given in Section 3.1.

3.8. Assume that regression through the origin is to be used and that there is a single regressor. (Recall that this was presented in Section 1.6). Determine the general form of the $\mathbf{X}$ matrix and then show that the use of matrices produces the same least squares estimator as was given in Section 1.6.

3.9. Show that the numerator of Eq. (3.7) equals $\sum (\hat{Y} - \overline{Y})^2$.

3.10. Use the appropriate matrix expressions in the chapter to show, after some additional algebra, that the application of Eq. (3.5) produces the same expressions for $\hat{\beta}_0$ and $\hat{\beta}_1$ that were given in Chapter 1.

3.11. Prove that $\mathbf{X}'\mathbf{e} = \mathbf{0}$, where $\mathbf{e}$ denotes the vector of residuals and the $\mathbf{X}$ matrix is for an intercept model.

CHAPTER 4

Introduction to Multiple Linear Regression

Most applications of regression analysis involve the use of more than one regressor. The model for *multiple linear regression* is

$$Y = \beta_0 + \beta_1 X_1 + \beta_2 X_2 + \cdots + \beta_m X_m + \epsilon \tag{4.1}$$

with the corresponding prediction equation

$$\hat{Y} = \hat{\beta}_0 + \hat{\beta}_1 X_1 + \hat{\beta}_2 X_2 + \cdots + \hat{\beta}_m X_m$$

As in Eq. (4.1), m will hereinafter denote the number of regressors.

Multiple regression is, unfortunately, sometimes referred to as *multivariate regression* to indicate that there is more than one regressor. The two terms are not synonymous, however. The word *multivariate* must be reserved for the case where there is more than one *dependent* variable. Although multivariate regression is indeed a field of study, a discussion of the topic is more appropriately given in a book on multivariate methods (e.g., Gnanadesikan, 1977), and the reader is referred to such books for additional information. (We will similarly eschew the potentially misleading term *multivariable regression*.)

Even though we may appropriately regard multiple regression as an extension of simple regression, there are some questions that the user of multiple regression must address that are not encountered in simple regression.

In particular, if data are available on, say, k variables that might seem to be related to the dependent variable, should all k variables be used? If not, which ones should be used? What is gained, if anything, by having $m < k$? Can we use scatter plots to determine which independent variables to include in the model? Can possible transformations of the regressors be determined simply by examining such scatter plots? Should alternatives to least squares be used under certain conditions? If so, under what conditions should they be used, and which ones should be considered? Specifically, should least squares still be used when there are high correlations among the regressors?

118

In addition to these questions, there are other issues that are common to both simple regression and multiple regression. For example, if the regressors are to have fixed values, what configuration of values should be used?

These and many other questions will be addressed in this chapter and in subsequent chapters. Snedecor and Cochran (1980, p. 334) stated that "multiple linear regression is a complex subject," and approaches to multiple regression have become increasingly more sophisticated since that statement was made.

Although increased sophistication usually entails increased complexity, there is also the potential of doing a better job of building regression models once these new methods are mastered.

4.1 AN EXAMPLE OF MULTIPLE LINEAR REGRESSION

In Chapter 1 the issue of fixed X versus random X was discussed, and the conclusions that were drawn there also apply to multiple regression. There are consequences that occur in multiple regression from having random variables as regressors that do not occur in simple regression, however. For example, in simple regression we can make $\text{Var}(\hat{\beta}_1)$ small by selecting the X values in such a way that there is a considerable spread, as will be seen in Chapter 14. In multiple regression, however, there is a distinct possibility that one or more $\text{Var}(\hat{\beta}_i)$ will be large when the regressors are random, even if the spread of values of X_i is large.

Another problem that results from having random regressors can be explained as follows. The definition of $\beta_i(i \neq 0)$ is that it is the change in Y that would be expected to occur per unit change in X_i when the values of the other regressors are held constant. But when the regressors are random, there is no "holding constant" the values of the other regressors, so the $\hat{\beta}_i$ are then not easily interpretable. As discussed in Section 4.3, the sign of some $\hat{\beta}_i$ may even appear to be wrong. (Technically, the definition of β_i also holds when the regressors are random and the values of the other regressors repeat by chance, but this is almost meaningless since the values are not directly fixed.)

Thus, it is highly desirable to obtain regression data from a designed experiment. Designs with desirable properties and methods for constructing such designs are discussed in Chapter 14.

In this section we will start with a two-regressor example in which the regressors will be assumed to be fixed. We will then rearrange the values of those regressors so as to create a configuration that is typical of what can occur when the regressors are highly correlated random variables. (We will not be concerned with checking assumptions for this example, as assumptions will be discussed in Chapter 5, and they were discussed extensively for simple regression in Chapter 2.)

The data for the fixed-regressor example are given in Table 4.1. As was shown in Chapter 3, the least squares estimators can be obtained as $\hat{\boldsymbol{\beta}} = (\mathbf{X}'\mathbf{X})^{-1}\mathbf{X}'\mathbf{Y}$. In simple regression $\mathbf{X}'\mathbf{X}$ is a 2×2 matrix, and the inversion

Table 4.1 Orthogonal Regressors

Y	X_1	X_2
23.3	5	17
24.5	6	14
27.2	8	14
27.1	9	17
24.1	7	13
23.4	5	17
24.3	6	14
24.1	7	13
27.2	9	17
27.3	8	14
27.4	8	14
27.3	9	17
24.3	6	14
23.4	5	17
24.1	7	13
27.0	9	17
23.5	5	17
24.3	6	14
27.3	8	14
23.7	7	13

of a 2×2 matrix was illustrated in Chapter 3. With two regressors a 3×3 matrix would have to be inverted if the calculations are performed in the usual manner.

Although the inversion of such a matrix by hand is not difficult, we can invert a 2×2 matrix by using a modified approach. There are two reasons for using such an approach at this point in the chapter. First, it will allow us to use an approach that was previously illustrated in Chapter 3 (and will be discussed in more detail in Section 4.2), and we also wish to avoid giving undue emphasis to hand computations since regression analyses are generally performed with a computer.

The *centering* of regression data was discussed in Section 3.2.4, with the term referring to the construction of a new random variable by subtracting the mean of a random variable from each of the values in the sample. For two regressors the prediction equation with centered data is

$$\hat{Y} - \overline{Y} = \hat{\beta}_1(X_1 - \overline{X}_1) + \hat{\beta}_2(X_2 - \overline{X}_2) \tag{4.2}$$

which is equivalent to $\hat{Y} = \hat{\beta}_0 + \hat{\beta}_1 X_1 + \hat{\beta}_2 X_2$ since $\hat{\beta}_0 = \overline{Y} - \hat{\beta}_1 \overline{X}_1 - \hat{\beta}_2 \overline{X}_2$. Thus, $\hat{\beta}_0$ is embedded in Eq. (4.2) rather than being written explicitly. Accordingly, only $\hat{\beta}_1$ and $\hat{\beta}_2$ are obtained by inverting $(\mathbf{X}^+)'\mathbf{X}^+$, with $\mathbf{X}^+$ denoting the matrix of centered regressors.

With $\overline{X}_1 = 7$, $\overline{X}_2 = 15$, and $\overline{Y} = 25.24$, we obtain

$$(\mathbf{X}^+)'\mathbf{X}^+ = \begin{bmatrix} 40 & 0 \\ 0 & 56 \end{bmatrix}$$

which is a diagonal matrix. The general form for centered regressors is

$$(\mathbf{X}^+)'\mathbf{X}^+ = \begin{bmatrix} \sum(X_1 - \overline{X}_1)^2 & \sum(X_1 - \overline{X}_1)(X_2 - \overline{X}_2) \\ \sum(X_1 - \overline{X}_1)(X_2 - \overline{X}_2) & \sum(X_2 - \overline{X}_2)^2 \end{bmatrix}$$

and in this example the off-diagonal elements are zero. The inverse of a diagonal matrix is obtained by taking the reciprocal of the diagonal elements, so

$$[(\mathbf{X}^+)'\mathbf{X}^+]^{-1} = \begin{bmatrix} \frac{1}{40} & 0 \\ 0 & \frac{1}{56} \end{bmatrix}$$

We then obtain

$$(\mathbf{X}^+)'\mathbf{Y}^+ = \begin{bmatrix} 41.80 \\ 5.80 \end{bmatrix}$$

with $\mathbf{Y}^+$ denoting the vector of centered values of $\mathbf{Y}$. In general,

$$(\mathbf{X}^+)'\mathbf{Y}^+ = \begin{bmatrix} \sum(X_1 - \overline{X}_1)(Y - \overline{Y}) \\ \sum(X_2 - \overline{X}_2)(Y - \overline{Y}) \end{bmatrix}$$

so that

$$\hat{\boldsymbol{\beta}} = [(\mathbf{X}^+)'\mathbf{X}^+]^{-1}(\mathbf{X}^+)'\mathbf{Y}^+$$

$$= \begin{bmatrix} \frac{1}{40} & 0 \\ 0 & \frac{1}{56} \end{bmatrix}\begin{bmatrix} 41.80 \\ 5.80 \end{bmatrix}$$

$$= \begin{bmatrix} 1.045 \\ 0.104 \end{bmatrix} = \begin{bmatrix} \hat{\beta}_1 \\ \hat{\beta}_2 \end{bmatrix}$$

Solving for $\hat{\beta}_0$ produces $\hat{\beta}_0 = 16.4$, so the prediction equation is

$$\hat{Y} = 16.4 + 1.045X_1 + 0.104X_2$$

4.1.1 Orthogonal Regressors

The fact that the off-diagonal elements of $(\mathbf{X}^+)'\mathbf{X}^+$ are zero means that the regressors are *orthogonal*; that is, $\sum (X_1 - \overline{X}_1)(X_2 - \overline{X}_2) = 0$. This is desirable because the results of a regression analysis can then be interpreted unambiguously.

In particular, the signs of the $\hat{\beta}_i$ will be the same as the signs of the $r_{X_i Y}$, the correlation between X_i and Y, and the $\hat{\beta}_i$ will have the same values regardless of whether X_i is used alone in the model or in tandem with the other regressors. (The t-statistics should differ, however, as the estimate of σ will almost certainly change as regressors are added to the model.) Furthermore, the contribution of a particular regressor toward R^2 is independent of the contributions of the other regressors in the prediction equation. Specifically, if Y is regressed against X_i, and $R^2 = 0.45$, the contribution that X_i makes toward the R^2 value that is obtained when other regressors are added to the prediction equation is 0.45. This would also be the contribution of X_i if it were the last regressor added to the model.

Because the regressor values are both fixed and orthogonal, we may interpret the $\hat{\beta}_i$ as "rates of change," as was discussed earlier in this section. For example, the average change in Y per unit change in X_1 when X_2 is held constant (i.e., when X_1 changes from 5 to 9 when $X_2 = 17$ and from 6 to 8 when $X_2 = 14$) is 1.21, which is very close to $\hat{\beta}_1 = 1.045$.

The summary statistics for the Table 4.1 data are given in Table 4.2. We can see that we have a large F-value, and hence a small *p-value*. Furthermore, since the F-value is more than four times both $F_{2, 17, .05} = 3.59$ and $F_{2, 17, .01} = 6.11$,

Table 4.2   Regression Results for Table 4.1 Data

Regressor	Coefficient	Standard Deviation	t-statistic	p-value
Constant	16.3710	1.798	9.10	0.000
X_1	1.0450	0.1236	8.46	0.000
X_2	0.1036	0.1045	0.99	0.335

$$s = 0.7816 \qquad R^2 = 0.81$$

ANOVA Table				
SOURCE	DF	SS	MS	F	p-value
Regression	2	44.282	22.141	36.24	0.000
Due to X_1	1	43.681			
Due to X_2	1	0.601			
Residual	17	10.386	0.611		
Total	19	54.668			

the 4-to-1 rule of Wetz (1964) is satisfied. [This rule, which is described in Draper and Smith (1981, p. 8), states that for the regression model to be useful the calculated F-value should be more than four times the tabular F-value for the chosen significance level. The t-statistic version of this rule was discussed in Section 1.5.4.]

The fact that the F-statistic is significant does not necessarily mean that both regressors should be used, however. That statistic is used for testing $H_0: \beta_1 = \beta_2 = 0$ against the alternative hypothesis H_a: not both β_1 and β_2 equal zero. Rejecting H_0 means simply that at least one regressor should be used.

We can see from the t-statistics that only X_1 should be used in the prediction equation. (Note: The 2-to-1 rule given in Section 1.5.4 applies only to simple regression; in multiple regression we use the tabular values as critical values.)

The ANOVA table shows that with just X_1 in the model we would have $R^2 = 43.681/54.668 = 0.799$, which differs only slightly from the R^2 value with both regressors. Since $\hat{\beta}_0 = \overline{Y} - \hat{\beta}_1\overline{X}_1$ when only X_1 is used and $\hat{\beta}_1$ remains unchanged it follows that the prediction equation is $\hat{Y} = 17.89 + 1.045X_1$.

One last point will be made concerning this example. A lack-of-fit test for simple regression was given in Section 1.5.8. In multiple regression we need to have at least one *set* of regressor values repeated in order to perform that test. In Table 4.1 there are five different (X_1, X_2) pairs, and each of these occurs four times. Therefore, the test can be performed, and the computations would be essentially the same as the computations illustrated in Section 1.5.8 for simple regression.

4.1.2 Correlated Regressors

When the regressors are random variables, some of them may be highly correlated. This makes the analysis of regression data difficult, as will be seen in this section.

The data in Table 4.3 will be used for illustration. These are the same data that were given in Table 4.1, except that the values of X_2 have been rearranged, and we will now assume that the regressors are random. With this rearrangement X_1 and X_2 are now highly correlated, as $r_{X_1X_2} = .93$.

The regression equation is $\hat{Y} = 9.26 - 0.261X_1 + 1.19X_2$, and we should note that $\hat{\beta}_1$ is now negative, even though neither Y nor X_1 has been changed. Specifically, for both Table 4.1 and Table 4.3 the correlation between Y and X_1 is .894—a high positive correlation. (Here we use the word *correlation* very loosely for the Table 4.1 data since we assumed the regressors to be fixed.)

This illustrates one of the nuances of correlated data: the signs of the $\hat{\beta}_i$ may not be the same as the r_{X_iY}. This problem is discussed in detail in Section 4.3. At this point we will simply note that $\hat{\beta}_1$ is negative because $r_{YX_1} - r_{X_1X_2}r_{YX_2}$ is negative. That is, $r_{YX_1} - r_{X_1X_2}r_{YX_2} = .894 - (.93)(.994) = -.03$.

The negative value for $\hat{\beta}_1$ is thus caused by two factors: the high correlation between X_1 and X_2 and the fact that X_2 is more highly correlated with Y than

Table 4.3 Correlated Regressors

Y	X_1	X_2
23.3	5	13
24.5	6	14
27.2	8	17
27.1	9	17
24.1	7	14
23.4	5	13
24.3	6	14
24.1	7	14
27.2	9	17
27.3	8	17
27.4	8	17
27.3	9	17
24.3	6	14
23.4	5	13
24.1	7	14
27.0	9	17
23.5	5	13
24.3	6	14
27.3	8	17
23.7	7	14

is X_1. The consequences of a regression coefficient having the "wrong" sign depends on how the regression equation is used, as is discussed in Section 4.3.

4.1.2.1 Partial-F Tests and t-Tests

The other summary statistics for Table 4.3 data are given in Table 4.4. The F-statistic is (highly) significant, and both t-statistics are significant. The latter suggests that both X_1 and X_2 should be used in the regression equation, but we might want to examine this more closely.

(As in Table 4.4, s will be used to denote $\hat{\sigma}$, the estimate of the standard deviation of the error term, and s_e will also be used to represent the standard deviation of the error term when the standard error of some other estimator is being simultaneously discussed, so that the two standard errors can be clearly distinguished.)

Those t-statistics give the contribution of X_i, given that the other regressor is in the equation. Since $s_{\hat{\beta}_i}$ will be small when s_e is small, the contribution of X_i need not be very large in order to produce a moderately large t-statistic. And that is just what happens here. Whereas $R^2 = .995$ for the full value, the value decreases only very slightly to .988 when just X_2 is in the model. Thus, the very slight improvement is magnified by a small residual standard deviation.

This raises the question as to whether t-statistics should determine the regressors that are to be used in the equation, and if not, how should the regressors

Table 4.4   Regression Results for Table 4.3 Data

Regressor	Coefficient	Standard Deviation	t-statistic	p-value
Constant	9.2611	0.3706	24.99	0.000
X_1	−0.26053	0.05525	−4.72	0.000
X_2	1.18684	0.04669	25.42	0.000

$$s = 0.1287 \qquad R^2 = 0.995$$

ANOVA Table

SOURCE	DF	SS	MS	F	p-value
Regression	2	54.386	27.193	1641.14	0.000
Due to X_1	1	43.681			
Due to $X_2\|X_1$	1	10.705			
Due to X_2	1	54.018			
Due to $X_1\|X_2$	1	0.368			
Residual	17	0.282	0.017		
Total	19	54.668			

be selected. The latter is considered in detail in Chapter 7, and we will use this example in illustrating the methods that are presented therein.

There are certain relationships in Table 4.4 that should be noted. In particular, since the regressors are correlated, the order in which they are entered determines their contribution to the regression sum of squares. For example, the sum of squares for X_2 given that X_1 is in the equation is 10.705. If this number is divided by the residual mean square, the result is a *partial-F test* with the value of F equal, within rounding, to the square of the t-statistic for X_2. Thus, the hypothesis $H_0: \beta_2 = 0$ can be tested using either a t-test or a partial-F test.

The computational formulas for partial-F tests are generally given in some detail in regression books. Formally, a partial-F test for determining if X_2 should be added to a model that already contains X_1 is performed by computing

$$F = \frac{\text{SS}_{\text{reg}}(X_1, X_2) - \text{SS}_{\text{reg}}(X_1)}{\text{MS}_{\text{res}}} \tag{4.3}$$

where $\text{SS}_{\text{reg}}(\cdot)$ denotes the regression sum of squares for the model with the indicated variable(s), and MS_{res} denotes the mean square residual ($= \hat{\sigma}^2$). Substituting the numbers from Table 4.4 we obtain $F = (54.386 - 43.681)/0.017 = 646.28$, which is the square of the t-statistic for X_2.

If we were to compute the components of Eq. (4.3) by hand, we could obtain the numerator using $\text{SS}(X_1, X_2) = \hat{\boldsymbol{\beta}}' \mathbf{X}' \mathbf{Y} - (\sum Y)^2/n$, and $\text{SS}(X_1) = \hat{\boldsymbol{\beta}}_1 \mathbf{X}_1' \mathbf{Y} - $

$(\sum Y)^2/n$, where $\mathbf{X}_1$ is just the column of values for X_2, and $\hat{\boldsymbol{\beta}}_1$ contains the two regression coefficients when only X_1 is used in the model.

Equation (4.3) could be generalized so that more than one additional variable is being tested for inclusion, as could be the case if we were testing a nested model against a larger model. For such a scenario the partial-F test would have to be used rather than a t-test, since the latter is applicable only in testing for inclusion of a single variable.

Either test should be used with caution, however, as the results could be misleading if not viewed in the proper perspective. This is especially true of t-tests with two very highly correlated regressors, as both t-statistics could be small even when each regressor is highly correlated with Y. The t-statistics would then be telling us that both regressors are not needed, but we should not interpret such statistics to mean that the regressor that corresponds to a small t-statistic is of no value. The regressor might indeed have a low correlation with Y, or the correlation could be quite high.

Although the ANOVA table in Table 4.4 provides complete information on the total and marginal contributions of the regressors for each of the two possible sequences, with r regressors in the equation there would be $r(r-1)(r-2)$ $\cdots 2 \cdot 1$ such sequences. Thus, it would be impractical to view such contributions if r were greater than 2 or 3.

Consequently, there is a need to determine a good subset of available regressors without having to look at every combination and every sequence of possible regressors, and such methods are discussed in Chapter 7.

4.1.3 Confidence Intervals and Prediction Intervals

Assume that we wish to construct a confidence interval for β_1 for Table 4.1 data. In general, a $(1 - \alpha)\%$ confidence interval for β_i is of the form

$$\hat{\beta}_i \pm t_{\alpha/2, \nu} s \sqrt{c_{ii}}$$

where ν is the degrees of freedom for estimating σ and c_{ii} is the ith diagonal element of $[(\mathbf{X}^+)'\mathbf{X}^+]^{-1}$, again assuming centered form, with $s \sqrt{c_{ii}} = s_{\hat{\beta}_i}$.

With $t_{.025, 17} = 2.11$, $s = 0.7816$, $c_{11} = \frac{1}{40}$, and $\hat{\beta}_1 = 1.045$, a 95% confidence interval for β_1 is obtained as $1.045 \pm 2.11(0.7816)(\frac{1}{40})^{1/2} = 1.045 \pm 0.26$, so that the lower limit is 0.78, and the upper limit is 1.30. Here we assume for purposes of illustration that both regressors are to be used in the model. As indicated in Section 4.1.1, an argument could be made for using only X_1 in the equation. If such a decision were subsequently made, the confidence interval would have to be recomputed, since s_e would almost certainly change. Since σ_ϵ is the standard deviation of the error term for the true model, we want our estimate, s_e, to be obtained using a model that is our best guess as to the true model. For this example there is hardly any change, however, as $s = 0.7813$

if X_2 is not used in the model, and with $t_{.025, 18} = 2.101$, the confidence limits would be the same to two decimal places.

It was stated in Section 1.5.6.2 that a confidence interval for $\mu_{Y|X}$ is less useful than a prediction interval. (We could go further and state that it has limited usefulness.) Therefore, it will not be illustrated for multiple regression. Instead, we will illustrate a prediction interval for Y given X_1 and X_2. Recall that the general form of the prediction interval in simple regression was given by Eq. (1.13). With an appropriate modification to reflect the difference in the degrees of freedom for multiple regression, we obtain

$$\hat{Y} \pm t_{\alpha/2, \nu}(\hat{\sigma}_\epsilon^2 + \hat{\sigma}_{\hat{Y}}^2)^{1/2} \tag{4.4}$$

Using the expression for $\text{Var}(\hat{Y})$ given in Eq. (3.13), and using $\hat{\sigma}_\epsilon^2 = s^2$, Eq. (4.4) can be written equivalently as

$$\hat{Y} \pm t_{\alpha/2, \nu} s \left\{ 1 + \frac{1}{n} + (\mathbf{x}^+)'[(\mathbf{X}^+)'\mathbf{X}^+]^{-1}(\mathbf{x}^+) \right\}^{1/2} \tag{4.5}$$

where $(\mathbf{x}^+)'$ may be (but is not necessarily) one of the rows of $\mathbf{X}^+$.

Using $(\mathbf{x}^+)' = [-2 \quad 2]$ (which is $x_1 = 5$ and $x_2 = 13$ in raw form from Table 4.3), a 95% prediction interval for Y is obtained as

$$23.382 \pm 2.11(0.1287)(1 + \tfrac{1}{20} + 0.105)^{1/2}$$
$$= 23.382 \pm 0.292$$

so the lower limit is 23.090 and the upper limit is 23.674.

It should be noted that a prediction interval is, strictly speaking, only valid for (X_1, X_2) within the region covered by X_1 and X_2 in the data set, and this is generally *not* the rectangular region formed by considering the range of X_1 and X_2 separately. In particular, for correlated data the (X_1, X_2) region may be small relative to the rectangular region. For two regressors it is easy to determine whether $\mathbf{x}$ lies within the experimental region, since the region can be graphed. There are various methods that can be used to make this determination when there are more than two regressors, two of which are given by Weisberg (1985).

We should also note that the *width* of the prediction interval depends on the distance that $\mathbf{x}^+$ is from the center of the data. In particular, since $(\mathbf{x}^+)' = [x_1 - \bar{x}_1 \quad x_2 - \bar{x}_2]$, $(\mathbf{x}^+)'[(\mathbf{X}^+)'\mathbf{X}^+]^{-1}\mathbf{x}^+$ is the *Mahalanobis distance* (Mahalanobis, 1936) that $\mathbf{x}$ is from $\bar{\mathbf{X}}$, where $\bar{\mathbf{X}}' = [\bar{X}_1 \quad \bar{X}_2]$. This also means, of course, that $\text{Var}(\hat{Y})$ is the greatest when $\mathbf{x}$ denotes a point on the boundary of the experimental region.

4.2 CENTERING AND SCALING

The centering of regression data was illustrated in Section 3.2.4 and was also used in Section 4.1. In this section centering is considered in greater detail, and the use of both centering and scaling is illustrated.

4.2.1 Centering

There are two reasons why centering could be desirable. If a regression model contains polynomial terms, the correlation between the linear and quadratic terms can be reduced (and perhaps reduced to zero) if X^+ and $(X^+)^2$ are used instead of X and X^2. [See, e.g., the discussions in Bradley and Srivastava (1979) and Snee and Marquardt (1984).]

The other reason involves the column of 1's that is the first column of the **X** matrix when the data are not centered. If the values for one (or more) of the regressors differ very little, then the values of that regressor will almost be equal to a multiple of the column of 1's. These two columns of the **X** matrix will then be highly "correlated," with the word being used loosely since the column of 1's does not represent a random variable.

An **X** matrix can be said to be *ill-conditioned* when there are near linear dependencies involving two or more columns, and as first emphasized by Longley (1967), ill-conditioned data can cause computational problems, even when a computer is used. This problem has been addressed during the past two decades for software used on mainframes, but the emergence of the microcomputer during the 1980s and statistical software written for it has led to new problems, some of which are discussed by Lesage and Simon (1985).

The centering of regression data was quite controversial during the 1980s. Papers in support of centering included Marquardt (1980) and Snee and Marquardt (1984), whereas opposition to centering was stated in papers such as Smith and Campbell (1980) and Belsley (1984). See also Belsley (1991, p. 195), who maintains that while centering can help foster statistical accuracy, the extreme conditions for which computational accuracy might not be possible without centering are those where the statistical results will not be reliable anyway. The discussion in Belsley (1984) and in the comments of the discussants of that paper was primarily in regard to how centering affects regression diagnostics for assessing possible near dependencies involving two or more regressors (such as the diagnostics that are given in Section 4.3.2). The main objections to centering raised by Belsley are (1) that the intercept is lost and (2) that measures of ill-conditioning are disturbed by the centering.

When the regressors (only) are centered, the intercept becomes $\overline{Y}$, but $\hat{\beta}_0$ can be easily restored, as was indicated in Section 4.1. Which is more meaningful, $\overline{Y}$ or $\hat{\beta}_0$? The latter gives us the predicted value of Y when the value of each regressor is equal to zero, whereas $\overline{Y} = \hat{Y}$ when the value of each regressor is set equal to its mean. Obviously, the latter is more meaningful, since we are generally not interested in predicting Y when each regressor is set equal to zero. This would also generally involve extrapolating beyond the range of the data. Thus, as Snee and Marquardt (1984) point out, β_0 is essentially a nuisance parameter.

The other point raised by Belsley involves the *condition number* of a matrix, which was discussed in Section 3.1.1. (Recall that the condition number is defined as the square root of the largest eigenvalue divided by the square root of the smallest eigenvalue.) The use of the condition number of $X'X$ in assessing multicollinearity (near dependencies involving two or more regressors; see Section 4.3) seems questionable, however, since this is essentially a numerical diagnostic rather than a statistical diagnostic. As Cook (1984) states in discussing Belsley (1984): "A meaningful interpretation of statistical stability—which seems to be the main issue in this paper—is lacking. Part of the problem may stem from trying to attach statistical interpretations to numerical diagnostics." Even if we were to accept the use of the condition number, it seems as though it would have to be applied to the centered-data form of $X'X$ since a column of 1's does not represent data.

The issue of centering is also discussed extensively by Myers (1990), in which the discussion is similar to that given here.

4.2.2 Scaling

In this section we discuss centering and scaling of Y and the regressors. Centering and scaling are accomplished simultaneously as follows. If we define Y_i^* as

$$Y_i^* = \frac{Y_i - \bar{Y}}{(S_{yy})^{1/2}} \tag{4.6}$$

with $S_{yy} = \sum (Y_i - \bar{Y})^2$, it should be apparent that $\bar{Y}^* = 0$ and $\sum Y_i^{*2} = 1$. We would transform each regressor in the same manner; that is, X_i is transformed as

$$X_{ij}^* = \frac{X_{ij} - \bar{X}_i}{(S_{x_i x_i})^{1/2}} \tag{4.7}$$

with $S_{x_i x_i} = \sum_{j=1}^{n} (X_{ij} - \bar{X}_i)^2$.

Transforming to Y^* and X_i^* causes $(X^*)'X^*$ and $(X^*)'Y^*$ to be in "correlation form." Specifically, $(X^*)'X^*$ is a correlation matrix whose elements are the correlations between the regressors, and $(X^*)'Y^*$ is a vector that contains the correlation between Y and each regressor. Accordingly, for two regressors

$$(X^*)'X^* = \begin{bmatrix} 1 & r_{12} \\ r_{12} & 1 \end{bmatrix}$$

and

$$(\mathbf{X}^*)'\mathbf{Y}^* = \begin{bmatrix} r_{1Y} \\ r_{2Y} \end{bmatrix}$$

With

$$[(\mathbf{X}^*)'\mathbf{X}^*]^{-1} = \frac{1}{1 - r_{12}^2} \begin{bmatrix} 1 & -r_{12} \\ -r_{12} & 1 \end{bmatrix} \tag{4.8}$$

it follows that

$$\hat{\beta}^* = \frac{1}{1 - r_{12}^2} \begin{bmatrix} 1 & -r_{12} \\ -r_{12} & 1 \end{bmatrix} \begin{bmatrix} r_{1Y} \\ r_{2Y} \end{bmatrix}$$

$$= \begin{bmatrix} \dfrac{r_{1Y} - r_{12}r_{2Y}}{1 - r_{12}^2} \\[2mm] \dfrac{r_{2Y} - r_{12}r_{1Y}}{1 - r_{12}^2} \end{bmatrix} \tag{4.9}$$

We can now see clearly what was mentioned in Section 4.1.2; namely, that $\hat{\beta}_1$ will be negative when $r_{1Y} - r_{12}r_{2Y}$ is negative.

We may also note that each component of $\hat{\beta}^*$ is the same as the numerator of the corresponding partial correlation coefficient, and will have the same sign as the latter. Specifically, $r_{2Y\cdot 1}$ denotes the correlation between X_2 and Y after each has been adjusted for its linear relationship (if such exists) with X_1, and is termed a *partial correlation coefficient*. If X_1 were orthogonal to both X_2 and Y, it could be easily seen that $r_{2Y\cdot 1} = r_{2Y}$, since $r_{2Y\cdot 1} = \hat{\beta}_2^*[(1 - r_{12}^2)/(1 - r_{1y}^2)]^{1/2}$. This follows from Eq. (4.9), since $\hat{\beta}_2^*$ would then equal r_{2Y}, and of course r_{12} and r_{1Y} would be zero.

Thus, from Eq. (4.9) we can see the desirability of having uncorrelated regressors (i.e., $r_{12} = 0$), as then the $\hat{\beta}_i^*$ will not only have the same sign as the r_{iY}, but will be equal to the correlation coefficients.

We can obtain the $\hat{\beta}_i$ as

$$\hat{\beta}_i = \left(\frac{S_{yy}}{S_{ii}} \right)^{1/2} \hat{\beta}_i^* \qquad i = 1, 2, \ldots, r \tag{4.10}$$

where S_{ii} is used to represent $S_{X_iX_i}$. The relationship in Eq. (4.10) should be intuitively apparent since the $(S_{ii})^{1/2}$ are used as divisors in producing the

$(\mathbf{X}^*)'\mathbf{X}^*$ matrix, which is then inverted, and $(S_{yy})^{1/2}$ is the divisor for Y. Thus, to return to the $\hat{\beta}_i$ we would have to "invert" the effect of these two components, as was indicated in Section 3.2.5.

4.3 MULTICOLLINEARITY AND THE "WRONG SIGNS" PROBLEM

The word *multicollinearity* has been used to represent a near exact relationship between two or more variables. If $a_1 X_1 + a_2 X_2 + a_3 X_3 + \cdots + a_u X_u \doteq c$, where c is some constant and $a_1, a_2, \ldots, a_u$ are also constants, some of which may be zero, then the regressors $X_1, X_2, \ldots, X_u$ with nonzero constants are multicollinear. [Note: Some writers have used the word *collinearity* to describe the condition that we have described here as multicollinearity. See, e.g., Belsley et al. (1980) and Belsley (1984). Technically, the two terms are misnomers since the literal definition of *collinear* is that there is an *exact* relationship between two or more variables. Nevertheless, we shall use the "statistical definition" of the terms.]

It is often stated that multicollinearity can cause the signs of the coefficients to be wrong (that is, the sign of $\hat{\beta}_i$ or $\hat{\beta}_i^*$ is different from the sign of r_{iY}). But the extent to which the r_{iY} differ will also be an important factor. For example, assume that $r_{1Y} = r_{2Y} = r_{12} = .99$. Thus, there is a very high degree of correlation between X_1 and X_2, yet we can see from Eq. (4.9) that the signs of $\hat{\beta}_1^*$ and $\hat{\beta}_2^*$ will be "right." Conversely, if $r_{1Y} = .3$, $r_{2Y} = .8$, and $r_{12} = .4$, the sign of $\hat{\beta}_1^*$ will be "wrong" even though there is only a moderate correlation between X_1 and X_2. Thus, the signs can be "right" even when there is a high degree of multicollinearity and "wrong" when there is essentially no multicollinearity.

The point is that the three correlations determine the signs of the regression coefficients in the two-regressor case, and in general there is really no such thing as a right or wrong sign for a coefficient.

It was indicated in Section 1.6.1 that the use of regression for control implies the existence of a cause-and-effect relationship. If the data were not collected from a controlled experiment, or if there are restrictions on the regressor values that cause the values of certain regressors to be highly correlated, then the manipulation of certain regressor values in an effort to control Y within specified bounds may not be successful.

Similarly, regression might appear to the uninitiated to be of dubious value in a scenario such as the following true story. Many years ago a prominent scholar mentioned a regression equation that someone had constructed for measuring overall intelligence of children. (Recall the discussion in Chapter 1 concerning the use of regression to predict GPA scores.) One of the regressors had a negative coefficient, which prompted this person to ask the regression user if his daughter's overall intelligence would decline if her score increased on the measure that was represented by the regressor with the negative coefficient.

Since measures of intelligence are apt to be highly correlated, negative coefficients can result simply because of certain combinations of values of correlation coefficients, as has been shown in this section.

The reader is referred to Mullet (1976) for additional reading on the wrong–signs problem.

4.3.1 Inflated Variances

A related problem is that the variances of the $\hat{\beta}_i$ are a function of the extent of the multicollinearity, provided that X_i is part of the multicollinearity, as is discussed in Section 4.3.2. Since $\text{Var}(\hat{\beta}_i) = \sigma^2 c_{ii}$ where c_{ii} is as previously defined, it follows that $\text{Var}(\hat{\beta}_i^*) = \sigma_*^2 c_{ii}^*$, where c_{ii}^* is the ith diagonal element of $[(\mathbf{X}^*)'\mathbf{X}^*]^{-1}$ and σ_*^2 is the error variance for the correlation-form model.

When correlation form is used, $\text{Var}(\hat{\beta}_i^*)$ is given by

$$\text{Var}(\hat{\beta}_i^*) = \frac{\sigma_*^2}{1 - r_{12}^2}$$

for the two-regressor case, as is apparent from Eq. (4.8). (Different expressions are used when there are more than two regressors, as shown in Section 4.3.2.)

Regardless of which form is used, $s^2(s_*^2)$ would be used in estimating $\sigma^2(\sigma_*^2)$ in obtaining estimates of the variances.

A consequence of having inflated variances is that the width of the confidence intervals for the β_i will also be inflated, perhaps even to the point of rendering one or more intervals useless.

4.3.2 Detecting Multicollinearity

Since multicollinearity inflates $\text{Var}(\hat{\beta}_i)$ and causes problems with the signs of the $\hat{\beta}_i$ and confidence intervals for the β_i, it is important to be able to detect it when it exists.

Multicollinearity is easy to detect in the two-regressor case; we need only look at the value of r_{12}. When there is more than two regressors, however, inspection of the r_{ij} is not sufficient, as the latter could only identify a two-regressor collinearity.

For example, assume that we have four regressors, and the population correlation coefficients, ρ_{ij}, (as defined in Section 1.8) are $\rho_{12} = \rho_{13} = \rho_{23} = 0$, with $\sigma_1^2 = \sigma_2^2 = \sigma_3^2$ and $X_4 = X_1 + X_2 + X_3$. It can be easily shown (as the reader is asked to do in Exercise 4.9) that $\rho_{14} = \rho_{24} = \rho_{34} = .577$. Thus, three of the pairwise correlations are zero and the other three are not especially large, yet we have the most extreme multicollinearity problem possible in that there is an exact linear relationship between the four regressors. Consequently, $\mathbf{X}'\mathbf{X}$ [or $(\mathbf{X}^*)'\mathbf{X}^*$] will not have a unique inverse, so the ordinary least squares estimators will not be obtainable.

Although we would not expect to encounter exact linear relationships involving three or more regressors with real data, near-exact relationships can occur in certain fields of application, such as with economic data, as is shown in Belsley et al. (1980).

Since multicollinearity can thus exist without large pairwise correlations, we must address the question "How can multicollinearity be detected."

Marquardt (1970) proposed looking at *variance inflation factors (VIF)*, where $VIF(i) = c_{ii}^*$. When the regressors are orthogonal, $(\mathbf{X}^*)'\mathbf{X}^*$ is an identity matrix (and thus so is $[(\mathbf{X}^*)'\mathbf{X}^*]^{-1}$), and since $Var(\hat{\beta}_i^*) = \sigma_*^2 c_{ii}^*$, the value of $VIF(i)$ is thus the multiple by which $Var(\hat{\beta}_i^*)$ has been inflated relative to what it would have been with orthogonal data. Multicollinearity is declared to exist whenever any VIF value is at least equal to 10.

For the two-regressor case, $VIF(1) = VIF(2)$, and the common value will be at least equal to 10 when $|r_{12}| \geq \sqrt{0.9} = .949$, a rather stringent requirement.

For the r-regressor case with equal pairwise correlations, a, the threshold value will be met when $a \geq .9325$ or $a \leq .1825$ for $r = 3$, and, for example, $a \geq .9164$ or $a \leq -.1964$ for $r = 6$. [Note: These requirements may seem to be much more stringent for a positive than for a negative, but this is not true since a must exceed $-1/(r-1)$ for the set of equal pairwise correlations to be feasible. Thus, for $r = 6$ the bounds for negative a are $-.2 < a \leq -.1964$ and $-.5 < a \leq -.4825$ for $r = 3$.]

Thus, using the equicorrelation case as a benchmark, we can see that the correlations, if positive, must be quite large in order to have VIF values at least equal to 10.

We have seen that VIF values can be inflated by large pairwise correlations, but that multicollinearity (and hence large VIF values) can exist without such correlations. A more intuitive view of $VIF(i)$ can be realized by considering the relationship

$$VIF(i) = \frac{1}{1 - R^2(i)} \tag{4.11}$$

where $R^2(i)$ is the R^2 value that results from regressing X_i on the other regressors in the prediction equation.

Notice that Eq. (4.11) is in the same general form as $VIF(i)$ for the two-regressor case, where the latter is

$$VIF(i) = \frac{1}{1 - r_{12}^2} \tag{4.12}$$

as can be seen from Eq. (4.8). Since $r_{12}^2 = R^2(1) = R^2(2)$, Eqs. (4.11) and (4.12) have the same form. If we define multicollinearity in terms of VIF values, we can thus see why r_{12} is both necessary and sufficient for detecting multicollinearity in the two-regressor case.

When there is more than two regressors, there are two immediate questions that should be addressed: (1) how can we determine which regressors are causing the variances to be inflated and (2) will the existence of multicollinearity cause all of the variances to be inflated, or only those involved in the multicollinearity.

We will address the second question first. Assume that two highly correlated regressors are combined with $r - 2$ regressors, with the latter being orthogonal to the former. The $r - 2 \operatorname{Var}(\hat{\beta}_i)$ will be the same with or without the other two highly correlated regressors [Belsley et al. 1980, p. 108).] This should be apparent from using Eq. (4.11) and recognizing that $R^2(i)$ will be unchanged if X_i is regressed upon two additional regressors that are orthogonal to both X_i and to the other $r - 3$ regressors that are already being used. Thus, variance inflation due to multicollinearity will occur only for estimators that are coefficients of regressors involved in the multicollinearity.

The first question may be answered by using, for example, a set of eigenvectors corresponding to the eigenvalues of $(\mathbf{X}^*)'\mathbf{X}^*$, or equivalently using $\mathbf{X}'\mathbf{X}$. (Eigenvalues and eigenvectors were covered in Section 3.1.1.)

When the regressors are orthogonal, all of the eigenvalues of $(\mathbf{X}^*)'\mathbf{X}^*$ are equal to 1.0, but when multicollinearity exists at least one of the eigenvalues will be close to zero. The elements of an eigenvector corresponding to a small eigenvalue will signify the regressors that are involved in the multicollinearity in that large elements in an eigenvector indicate that the corresponding regressors are involved in the multicollinearity.

To illustrate, consider the example given earlier in this section in which three of the regressors are orthogonal, and the fourth, X_4, is defined as $X_4 = X_1 + X_2 + X_3$. So as to not have an exact linear relationship, we will define X_4 as $X_4 = X_1 + X_2 + X_3 + w$, where $w_i \sim \operatorname{NID}(0, \sigma_w^2 = 0.0001)$.

We will use X_1 and X_2 from Table 4.1, and then center and scale those values in addition to the values of X_3 and X_4, with X_3 selected so as to be orthogonal to X_1 and X_2. The matrix $\mathbf{X}^*$ is then as given in Table 4.5, and

$$(\mathbf{X}^*)'\mathbf{X}^* = \begin{bmatrix} 1 & 0 & 0 & 0.57957 \\ 0 & 1 & 0 & 0.57282 \\ 0 & 0 & 1 & 0.57917 \\ 0.57957 & 0.57282 & 0.57917 & 1 \end{bmatrix}$$

Since $(\mathbf{X}^*)'\mathbf{X}^*$ is a correlation matrix, as was stated in Section 4.2.2, the last row (or equivalently the last column) gives the correlations between each of the first three regressors and X_4, and the submatrix that is a 3×3 identity matrix represents the fact that X_1, X_2, and X_3 are pairwise uncorrelated. Notice that the nonzero sample correlations are very close to the corresponding population correlation values that were given earlier in this section. This results from the fact that σ^2 was chosen to be very small.

Table 4.5   Correlation-form Regressor Values

X_1	X_2	X_3	X_4
−0.316228	0.267261	−0.111803	−0.093689
−0.158114	−0.133631	0.167705	−0.071460
0.158114	−0.133631	−0.391312	−0.218343
0.316228	0.267261	0.167705	0.431520
0.000000	−0.267261	0.167705	−0.055130
−0.316228	0.267261	−0.111803	−0.088172
−0.158114	−0.133631	0.167705	−0.064976
0.000000	−0.267261	0.167705	−0.065264
0.316228	0.267261	0.167705	0.433429
0.158114	−0.133631	−0.391312	−0.207645
0.158114	−0.133631	−0.391312	−0.211656
0.316228	0.267261	0.167705	0.426754
−0.158114	−0.133631	0.167705	−0.063471
0.316228	0.267261	−0.111803	−0.101922
0.000000	−0.267261	0.167705	−0.054215
0.316228	0.267261	0.167705	0.442818
−0.316228	0.267261	−0.111803	−0.099139
−0.158114	−0.133631	0.167705	−0.075339
0.158114	−0.133631	−0.391312	−0.206952
0.000000	−0.267261	0.167705	−0.057150

The eigenvalues and a set of corresponding eigenvectors for this matrix are given in Table 4.6. As indicated earlier in this section, the elements of the eigenvector that corresponds to the smallest eigenvalue should give some clue as to the nature of the multicollinearity. Here the presence of multicollinearity is indicated by the fact that the smallest eigenvalue is so close to zero. Notice that the first three elements of the eigenvector are almost identical, suggesting that X_1, X_2, and X_3 are contributing equally to the multicollinearity. (And, of course, we know that to be true since $X_4 \doteq X_1 + X_2 + X_3$.)

More specifically, as was shown in Chapter 3, $V'AV = E$, where E is a diagonal matrix that contains the eigenvalues of an arbitrary matrix A, and V is the matrix of eigenvectors. Letting $A = X'X$ [or we could equivalently use

Table 4.6   Eigenvalues and Corresponding Eigenvectors for the Regressors in Table 4.5

Eigenvalue	Eigenvector			
1.99973	−0.409929	−0.405151	−0.409647	−0.707107
1.00000	0.814810	−0.407661	−0.412185	0.000000
1.00000	0.000000	0.710998	−0.703194	0.000000
0.00027	0.409929	0.405151	0.409647	−0.707107

$(\mathbf{X}^*)'\mathbf{X}^*]$, it follows that we may write $\mathbf{E} = (\mathbf{XV})'\mathbf{XV}$, and so a small eigenvalue will result when the rows of $\mathbf{X}$ are almost orthogonal to a column in $\mathbf{V}$ (and recall that each column in $\mathbf{V}$ is an eigenvector). Since the ith row of $\mathbf{X}$ gives the values of each regressor for the ith observation, it follows that with four regressors we should consider the expression $X_1 v_{1j} + X_2 v_{2j} + X_3 v_{3j} + X_4 v_{4j}$ when the jth eigenvalue is close to zero, where v_{ij} denotes the ith element of the jth eigenvector.

As can be seen from Table 4.6, the first three elements of the eigenvector that corresponds to the near-zero eigenvalue are almost identical, and this suggests that X_1, X_2, and X_3 contribute almost equally to the multicollinearity, as stated previously. If we let that eigenvalue be set to zero, we obtain $0.707X_4 = 0.4099X_1 + 0.405X_2 + 0.4096X_3$, so that $X_4 \doteq 0.58(X_1 + X_2 + X_3)$. Thus, the general form of the mutlicollinearity is identified. [The use of eigenvalues and eigenvectors in examining multicollinearities is discussed extensively by Gunst and Mason (1980), and they give an interesting application on page 298.]

The VIF values for this example are: VIF(1) = 626.32, VIF(2) = 611.83, VIF(3) = 625.46, and VIF(4) = 1861.62. Thus, the variances of the estimators are unequally affected. This can be explained by using Eq. (4.11). Since X_4 is an almost exact linear combination of X_1, X_2, and X_3, it follows that $R^2(4)$ must be very close to 1. Specifically, $R^2(4) = .99946$, and substituting this value in Eq. (4.11) produces, within rounding, the indicated value of VIF(4). The other VIF values are considerably less because $R^2(i) \doteq .998$ for $i = 1, 2, 3$.

Specifically, when all four regressors are used, the R^2 value is .932, yet each of the four t-statistics for the regressors is less than 1 in absolute value. This illustrates the statement made in Section 4.1.2.1 regarding how t-statistics should be interpreted and the care that must be exercised in interpreting t-statistics in the presence of multicollinearity. Much more startling is the fact that R^2 is close to 1 when one of the orthogonal regressors is regressed against the other two orthogonal regressors and X_4, as one might expect R^2 to be equal to the square of the correlation between X_4 and the regressor that is serving as the dependent variable. This and other seeming contradictions are discussed in detail in Section 5.7.

4.3.3   Variance Proportions

A useful diagnostic that is discussed in detail by Myers (1990) and Belsley et al. (1980) is the proportion of $\text{Var}(\hat{\beta}_i)$ that results from the multicollinearity. These are termed *variance proportions* by the first source, and *variance-decomposition proportions* by the latter.

A variance proportion is defined as follows. Since $\mathbf{V}'(\mathbf{X}'\mathbf{X})\mathbf{V} = \mathbf{E}$, we may write $\mathbf{VV}'(\mathbf{X}'\mathbf{X})\mathbf{VV}' = \mathbf{VEV}'$, so that $\mathbf{X}'\mathbf{X} = \mathbf{VEV}'$ since $\mathbf{V}'\mathbf{V} = \mathbf{I}$. Then, $(\mathbf{X}'\mathbf{X})^{-1} = (\mathbf{VEV}')^{-1} = (\mathbf{V}')^{-1}\mathbf{E}^{-1}\mathbf{V}^{-1} = \mathbf{VE}^{-1}\mathbf{V}'$, since $\mathbf{V}' = \mathbf{V}^{-1}$, which follows from the fact that $\mathbf{V}'\mathbf{V} = \mathbf{I}$ and $\mathbf{V}^{-1}\mathbf{V} = \mathbf{I}$.

Since the inverse of a diagonal matrix is obtained by inverting the diagonal elements of that matrix, if follows that c_{ii}, the ith diagonal element of $(\mathbf{X}'\mathbf{X})^{-1}$, is given by

$$c_{ii} = \sum_{j=1}^{p} (v_{ij}^2/\lambda_j)$$

where the λ_j^{-1} are the diagonal elements of $\mathbf{E}^{-1}$. A variance proportion is then defined as

$$p_{ji} = \frac{v_{ij}^2/\lambda_j}{\sum_{j=1}^{p} (v_{ij}^2/\lambda_j)}$$

where p_{ji} represents the proportion of VIF(i) that results from the multicollinearity (if one exists) represented by λ_j. For example, using the Table 4.6 data we can show that $p_{44} = .9999$, so virtually the entire VIF(4) value is due to the multicollinearity that corresponds to the smallest eigenvalue.

These variance proportions thus show us the extent to which VIF(i), and consequently Var($\hat{\beta}_i$), are inflated by the multicollinearity corresponding to a small eigenvalue. Although the nature of the multicollinearity is not indicated by the variance proportion, it is indicated roughly by the eigenvector that corresponds to the small eigenvalue. Accordingly, eigenvectors and variance proportions can be used together to show how certain forms of multicollinearity inflate Var($\hat{\beta}_i$). [See Hocking (1984) for a related discussion.]

4.3.4   What To Do about Multicollinearity?

Once multicollinearity has been detected and the regressors that cause it have been identified, what can be done about it? Or should anything be done about it? We should remember that multicollinearity is a problem if we are using a regression equation for description, estimation, or control, but it may or may not be a problem if we are using it for prediction (as will be seen in Chapter 12).

Furthermore, multicollinearity need not be harmful. Belsley (1991) distinguishes between degrading multicollinearity and harmful multicollinearity and states that "degrading collinearity need not be harmful, and while the collinearity diagnostics can determine degrading collinearity, they cannot alone determine whether the collinearity is degrading enough also to be harmful." Belsley (1991, Chapter 7) provides "signal-to-noise" ratios for determining when multicollinearity is harmful. Briefly, the argument is made that variance inflation may not be harmful if a parameter that is to be estimated is large.

One suggestion that has been frequently made in trying to overcome multicollinearity is to collect new data. We must remember, however, that multicollinearity occurs with random regressors, and if we collect new data with the values of the regressors occurring at random, we would expect the new data to resemble the old data if both sets are representative of the populations that have been sampled.

One obvious solution is to eliminate one or more of the regressors that are causing the multicollinearity. But which ones do we eliminate? In the example given in Section 4.3.2 it was known from the definition of X_4 that this regressor was the one causing the multicollinearity. Therefore, X_4 could be deleted from the model. Not only would R^2 be virtually the same, but the regression coefficients would be more easily interpretable since the remaining regressors are orthogonal.

Deleting X_4 causes R^2 to change from .932 to .929, and the estimated standard deviations of $\hat{\beta}_1$, $\hat{\beta}_2$, and $\hat{\beta}_3$ change from 1.976, 1.651, and 0.6982 to 0.078, 0.066, and 0.0276, respectively. Thus, removing the multicollinearity affects the estimated standard deviations of the remaining estimators by an order of magnitude.

In general, selecting regressors to delete for the purpose of removing or reducing multicollinearity is not as straightforward as it might seem here. Even with extensive examination of different subsets of the available regressors, we still might select a subset of regressors that is far from optimal (in some sense), as with multicollinear data a small amount of sampling variability in the regressors and/or Y can result in a different subset being selected. Consequently, deleting regressors is not a safe strategy with multicollinear data.

An alternative to regressor deletion is to retain all of the regressors, but to use them in a manner that is different from the way that they are used with ordinary least squares. One such technique is ridge regression, which can be an effective alternative to ordinary least squares for multicollinear data. Ridge regression is discussed in Chapter 12.

4.4 SOFTWARE

Software for multiple regression is plentiful. Some popular statistical software packages were introduced in Section 2.7, and each of those can be used for multiple regression, including the detection of multicollinearity.

For example, SAS Software provides several options within the MODEL statement in the REG Procedure. These include COLLIN, which gives eigenvalues and eigenvectors, and VIF, which produces the variance inflation factors. Variance-decomposition proportions can also be produced. VIFs may be obtained in MINITAB as a subcommand to the Regress command, and both VIFs and variance-decomposition proportions may be obtained with SPSS by specifying the option collin in the statistics subcommand to the main regression command. Eigenvalues, eigenvectors, and variance-decomposition proportions can also be obtained with SYSTAT but not variance inflation factors. Variance inflation factors and variance-decomposition proportions are not produced directly in BMDP.

These software packages also have extensive capabilities in regard to other parts of regression analyses not discussed in this chapter, and these capabilities are discussed in subsequent chapters.

SUMMARY

An introduction to fundamental concepts of multiple linear regression has been given that has included orthogonal and correlated regressors, multicollinearity, the signs of regression coefficients, and centering and scaling. The problems caused by multicollinearity were discussed in detail. The frequently serious problems caused by multicollinearity mandate that orthogonal regressors be used whenever possible. Experimental designs for producing the latter are discussed in Chapter 14.

There are many other aspects to multiple regression, and these are covered in subsequent chapters.

REFERENCES

Belsley, D. A. (1984). Demeaning conditioning diagnostics through centering. *The American Statistician*, **38,** 73–77 (discussion: 78–93).

Belsley, D. A. (1991). *Conditioning Diagnostics: Collinearity and Weak Data in Regression.* New York: Wiley.

Belsley, D. A., E. Kuh, and R. E. Welsch (1980). *Regression Diagnostics: Identifying Influential Data and Sources of Collinearity.* New York: Wiley.

Bradley, R. A. and S. S. Srivastava (1979). Correlation in polynominal regression. *The American Statistician* **33,** 11–14.

Cook, R. D. (1984). Comment on "Demeaning conditioning diagnostics through centering." *The American Statistician*, **38,** 78–79.

Gnanadesikan, R. (1977). *Methods for Statistical Data Analysis of Multivariate Observations.* New York: Wiley.

Gunst, R. F. and R. L. Mason (1980). *Regression Analysis and Its Application: A Data-Oriented Approach.* New York: Dekker.

Hocking, R. R. (1984). Response to David A. Belsley. *Technometrics*, **26,** 299–301.

Lesage, J. P. and S. D. Simon (1985). Numerical accuracy of statistical algorithms for microcomputers. *Computational Statistics and Data Analysis*, **3,** 47–57.

Longley, J. W. (1967). An appraisal of least squares programs for the electronic computer from the point of view of use. *Journal of the American Statistical Association*, **62,** 819–841.

Mahalanobis, P. C. (1936). On the generalized distance in statistics. *Proceedings of the National Institute of Sciences of India*, **12,** 49–55.

Marquardt, D. W. (1970). Generalized inverses, ridge regression, biased linear estimation, and nonlinear estimation. *Technometrics*, **12,** 591–612.

Marquardt (1980). You should standardize the predictor variables in your regression equation (discussion of a paper by Smith and Campbell). *Journal of the American Statistical Association*, **75**, 87–91.

Mullet, G. M. (1976). Why regression coefficients have the wrong sign. *Journal of Quality Technology*, **8**, 121–126.

Myers, R. H. (1990). *Classical and Modern Regression with Applications*, 2nd edition. Boston: PWS-Kent.

Smith, G. and F. Campbell (1980). A critique of some ridge regression methods. *Journal of the American Statistical Association*, **75**, 74–81 (discussion, 81–103).

Snedecor, G. W. and W. G. Cochran (1980). *Statistical Methods*, 7th edition. Ames, IA: Iowa State University Press.

Snee, R. D. and D. W. Marquardt (1984). Comment on "Demeaning conditioning diagnostics through centering." *The American Statistician*, **38**, 83–87.

Weisberg, S. (1985). *Applied Linear Regression*, 2nd edition. New York: Wiley.

Wetz, J. M. (1964). Criteria for judging adequacy of estimation by an approximating response. Ph.D. thesis, Dept. of Statistics, University of Wisconsin.

EXERCISES

4.1. Consider the following regression data.

Y	X_1	X_2
14.3	8.1	7.2
13.2	7.8	6.4
14.2	8.6	7.1
12.5	7.6	6.1
14.0	8.4	7.5
12.6	7.7	6.3
14.9	9.0	7.7
13.4	7.9	6.4
13.6	9.1	7.8
14.8	8.3	7.5
13.7	8.4	6.6
15.4	7.9	7.9
14.8	8.2	7.4
12.9	7.5	6.2
14.2	8.5	7.0
13.0	7.6	6.5
15.1	8.8	7.8
14.2	8.2	7.2
13.2	7.1	6.5
15.0	7.4	8.0

    a. Compute the regression equation.

    b. Compute r_{1Y}, the correlation between X_1 and Y, and explain why the sign differs from the sign of $\hat{\beta}_1$.

    c. Construct a 95% confidence interval for β_2.

    d. Determine if both regressors should be used in the model. Would it be necessary to recompute the confidence interval in (c) if both regressors are not used in the model? Explain.

    e. Construct a 99% prediction interval for Y when $X_1 = 8.6$ and $X_2 = 7.0$, after first determining that this combination of X_1 and X_2 is within the experimental region.

4.2. What is the value of VIF(i) if $R^2(i) = 0.93$? Should an experimenter be concerned about such a large VIF value if he intends to use the regression equation for estimation or control?

4.3. What must be the values of VIF(1) and VIF(2) for the data in Table 4.2?

4.4. If we apply the rough rule-of-thumb given in Section 1.5.4 for determining the minimum number of data points for computing a regression equation, what is that minimum number when there are two regressors?

4.5. It was indicated at the beginning of Chapter 1 that regression can be used to predict body fat percentage. Obviously, this could be of interest to people who have an aversion to underwater weighing (which is necessary to determine the percentage accurately) or who simply do not have easy access to the proper equipment. One prediction equation that has been claimed to be reasonably accurate is

$$\text{Body fat \%} = \frac{(\text{height})^2 \times (\text{waist})^2}{\text{bodyweight} \times c}$$

where height and waist measurements are in inches, bodyweight is in pounds, and c is a constant that is equal to 970 for women and 2304 for men.

    Assume that you are given the task of verifying this equation for women. How would you proceed? In particular, is it possible to obtain a multiple linear regression model after making an appropriate transformation? If so, outline the steps that you would follow in a study designed to check the equation.

4.6. Assume that college GPA is to be predicted using a regression equation that contains two regressors: the total score on the Scholastic Aptitude Test (X_1) and high school GPA (X_2).

Describe the steps that you would take to produce the regression equation for a particular university. Will the two regressors logically be fixed or random? Assume that you have obtained the following data, where the Y values are numbers that are obtained after four years in college.

Y	X_1	X_2
2.8	1065	3.6
2.7	1050	3.3
3.0	1120	3.5
2.2	1005	3.0
2.0	970	2.7
2.4	1010	2.9
3.2	1205	3.5
3.4	1220	3.8
2.9	1090	3.5
3.5	1280	3.9
3.1	1130	3.6
1.9	950	2.6
2.4	990	2.9
1.9	970	2.7
2.0	980	2.6
3.6	1320	4.0
3.2	1140	3.8
2.1	1015	2.8
2.3	1020	2.8
3.0	1100	3.6
3.5	1270	3.9
3.1	1220	3.4

a. Compute the regression equation and determine R^2.

b. Determine the approximate (X_1, X_2) region for which the equation is valid. In particular, can the regression equation be used for $X_1 = 1250$ and $X_2 = 2.9$?

c. Would you recommend that one of the regressors be deleted from the equation or that they both be retained? Explain.

d. Would confidence intervals for β_1 and β_2 be of very much value considering the nature of the data?

e. Which would be of more value in trying to reach a decision regarding the possible admittance of a particular student: a confidence interval for the mean of Y given X_1 and X_2 or a prediction interval for Y given X_1 and X_2?

4.7. Assume that a regression equation has three regressors. What statistic would be used to test $H_0: \beta_1 = \beta_2 = \beta_3 = 0$? If H_0 is rejected, would this mean that the regression equation has value for prediction?

4.8 Consider the data in Exercise 4.6. Could the lack-of-fit test described in Section 1.5.8 be performed if extended to multiple regression? In general, is it likely that a (pure) lack-of-fit test based on exact replicates can be performed when the regressors are random, especially if there is more than just a few regressors?

4.9. Use the definition of ρ_{ij} to show that $\rho_{14} = \rho_{24} = \rho_{34} = .577$ for the example given at the beginning of Section 4.3.2. (Recall that ρ_{ij} was defined in Section 1.7.)

CHAPTER 5

Plots in Multiple Regression

Various diagnostic procedures were introduced in Chapter 2 for use with simple linear regression. Additional procedures are necessary for multiple linear regression, however, and such procedures are presented in this chapter. Specifically, scatter plots and standardized residual plots are sufficient when there is a single regressor, but the higher dimensions that result from the use of multiple regressors introduce some complexities. Accordingly, variations of standardized residual plots that can be used to detect the need for nonlinear terms are introduced. Similarly, an added variable plot is presented as an aid in detecting outliers.

5.1 BEYOND STANDARDIZED RESIDUAL PLOTS

In Chapter 2 the use of standardized residual plots was illustrated for the purpose of checking the assumptions of the simple linear regression model, including the assumption of a constant error variance. In multiple regression the error variance must be constant in regard to each regressor, so there is more checking that must be done.

There is also a need for some procedure that can be used to identify the form of each regressor (linear, quadratic, etc.) that is to be used in the regression equation. In simple regression we usually need only plot Y against X to determine an appropriate form of X to use.

It might seem that we should similarly plot Y against each regressor in multiple regression. Assume that if X_i belongs in the model, then X_i should be a linear term. Looking at scatter plots of Y versus X_i is then essentially tantamount to trying to determine if X_i should be included in the model based on the extent to which $|r_{YX_i}|$ differs from zero. As was shown in Chapter 4, the strength of the linear relationship between Y and X_i, given that other regressors are in the model, can vary greatly (see, e.g., Table 4.4). Thus, scatter plots of Y against each X_i will usually not be informative unless the correlation between the regressors is rather low. Moreover, such plots can be misleading, as is discussed and illustrated by Daniel and Wood (1980, p. 407).

144

Therefore, what is needed is one or more graphical devices that will allow a regression user to see the relationship between Y and X_i when the other regressors are in the model. One such device is described in the next section. (Additionally, *coplots* can be used; these are discussed briefly in Chapter 10.)

5.1.1 Partial Residual Plots

Although standardized residual plots can be used in multiple regression for detecting outliers and, as mentioned, for checking the assumption of a constant error variance, they will generally be less than effective in indicating the appropriate form of X_i, especially when a nonlinear term in X_i is needed. This is due to the fact that the residuals are orthogonal to each regressor, as was indicated in Exercise 3.11 in Chapter 3. It follows that if we fit the points in a residuals plot using regression through the origin, the slope of the line must be zero. Consequently, a residuals plot will not necessarily identify a needed transformation for which the slope of the transformed regressor plotted against the regressor would not be zero. Most transformations would be included in this category, including a logarithmic transformation and a reciprocal transformation. The need for a logarithmic or reciprocal term will thus not be indicated by a strong signal because such a signal would result from a configuration of points with a linear correlation much different from zero. Since a standardized residual plot will usually not differ greatly from the corresponding residual plot, the standardized residual plot will have essentially the same shortcoming.

A standardized residual plot could give the proper message when a quadratic term is needed, however, because a quadratic configuration would not violate the "zero correlation requirement." Therefore, it is not surprising that the standardized residual plot used in the third example in Section 5.2 *does* detect the need for the appropriate quadratic term.

A *partial residual plot* will usually be more valuable than a standardized residual plot for detecting the need for a nonlinear term in a regressor, although there are conditions under which they should be used together.

We will use the notation $\mathbf{e}^*(X_i)$ to denote the set of n partial residuals that are to be plotted against X_i. These are defined as

$$\mathbf{e}^*(X_i) = \mathbf{Y} - \mathbf{X}\hat{\boldsymbol{\beta}} + \hat{\beta}_i \mathbf{X}_i$$
$$= \mathbf{e} + \hat{\beta}_i \mathbf{X_i} \tag{5.1}$$

where $\mathbf{X_i}$ is the column vector in $\mathbf{X}$ that contains the observations for regressor X_i, and $\mathbf{e}$ is the vector of (ordinary) residuals. It should be noted that the $\mathbf{e}^*(X_i)$ that are plotted against X_i are actually pseudo-residuals in that they are not residuals obtained from either using X_i or not using X_i.

Ezekiel (1924) was apparently the first to use such a plot to determine if a regressor should be transformed. It was later given in Ezekiel and Fox (1959) and "rediscovered" by Larsen and McCleary (1972), who referred to it as a partial residual plot. We will later refer to this plot as the PR plot.

There are also some properties of partial residual plots that need to be examined closely. One feature is that fitting a least-squares line with no intercept to the points in the scatter plot of $e*(X_i)$ against X_i would have a slope of $\hat{\beta}_i$. That is, if the partial residuals were used as the dependent variable instead of Y, the coefficient of X_i would be the same as the coefficient of X_i in the multiple regression equation.

This result can be easily shown as follows. As discussed in Section 1.6, the slope estimate in regression through the origin is given by $\sum XY / \sum X^2$. Here the partial residual is the dependent variable, so we immediately see that the slope is given by $\sum (X_i)(e + \hat{\beta}_i X_i) / \sum X_i^2 = \hat{\beta}_i$, since e is orthogonal to X_i.

Unfortunately, however, a partial residual plot does not give the correct picture regarding the strength of the linear relationship between Y and X_i. This is due to the fact that the variability of the points about the regression line corresponds to

$$\text{Var}(\hat{\beta}_i) = \frac{\sigma^2}{\sum X_i^2} \tag{5.2}$$

which would be the correct variance only if the regressors were orthogonal. The correct variance is

$$\text{Var}(\hat{\beta}_i) = \frac{\sigma^2}{\mathbf{X}_i'[\mathbf{I} - \mathbf{X}(\mathbf{X}'\mathbf{X})^{-1}\mathbf{X}']\mathbf{X}_i} \tag{5.3}$$

where $\mathbf{X}_i$ and $\mathbf{X}$ are as previously defined. Since Eq. (5.3) may be written as

$$\text{Var}(\hat{\beta}_i) = \frac{\sigma^2}{\mathbf{X}_i'\mathbf{X}_i - \mathbf{X}_i'\mathbf{X}(\mathbf{X}'\mathbf{X})^{-1}\mathbf{X}'\mathbf{X}_i} \tag{5.4}$$

the actual variance is larger than the apparent variance from the plot since the first term in the denominator of Eq. (5.4) is the same as the denominator of Eq. (5.2), and the second term in the denominator of Eq. (5.4) must be positive. As pointed out by Atkinson (1985, p. 75), the difference in the two denominators could be great if X_i is highly correlated with any of the other regressors.

It is worth noting that no one has proposed the use of standardized partial residual plots. We saw the need for standardized residual plots in Chapter 2, but as Montgomery and Peck (1992, p. 162) and Larsen and McCleary (1972) point out, scaling the partial residuals would exaggerate any nonlinearity that is present.

[Note: Mallows (1986) believes that it is slightly preferable to define the partial residuals as $\mathbf{e} + \hat{\beta}_i(\mathbf{X_i} - \overline{\mathbf{X}}_i) + \overline{\mathbf{Y}}_i$, where $\overline{\mathbf{X}}_i = (\overline{X}_i, \overline{X}_i, \ldots, \overline{X}_i)'$ and $\overline{\mathbf{Y}}_i$ is similarly defined. We will not use this form herein, however.]

5.1.2 CCPR Plot

Wood (1973) and Daniel and Wood (1980, p. 124) present a similar plot, which differs from a partial residual plot in that $\hat{\beta}_i X_i$ is also plotted against X_i on the same plot. Accordingly, the plot was called a *component and component-plus-residual (CCPR) plot* with $\hat{\beta}_i X_i$ being their "component." Daniel and Wood argue (p. 124) that "without the component for reference, it is impossible to judge exactly where the fitted line would lie." (The "fitted line" is the set of fitted values for the regression of the partial residual against X_i. We will see in Section 5.2 how important the fitted line can be.)

The component obviously does not provide the fitted Y at each value of the regressor, however. Rather, by additionally plotting the component, we can (sometimes) practically see the magnitude of the residual at each value of the regressor. Specifically, $e + \hat{\beta}_i X_i - \hat{\beta}_i X_i \equiv e$, so the residual at each point is the difference between the partial residual and the component. When viewed in this manner, the plot described by Daniel and Wood is virtually a combination of a partial residual plot and a residual plot, in that the latter can almost be seen from their plot. This "two-for-one" feature can be important because, as was indicated earlier in this section, a residuals plot is useful only for detecting needed transformations for which the linear correlation is zero, whereas the partial residuals plot can detect a needed transformation when the correlation is nonzero. We will see examples of this in the next section. In particular, we will see an example in which the use of the CCPR plot makes it much easier to identify a needed nonlinear term than when the PR plot is used.

One potential problem with the CCPR plot is that it may not be possible to "see" clearly the residuals because the component may be much larger than the residual. If the residuals are much smaller than the component, the necessary scaling of the vertical axis will make it difficult to see the difference between the component and the component plus the residual. Another potential problem is that the raw residuals can have limited usefulness for detecting model inadequacy. In addition to the problems previously cited, the raw residuals may not be informative when there are large leverage values and the model inadequacy at those data points is not great [see Cook and Weisberg (1982, p. 15)]. Thus, although a CCPR plot is potentially more useful than a PR plot, it will not necessarily be superior.

5.1.3 Augmented Partial Residual Plot

Mallows (1986) proposed an *augmented partial residual (APR) plot* as an improvement on a partial residual plot, claiming that when a transformation is needed, the former often provides a much clearer signal than the latter. The author provided examples to support his claim.

The APR plot results from fitting a quadratic term in each regressor for which the need for a transformation is to be checked. The coefficient of the quadratic term is then multiplied times that term, and the product added to the expression for the partial residual that was given in Eq. (5.1).

As with the partial residual plot, Mallows (1986) indicates that it is slightly preferable to use a mean-centered form for the APR. If this suggestion is followed, the APR would be defined as

$$\text{APR}(X_i) = \mathbf{e} + \hat{\beta}_i(\mathbf{X_i} - \overline{\mathbf{X}_i}) + \hat{\beta}_{ii}[(\mathbf{X_i} - \overline{\mathbf{X}_i})^2 - \text{ave}] + \overline{\mathbf{Y}} \qquad (5.5)$$

where $\mathbf{e}$ and $\overline{\mathbf{X}_i}$ are as defined in Section 5.1.1, $\overline{\mathbf{Y}}$ is a vector in which each element is $\overline{Y}$, and ave is the average of the $(\mathbf{X_i} - \overline{\mathbf{X}_i})^2$ values.

5.1.4 CERES Plots

Cook (1993, 1994) examined PR and APR plots and presented a new type of plot. Partial residual and APR plots were shown to be expectedly inadequate under certain conditions. Cook assumed that the joint distribution of Y and the regressors is multivariate normal and introduced a *combining conditional expectations and residuals (CERES)* plot.

To see how the need for a CERES plot might arise, assume that the true model contains X_1 and the unknown function $g(X_2)$. That is, $Y = \beta_0 + \beta_1 X_1 + g(X_2) + \epsilon$. The PR plot for X_2 when X_2 appears only as a linear term in the fitted model can be viewed as a plot of $Y - \hat{Y} + \hat{\beta}_2 X_2 = (\beta_0 - \hat{\beta}_0) + (\beta_1 - \hat{\beta}_1)X_1 + g(X_2) + \epsilon$ versus X_2. If the true model does contain just linear terms in X_1 and X_2, the PR plot is then a plot of $\epsilon + \beta_2 X_2$ versus X_2, which can be expected to show the need for the linear term in X_2. If $g(X_2) \neq X_2$, however, and there is a nonlinear relationship between X_1 and X_2, the PR plot can fail. To see this, let $E(\hat{\beta}_0) = \beta_0^*$ and $E(\hat{\beta}_1) = \beta_1^*$. Then, conditioning on X_2, the expected value of the ordinate for the PR plot is $(\beta_0 - \beta_0^*) + (\beta_1 - \beta_1^*)E(X_1|X_2) + g(X_2)$. The $E(\hat{\beta}_1)$ will generally not equal β_1 when the wrong model is fit, so the relationship between X_1 and X_2 must be considered. Using an asymptotic argument, Cook (1993) showed that there is no problem if the relationship between X_1 and X_2 is linear. Without relying upon asymptotics, however, it follows that if a scatter plot of X_1 versus X_2 showed a strong linear relationship, the PR plot for X_2 would expectedly have roughly the same general shape as it would if the second term in the expected value of the ordinate were zero, provided that $g(X_2)$ contained the linear term X_2. If, however, there was a reciprocal relationship between X_1 and X_2, the expected value of the ordinate would have a functional form that would differ from $g(X_2)$. Then the PR plot could be quite misleading.

Would the APR be any better? If we assume, as previously, that the model for Y is $Y = \beta_0 + \beta_1 X_1 + g(X_2) + \epsilon$ and let $\tilde{\beta}_0$ and $\tilde{\beta}_1$ denote the parameter estimates of β_0 and β_1, respectively, when the fitted equation contains a quadratic term in X_2, the APR plot for X_2 is then a plot of $(\beta_0 - \tilde{\beta}_0) + (\beta_1 - \tilde{\beta}_1)X_1 + g(X_2) + \epsilon$ versus X_2. Notice that this is essentially the same general form as the ordinate of the PR plot, the only difference is that the estimates of β_0 and β_1 will not be the same as for the PR plot since a different fitted equation is used.

If the quadratic model is the correct model, then the ordinate of the APR plot is $\hat{g}(X_2) + e$, with $\hat{g}(X_2) = \hat{\beta}_2 X_2 + \hat{\beta}_{22} X_2^2$ and $\hat{\beta}_{22}$ denoting the coefficient of X_2 in the fitted model. Thus, the APR plot could be expected to perform as desired. Notice that this will be true even if there is some nonlinear relationship between X_1 and X_2. (We should remember, however, that what is "expected," in terms of expected values, is likely to differ from what eventuates.)

If X_2^2 is not a term in the true model, however, and $\tilde{\beta}_2 X_2 + \tilde{\beta}_{22} X_2^2$ is not a suitable proxy for $g(X_2)$, the APR plot can fail completely. An example of this is given in Section 5.2. One obvious potential problem is that the values of X_1 might be considerably larger than the values of X_2, and fitting the wrong model could cause $\tilde{\beta}_1$ to differ so much from β_1 that the values of X_1 will tend to obscure the plot. Notice that this could happen regardless of the relationship between X_1 and X_2.

Thus, the PR and APR plots can easily fail. Cook (1993) claims that the fitted equation should be obtained using the model $Y = \beta_0^+ + \beta_1^+ X_1 + \beta_2^+ m(X_2) + \epsilon$, with $m(X_2) = E(X_1|X_2) - E(X_1)$. As shown by Cook, β_0^+ will estimate $\beta_0 + E[g(X_2)]$, and β_1^+ will estimate β_1. If a CERES plot is constructed using an ordinate of $e + \hat{\beta}_2^+ \hat{m}(X_2)$, it follows that this represents $(\beta_0 - \beta_0^+) + (\beta_1 - \beta_1^+)X_1 + g(X_2) + \epsilon$. Thus, for a large sample the ordinate is essentially $g(X_2) - E[g(X_2)] + \epsilon$.

Estimates of $E(X_1)$ and $E(X_1|X_2)$ must be obtained. The former would logically be estimated by the average of X_1 in the sample. There are several options for the latter. Cook (1993) suggests the use of prior information, if available, or the use of a nonparametric approach such as *loess* if prior information is not available. (Nonparametric regression is covered in Chapter 10.) Alternatively, the Box–Tidwell approach (see Section 2.5.2.1) or ACE (see Section 6.7) could be used.

What if X_1 and X_2 are unrelated? Then we would expect $\hat{\beta}_2^+$ to be close to zero and $\hat{m}(X_2)$ to not differ greatly from a null vector. Then the CERES plot would be expected to differ very little from an ordinary residuals plot against an omitted variable. Such plots can easily fail, just as an ordinary residuals plot against an included variable can fail. But the ordinate of the plot has the same expected value regardless of the value of $\hat{\beta}_2^+$, so if the plot fails it may be a result of having used a small-to-moderate sample.

5.2 SOME EXAMPLES

Partial residual plots can be very useful under certain conditions, but there are also conditions under which they are of no value. An example of each type of scenario is given in this section.

We will assume that the true model is

$$Y = \beta_0 + \beta_1 X_1 + \beta_2 X_2^{-1} + \epsilon \tag{5.6}$$

The fitted model will be

$$Y = \beta_0 + \beta_1 X_1 + \beta_2 X_2 + \epsilon \tag{5.7}$$

The data in Table 5.1 were constructed using the form of Eq. (5.6). It was mentioned in Section 5.1 that scatter plots of Y against individual regressors can frequently be very misleading. Figures 5.1(a) and 5.1(b) are examples of this. Notice that each plot suggests that there is no relationship between Y and each regressor. Yet when both regressors are used (using more decimal places than are displayed in Table 5.1), there is virtually an exact fit, as Table 5.2 shows that $R^2 = 1.00$. (Actually R^2 rounds to 1.0; it is less than 1.0 by an extremely small amount. This paradoxical result is explained in Section 5.7, using the data in Table 5.6 that also possesses this property.)

At this point we simply wish to emphasize how misleading scatter plots of Y against individual regressors can sometimes be. Figure 5.2 shows how unin-

Table 5.1   Simulated Data

Y	X_1	X_2^{-1}	X_2
11.3141	1.35	6.2099	0.1610
12.0623	2.34	9.8280	0.1017
10.2527	2.85	11.3543	0.0881
10.3557	1.55	6.7701	0.1477
7.5074	2.50	9.6935	0.1032
10.6675	1.48	6.5704	0.1522
7.1506	1.79	7.1262	0.1403
4.5246	1.84	6.8992	0.1449
3.4678	3.50	12.6104	0.0793
−1.5040	1.67	5.3701	0.1862
18.3680	1.99	9.5597	0.1046
8.0786	2.25	8.8967	0.1124
15.5961	3.04	12.8486	0.0778
20.0016	2.59	11.9341	0.0838
14.5630	3.56	14.5297	0.0688
13.4005	3.87	15.4478	0.0647
14.3493	2.05	9.1537	0.1092
5.2692	1.68	6.4476	0.1551
16.6721	1.45	7.3880	0.1354
4.4787	2.12	7.8829	0.1269

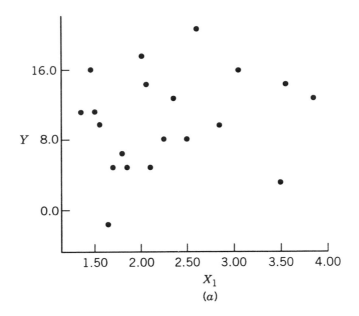

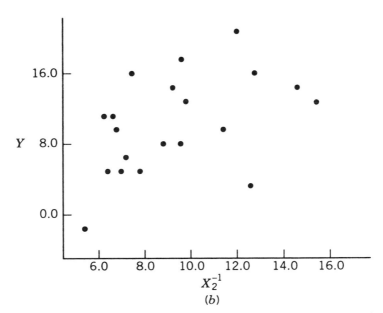

Figure 5.1 (a) Scatter plot of Y vs. X_1 for Table 5.1 data. (b) Scatter plot of Y vs. X_2^{-1} for Table 5.1 data.

Table 5.2   Analysis of the Table 5.1 Data Using X_2^{-1}

The regression equation is: $\hat{Y} = 2.00 - 23.0X_1 + 6.50X_2^{-1}$

Predictor	Coef	Stdev	t-ratio	p
Constant	2.00005	0.00017	11655.65	0.000
X_1	−23.0000	0.0002	−95384.49	0.000
X_2^{-1}	6.49999	0.00006	105127.16	0.000

$s = 0.0002250$ R-sq = 100.0% R-sq(adj) = 100.0%

Analysis of Variance

SOURCE	DF	SS	MS	F	p
Regression	2	577.52	288.76	5.706E+09	0.000
Error	17	0.00	0.00		
Total	19	577.52			

SOURCE	DF	SEQ SS
X_1	1	18.25
X_2^{-1}	1	559.27

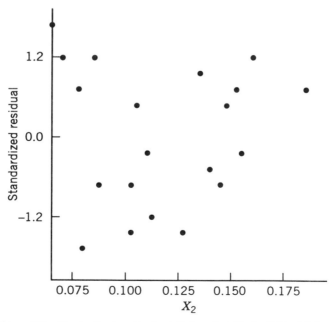

Figure 5.2   Standardized residuals plotted against X_2 for Table 5.1 data.

formative a standardized residual plot can also be when a transformation of a regressor is needed, as this plot was produced from the standardized residuals that were obtained from fitting the model in Eq. (5.7). We should compare this against the partial residual plot in Figure 5.3. Notice that the latter does, upon close inspection, show the need for a reciprocal term in X_2.

Table 5.3 gives the particulars for the regression equation that contains the linear term in X_2 (as given in Table 5.1). Notice that the R^2 value is rather low, which shows that in this instance the linear term is not an acceptable surrogate for the reciprocal term. This will generally be the case. We need to keep in mind what was discussed in Section 2.3.2 regarding the potential for improvement when a nonlinear term is substituted for a linear term.

It was stated in Section 2.3 that finding a transformation that causes a marked improvement can often be difficult. A partial residual plot will frequently enable an experimenter to avoid trial and error (as well as being a substitute for the Box–Tidwell approach), but it is important to recognize that partial residual plots will often be uninformative, just as the standardized residual plot did not suggest the appropriate transformation for the example in this section.

It has been argued that one of the shortcomings of partial residual plots is that they can fail to indicate the appropriate transformation when the appropriately transformed regressor is highly correlated with one of the regressors in

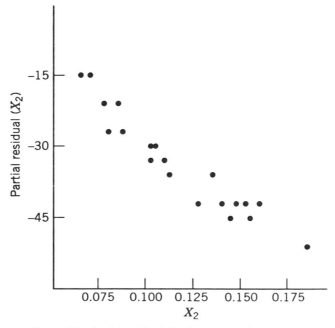

Figure 5.3 Partial residual plot for X_2 for Table 5.1 data.

Table 5.3   Analysis of the Table 5.1 Data Using X_2

The regression equation is: $\hat{Y} = 68.1 - 10.6X_1 - 288X_2$

Predictor	Coef	Stdev	t-ratio	p
Constant	68.07	10.67	6.38	0.000
X_1	-10.553	2.267	-4.65	0.000
X_2	-288.02	49.33	-5.84	0.000

$s = 3.309$ R-sq $= 67.8\%$ R-sq(adj) $= 64.0\%$

Analysis of Variance

SOURCE	DF	SS	MS	F	p
Regression	2	391.40	195.70	17.88	0.000
Error	17	186.12	10.95		
Total	19	577.52			

SOURCE	DF	SEQ SS
X_1	1	18.25
X_2	1	373.15

Unusual Observations

Obs.	X_1	Y	Fit	Stdev. Fit	Residual	St. Resid
10	1.67	-1.504	-3.186	2.382	1.682	0.73X

X denotes an obs. whose X value gives it large influence.

the model. This won't always happen, however, because in the present example the correlation between X_1 and $1/X_2$ is .959, and the correlation between X_2 and $1/X_2$ is $-.962$. Thus, the transformed regressor is very highly correlated with the two regressors that are in the model, yet the signal from the partial residual plot is quite clear.

Since the partial residual plot was informative but the standardized residual plot was uninformative, it follows that the CCPR plot must be informative since for this example it is a combination of a plot that is informative and a plot that is uninformative. The CCPR plot is given in Figure 5.4. The dotted line is the "component" and the plotted points are connected to highlight their relationship to the component. We can think of the CCPR plot as illuminating and magnifying the somewhat weak message from Figure 5.3. Clearly if we fit a line to points that form a reciprocal relationship the points in the middle of the graph should lie below the line and the points at each extreme should lie above the line. Since this is what happens in Figure 5.4, the latter helps us see the signal for a reciprocal term that Figure 5.3 emits.

Berk and Booth (1995) state that the partial residual plot, the augmented partial residual plot, and the CERES plot can all fail in the presence of multi-

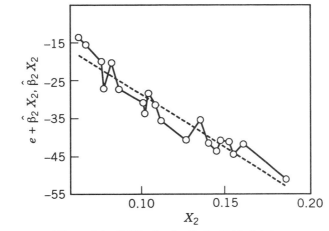

Figure 5.4 CCPR plot for X_2 for Table 5.1 data.

collinearity. Here the correlation between X_1 and X_2 is $-.896$, but we observed that the partial residual plot did not fail, although the signal was weak. Using loess (see Section 10.3.3) to estimate $E(X_{1i}|X_{2i})$, it can be shown that the CERES plot does fail, however.

For a second example, we consider the data in Table 5.4. As in the preceding example, the Y-values are generated using Eq. (5.6), and the fitted model is given by Eq. (5.7). The particulars for the correct and incorrect models are given in Table 5.5. Notice the very large difference in the R^2 values. Because of this difference, we would hope that one of the plots identifies the appropriate model.

Table 5.4 Simulated Data

Y	X_1	X_2	X_2^{-1}
4.0181	1.35	1.8236	0.5484
4.2717	1.52	0.9388	1.0652
4.6949	1.00	0.7013	1.4258
5.7268	1.27	0.5040	1.9843
6.6580	1.86	0.3828	2.6126
6.9979	2.36	0.3926	2.5473
5.3859	2.17	0.9137	1.0945
4.8093	1.92	1.4290	0.6998
7.0720	2.67	0.3847	2.5996
7.3504	2.84	0.4059	2.4635
6.1506	2.99	0.6137	1.6294

Table 5.5   Regression Output for Table 5.4 Data

(a) Using X_2^{-1}: The regression equation is $\hat{Y} = 2.45 + 0.696X_1 + 1.12X_2^{-1}$

Predictor	Coef	Stdev	t-ratio	p
Constant	2.4504	0.2360	10.38	0.000
X_1	0.6958	0.1222	5.69	0.000
X_2	1.1202	0.1043	10.74	0.000

$s = 0.2300$ $R\text{-sq} = 97.0\%$ $R\text{-sq(adj)} = 96.3\%$

Analysis of Variance

SOURCE	DF	SS	MS	F	p
Regression	2	13.7493	6.8746	129.98	0.000
Error	8	0.4231	0.0529		
Total	10	14.1724			

SOURCE	DF	SEQ SS
X_1	1	7.6491
X_2	1	6.1001

Unusual Observations

Obs.	C51	C50	Fit	Stdev. Fit	Residual	St. Resid
2	1.52	4.2700	4.7066	0.0946	−0.4366	−2.08R

R denotes an obs. with a large st. resid.

(b) Using X_2: The regression equation is $\hat{Y} = 5.25 + 0.845X_1 - 1.55X_2$

Predictor	Coef	Stdev	t-ratio	p
Constant	5.2493	0.7011	7.49	0.000
X_1	0.8448	0.2595	3.26	0.012
X_2	−1.5520	0.3657	−4.24	0.003

$s = 0.5008$ $R\text{-sq} = 85.8\%$ $R\text{-sq(adj)} = 82.3\%$

Analysis of Variance

SOURCE	DF	SS	MS	F	p
Regression	2	12.1662	6.0831	24.26	0.000
Error	8	2.0062	0.2508		
Total	10	14.1724			

SOURCE	DF	SEQ SS
X_1	1	7.6491
X_2	1	4.5170

Figure 5.5(a) shows that in this instance the scatter plot of Y against X_2 correctly indicates the need for the reciprocal transformation. It was stated in Section 5.1 that such scatter plots will generally be uninformative unless the correlation between the regressors is fairly low. Here $r_{X_1X_2} = -.417$, so the appropriate signal being given by the scatter plot should not be surprising. Figures 5.5(b) and 5.5(c) give the standardized residual plot and the partial residual plot, respectively. Notice that the former again fails to indicate the need for a transformation, but the correct signal is given by the latter, and the signal is somewhat stronger than that given by the scatter plot. The CERES plot also gives the correct signal.

As a final example illustrating the performance of the different types of plots, we consider an example given by Wood (1973), which was also used in Daniel and Wood (1980, p. 406). The data were generated as $Y = 20 + 20X_1 - 3X_2 - X_1^2 + \epsilon$, where $\epsilon \sim N(0, 1)$, and are reproduced here in Table 5.6. The example was originally given for the primary purpose of demonstrating how misleading scatter plots can be. In that respect this data set is similar to the first example given in this section. Therefore, the scatter plots will not be shown or discussed here. (The reader is asked to show in Exercise 5.1 that the scatter plots are indeed uninformative and to explain why.)

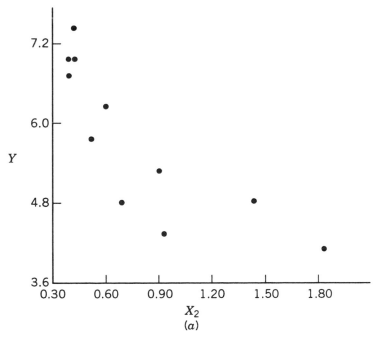

Figure 5.5 (a) Scatter plot of Y against X_2 for Table 5.4 data. (b) Plot of standardized residuals against X_2 for Table 5.4 data. (c) Plot of the partial residuals for X_2 against X_2 for Table 5.4 data.

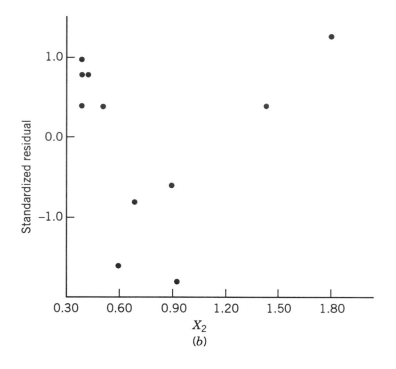

(b)

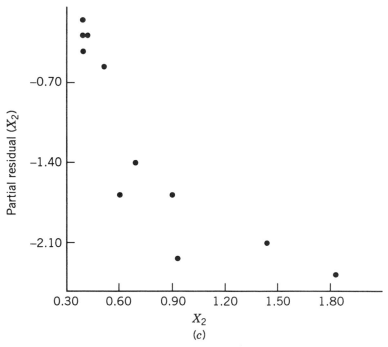

(c)

Figure 5.5   (*Continued*)

Table 5.6   Data from Wood (1973)

Y	X_1	X_2
20.164	0.0	0
18.847	0.2	2
6.678	0.5	7
8.709	1.0	10
36.934	1.5	4
8.397	2.0	16
−1.778	2.5	22
53.136	3.0	6
51.883	3.5	9
31.564	4.0	18
5.739	4.5	28
−6.466	5.0	34
56.494	5.3	14
20.503	5.5	24
47.498	6.0	19
−4.575	6.5	38
10.730	7.0	33
8.214	7.5	35
23.790	7.8	30
−3.518	8.0	40
20.283	0.0	0
16.970	0.2	2
8.379	0.5	7
9.876	1.0	10
37.719	1.5	4
8.445	2.0	16
−1.841	2.5	22
54.339	3.0	6
51.646	3.5	9
30.738	4.0	18
5.373	4.5	28
−7.537	5.0	34
53.974	5.3	14
28.636	5.5	24
46.065	6.0	19
−6.327	6.5	38
13.001	7.0	33
7.855	7.5	35
26.869	7.8	30
−3.659	8.0	40

We will go somewhat further and examine the performance of the standard-ized residual plot, the partial residual plot, the augmented partial residual plot, and the CERES plot. The first two are given in Figures 5.6(a) and 5.6(b). Notice

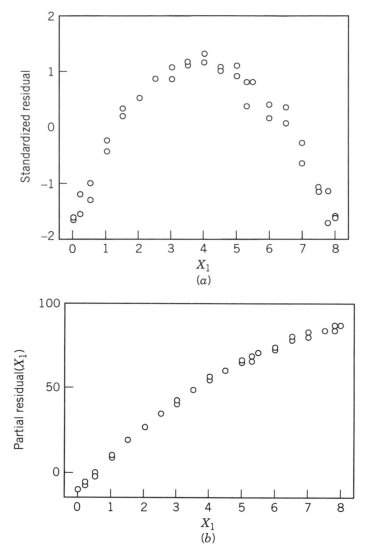

Figure 5.6 (*a*) Standardized residuals plot for Table 5.6 data. (*b*) Partial residuals plot for Table 5.6 data. (*c*) Augmented partial residuals plot for Table 5.6 data.

that the results are the opposite of those for the preceding examples. That is, the standardized residual plot very clearly indicates the need for the quadratic term in X_1, but the partial residual plot fails to do so.

The APR plot for this example is given in Figure 5.6(*c*), for which the APR is defined as in Eq. (5.5). We can see that there is hardly any difference between the partial residual plot and the APR plot. Although there is nonlinearity in each

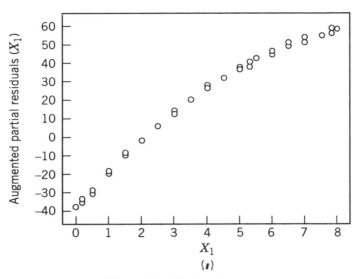

Figure 5.6   (*Continued*)

plot, the nonlinearity is suggestive of the possible need for a logarithmic term rather than a squared term.

The CERES plot also fails. Since there is a strong linear relationship between X_1 and X_2, $\hat{\beta}_1^+ \hat{m}(X_1)$ dominates the residuals, with the result that the correlation between the ordinate and abscissa of the plot is $-.964$. Although the example given by Daniel and Wood (1980) is unusual, it does show that the CERES plot can fail if there is a strong linear relationship between the two regressors and there is also a high R^2 value when the regression is performed using the term that contains the loess fits. It is worth noting that the plot of the residuals from the prediction equation $\hat{Y} = \hat{\beta}_0^+ + \hat{\beta}_1^+ m(X_1) + \hat{\beta}_2^+ X_2$ against X_1 does indicate the need for the quadratic term. This illustrates that a CERES plot can fail when there is a linear relationship between X_1 and X_2 and the residuals from the model with the estimated conditional expectations are small.

Daniel and Wood (1980, p. 413) give the CCPR plot for this example, and the need for the quadratic term can be seen from the quadratic-type curvature in the partial residuals relative to the component. Although they have not suggested this possible use of the CCPR plot, here we have an example in which it can be used to determine the appropriate transformation, and the necessary scaling of the vertical axis to accommodate the component does not obfuscate the signal.

When scaling of the vertical axis is a problem, one possible remedy would be to fit a regression line (through the origin) for the points that are plotted, and then plot the residuals from this regression against the regressor. This should, for example, enable the need for a quadratic transformation to be identified, just as it could be identified when there is no scaling problem.

Although in this example the CCPR plot was sufficient, in many if not most other cases a scatter plot or a residuals plot will be needed as a supplementary plot.

The properties of the APR plot are somewhat different from those of the PR plot. For example, when the partial residuals for X_i are regressed against X_i, the regression line must go through the origin since the Y-intercept is zero (as the reader is asked to show in Exercise 5.2). The regression line will not go through the origin when the augmented partial residuals are used, however, nor will the slope of the line be equal to $\hat{\beta}_i$.

Rather than using a quadratic term in the APR plot, it might be better to use the term $X_i \ln(X_i)$, as this is what is used in the Box–Tidwell transformation approach. This would be a much more easily defensible choice, as a quadratic term is not a reasonable substitute for all nonlinear terms.

When a nonlinear term is needed without the corresponding linear term, we might suspect that this modification will not work because there is then no linear trend that needs to be removed. Therefore, the plot can be uninformative, and it can be shown that this is indeed what happens for Tables 5.2 and 5.4 data, for which a linear term in the regressor to be transformed is not in the true model.

In the next section we similarly examine when a scatter plot and a partial residual plot can be expected to be informative.

5.3 WHICH PLOT?

We have seen three examples in which the results differed considerably, thus suggesting that no one plot can always be relied upon to provide the appropriate signal. It is instructive to examine how a false signal or no signal can result.

Consider the use of an ordinary residuals plot or a standardized residuals plot. To reiterate what was stated in Section 5.1.1, the residuals are orthogonal to each regressor. Therefore, we should not rule out an ordinary residuals plot or a standardized residuals plot when attempting to determine if a transformation is needed, as this will frequently be the appropriate plot to use, as was seen in the third example in Section 5.2. (We note for that example that the leverages do not differ greatly. When that occurs, the correlation between the residuals and the standardized residuals will be very close to 1, so the standardized residuals will be nearly orthogonal to each regressor.)

Although a CERES plot can be informative, as it was in the second example, it can frequently fail. In particular, it can fail when the standardized residual plot fails and there is essentially no relationship between X_1 and X_2. It can also fail when an important regressor is nearly a nonlinear function of the regressor that is being examined, as discussed by Berk and Booth (1995).

5.3.1 Relationships Between Plots

We will assume initially that the true model contains only linear terms in the regressors. A simple way to view the relationship between scatter plots and partial residual plots is as follows. Let Y denote the response variable after it has been mean centered, and assume that the correct model is the fitted model $Y = \sum \beta_i X_i + \epsilon$. With a scatter plot of Y against X_j (which here denotes an arbitrary regressor), we are implicitly assuming that the correct model is $Y = \beta_j f(X_j) + \epsilon$ for some function f, whereas the correct model, relative to X_j, is

$$Y - \sum_{\substack{i=1 \\ i \neq j}}^{r} \beta_i X_i = \beta_j X_j + \epsilon \tag{5.8}$$

When the β_j are estimated, we obtain

$$Y - \sum_{\substack{i=1 \\ i \neq j}}^{r} \hat{\beta}_i X_i = \hat{\beta}_j X_j + e \tag{5.9}$$

where the right side of Eq. (5.9) is, by definition, the partial residual for X_j. The extent to which a scatter plot is uninformative can thus be influenced by the correlation between the partial residuals and Y. Conversely, if the partial residual plot is informative in terms of correctly indicating a linear relationship between Y and X_j, then the scatter plot of Y against X_j should also be informative when there is a high correlation between Y and the partial residual. When will this occur?

To facilitate the derivation of a result that we wish to show, we will now assume that the regressors are also centered, in addition to Y. Using the definition of a partial residual given in Eq. (5.1), it can be easily shown (as the reader is asked to do in Exercise 5.10) that

$$r_{\mathrm{PR}_j Y} = \frac{\sum e^2 + \hat{\beta}_j \sum X_j Y}{\left(\sum Y^2\right)^{1/2} \left(\sum e^2 + \hat{\beta}_j^2 \sum X_j^2\right)^{1/2}} \tag{5.10}$$

where PR_j denotes the partial residual for (the centered) X_j. (The first term in the numerator results from $\sum eY = \sum e^2$.) Since $\sum Y^2 = S_{yy}$ when the response values have been centered (and similarly for X_j), we observe that $r_{\mathrm{PR}_j Y}$ will be close to $r_{X_j Y}$ when the fitted model is quite good, so that $\sum e^2$ is small relative to the other terms in the expression for $r_{\mathrm{PR}_j Y}$. Thus, when there is very little error in the fitted model, we would expect the scatter plot to perform similar to the partial residual plot.

If, for example, X_j is *not* needed in the model, then we would generally expect $\hat{\beta}_j$ to be relatively small. Then $r_{\mathrm{PR}_j Y}$ would be approximately equal to r_{eY}, and the latter is known (Draper and Smith, 1981, p. 147) to be equal to $(1 - R^2)^{1/2}$. Thus, if R^2 is close to 1 and X_j is not part of the true model, the scatter plot may perform quite differently from the partial residual plot. Since the latter will not differ greatly from a plot of the ordinary residuals against X_j when $\hat{\beta}_j X_j$ is small, which of course will show no linear relationship, it follows that the scatter plot may not correctly show that X_j does not belong in the model. Conversely, when R^2 is quite small, the scatter plot may give the appropriate message.

(These results are quite intuitive. The last result certainly makes sense. If R^2 is close to zero, then obviously no regressor is important, so we would not expect to see a message indicating that a regressor is important. The first result means that a regressor that is unimportant when used in combination with other regressors may not appear unimportant in the scatter plot. Certainly this is well known.)

It is important to note that these observations say nothing about whether the partial residual plot will be informative under these conditions. Remembering that a partial residual is given by $e + \hat{\beta}_j X_j$, when there is almost an exact fit a partial residual plot should have a very small amount of random scatter about a zero slope line when X_j is not part of the correct model, and very little scatter about a nonzero slope line when X_j should be in the model. Notice that these results do not depend on the correlation between X_j and the other regressors in the model.

The data in Table 5.1 serve to illustrate some of these points. Assume that we have initially fit the correct form of X_2 (i.e., the reciprocal term), and we will now refer to that term as X_2^*. Then $R^2 \doteq 1.00$, and the scatter plot of Y against X_2^* is misleading because $r_{PR_2 * Y} = r_{X_2^* Y} = 0.45$. The partial residual plot is just the opposite, however, because for all practical purposes all of the points lie on a line with a slope of 6.5 ($= \hat{\beta}_2$). Thus, the partial residual plot could not have been more informative, nor could the scatter plot be less informative.

If we similarly designate the reciprocal term in Table 5.5 as X_2^*, we find that $r_{PR_2 * Y} = .937$, which explains why the scatter plot was informative. Since R^2 was .971, it follows that $r_{X_2^* Y}$ must be close to .937, and in fact the correlation is .922. Recall that the partial residual plot was also informative, and this is due to the fact that R^2 is close to 1, coupled with the fact that the coefficient of the reciprocal term differs significantly from zero.

The augmented partial residual plot should, under the assumption of linearity, be virtually the same as the partial residual plot. This is due to the fact that $\hat{\beta}_{ii}$ should be close to zero, since a quadratic term is unnecessary. Therefore, the augmented partial residuals should be almost the same as the partial residuals.

We have seen what should happen with a scatter plot, in particular, under certain conditions when the true model is linear. It is of somewhat more interest, however, to determine what should happen with a scatter plot, residual plot, and augmented partial residual plot when the true model is *nonlinear*. This we consider in the next section.

Mansfield and Conerly (1987) discuss residual and partial residual plots and indicate conditions under which the latter is apt to give appropriate and inappropriate signals. They also suggest using the CCPR plot so that the component can be used as a reference line in assessing a departure from linearity.

5.3.2   True Model Contains Nonlinear Terms

Assume that the actual model has a squared term in X_j, so that the right side of Eq. (5.8) should additionally contain the term $\hat{\beta}_{jj}X_j^2$. It should be apparent that the main argument given in Section 5.3.1 still applies. That is, the ability of a scatter plot to detect the need for (and identify the form of) a transformation will essentially depend on $r_{\mathrm{PR}_j,Y}$.

To see this numerically, we will again use the numerical examples introduced in Section 5.2. Recall that the scatter plots were totally uninformative in the first example (Table 5.1 data). Those plots were constructed using the true form of each regressor. In particular, the appropriate transformation of X_2 was used. The scatter plot using X_2 itself was not shown, but the plot would have shown linearity, rather than the need for the reciprocal term. This should not be surprising in view of the fact that $r_{\mathrm{PR}_2 Y} = .663$. In the second example the scatter plot gave a strong signal, which would be expected since $r_{\mathrm{PR}_2 Y} = .896$.

In the last example (Table 5.6 data) it was stated that the scatter plot for X_2 was uninformative, and that would be expected since $r_{\mathrm{PR}_2 Y} = .022$. It was shown that the partial residual plot and augmented partial residual plot did not suggest the appropriate transformation, but the standardized residual plot gave a very strong signal. The plot of the ordinary residuals against X_1 was not shown, but it was almost identical to the standardized residual plot.

It is interesting that the ordinary residuals are so informative and the partial residuals are so uninformative. This is the opposite of what we might expect, as a partial residual plot is generally considered to be an improvement over a plot of the ordinary residuals in identifying a needed transformation. Cook and Weisberg (1982, p. 15) discuss when the ordinary residuals are likely to be informative or uninformative. Specifically, they give the result

$$E(e_i) = (1 - h_{ii})B_i - \sum_{j \neq i} h_{ij}B_j \qquad (5.11)$$

where h_{ij} denotes the i,jth element of the hat matrix given in Section 3.2.6, and B_i denotes the model bias at data point i. [Notice the similarity between Eq. (5.11) and (2.6) in Section 2.1.2. Equation (5.11) follows from Eq. (2.6) because $E(\epsilon_i) \neq 0$ when the fitted model is not the true model; that is, when there is model bias.] Of particular interest is the first term on the right side of Eq. (5.11), as a large leverage value (h_{ii}) can obscure the bias that exists at that point.

There are not any large leverage values in the last example, however. Furthermore, the bias is larger at the most extreme X_1 values than in the center, and the sign of the bias differs at the extremes from the sign of the bias for the central X_1 values. Thus the model bias at the extremes stands out, which is why the residuals plot (and standardized residuals plot) are informative.

In the first two examples there were high leverage points, and the failure of the residuals plot to give a clear signal in the second example was due largely to the value of the residual at the single high leverage value. The residuals plot for the first example gave no signal whatsoever, which is not surprising since there are two high leverage points and the entire set of leverage values differ considerably.

Unlike the first two examples, in the last example there is a need for both a linear term and a nonlinear term, and further analysis shows that the linear term is the far more important of the two. But the quadratic term is statistically significant, and adding it to the model causes R^2 to increase from .911 and .998. Adding a strong linear term component to the residuals causes the quadratic configuration of the residuals plot to virtually disappear. We should be concerned that a strong linear term can virtually obscure the need for a nonlinear term. If we knew that a quadratic term is needed, we can then virtually see half of a parabola in the partial residuals plot, but if we didn't know that a quadratic term was needed, we would be hard pressed to identify correctly the appropriate transformation from the plot. We would rather see a less obscure signal. Thus, although Mansfield and Conerly (1987, p. 113) would likely conclude that the PR plot has performed satisfactorily, we should not be satisfied with the result. The APR plot is very similar to the PR plot because the linear component dominates the quadratic component in the APR.

We should not be surprised by the results for the PR and APR plots because the correlation between X_1 and X_1^2 is .964. Thus, the linear term is serving as a proxy for the quadratic term to a considerable extent, and this affects the appearance of the PR and APR plots.

What if the equation that was used to generate the Y values had contained X_1^2 but not X_1? If the fitted model had contained X_1 and X_2, the PR plot would have a strong quadratic element, but the linear term serving as a good proxy for the quadratic term would prevent the PR plot from being parabolic.

The discussion in Section 2.5.2 regarding when a transformation is likely to be helpful is germane here because a linear term in a regressor can dominate a nonlinear term in that regressor and thus camouflage somewhat the need for the nonlinear term.

5.4 RECOMMENDATIONS

We have seen that a scatter plot, residual plot, standardized residual plot, partial residual plot, augmented partial residual plot, and CERES plot have all failed at least once to indicate the appropriate transformation or to show the need for a linear term. This suggests that combinations of plots, such as the CCPR plot, should be used.

Although other writers have warned that scatter plots are unreliable, there is no harm in using them as part of a battery of plots that are to be examined. There are conditions under which a scatter plot will provide the appropriate signal, and that signal might be used to reinforce the signal received from other plots (as was the case for the data in Table 5.5).

A useful way of displaying the scatter plots is in the form of a *scatterplot matrix*, which contains all of the scatter plots using $Y, X_1, X_2, \ldots, X_p$. The scatterplot matrix for the Table 5.1 data is given in Figure 5.7. This is a convenient form for displaying the individual scatter plots and is especially useful when there is more than a few regressors.

Our most difficult task is to detect the need for a nonlinear form of a regressor to be used in addition to the linear term when there is a high correlation between the two terms. That is, the nonlinear term may be reasonably well represented by the linear term, but still make a significant contribution when used in addition to the linear term (as was true for the Table 5.6 data). For this scenario the PR and APR plots will frequently not give strong signals, so it will be necessary to rely upon other plots.

Although not suggested by Mallows (1986), an APR plot could be converted into a counterpart to the CCPR plot. Specifically, the component would be $\beta_j X_j$ if we wished to use a reference line as an aid in judging a departure from linearity.

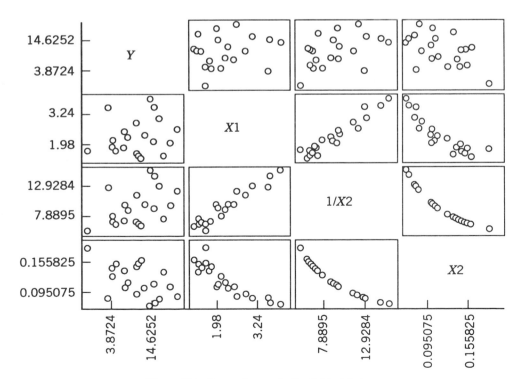

Figure 5.7 Scatterplot matrix for Table 5.1 data.

5.5 PARTIAL REGRESSION PLOTS

Whereas the various modifications of the ordinary residual plot are used for detecting the need for nonlinear terms in the regressors, a partial regression plot is used for different purposes.

The term *partial regression plot* is due to Belsley et al. (1980), who originally gave it the name *partial regression leverage plot*. The partial regression plot for regressor X_j is constructed as follows. The residuals that result from regressing Y on all of the other regressors in the model are computed, as are the residuals that are produced when X_j is regressed on all of the other regressors. The partial regression plot is defined as the scatter plot of the first set of residuals against the second set.

As Cook and Weisberg (1982, p. 46) have shown, the first set of residuals can be expressed in terms of the second set. Specifically, the first set is equivalent to $e + \hat{\beta}_j e_j^*$, where e is the set of residuals from the regression of Y on all of the regressors, and e_j^* is the second set of residuals. Using this result, it can be easily shown that regressing the first set against the second set produces a regression line (through the origin) that has a slope of $\hat{\beta}_j$.

We may demonstrate this by following the same general approach that was used to show that the slope of the regression line through the points on a partial residual plot is also $\hat{\beta}_j$. Again using the fact that the slope for regression through the origin is $\sum XY / \sum X^2$, here we have $e + \hat{\beta}_j e_j^*$ playing the role of Y and e_j^* playing the role of X. To obtain the desired result we need only show that $\sum e e_j^* = 0$. This we may do by using the result that was stated in Exercise 3.11. Let $\mathbf{X}^*$ denote the matrix for the regressors upon which X_j is regressed and $\hat{\boldsymbol{\beta}}^*$ represent the corresponding vector of parameter estimates. Then $\sum e e_j^* = \mathbf{e}'(\mathbf{X_j} - \mathbf{X}^* \hat{\boldsymbol{\beta}}^*) = 0$ because $\mathbf{e}$ is orthogonal, by definition, to $\mathbf{X_j}$ and $\mathbf{X}^*$.

Thus, partial regression plots and partial residual plots have this result in common, and the residuals about the regression line for the partial regression plot are the same as the residuals from the full multiple regression equation (this is also true for the partial residual plot). A dissimilarity is that in the partial regression plot the vertical and horizontal axis labels are true residuals, whereas model residuals are not used directly in the partial residual plot.

One relative advantage of the partial regression plot is that, unlike the partial residual plot, the partial regression plot will, apart from a constant, show the correct strength of the linear relationship between Y and X_j when the other regressors are used. This is because the simple correlation between the two sets of residuals is equal to the partial correlation between Y and X_j. (Partial correlation was discussed briefly in Section 4.2.2 and is also discussed in Section 5.7.)

It is helpful to see that the two types of plots have some common (and some distinct) features, but it is important to recognize that they should be used for different purposes. A partial regression plot is most often used for detecting leverage (potentially influential) points and influential data points that might not be leverage points.

Belsley et al. (1980) give several examples illustrating the use of these plots (including a plot for the intercept term), for detecting influential observations. They term the two sets of residuals the *partial Y residual* (on the vertical axis) and the *partial X residual*.

Hamilton (1992) gives an interesting variation of the standard plot, where the size of the symbol that is used for each point is proportional to the value of DFBETAS at that point. This is termed a *proportional leverage plot*, and such a plot could be particularly useful when the sample size is so large that it may not be apparent whether an outlying point is influential. This can also be an aid in identifying points, as point identification is potentially a problem since the axis values are not coordinates of Y or any of the regressors.

Before illustrating partial regression plots, it should be noted that these plots have also been referred to as *added variable plots*. [See, e.g., Cook and Weisberg (1982) or Atkinson (1985).] Other names that have been used include *adjusted variable plot* (Chambers et al. (1983) and *individual coefficient plot* (Lawrence, 1986). Cook and Weisberg support their use of the term by contending that the plot "is designed to measure the effect of adding a variable to the model," while also stating that the plot can be used for a regressor already in the model. The latter use obviously makes the term *added variable plot* potentially misleading, however. Atkinson (1985, p. 68) states that "an added variable plot ... provides information about the addition of one further carrier to the model." [The term *carrier*, which is also used by Mosteller and Tukey (1977) and others, is synonymous with *regressor*.] Atkinson refers to a partial regression leverage plot as one that is used for a regressor already in the model and states that the two plots have a different starting point but produce the same end result. Thus, Atkinson's definition of an added variable plot differs somewhat from the definition given by Cook and Weisberg but is consistent with the definition given by Belsley et al. (1980).

Belsley et al. (1980) claim that a partial regression plot "should supplant the traditional plots of residuals against explanatory variables" as a means of detecting leverage points and influential data.

Using the hat matrix $\mathbf{H}$, if we wish to construct the plot for a variable W that is *not* presently in the model, we would plot $(\mathbf{I} - \mathbf{H})\mathbf{Y}$ against $(\mathbf{I} - \mathbf{H})\mathbf{W}$. Recalling that $\mathbf{H} = \mathbf{X}(\mathbf{X}^T\mathbf{X})^{-1}\mathbf{X}^T$, it should be apparent that this plot is a plot of the two sets of residuals that were described earlier.

If the variable (say, X_j) *is* presently in the model, then the plot for X_j is a plot of $(\mathbf{I} - \mathbf{H}^*)\mathbf{Y}$ against $(\mathbf{I} - \mathbf{H}^*)\mathbf{X_j}$, where $\mathbf{H}^*$ is computed in the usual way after first deleting the column that contains the values of X_j from the $\mathbf{X}$ matrix. Thus, as Atkinson has stated, the plots are defined similarly; the only real difference is the starting point.

One obvious problem with a partial regression plot used to detect the need for a transformation, as is often done, is that the quantity plotted on the horizontal axis is not X_j, as it is with the partial residual plot. Accordingly, we could not appeal to an argument such as was used in determining when a scatter plot might work to similarly determine when a partial regression plot could be successful in detecting a necessary transformation.

Attention has been focused on partial residual plots and partial regression plots as an improvement on a residuals plot because of the properties that the partial plots have when the correct model is fit, as discussed in Atkinson (1985, p. 67). But is is probably more important to have one or more plots that we can rely upon for guidance when the fitted model is *not* the correct model, remembering that all models are wrong, anyway.

The examples given in the next section show that the partial regression plot performs somewhat worse than the partial residual plot in terms of identifying a needed transformation, and we should expect this to be the general rule. Johnson and McCulloch (1987) consider various types of plots, including an added variable plot, and conclude that the latter "may cause undesirable distortion in general" when the proper form of the added variable is not linear.

5.5.1 Examples

Figure 5.8 exemplifies the frequent inability of a partial regression plot to signal a needed transformation, as this is the plot for X_1 for the Table 5.6 data. Notice that there is no hint of nonlinearity, even though a quadratic term was needed. Thus, the partial regression plot gives an even more misleading signal than did the partial residual plot, which at least showed nonlinearity without suggesting the appropriate transformation.

It could also be shown that the partial regression plot for the Table 5.1 data does not identify the needed nonlinear term, and although the identification is made for the data in Table 5.5, the signal is not as strong as we would prefer.

When linearity holds, it is desirable that the strength of the linear relationship not be exaggerated, so in that respect a partial regression plot has an advantage over a partial residual plot. But when a nonlinear term is needed, we want the

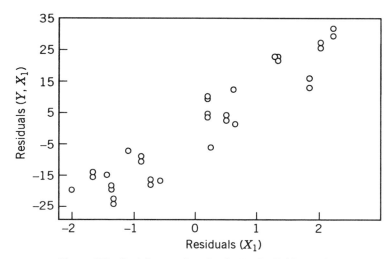

Figure 5.8 Partial regression plot for X_1 for Table 5.6 data.

signal to be as strong as possible, even if the signal may exaggerate the strength of the nonlinear relationship.

5.5.2 Detrended Added Variable Plot

Although the use of a partial regression plot to identify model deficiencies is highly questionable in view of the quantity that is plotted on the horizontal axis, Cook (1986) proposed a *detrended added variable plot* as an improvement on the standard added variable plot when a fitted model has a strong linear term in a particular regressor, but a nonlinear term in that regressor is also needed. Since this is what occurred with the Table 5.6 data, we will see if the detrended plot is more informative than the regular plot. [Examples showing the value of the detrended plot were given by Cook (1986).]

The detrended plot is obtained by first removing the second term in the ordinate for the plot. That is, e is plotted against e_j^*, rather than using $e + \hat{\beta}_j e_j^*$ as the ordinate. The detrended plot for X_1 for the Table 5.6 data is given in Figure 5.9. Obviously the detrended plot fails completely. This will happen frequently because e is orthogonal to e_j^*, as was shown in Section 5.5. As discussed in Section 5.1.1, any plot in which the two plotted components are orthogonal cannot be relied upon to detect the need for many types of nonlinear terms. Although a signal that a quadratic term is needed could still occur in the presence of orthogonality, this does not happen for the Table 5.6 data.

Notice that this result is independent of whether the true model is linear or not, so a detrended added variable plot has the same shortcoming as an ordinary residuals plot in that it cannot be expected to show the need for most transformation types. The argument given by Cook and Weisberg (1989) is actually an

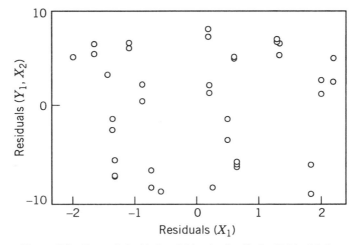

Figure 5.9 Detrended added variable plot for X_1 for Table 5.6 data.

argument that favors a plot of the ordinary residuals against X_j, and we saw for the Table 5.6 data that this plot gave a practically perfect signal regarding the transformation that was needed.

But a detrended added variable plot has an additional weakness. Assume that, as in the third example, a quadratic term is needed. There is no reason why the quadrature in a plot of the residuals against a regressor should also exist when these residuals are plotted against the residuals that are obtained when that regressor is regressed against all of the other regressors. Clearly, there is no reason to believe that there should be any relationship between the first set of residuals and the second set, relative to the necessary transformation. Certainly the second set will be unrelated to the needed transformation because Y is not involved.

Thus, a detrended added variable plot is the weakest of all of the plots that have been considered in this chapter for detecting a needed transformation.

These results apply to a variable X_j that has been included in the prediction equation. What if X_j was not used in the fitted equation so that the plot is truly a (detrended) added variable plot? As indicated earlier, the end result is the same for an added variable plot regardless of whether the regressor is in the initial fitted model or not, and obviously this must also be true for a detrended added variable plot.

Until better plots are developed (or a well-conceived strategy is put forth), the regression user is advised to construct all of the plots that were previously recommended and not to rely upon added variable plots or detrended added variable plots in attempting to determine the need for nonlinear terms.

5.5.3   Partial Regression Plots Used to Detect Influential Observations

We illustrate the use of a partial regression plot to detect influential observations with the following example. The data in Table 5.7 were generated as follows: $X_1 \sim N(0, 1), X_2 = X_1 + 4a$, with $a \sim N(0, 0.1), X_3 = X_2 + 6b$, with $b \sim N(0, 0.2)$, and $Y = 2 + X_1 + X_2 + X_3 + c$, where $c \sim N(0, 1)$.

Thus, although the regressor values are generated randomly, having X_2 and X_3 each obtained from X_1 creates the distinct possibility that a point that is extreme on X_1 will also be extreme on X_2 and X_3. Thus, the point might be a high leverage point, with the Y-value determining whether or not the point is influential.

The partial regression plots for X_1, X_2, and X_3 are given in Figures 5.10(a), (b), and (c), respectively. The convention that is adopted here is a variation of the proportional leverage plot of Hamilton (1992) in that points that have values of DFBETAS in excess of the threshold value are identified, with the point number and the value of DFBETAS given on the plot.

This can be a helpful addition to a partial regression plot because the position of the regression line won't always be obvious from the plot, and it will frequently be necessary to make that determination before influential observations can be identified. Figure 5.10(a) is a good example of this, as it isn't obvi-

Table 5.7 Simulated Data

Y	X_1	X_2	X_3
7.62	2.37	2.01	1.34
2.43	0.01	0.27	0.98
12.75	3.31	3.54	4.06
3.65	0.43	1.15	−0.41
1.68	0.86	0.39	−0.81
−0.39	−0.27	−0.11	−2.21
0.63	0.01	0.01	−0.36
2.94	0.21	0.65	0.41
−3.97	−1.33	−1.90	−2.48
4.39	1.70	0.75	0.22
2.14	0.30	0.06	−0.42
1.36	0.72	−0.05	−1.20
2.59	0.13	0.29	0.42
2.30	0.49	0.63	−0.68
6.14	1.43	1.23	1.28
−3.59	−0.81	−0.52	−2.95
0.78	−0.63	−0.02	0.33
−2.13	−0.92	−0.80	−0.33
5.19	1.05	1.90	−0.28
5.81	0.10	0.01	1.76
3.59	−0.51	0.14	0.20
−2.78	−2.49	−2.49	−0.62
5.58	0.89	1.39	0.73
0.41	−0.65	−1.10	−2.19
−4.70	−2.25	−2.41	−1.97
0.47	0.10	−0.00	−0.92
7.79	2.01	1.73	2.01
4.61	0.91	0.57	1.01
4.92	0.75	1.03	1.54
4.83	0.93	0.76	1.03

ous that the slope of the regression line that would be fit through the points is approximately 1. [Of course, we know what the slope is from the full regression model, but unless the slope is drawn on the graph, influential data points may not be apparent. The suggested use of DFBETAS practically renders unnecessary the construction of the line.]

Some care must be exercised in interpreting these three plots, as there appears to be virtually no linear relationship between Y and X_1 or between Y and X_2. Indeed, the correlation between each pair of residuals is only about .4, even though r_{YX_1} and r_{YX_2} are each greater than .9. As was stated in Section 5.5, the correlation between a pair of plotted residuals is equal to the *partial* correlation between Y and the regressor that is used for the plot.

When the true model is linear, as in this example, viewing the partial regres-

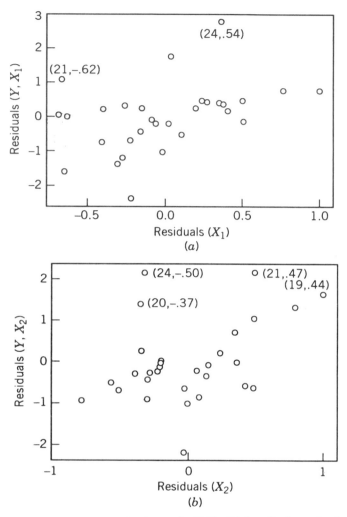

Figure 5.10 (*a*) Partial regression plot for X_1 for Table 5.7 data. Designated points are point numbers and values of DFBETAS for points where the threshold value of .365 is exceeded. (*b*) Partial regression plot for X_2 for Table 5.7 data. Points whose DFBETAS value exceeds the threshold value of .365 are identified with the point number and the value of DFBETAS. (*c*) Partial regression plot for X_3 for Table 5.7 data. Points at which the value of DFBETAS exceeds the threshold value of .365 are identified by point number and the value of DFBETAS.

sion plots is similar to looking at the *t*-statistics from the multiple regression. [For this example it can be shown that the *t*-statistics are very close to the individual *t*-statistics for regression lines fit for each of the partial regression plots. We would expect this to generally occur when the number of regressors is small. See also Cook and Weisberg (1982, p. 51).] Consequently, the par-

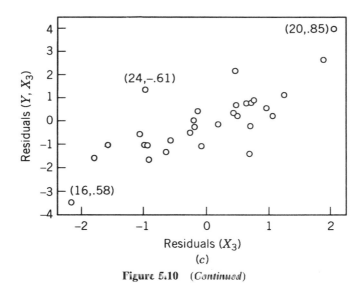

Figure 5.10 (Continued)

tial regression plots can be fallible in the same way that t-statistics in multiple regression can be fallible, and in this example the culprit is multicollinearity.

The results regarding the influential data points are summarized in Table 5.8. Point 24 significantly affects all of the regression coefficients, especially $\hat{\beta}_1$ and $\hat{\beta}_2$, as can be seen by comparing the regression equation using all 30 points, $\hat{Y} = 2.02 + 0.954X_1 + 0.93X_2 + 1.11X_3$, with the regression equation obtained after first omitting point 24, $\hat{Y} = 1.93 + 0.705X_1 + 1.11X_2 + 1.20X_3$. As a reference point, we can compare this change with the change that results when one of the other points is deleted at random. When point 9 is deleted, the regression equation is $\hat{Y} = 2.03 + 0.979X_1 + 0.904X_2 + 1.10X_3$. Clearly the change in the coefficients is much less when one of the other points is randomly deleted than when one of the spotlighted points is deleted. Points 20 and 21 are also seen as being considerably influential.

Once influential data points have been identified, there is the obvious question of what to do with them. An argument can be made for not deleting them unless they can be identified as being bad data. A counter argument, however,

Table 5.8 Summary of Influential Points[a] for Table 5.7 Data

		Point Number (Value of DFBETAS)			
Coefficient	$\hat{\beta}_1$			21 (−0.62)	24 (0.54)
Affected	$\hat{\beta}_2$		19 (0.44)	20 (−0.37) 21 (0.47)	24 (−0.50)
	$\hat{\beta}_3$	16 (0.58)		20 (0.85)	24 (−0.61)

[a]The cutoff value for DFBETAS is $2/\sqrt{n} = .365$.

is that even good data can sometimes be configured in such a way as to cause some data points to be influential. If we accept the last argument, and also recognize that data sets routinely contain a small percentage of bad data, we may then argue for an alternative to ordinary least squares that either downweights certain observations or else does not use them. Such alternatives are given in Chapter 11.

Although using DFBETAS in conjunction with a partial regression plot does enhance the latter considerably, we should keep in mind that DFBETAS is a single-point deletion diagnostic. That is, the standardized change in a regression coefficient is determined where the change is that which results when a single point is deleted. A cluster of influential data points may exist such that deleting a single point will cause little, if any, change in a regression coefficient. This is called *masking*, and is frequently a difficult problem to detect. Another potentially difficult problem is *swamping*, the false identification of good data points as being suspicious observations.

The problems of masking and swamping and suggested solutions are discussed in detail in Chapter 11. Additional approaches include the use of multiple-point generalizations of such statistics as Cook's-D, as given by Takeuchi (1991) and the MDFFIT statistic of Belsley et al. (1980, p. 32). These are not graphical procedures, although like DFBETAS, they might be used in conjunction with some graphical procedures.

5.6 OTHER PLOTS FOR DETECTING INFLUENTIAL OBSERVATIONS

Other types of plots have been suggested for detecting influential observations. Not surprisingly, these are based on some of the influence measures that are in use. For example, Atkinson (1985) illustrates the use of half-normal plots of the modified Cook statistic (see Section 2.4.4). Cook and Weisberg (1982) illustrate a half-normal plot of the square root of Cook's-D statistic, and Johnson and McCulloch (1987) proposed a method that involves partitioning the data into clusters.

5.7 LURKING VARIABLES

Plots of regression data are also used for additional purposes. As defined by Box (1966), and discussed and illustrated extensively by Joiner (1981), a *lurking variable* is a variable that has an important effect but has not been included in the model either because its existence is unknown or because it is thought to be unimportant.

The plots that have been discussed previously are for detecting influential and/or bad observations, determining the need for a transformed regressor, and (in Chapter 2) for detecting heteroscedasticity. So how do we detect the need for a regressor that has not been included in the analysis? Joiner (1981) indicates that a lurking variable is frequently a variable that varies over time in a systematic manner. Accordingly, if the data have been collected over time, a plot of Y against time that has a definite pattern (such as a downward trend) might lead to the discovery of a lurking variable, such as an environmental variable, that is highly correlated with Y.

Recall the discussion on regression for control in Section 1.8.1, in which the possible effect of one such variable, humidity, was illustrated. For that scenario the lurking variable was related to Y but was not related to the single regressor in the model. Therefore, a plot of Y against time might, with some detective work, lead to the identity of the lurking variable.

In other scenarios the lurking variable might also be related to one or more of the regressors, so plots of the regressors against time can also be helpful.

5.0 EXPLANATION OF TWO DATA SETS RELATIVE TO R^2

Two of the three data sets that were used in Section 5.2 illustrate an interesting phenomenon that is not well known. When the regressors are orthogonal, it can be easily shown that $R^2 = r_{YX_1}^2 + r_{YX_2}^2 + \cdots + r_{YX_p}^2$. When the regressors are highly correlated, the usual scenario is that R^2 is much less than $\sum r_{YX_i}^2$. It is possible, however, that R^2 could be much greater than this sum, although it is conjectural how frequently this will occur with real data.

For the Table 5.6 data, $r_{YX_1} = -.0785$, so R^2 would equal .0062 if only X_1 were used in the regression equation. As before, we will let X_2^* denote the transformed second regressor, which we recall was the reciprocal of the original regressor. Then $r_{YX_2^*} = -.5400$, which produces an R^2 value of .2916. When both regressors are used, however, $R^2 = .911$, which far exceeds the sum of the two individual R^2 values.

For the two-regressor case with regressors X_1 and X_2^*, we have the general result

$$R^2 = 1 - (1 - r_{YX_2^*}^2)(1 - r_{YX_1 \cdot X_2^*}^2)$$
$$= r_{YX_2^*}^2 + r_{YX_1 \cdot X_2^*}^2 - r_{YX_2^*}^2 r_{YX_1 \cdot X_2^*}^2 \tag{5.12}$$

where $r_{YX_1 \cdot X_2^*}$ is the partial correlation between Y and X_1, adjusting each variable for its relationship with X_2^*. This partial correlation is defined by

$$r_{YX_1 \cdot X_2^*} = \frac{r_{YX_1} - r_{X_1 X_2^*} r_{YX_2^*}}{\sqrt{1 - r_{X_1 X_2^*}^2}\sqrt{1 - r_{YX_2^*}^2}}$$

With $r_{X_1 X_2^*} = .8683$, we obtain

$$r_{YX_1 \cdot X_2^*} = \frac{-.0785 - (.8683)(-.5400)}{\sqrt{1 - (.8683)^2} \sqrt{1 - (-.5400)^2}}$$

$$= 0.9350$$

Notice the very large difference between r_{YX_1} and $r_{YX_1 \cdot X_2^*}$; it is this difference that causes R^2 to be much larger than might be anticipated. Using Eq. (5.12), we obtain $R^2 = (-.5400)^2 + (.9350)^2 - (-.5400)^2(.9350)^2 = .911$.

This relationship involving R^2 has been discussed by various writers including Hamilton (1987) and Bertrand and Holder (1988), and readers are referred to these papers for additional details.

5.9   SOFTWARE

PROC REG in SAS Software has an option to produce partial regression plots, but there is no option for partial residual plots. Both partial residual plots and added variable plots are available in BMDP Program 2R, and partial residual plots are available in the REGRESSION program of SPSS, but partial regression plots are not available. Neither of these two plots can be obtained as a subcommand in Minitab, but both plots can be produced with minimal additional effort. Similarly, neither of the two plots may be obtained directly with SYSTAT. It is claimed that an option allows partial residuals to be saved, but these are actually the residuals that would be used in constructing an added variable plot, not the pseudo-residuals that are used in constructing a partial residuals plot. S-Plus provides the capability for *brushing* and *spinning* (S-Plus is the registered trademark of Statistical Sciences, Inc.). The former can be helpful in identifying outliers and possibly influential data points, since a selected point in one scatter plot in a scatter plot matrix can be highlighted in the other scatter plots. Spinning entails rotating a three-dimensional scatter plot. This is helpful in examining structure, although its use is limited to the two-regressor case. Brushing and spinning can also be performed using the *R-code* that accompanies Cook and Weisberg (1994). Discussions of brushing and spinning are also given in Cook and Weisberg (1989), and in Chapter 5 of Statistical Sciences, Inc. (1991).

SUMMARY

We have looked at the use of various types of regression plots for detecting influential observations and for detecting the possible need to transform one or more regressors. Emphasis has been given to partial residual plots and partial regression plots, and variations of each. We have seen that no one type of plot can be trusted to be suggestive of a needed regressor transformation. We have also seen that scatter plots can succeed when the other plots fail, and conditions under which each type of plot will be likely to succeed or fail were given. Berk and Booth (1995) provide a good discussion of these plots, including conditions under which they are likely to succeed or fail.

A CCPR plot will frequently be superior to the other plots simply because it is a combination of an ordinary residuals plot and a partial residuals plot. Possible modifications to standard plots include residual plots constructed using residuals from a regression line fit through points on other plots.

We need not rely solely on plots for determining needed transformations, however, as numerical procedures may also be used. A strategy for using modifications of two of the standard numerical procedures is discussed in Chapter 6.

REFERENCES

Atkinson, A. C. (1985). *Plots, Transformations, and Regression*. Oxford, England: Clarendon Press.

Belsley, D. A., E. Kuh, and R. E. Welsch (1980). *Regression Diagnostics: Identifying Influential Data and Sources of Collinearity*. New York: Wiley.

Berk, K. N. and D. E. Booth (1995). Seeing a curve in multiple regression. *Technometrics*, **37**, 385–398.

Bertrand, P. V. and R. L. Holder (1988). A quirk in multiple regression: The whole regression can be greater than the sum of its parts. *The Statistician*, **37**, 371–374.

Box, G. E. P. (1966). Use and abuse of regression. *Technometrics*, **8**, 625–629.

Chambers, J. M., W. S. Cleveland, B. Kleiner, and P. A. Tukey (1983). *Graphical Methods for Data Analysis*. Belmont, CA: Wadsworth.

Cook, R. D. (1986). Assessment of local influence. *Journal of the Royal Statistical Society, Series B*, **48**, 133–155 (discussion: 156–169).

Cook, R. D. (1991). Added Variable Plots in Linear Regression, in *Directions in Robust Statistics and Diagnostics*, W. Stahel and S. Weisberg, eds. New York: Springer-Verlag, pp. 47–60.

Cook, R. D. (1993). Exploring partial residual plots. *Technometrics*, **35**, 351–362.

Cook, R. D. (1994). On the interpretation of regression plots. *Journal of the American Statistical Association*, **89**, 177–189.

Cook, R. D. and S. Weisberg (1982). *Residuals and Influence in Regression*. New York: Chapman and Hall.

Cook, R. D. and S. Weisberg (1989). Regression diagnostics with dynamic graphics. *Technometrics*, **31**, 277–291 (discussion: 293–311).

Cook, R. D. and S. Weisberg (1994). *An Introduction to Regression Graphics*. New York: Wiley.

Daniel, C. and F. S. Wood (1980). *Fitting Equations to Data*, 2nd edition. New York: Wiley.

Draper, N. R. and H. Smith (1981). *Applied Regression Analysis*, 2nd edition. New York: Wiley.

Ezekiel, M. (1924). A method of handling curvilinear correlation for any number of variables. *Journal of the American Statistical Association*, **19**, 431–453.

Ezekiel, M. and K. Fox (1959). *Methods of Correlation and Regression Analysis*, 3rd edition. New York: Wiley.

Hamilton, D. (1987). Sometimes $R^2 > r_{YX_1}^2 + r_{YX_2}^2$. *The American Statistican*, **41**, 129–132.

Hamilton, L. C. (1992). *Regression with Graphics: A Second Course in Applied Statistics*. Pacific Grove, CA: Brooks/Cole.

Johnson, B. W. and R. E. McCulloch (1987). Added-variable plots in linear regression. *Technometrics*, **29**, 427–433.

Joiner, B. L. (1981). Lurking variables: some examples. *The American Statistician*, **35**, 227–233.

Larsen, W. A. and S. J. McCleary (1972). The use of partial residual plots in regression analysis. *Technometrics*, **14**, 781–790.

Lawrance, A. (1986). Discussion of "Assessment of local influence" by R. D. Cook. *Journal of the Royal Statistical Society, Series B*, **48**, 157–159.

Mallows, C. L. (1986). Augmented partial residual plots. *Technometrics*, **28**, 313–319.

Mansfield, E. R. and M. D. Conerly (1987). Diagnostic value of residual and partial residual plots. *The American Statistician*, **41**, 107–116.

Montgomery, D. C. and E. A. Peck (1992). *Introduction to Linear Regression Analysis*, 2nd edition. New York: Wiley.

Mosteller, F. and J. W. Tukey (1977). *Data Analysis and Regression*. Reading, MA: Addison-Wesley.

Statistical Sciences, Inc. (1991). S-Plus for DOS User's Manual, Vol. 1. Seattle: Statistical Sciences, Inc.

Takeuchi, H. (1991). Detecting influential observations by using a new expression of Cook's distance. *Communications in Statistics*, **A20**, 261–274.

Wood, F. S. (1973). The use of individual effects and residuals in fitting equations to data. *Technometrics*, **15**, 677–695.

EXERCISES

5.1. Construct the scatter plots for the data in Table 5.6 and explain why they are misleading.

5.2. Show that the regression equation constructed from the points in a partial residual plot will be the same whether an intercept or no-intercept equation is used.

5.3. Assume that there are three orthogonal regressors under consideration for inclusion in a regression model. Will scatter plots of Y against each potential regressor be informative in terms of indicating whether a linear term in each regressor is likely to be significant? Explain.

5.4. Two regressors are used in a regression equation, and a detrended added variable plot is used to determine if a transformation of either regressor is necessary. Assume that a reciprocal transformation of one regressor is needed. Could the detrended added variable plot signal the need for this transformation? Which plot(s) discussed in the chapter would most likely produce a signal if the regressors are moderately correlated?

5.5. Assume that there are two regressors, with $r_{YX_1} = .6$ and $r_{YX_2} = .4$. What will be the value of R^2 if (a) $r_{X_1X_2} = 0$, (b) $r_{X_1X_2} = .3$?

5.6. Using Eq. (5.12), determine the general expression for R^2 in terms of the appropriate simple correlation coefficients.

5.7. (a) Assume that one of the regressors in a two-regressor model has been misspecified, and a logarithmic term should be used rather than a linear term. If the values of that regressor in the data set range from 8 to 14, would a partial residual plot likely indicate the need for the logarithmic term? (Hint: Assume that $n = 25$ and that the regressor values are equally spaced. What is the correlation between the linear term and the logarithmic term?) Should we really be concerned whether any plot identifies or fails to identify the need for the term? In particular, recall the rule-of-thumb from Mosteller and Tukey (1977) that was given in Section 2.3.2.

(b) Now assume that the regressor values range from 8 to 200, and n stays the same. Again compute the correlation coefficient. Explain why a residuals plot, standardized residuals plot, or detrended added variable plot will not provide the appropriate signal. The following data were constructed using the equation $Y = 4 + 3X_1 + 4\ln(X_2) + \epsilon$.

Y	X_1	X_2
18.0168	2	8
26.1803	4	16
34.1385	5	24
37.4684	6	32
40.4111	7	40
25.7534	2	48
28.7795	3	56
33.1376	4	64
38.3942	6	72
28.6246	3	80
34.2770	4	88
49.7442	9	96
45.7326	8	104

Y	X_1	X_2
28.7993	2	112
36.7360	4	120
38.0868	5	128
44.0260	7	136
41.4235	6	144
32.5727	3	152
36.5755	4	160
40.1900	5	168
42.2187	6	176
44.2683	7	184
41.4570	6	192
37.3367	4	200

Construct the residuals plot, standardized residuals plot, and detrended added variable plot for X_2. Now construct the partial residual plot for X_2. Then use the logarithmic term in place of the linear term and compare the two R^2 values. Does this difference suggest that it is worthwhile to identify the correct term? What would have been the approximate difference in the two R^2 values if X_2 had ranged from 8 to 14, as in part (a)?

5.8. Assume that an experimenter intends to use only one plot in checking to see if any of the four regressors in the regression equation need to be transformed. Would you recommend (a) scatter plots, (b) standardized residual plots, (c) partial residual plots, or (d) CCPR plots?

5.9. Compute, in symbols, the sample correlation between X_j and the partial residual for X_j. What does your expression suggest regarding the appearance of the partial residual plot when the R^2 value for the model with linear terms is very close to one? Does this result make sense?

5.10. Derive Eq. (5.10).

5.11. What plots might be used to detect a lurking variable?

5.12. Assume that an experimenter inadvertently constructs a partial residual plot when there is only a single regressor. Describe the configuration of points that will always result, after first solving for the ordinate for the plot.

CHAPTER 6

Transformations in Multiple Regression

The subject of transformations was introduced in Section 2.5, and in this chapter we consider the subject in more detail.

6.1 TRANSFORMING REGRESSORS

The reasons for making transformations were given at the beginning of Section 2.5; one of the stated reasons was to improve the fit of the model. Plots such as the CCPR plot given in Chapter 5 can be used to identify a needed nonlinear term in one or more regressors. In this section we consider other methods for identifying necessary transformations, including diagnostics. We will use the term *transformation* to represent the situation when the variable is transformed and the linear form that existed originally is not retained (as occurs when Y is transformed), and also to mean that a variable is transformed but the linear term *is* retained, as will generally be the case when regressors are transformed.

In particular, it is important to identify the number of observations that are signaling a needed transformation. For example, Daniel and Wood (1980, p. 81) fit a regression equation with a quadratic term to the stack loss data of Brownlee (1960), but as discovered by Cook (1979), there is only one observation that suggests the inclusion of the quadratic term. In this section we consider diagnostics for identifying such observations.

It is also important to recognize that observations that are influential (not influential) when only linear terms are used may be noninfluential (influential) after one or more transformations have been made. Atkinson (1985) and Cook and Weisberg (1982) illustrate the use of an added variable plot for a *constructed variable* (Box, 1980) in determining not only the need for a transformation but also to determine if the signal for a transformation is coming from all of the observations or only a few.

A constructed variable is simply a new regressor variable. As presented in the literature, it arises in the following way. Let

183

$$Z(\lambda) = \begin{cases} \dfrac{y^\lambda - 1}{\lambda \dot{y}^{\lambda-1}} & \lambda \neq 0 \\[2em] \dot{y} \log(y) & \lambda = 0 \end{cases}$$

with $\dot{y}$ denoting the geometric mean of y. The objective is to obtain $Z(\lambda) = X\beta + \epsilon$ as an adequate model for some value of λ. Expanding $Z(\lambda)$ in a Taylor series about a stated value λ_0 produces

$$Z(\lambda) \doteq Z(\lambda_0) + (\lambda - \lambda_0)W(\lambda_0) + \epsilon$$

so that

$$Z(\lambda_0) \doteq X\beta + \theta W + \epsilon$$

with $\theta = \lambda - \lambda_0$ and $W = \partial Z(\lambda)/\partial \lambda$ evaluated at $\lambda = \lambda_0$ being the constructed variable.

We could use $X \ln(X)$ as the form of the constructed variable, which is actually what occurs when the Box–Tidwell transformation approach is used (see Section 2.3.2.1), as is indicated by Eq. (2.15). The steps that would be followed in producing the added variable plot would be essentially the same as the steps followed when one of the included variables is used. [Notice that such a plot for a constructed variable is truly an "added variable" plot.]

The stack loss data, given in Table 6.1, will be used for illustration. (The data and the objective of the study are discussed in detail in Section 6.6.5). Daniel and Wood (1980, p. 81) concluded that observations 1, 3, 4, and 21 do not fit with the other 17 data points and indicate that a model with linear terms in X_1 and X_2 and a quadratic term in X_1 provides a satisfactory fit.

In his primary analysis of these data, Atkinson (1985, pp. 129–136) elects to use all of the data. His added variable plot [using a constructed variable that differs slightly from $X \ln(X)$] shows the need for a transformation of X_1 and also shows that evidence of the need to transform is spread throughout the data.

Figure 6.1 is the added variable plot for the constructed variable $X_1 \ln(X_1)$, where only the 17 "good" data points are used, and X_3 is not used in the computations. Notice that the rightmost point, which is point 2 of the original 21, seems influential in helping to create some impression of linearity if we think of a regression line fit to these points going through the point (0,0), as it must. The influence of point 2 becomes apparent when we compute the correlation between the two residuals that comprise each point. [Recall from Section 5.4 that an added variable (partial regression) plot is a plot of two sets of residuals.] The correlation is .509 *with* point 2 and .410 without it. When this point is removed, the R^2 values with and without the quadratic term are both .942,

Table 6.1   Stack Loss Data

Obs.	Y (stack loss)	X_1 (airflow)	X_2 (water temp.)	X_3 (acid)
1	42	80	27	89
2	37	80	27	88
3	37	75	25	90
4	28	62	24	87
5	18	62	22	87
6	18	62	23	87
7	19	62	24	93
8	20	62	24	93
9	15	58	23	87
10	14	58	18	80
11	14	58	18	89
12	13	58	17	88
13	11	58	18	82
14	12	58	19	93
15	8	50	18	89
16	7	50	18	86
17	8	50	19	72
18	8	50	19	79
19	9	50	20	80
20	15	56	20	82
21	15	70	20	91

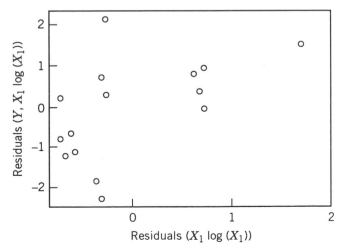

Figure 6.1   Added variable plot for $X_1 \log(X_1)$ for Table 6.1 data.

whereas when all 17 points are used, $R^2 = .980$ with the quadratic term and $R^2 = .973$ without it. Thus, one point is signaling the need for the quadratic term [as discovered also by Cook (1979)], and although the value of the t-statistic for the quadratic term is 2.15 (so that the p-value equals .051), we will generally not want to make a transformation that seems to be supported by only one observation. (With a small sample, however, we might have only one or two points that signal the need for a transformation, thus possibly presenting us with a difficult decision.)

Since the strength of the linear relationship between the two types of residuals in the added variable plot for the constructed variable is also the strength of the signal to add a nonlinear term, we might use the type of influence plot that was used in Figure 2.13 in Section 2.4.1 in lieu of or in conjunction with the numerical analysis described in the preceding paragraph. Figure 6.2 is a plot of the points in Figure 6.1, with the added feature that the size of the plotting symbol indicates the magnitude of the influence that each point has on the correlation between the two sets of residuals. The graph indicates that the correlation is increased approximately .1 by the presence of point 2, as was determined by the preceding numerical analysis. What may not have been apparent from Figure 6.1, however, is that there is another point, point 20 in Table 6.1, that has an even larger (negative) effect on the correlation coefficient. We might consider deleting that point and producing the plot with the remaining 16 points to see if point 2 still stands out as being the only point that increases the correlation to any extent. The configuration of points suggests that

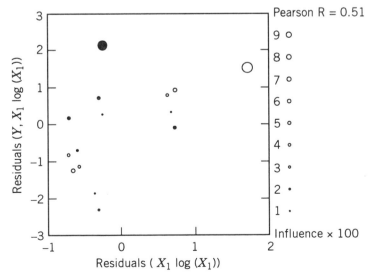

Figure 6.2   Added variable plot for the constructed variable $X_1 \log(X_1)$ for Table 6.1 data, with the size of the plotting symbol showing the influence of each point on the correlation coefficient.

deleting this point will not cause the appearance of any additional large circles, however, and this could be seen to be true.

In addition to illustrating an added variable plot of a constructed variable, this example also shows that the signal indicating a transformation must be carefully checked.

6.2 TRANSFORMING Y

There are some subtle points that must be kept in mind regarding transformations to correct nonnormality and/or heteroscedasticity. It was emphasized in Section 2.1.2 that errors and the corresponding residuals will generally not have the same distribution when the errors are nonnormal. Furthermore, the *empirical* distributions of ϵ and e may differ considerably, as we will see in the next section. (Recall that ϵ is the model error term and e is the difference between the observed and predicted values.) This creates a somewhat thorny problem because we are transforming Y in an effort to induce normality in ϵ, and we are using e to gauge whether or not we have been successful.

In the examples that follow, we will use a known distribution for ϵ, and we will see how an apparently appropriate transformation of Y translates (or does not translate) into a "more normal" distribution of e. We will compute the correlations using the standardized residuals, rather than the residuals, since the variances of the residuals will differ, in general. Therefore, e'_a will be used to denote the standardized residuals in the examples that follow, after a transformation has been made.

6.2.1 Transformation Needed But Not Suggested

The data in Table 6.2 will be used for the first example. The Y values were generated as $Y = 3 + X_1 + X_2 + \epsilon$, where ϵ has a chi-square distribution with 2 degrees of freedom, and the values for X_1 and X_2 were generated using the same distribution. Thus, Y has a chi-square distribution with 6 degrees of freedom.

We may transform a chi-square random variable to approximate normality by using the Wilson–Hilferty (1931) transformation. The transformation is given by

$$Y^* = [(Y/\gamma)^{1/3} - 1 + \tfrac{2}{9}\gamma^{-1}](\sqrt{9\gamma/2}) \tag{6.1}$$

where γ is the degrees of freedom of the chi-square random variable. For all practical purposes we can simply use $Y^* = Y^{1/3}$, since the other components of Eq. (6.1) will not affect the quality of the fit.

We will use the correlation between a random variable and the normalized random variable to show the effect of the transformation on each of the random

Table 6.2   Simulated Data: Chi-square Errors

Y	X_1	X_2
11.68	0.20	5.26
12.68	4.60	1.66
6.15	0.00	2.79
11.67	3.43	3.64
8.22	4.41	0.40
9.02	3.81	0.72
13.42	5.54	4.40
9.56	4.56	0.72
12.41	5.12	1.58
4.67	0.34	1.31
4.31	0.70	0.12
5.76	1.08	1.05
13.41	3.38	3.05
7.00	1.86	1.84
5.85	1.65	0.97
3.91	0.34	0.18
5.08	1.43	0.05
5.56	1.04	1.01
5.54	1.36	1.03
10.44	2.47	2.84

variables. (The normalized values are obtained by fitting a normal distribution to the observed values.) We will let e'_b denote the standardized residuals before any transformation is made, and as stated previously, e'_a will denote the standardized residuals after Y has been transformed. Similarly, we will let ϵ_a denote the transformed pseudo-errors, which are obtained by direct transformation of the errors.

The pseudo-errors will be used simply as a benchmark, as the actual transformed errors cannot be computed. The objective will be to show how the appropriate transformation affects the original errors relative to the effect on the residuals. A subscript n attached to each of these variables will denote the normalized form of each.

Before any transformation is made, $r_{\epsilon \epsilon_n} = .908$ and $r_{e'_b e'_{bn}} = .956$. Since the critical value of the test statistic (i.e., the correlation coefficient) is .9503 using a .05 significance level, we conclude that the residuals are (approximately) normally distributed, whereas the hypothesis of normality is rejected for the errors. Thus, we would reach the wrong conclusion if we relied upon the residuals in trying to determine the distribution of the error term (and forgot about the supernormality tendency that was discussed in Section 2.1.2.1), which means we would decide that a transformation is not necessary, even though the distribution of the error term is highly skewed. In general, the empirical distribution of the residuals need not even remotely resemble the distribution of the errors, which is essentially what we have in this example. This is why we need to use methods such as those given in Section 2.1.2.3.

We should also keep in mind the fact that there is a major difference between transforming empirical distributions and transforming random variables. Specifically, we cannot rely upon results for the latter to tell us what should happen with the former.

6.2.2 Transformation Needed and Suggested

We have just seen an example in which a transformation was strongly needed, but there was no signal indicating the need for a transformation. In this section we consider an example where a transformation is both needed *and* suggested. We will again simulate the data so that we will know that a transformation is needed.

We will use the data in Table 6.3, which were generated as $Y = 2 + X_1 + .4X_2 + \epsilon$, where ϵ has the standard lognormal distribution. This means that we

Table 6.3 Simulated Data:
Lognormal Errors

Y	X_1	X_2
11.51	3.20	12.50
22.25	4.30	38.70
15.71	2.20	22.60
17.90	3.40	19.50
12.18	2.50	17.80
12.81	2.10	19.70
15.84	2.70	26.30
17.93	3.40	27.90
14.99	1.90	20.10
15.04	3.00	22.30
15.94	2.20	22.40
12.93	4.30	14.80
12.31	2.00	16.90
19.20	3.60	21.60
18.91	6.20	24.30
15.25	3.80	22.30
14.09	3.90	17.80
11.27	1.90	17.40
12.89	3.50	17.90
11.57	1.70	15.70
28.76	6.70	38.20
23.44	6.80	36.20

should consider a log transformation, as this would transform the errors to a standard normal distribution. Before the transformation we have $r_{\epsilon\epsilon_n} = .905$ and $r_{e'_b e'_{bn}} = .930$. Thus, the residuals suggest that a transformation is necessary.

After the log transformation, applied to the left side only, we have $r_{\epsilon_a \epsilon_{an}} = .990$ and $r_{e'_a e'_{an}} = .949$. Thus, the transformation has the desired effect on the errors, but not on the residuals, since the .05 critical value for the latter is .954. Remember that the empirical distribution of the residuals depends on various factors, such as the leverages, that can have a prominent effect, and here there are several leverage values that exceed $2p/n$. Consequently, the correlation between the residuals and the normalized residuals should be used only as a rough guideline. Here we see that the residuals are considerably "less normal" than the pseudo-errors.

Notice that the manner in which X_1 and X_2 were generated was not stated, thus leaving open the possibility that the distribution of Y may be considerably different from the distribution of ϵ. This example shows that transforming Y in accordance with the way ϵ would have to be transformed directly will not necessarily be the appropriate transformation.

Since the residuals failed the normality test, we would assume that the errors for the transformed model were nonnormal, thinking about the supernormality tendency of the residuals. Therefore, we will not complete the analysis, saving a thorough analysis for the next section.

6.2.3 Transformation Apparently Successful

Our last example will be one in which the residuals appropriately signal the need for a transformation, and the correct transformation produces residuals that are approximately normal. The data in Table 6.4 were generated as $Y = 2 + X_1 + 2X_2 + \epsilon$, where ϵ has an exponential distribution with a mean of 2.

Before a transformation is performed, we have $r_{\epsilon\epsilon_n} = .904$ and $r_{e'_b e'_{bn}} = .945$. Both numbers suggest that a transformation is needed, since the .05 critical value for the test statistic is, as before, .9503. If we use a log transformation, we obtain $r_{\epsilon_a \epsilon_{an}} = .993$ and $r_{e'_a e'_{an}} = .985$. Thus, the transformation is judged to be successful, as evidenced by the second correlation being close to 1.

We have thus seen three examples in which the error term was highly nonnormal, but the results varied greatly when the residuals were examined. In the first example, the supernormality tendency of the residuals resulted in the residuals suggesting that the errors may be normally distributed. In the second example the residuals suggested that the transformation was not successful, whereas in the third example success was apparently achieved.

In Section 6.6 we will use the standardized skewness coefficient of the residuals as a second measure of possible nonnormality. This will be denoted as α_3 and is defined as $\alpha_3 = \sqrt{n}(\sum e_i^3 / (\sum e_i^2)^{3/2})$.

**Table 6.4   Simulated Data:
Exponential Errors**

Y	X_1	X_2
86.482	13	35
106.435	19	41
75.398	13	30
99.996	21	38
84.095	17	31
78.985	16	30
89.254	15	36
101.437	24	37
103.179	12	44
108.792	10	48
87.751	17	34
87.394	22	31
96.055	26	33
97.329	18	38
98.512	22	37
90.598	10	38
113.413	16	47
126.788	28	47
119.715	21	46
120.787	20	46

6.3   FURTHER COMMENTS ON THE NORMALITY ISSUE

We have seen in the examples in Section 6.2 that there can be a considerable difference in the distributions of ϵ and e. Since ϵ and its distribution are unobservable, should we focus most of our attention on the distribution of Y_i or the distribution of e_i?

Both Y_i and ϵ_i need to be normally distributed, but for different reasons. The confidence intervals, prediction intervals, and hypothesis tests that are used in regression are based on the assumption that Y_i has a normal distribution. (Recall the discussion in Section 1.5.4 that $\hat{\beta}_1$ in simple regression has a normal distribution if Y_i is normally distributed, assuming that the values of the regressor are fixed.)

What if Y_i has a nonnormal distribution? As shown by Box and Watson (1962), and discussed by Seber (1977, p. 190), the effect of nonnormality in Y on the F-test that all of the regression parameters are zero depends on the extent of nonnormality in the regressors (here making no distributional assumption on ϵ). More specifically, whether the regressor values are preselected or not, if the empirical distribution of each regressor is approximately normal, then the F-test will be insensitive to nonnormality in Y, but nonnormality in the regressors will tend to undermine the test. Thus, the effect of nonnormality in Y depends on more than just the extent to which Y is nonnormal.

We should view the normality requirement on ϵ as being a necessary condition for use of the method of least squares. Least squares corresponds to a squared error loss function, which would be appropriate only if the error term had a symmetric distribution. If ϵ had a highly asymmetric distribution, it would be illogical to weight equally positive and negative errors that have the same absolute values. Yet this is what the method of least squares does. We may also argue that least squares would be inappropriate, or at least suboptimal, if the error distribution were symmetric but nonnormal [see also, e.g., Montgomery and Peck (1992, p. 383)].

Thus, approximate normality for ϵ is necessary for the proper use of least squares, but the effect of nonnormality in Y depends on the empirical distribution of the regressor values.

6.4 BOX–COX TRANSFORMATION

Discussed briefly in Section 2.3.4, the Box–Cox transformation approach is one of the most widely used transformations. As originally presented by Box and Cox (1964), the objective is to transform Y to $Y^{(\lambda)}$ in such a way that the transformed Y is a linear function of the regressors, with approximate normality for $Y_i^{(\lambda)}$ and a constant error variance.

We should keep in mind that we do pay a price when we transform Y. In particular, Y may be the natural response variable, so we would have to later transform back to Y in order to have results that are in a meaningful form. In general, we should not try to transform Y solely for the purpose of improving the fit without first trying to transform one or more of the regressors.

The Box–Cox family of transformations is generally presented in textbooks as

$$Y^{(\lambda)} = \begin{cases} \dfrac{Y^\lambda - 1}{\lambda} & \lambda \neq 0 \\[2mm] \log(Y) & \lambda = 0 \end{cases} \tag{6.2}$$

although it was presented somewhat differently in Box and Cox (1964). When we use a model with an intercept, we may simply use Y^λ in place of $(Y^\lambda - 1)/\lambda$. With this substitution, the Box–Cox transformation approach is then the same as that given by Tukey (1957).

There are two methods available for estimating λ, and these are explained in detail by Draper and Smith (1981, pp. 225–235). One method involves computing

$$L(\lambda) = -\tfrac{1}{2}n \log(\text{SSE}/n) + (\lambda - 1) \sum \log(Y) \tag{6.3}$$

for a range of possible values for λ, and then selecting the value of λ that maximizes $L(\lambda)$. This assumes that the transformed dependent variable has a normal distribution, since Eq. (6.3) is a simplified form of the likelihood function (i.e., joint probability density function) of $Y_1^{(\lambda)}, Y_2^{(\lambda)}, \ldots, Y_n^{(\lambda)}$ that results when $Y_i^{(\lambda)}$ is assumed to have a normal distribution. The use of the likelihood function for the normal distribution will not ensure that $Y^{(\lambda)}$ will have a normal distribution, however [see Hernandez and Johnson (1980) or Cook and Weisberg (1982, p. 64)].

The other (equivalent) method is to use the standardized form of Eq. (6.2) and select the value of λ that minimizes the residual sum of squares for the transformed observations. (The use of the standardized form makes the residual sum of squares comparable.) There are two potential problems: The Box–Cox transformation will not guarantee (approximate) normality, and either method of selecting λ could be undermined if the transformed Y is not at least approximately normally distributed.

Cook and Weisberg (1994a,b) suggest an *inverse response plot* (a scatter plot of Y against $\hat{Y}$) for determining the appropriate value of λ and give conditions under which the plot should be reliable. As with any graphical method, however, the inverse response plot may be more valuable as a device for suggesting that a transformation is needed, rather than providing a clear signal as to the form of the appropriate transformation. Cook and Weisberg (1994b, p. 157) essentially say the same thing when they indicate that the appropriate transformation can be estimated by fitting a curve to the plot. The curve might suggest a common transformation to try. If not, a Box–Tidwell transformation (see Sections 2.3.2.1 and 6.5) might be used, with $\hat{Y}$ playing the role of a regressor to be transformed.

Cook and Weisberg (1994b, p. 156) give an example in which the appropriate transformation was to transform Y as $Y^{1/3}$ and show how the inverse normal plot gives the correct signal that a transformation is needed. Since the same transformation was needed for the data in Table 6.2, it is of interest to see if the plot provides the proper signal for these data. It can be shown that the plot does exhibit nonlinearity, as the reader is asked to show in Exercise 6.10. The appropriate transformation has virtually no effect on the fit, however. Why?

Another possible approach would be to use a nonparametric alternative to these methods. Specifically, R^2 could be computed, *on the original scale*, for a range of possible values of λ, with $r_{e'_a e'_{an}}$ and α_3 also computed for each λ. In general, R^2 values computed before and after a transformation of Y will not be comparable. This point has been emphasized by Scott and Wild (1991) and Kvålseth (1985). The latter recommends that R^2 be computed as

$$R_{\text{raw}}^2 = 1 - \frac{\sum (Y - \hat{Y}_{\text{raw}})^2}{\sum (Y - \overline{Y})^2}$$

where $\hat{Y}_{\text{raw}}$ denotes the predicted values transformed back to the original scale. The value of R_{raw}^2 will frequently be negative, however, and this will particularly happen when a predicted value on the transformed scale is very close to zero, so that the predicted value "blows up" when transformed back to the original (raw) scale. Consequently, we shall henceforth use $r^2_{Y\hat{Y}_{raw}}$ to represent the square of the correlation between the Y values and the predicted Y values obtained after converting back to the original (raw) scale, and we shall use this instead of R^2_{raw}. (There will usually be close agreement between the two statistics for good models, however.)

The suggested method has some obvious advantages over the standard approach. First, the results do not depend on the assumption of normality, and normality could be checked roughly by computing $r_{e'_a e'_{an}}$ or by using simulated envelopes for the standardized residuals (see Section 2.1.2.3). Also, the maximum value of the likelihood function does not have any intrinsic meaning, whereas $r^2_{Y\hat{Y}_{raw}}$ obviously does have intrinsic value.

We note that there is a potential problem in computing $r^2_{Y\hat{Y}_{raw}}$, however. If there are any negative $\hat{Y}$ values on the transformed scale, it will not be possible to obtain the corresponding $\hat{Y}$ values on the raw scale for most values of λ. (E.g., if $\hat{Y} = -0.62$ on the transformed scale, it is not possible to obtain $\hat{Y}$ on the raw scale for $\lambda = 0.5$, or -0.5, or most other values.) Unfortunately this problem cannot be easily resolved. In particular, if we tried to alter the raw scale Y-values by using an additive or multiplicative constant, we would not be able to directly retrieve the statistics that we will use on the raw scale because we would be subsequently making a nonlinear transformation of a variable that has already been transformed linearly. For example, if we used $Y^* = Y + 100$ and then computed the predicted values for $(Y^*)^\lambda$, we would not be able to convert directly back to $\hat{Y}$ for the purpose of computing R^2_{raw} at a given value of λ.

The simplest way to avoid this problem is not to use data points in the computation of $r^2_{Y\hat{Y}_{raw}}$ at which $\hat{Y}$ values are negative. This is the approach that we will follow.

It is important to remember the homoscedasticity assumption when transforming Y (or the regressors). We will check this assumption in the examples that are given in Sections 6.6.1–6.6.5.

As was mentioned in Section 2.3.4, Ruppert and Aldershof (1989) presented a method of selecting λ so as to remove both skewness and heteroscedasticity. Certainly both objectives will not always be accomplishable with the same value of λ, which they have designated as λ_{sh}. The approach taken by Ruppert and Aldershof is to test the hypothesis that $\lambda_s = \lambda_h$, and hope that the hypothesis is not rejected. If it is not rejected, they use a weighted function of the equations that are used to produce $\hat{\lambda}_s$ and $\hat{\lambda}_h$, with the weights, w and $1 - w$, chosen so that w minimizes the asymptotic variance of $\hat{\lambda}_{sh}$. Their approach assumes that there is at least tentative identification of the model that relates Y to the regressors.

We will not illustrate that approach here, but we will use as a measure of heteroscedasticity the correlation between the squared standardized residuals and $\log(\hat{Y})$, which is a slight deviation from what Ruppert and Aldershof used.

This measure will be designated as r_{H_1}. As a second measure of possible heteroscedasticity, we will use the correlation between the log of the absolute value of the residuals and $\log(\hat{Y})$. This measure can be motivated by the assertion of Carroll and Ruppert (1988, p. 30), who state that a scatter plot of these two statistics will sometimes show a linear trend when σ_ϵ is proportional to a power of the mean response. We will use r_{H_2} to represent this second measure.

Unfortunately, the value of each of these statistics is susceptible to influential data points, and r_{H_1} could be approximately zero if the spread of the residuals first increases and then decreases so as to create, for example, a plot that looks like a football. Therefore, it is desirable to add a third measure. The latter will be called SPREAD-RATIO, and will be computed in the following manner. Six equally spaced values will be selected within the range of observed $\hat{Y}$ values, with each $\hat{Y}$ value changed to the boundary value to which it is closest. This produces six sets of standardized residuals, with SPREAD-RATIO defined as the sum of the two largest ranges of standardized residuals divided by the sum of the two smallest ranges. It is quite possible that one of the latter will be zero, as when only a single $\hat{Y}$ value is assigned to a particular boundary value. If this occurs, the second smallest range will be used in its place, with the statistic not defined if both of the two smallest ranges are zero. Since σ may be estimated as the range divided by 6 (or 4) for normally distributed data, this is equivalent to using the sum of the two largest estimates of σ divided by the sum of the two smallest estimates. Carroll and Ruppert (1988, p. 36) suggest that many residual plots will show some evidence of heteroscedasticity, and recommend that the ratio of the largest to the smallest estimate of σ exceed three before considering the use of weighted least squares (the latter was covered in Section 2.1.3.1). Here we are using a variation of their rule-of-thumb with the intention of having some resistance to standardized residual outliers.

We will use the auxiliary statistics that have been presented in this section in conjunction with the Box–Cox transformation. This will comprise the first stage of a two-stage procedure, and we will not illustrate the first stage until the second stage has been discussed. Numerical examples of the full two-stage procedure are given in Sections 6.6.1–6.6.5.

6.5 BOX–TIDWELL REVISITED

The Box–Tidwell transformation method was discussed in detail and illustrated in Section 2.3.2.1. In this section we discuss some problems that can occur when the technique is used in multiple regression, and in Section 6.6 we illustrate how it can be used in conjunction with the Box–Cox procedure.

One such problem will occur when the range of a regressor is limited. Recall that this will make it likely that most possible transformations of the regressor will not be helpful, and there is also a strong possibility that tolerance problems will result for whatever software is being used. For example, if X_1 has a limited range, there will probably be a very high correlation between X_1 and $X_1 \ln(X_1)$, as the values of $\ln(X_1)$ will differ by considerably less than do the values of X_1. The multicollinearity problem will, of course, be exacerbated if there are additionally high correlations among the regressors.

This could cause transformed or untransformed regressors to be deleted automatically from the regression equation, thus making the necessary computations impossible. Therefore, it will sometimes be necessary to adjust the "tolerance" factor in regression programs so that one or more regressors will not automatically be deleted by the software. Another potential problem is the fact that convergence problems can often occur when the Box–Tidwell approach is used, especially when the Box–Tidwell approach is used in conjunction with an inappropriate transformation of Y. Therefore, it is important that the transformation of Y be selected wisely.

6.6   COMBINED BOX–COX AND BOX–TIDWELL APPROACH

Some data sets will require a transformation of Y and the regressors. This might happen, for example, when transformations of the regressors have been identified that produce a model that fits the data very well, but there is evidence of nonnormality and/or heteroscedasticity.

This type of problem might be avoided by using a two-stage procedure that results from combining the two approaches suggested in Sections 6.4 and 6.5. Specifically, we will use the various indicators to decide how Y might be transformed, if at all. We will then apply the Box–Tidwell approach for the selected transformation of Y, using five iterations, and will employ the same five nonnormality and heteroscedasticity indicators that were suggested for use with the Box–Cox procedure.

We will illustrate this approach on five data sets: the three hypothetical data sets that were used in Sections 6.2.1.1, 6.2.1.2, and 6.2.1.3, and two real data sets. For the latter the error distributions are unknown, so we can contrast the results for these two data sets with the results obtained when the error distribution is known to be nonnormal.

6.6.1   Table 6.2 Data

Since we are using two types of transformations in the two-stage procedure, it is necessary to use additional notation in the tables that follow. To simplify the notation as much as possible, we will use BT (generally as a subscript) to indicate that the statistic is computed after both the Box–Cox approach and the Box–Tidwell method have been applied, and the absence of BT will mean that only the Box–Cox transformation has been used.

The application of the modified Box–Cox approach to the Table 6.2 data produces the results given in Table 6.5. Notice that three of the values of R_{raw}^2 are negative. As noted in Section 6.4, negative values can occur when predicted values on the transformed scale are very close to zero, so that the predicted val-

Table 6.5 Modified Box–Cox Approach Applied to the Table 6.2 Data

$r^2_{Y\hat{Y}_{\text{raw}}}$	R^2_{raw}	α_3	$r_{e'e'_n}$	λ	log-likelihood	r_{H_1}	r_{H_2}	SPREAD-RATIO
0.755	0.710	1.517	0.883	-2.0	-11.59	0.343	0.397	a
0.802	0.762	1.505	0.888	-1.8	-9.83	0.335	0.262	a
0.838	0.795	1.494	0.893	-1.6	-8.14	0.316	0.104	a
0.865	0.819	1.483	0.901	-1.4	-6.52	0.283	-0.084	a
0.887	0.838	1.468	0.909	-1.2	-4.98	0.231	-0.119	a
0.269	-13.063	1.447	0.915	-1.0	-3.55	-0.764	-0.458	a
0.436	-1.612	1.412	0.921	-0.8	-2.25	-0.704	-0.458	a
0.564	-0.082	1.359	0.925	-0.6	-1.10	-0.674	-0.473	a
0.660	0.410	1.280	0.930	-0.4	-0.12	-0.657	-0.376	a
0.732	0.629	1.169	0.931	-0.2	0.66	-0.647	-0.542	57.17
0.787	0.744	-1.024	0.927	0.0	1.22	0.577	0.539	20.38
0.828	0.811	-0.846	0.923	0.2	1.54	0.641	0.640	2.12
0.859	0.853	-0.643	0.922	0.4	1.62	0.641	0.708	2.46
0.882	0.880	-0.425	0.932	0.6	1.47	0.641	0.715	2.69
0.899	0.898	-0.204	0.942	0.8	1.09	0.639	0.660	2.93
0.910	0.910	0.007	0.956	1.0	0.51	0.632	0.604	3.01
0.916	0.916	0.200	0.967	1.2	-0.25	0.617	0.574	2.72
0.917	0.917	0.369	0.978	1.4	-1.17	0.590	0.465	2.45
0.913	0.913	0.510	0.983	1.6	-2.22	0.546	0.286	2.58
0.901	0.899	0.625	0.984	1.8	-3.38	0.480	0.291	2.73
0.892	0.891	0.717	0.980	2.0	-4.63	0.484	0.258	2.90

[a]The smoothed data configuration precluded the computation at the indicated value of λ.

ues are extremely large when converted back to the original scale. Notice also that SPREAD-RATIO is not defined for the first nine values of λ. This would likely be due to extreme values of $\hat{Y}$ that cause sparseness in the plot of the standardized residuals against $\hat{Y}$. Since there is evidence of heteroscedasticity at every value of λ, we need to apply the Box–Tidwell approach as a second-stage procedure to see if the heteroscedasticity can be removed.

It is desirable to use as a starting point for the Box–Tidwell procedure the point at which $r^2_{Y\hat{Y}_{\text{raw}}}$ and R^2_{raw} begin to agree. Starting at a smaller value of λ will generally cause convergence problems for the Box–Tidwell procedure. Consequently, considering this and the values of the other indicators ($r_{e'e'_n}$ in particular), it seems judicious to use $1.0 \le \lambda \le 1.8$ in the second stage.

The results for the second stage are given in Table 6.6, with α_1 and α_2 denoting the exponents of X_1 and X_2, respectively. The Box–Tidwell results essentially tell us that we should not transform the regressors, which is correct since the regressors are not transformed in the true model. We see evidence of a heteroscedasticity problem at each value of λ, and the values $r_{e'e'_n(\text{BT})}$ are also smaller than we would prefer. Therefore, normality would be questionable at each value of λ, remembering the supernormality tendency.

Thus, the severe nonnormality of the error term apparently cannot be removed, and plots of the standardized residuals against $\hat{Y}$ confirm that there is evidence of a heteroscedasticity problem. (Recall that the error term is a chi-square random variable with 2 degrees of freedom.) Consequently, an estimation technique other than ordinary least squares should be considered.

6.6.2 Table 6.3 Data

Recall that for the Table 6.3 data the appropriate transformation for the errors did not have the desired effect on the residuals, which suggests that the two-stage procedure may not produce acceptable results in regard to normality. The results for the first stage are given in Table 6.7. As suspected, the results are discouraging, not only in regard to normality but also in regard to homoscedasticity. Selecting the interval for λ in the second stage is not easily done, since there is not an obvious choice. Using $-0.6 \le \lambda \le 0.1$ produced results that indicated the normality assumption might be met but not the homoscedasticity assumption. Thus, weighted least squares seems unavoidable, so the second-stage results will not be given. Thus, we see another case in which attempting to overcome nonnormality of the error term creates a problem with heteroscedasticity.

6.6.3 Table 6.4 Data

The results for stage one with the Table 6.4 data are given in Table 6.8. Using $-0.5 \le \lambda \le 0.4$ for the second stage produces the results given in Table 6.9. The numbers suggest that we might consider using $Y^{(-0.3)}$, $X_1^{0.6}$, and $\ln(X_2)$. The only "red flag" is the value of SPREAD-RATIO. When one heteroscedasticity indicator is out of line with the other two, we should try to determine why that

Table 6.6 Modified Box–Cox and Box–Tidwell Procedure Applied to the Table 6.2 Data

$r^2_{Y\hat{Y}_{raw}}$	$r^2_{Y\hat{Y}_{raw}}(BT)$	λ	α_1	α_2	$r_{H_1}(BT)$	$r_{H_2}(BT)$	SPREAD-RATIO(BT)	$\alpha_3(BT)$	$r_{e'e'_n}(BT)$
0.910	0.912	1.0	0.83	1.00	0.644	0.470	a	0.070	0.960
0.913	0.915	1.1	0.84	1.01	0.638	0.462	a	0.158	0.966
0.916	0.917	1.2	0.84	1.02	0.629	0.499	a	0.242	0.971
0.917	0.918	1.3	0.85	1.03	0.617	0.481	a	0.323	0.974
0.917	0.918	1.4	0.86	1.04	0.601	0.429	a	0.399	0.976
0.916	0.917	1.5	0.87	1.05	0.581	0.314	2.384	0.471	0.977
0.913	0.915	1.6	0.87	1.06	0.557	0.335	2.480	0.537	0.978
0.908	0.911	1.7	0.88	1.07	0.528	0.335	2.582	0.598	0.978
0.901	0.904	1.8	0.89	1.08	0.492	0.289	2.691	0.654	0.979

[a]The smoothed data configuration precluded the computation at the indicated value of λ.

Table 6.7 Modified Box–Cox Procedure Applied to the Table 6.3 Data

$r^2_{Y\hat{Y}_{raw}}$	R^2_{raw}	α_3	$r_{e'e'_n}$	λ	log-likelihood	r_{H_1}	r_{H_2}	SPREAD-RATIO
0.822	0.335	−0.394	0.982	−2.0	−13.38	0.148	−0.011	6.58
0.832	0.579	−0.471	0.979	−1.8	−12.63	0.151	0.011	9.45
0.839	0.696	−0.550	0.976	−1.6	−11.91	0.151	0.032	8.32
0.846	0.762	−0.632	0.974	−1.4	−11.19	0.148	0.058	8.76
0.852	0.802	−0.715	0.971	−1.2	−10.51	0.141	0.087	9.54
0.857	0.827	−0.797	0.969	−1.0	−9.85	0.131	0.154	5.08
0.862	0.844	−0.877	0.965	−0.8	−9.23	0.115	0.375	4.54
0.866	0.856	−0.953	0.962	−0.6	−8.66	0.094	0.102	4.43
0.870	0.865	−1.020	0.959	−0.4	−8.15	0.065	−0.028	4.72
0.874	0.871	−1.076	0.956	−0.2	−7.74	0.028	−0.037	5.07
0.877	0.876	1.118	0.949	0.0	−7.42	0.025	0.045	2.95
0.880	0.879	1.144	0.944	0.2	−7.24	0.084	0.027	202.24
0.882	0.882	1.154	0.941	0.4	−7.22	0.163	0.112	114.07
0.884	0.883	1.151	0.936	0.6	−7.39	0.256	0.317	41.50
0.885	0.885	1.140	0.933	0.8	−7.75	0.358	0.475	25.18
0.885	0.885	1.129	0.930	1.0	−8.35	0.454	0.535	18.53
0.885	0.885	1.125	0.934	1.2	−9.17	0.529	0.612	14.90
0.884	0.884	1.136	0.935	1.4	−10.22	0.575	0.695	12.58
0.882	0.882	1.166	0.935	1.6	−11.48	0.593	0.749	10.97
0.879	0.878	1.213	0.935	1.8	−12.94	0.592	0.774	9.79
0.874	0.872	1.276	0.933	2.0	−14.57	0.579	0.766	8.90

Table 6.8 Modified Box–Cox Procedure Applied to the Table 6.4 Data

$r^2_{Y\hat{Y}_{raw}}$	R^2_{raw}	α_3	$r_{e'e'_{ii}}$	λ	log-likelihood	r_{H_1}	r_{H_2}	SPREAD-RATIO
0.929	0.891	1.535	0.886	-2.0	-22.46	-0.076	-0.030	7.12
0.940	0.915	1.516	0.887	-1.8	-21.08	-0.076	-0.064	6.05
0.950	0.934	1.490	0.889	-1.6	-19.64	-0.079	-0.097	5.20
0.958	0.948	1.454	0.893	-1.4	-18.14	-0.084	-0.117	4.51
0.965	0.958	1.402	0.899	-1.2	-16.59	-0.093	-0.091	3.94
0.971	0.967	1.327	0.910	-1.0	-14.99	-0.106	0.003	3.46
0.976	0.973	1.220	0.921	-0.8	-13.36	-0.124	-0.040	3.10
0.980	0.978	1.066	0.933	-0.6	-11.71	-0.151	-0.068	2.82
0.983	0.982	0.852	0.952	-0.4	-10.09	-0.188	-0.109	2.57
0.986	0.985	0.564	0.972	-0.2	-8.54	-0.240	-0.099	2.66
0.988	0.987	-0.202	0.985	0.0	-7.17	0.294	0.106	7.75
0.989	0.989	0.211	0.991	0.2	-6.08	0.376	0.107	a
0.990	0.990	0.618	0.986	0.4	-5.39	0.421	0.229	a
0.991	0.991	0.943	0.966	0.6	-5.18	0.424	0.397	a
0.991	0.991	1.130	0.947	0.8	-5.50	0.403	0.452	a
0.990	0.990	1.175	0.945	1.0	-6.30	0.377	0.536	a
0.989	0.989	1.118	0.951	1.2	-7.49	0.356	0.459	5.65
0.988	0.988	1.012	0.956	1.4	-8.94	0.340	0.005	4.94
0.986	0.986	0.899	0.960	1.6	-10.54	0.329	0.257	4.44
0.984	0.984	0.798	0.963	1.8	-12.22	0.319	0.281	4.05
0.981	0.980	0.717	0.964	2.0	-13.92	0.309	0.250	3.75

[a]The smoothed data configuration precluded the computation at the indicated value of λ.

Table 6.9 Modified Box–Cox and Box–Tidwell Procedure Applied to the Table 6.4 Data

$r^2_{Y\hat{Y}_{raw}}$	$r^2_{Y\hat{Y}_{raw}(BT)}$	λ	α_1	α_2	$r_{H_1}(BT)$	$r_{H_2}(BT)$	SPREAD-RATIO(BT)	$\alpha_3(BT)$	$r_{e'e'_n}(BT)$
0.981	0.988	−0.5	0.63	−0.12	−0.052	0.075	4.874	−0.057	0.986
0.983	0.989	−0.4	0.63	−0.03	−0.074	0.072	5.976	−0.128	0.989
0.984	0.989	−0.3	0.63	0.06	−0.098	0.033	5.424	−0.205	0.991
0.986	0.990	−0.2	0.63	0.16	−0.123	0.001	4.939	−0.288	0.991
0.987	0.990	−0.1	0.62	0.25	−0.149	−0.023	4.509	−0.375	0.989
0.988	0.991	−0.0	0.62	0.34	0.167	−0.029	4.088	0.463	0.986
0.988	0.991	0.1	0.62	0.43	0.199	0.126	4.082	0.552	0.982
0.989	0.991	0.2	0.62	0.52	0.220	0.158	4.053	0.638	0.975
0.990	0.991	0.3	0.62	0.61	0.239	0.206	4.028	0.718	0.966
0.990	0.991	0.4	0.62	0.70	0.255	0.259	4.088	0.789	0.960

is the case. Although the plot of the standardized residuals against X_2 does look slightly unusual for the set of suggested transformations, there is not a strong signal of heteroscedasticity. Thus, the suggested model seems to be a good choice.

6.6.4 Minitab Tree Data

We will use the Minitab tree data (Ryan et al., 1985) for illustration, as plausible models that have been proposed involved the transformation of either Y or at least one regressor. The data are given in Table 6.10. The objective is to develop

Table 6.10 Minitab Tree Data

Obs.	Y (Volume)	X_1 (Diameter)	X_2 (Height)
1	10.3	8.3	70
2	10.3	8.6	65
3	10.2	8.8	63
4	16.4	10.5	72
5	18.8	10.7	81
6	19.7	10.8	83
7	15.6	11.0	66
8	18.2	11.0	75
9	22.6	11.1	80
10	19.9	11.2	75
11	24.2	11.3	79
12	21.0	11.4	76
13	21.4	11.4	76
14	21.3	11.7	69
15	19.1	12.0	75
16	22.2	12.9	74
17	33.8	12.9	85
18	27.4	13.3	86
19	25.7	13.7	71
20	24.9	13.8	64
21	34.5	14.0	78
22	31.7	14.2	80
23	36.3	14.5	74
24	38.3	16.0	72
25	42.6	16.3	77
26	55.4	17.3	81
27	55.7	17.5	82
28	58.3	17.9	80
29	51.5	18.0	80
30	51.0	18.0	80
31	77.0	20.6	87

a regression equation for estimating (predicting) the volume of a tree so that the amount of timber in a given area of a forest may be estimated. Therefore, what is needed is a regression equation for estimating volume (Y) using diameter (X_1) and/or height (X_2). The results for the first stage are given in Table 6.11.

Notice that all of the indicators have satisfactory values without any transformations, so the only motivation for considering transformations would be to see if $r^2_{Y\hat{Y}_{raw}}$ can be increased from .948. We will use $-0.2 \le \lambda \le 1.0$ for the second stage, and the results are shown in Table 6.12. Notice that there is a sign of a heteroscedasticity problem at each value of λ, so it appears as though the use of weighted least squares would be necessary for any of the models in Table 6.12. Since it would be somewhat difficult to obtain good estimates of the weights for this (and most) data sets, it might be best to use the model with no transformations and settle for the smaller value of $r^2_{Y\hat{Y}_{raw}}$.

6.6.4.1 Other Analyses of the Tree Data

This data set has been analyzed by many other writers including Atkinson (1985) and Cook and Weisberg (1982). As discussed by Cook and Weisberg (1982, p. 69), the volume of a cylinder or cone is proportional to the diameter (X_1) squared times the height (X_2). This suggests the possible use of an additive model with terms of $\log(Y)$, $\log(X_1)$, and $\log(X_2)$, as suggested by Cook and Weisberg, although they prefer a model that is discussed later in this section.

Why not just use the single term $X_1^2 X_2$, rather than using logarithmic terms? Some reflection should indicate that this would almost certainly create a heteroscedasticity problem, as we would expect the prediction error to increase with the size of the tree, since a tree is neither a cylinder nor a cone. Thus, although $r^2_{Y\hat{Y}_{raw}} = .978$ using either an intercept or no-intercept model, it can be shown that the heteroscedasticity indicators are extremely large.

Our two-stage procedure suggests that if $\log(Y)$ is used, then $\log(X_1)$ is plausible (since α_1 is close to zero), but a reciprocal transformation is suggested for X_2. Since the range of X_2 is small relative to the order of magnitude, we would expect different transformations of X_2 to produce essentially the same results. Therefore, the heteroscedasticity measures may not change much when $\log(X_2)$ is used.

When the logarithmic transformation is applied to all three variables, we obtain $r^2_{Y\hat{Y}_{raw}} = .978$, $r_{H_1} = .133$, $r_{H_2} = .081$, and SPREAD-RATIO = 4.2. These values suggest that we may need to consider using weighted least squares if we use this model. A plot of the standardized residuals against $\hat{Y}$ also suggests doing so, and the signal is quite strong when the standardized residuals are plotted against X_2, as a pronounced fan-shaped configuration results. (The use of a no-intercept model with these variables produces acceptably small values of the heteroscedasticity indicators, but $r^2_{Y\hat{Y}_{raw}}$ drops off to .936. This is less than the R^2 value that resulted when no transformations are used and the heteroscedasticity indicators are satisfactory.)

The model with all logarithmic terms is also one of the two models suggested

Table 6.11 Modified Box–Cox Results for Tree Data

$r^2_{Y\hat{Y}_{raw}}$	R^2_{raw}	α_3	$r_{e'e'_n}$	λ	log-likelihood	r_{H_1}	r_{H_2}	SPREAD-RATIO
0.728	0.516	1.236	0.922	-2.0	-81.05	0.589	0.379	3.28
0.767	0.571	1.213	0.924	-1.8	-76.35	0.593	0.403	3.29
0.401	-79.362	1.183	0.927	-1.6	-71.58	0.320	0.155	3.88
0.762	-2.036	1.145	0.931	-1.4	-66.72	0.381	0.116	4.08
0.837	0.254	1.094	0.936	-1.2	-61.73	0.423	0.152	4.36
0.880	0.700	1.023	0.942	-1.0	-56.56	0.451	0.187	4.76
0.363	-107.365	0.923	0.951	-0.8	-51.16	-0.256	-0.134	5.62
0.690	-1.160	0.775	0.963	-0.6	-45.47	-0.125	-0.068	7.16
0.851	0.560	0.557	0.978	-0.4	-39.48	-0.107	0.134	6.29
0.924	0.863	0.253	0.992	-0.2	-33.29	-0.136	0.083	7.16
0.959	0.947	0.069	0.991	0.0	-27.48	0.184	0.054	2.82
0.974	0.972	0.129	0.990	0.2	-23.49	0.262	0.236	3.26
0.978	0.978	-0.190	0.991	0.4	-23.36	0.248	0.270	2.43
0.974	0.974	-0.320	0.976	0.6	-27.35	0.221	0.261	1.96
0.965	0.964	-0.062	0.991	0.8	-33.61	0.135	0.206	1.46
0.948	0.948	0.310	0.989	1.0	-40.47	0.022	0.061	1.25
0.930	0.929	0.650	0.972	1.2	-47.22	-0.039	-0.049	1.17
0.937	0.925	0.934	0.954	1.4	-53.68	0.480	0.724	1.45
0.923	0.910	1.173	0.939	1.6	-59.87	0.448	0.661	1.72
0.906	0.893	1.381	0.926	1.8	-65.83	0.417	0.733	1.97
0.886	0.871	1.567	0.913	2.0	-71.62	0.384	0.712	2.23

Table 6.12 Modified Box–Cox and Box–Tidwell Results for Tree Data

$r^2_{Y\hat{Y}_{raw}}$	$r^2_{Y\hat{Y}_{raw}(BT)}$	λ	α_1	α_2	$r_{H_1(BT)}$	$r_{H_2(BT)}$	SPREAD-RATIO(BT)	$\alpha_{3(BT)}$	$r_{e'e'_n(BT)}$
0.924	0.977	−0.2	−0.30	−1.57	−0.053	−0.323	3.676	0.269	0.988
0.945	0.977	−0.1	−0.06	−1.29	−0.081	−0.308	3.478	0.232	0.989
0.959	0.977	0.0	0.19	−1.01	0.150	0.281	4.692	−0.194	0.989
0.968	0.977	0.1	0.44	−0.74	0.148	0.248	72.845	−0.155	0.988
0.974	0.978	0.2	0.68	−0.47	0.187	0.296	607.169	−0.116	0.990
0.977	0.978	0.3	0.93	−0.21	0.229	0.293	4.781	−0.079	0.991
0.978	0.978	0.4	1.17	0.06	0.274	0.366	2.333	−0.042	0.991
0.977	0.977	0.5	1.41	0.33	0.321	0.405	2.107	−0.009	0.991
0.974	0.977	0.6	1.65	0.60	0.368	0.436	1.628	0.022	0.990
0.970	0.977	0.7	1.88	0.88	0.415	0.463	6.718	0.050	0.989
0.965	0.977	0.8	2.12	1.16	0.458	0.485	14.044	0.074	0.989
0.958	0.977	0.9	2.35	1.45	0.498	0.504	12.423	0.094	0.989
0.948	0.977	1.0	2.58	1.74	0.531	0.520	11.234	0.110	0.991

by Atkinson (1985, p. 128), the other model being $Y^{1/3}$ regressed against X_1 and X_2. [This second model is the model that is preferred by Cook and Weisberg (1982).] Table 6.11 suggests that this model may be reasonable, although there would apparently be a moderate heteroscedasticity problem. Table 6.12 suggests that a linear term in X_1 is indeed a good choice in terms of fit, although a logarithmic transformation is suggested for X_2. Again, however, the limited range of X_2 relative to its order of magnitude suggests that alternative transformations of X_2 should be virtually as good in terms of fit. Therefore, we would expect the heteroscedasticity indicators in Table 6.12 not to differ very much from the values of the indicators when the suggested model is fit.

When this second model is used, we obtain $r^2_{Y\hat{Y}_{raw}} = .977$, $r_{H_1} = .255$, $r_{H_2} = .395$, and SPREAD-RATIO = 2.56. Thus, two of the three indicators suggest considerable heteroscedasticity, and when the standardized residuals are plotted against X_2, we again observe a fan-shaped configuration.

6.6.5 Stack Loss Data

The stack loss data set given by Brownlee (1965, p. 454), which was mentioned briefly in Section 6.1, has perhaps been analyzed more than any other regression data set, but it has not been analyzed in terms of the indicators used in this chapter. Extensive analyses have been given by Daniel and Wood (1980, pp. 60–82) and Atkinson (1985, pp. 129–136), and the reader is referred to these sources for descriptive information not given here. The data set draws its name from the fact that the dependent variable is 10 times the percentage of ingoing ammonia that escapes from an absorption tower. The objective is to predict this variable in a plant setting, with the following possible regressors to choose from: airflow (X_1), cooling water inlet temperature (X_2), and acid concentrate (X_3). As indicated previously, the data are in Table 6.1.

Daniel and Wood (1980) identify 4 of the 21 data points (numbers 1, 3, 4, and 21) as corresponding to transitional plant states, and elect not to use those points. We shall adopt the same position here. The consensus of opinion is that X_3 is not of value, so we shall exclude it from our analysis. The analysis of the modified Box–Cox procedure produces the results that are shown in Table 6.13.

All of the indicators provide a strong signal concerning the range of λ that should be used in the second stage. Accordingly, we shall use $0.2 \leq \lambda \leq 1.0$ for that stage, while recognizing that we may encounter some heteroscedasticity problems. (Notice the considerable differences between $r^2_{Y\hat{Y}_{raw}}$ and R^2_{raw} when λ is close to -2. This shows how we could be misled by relying solely on the R^2_{raw} values.)

The results for the second stage are shown in Table 6.14. There is not an obvious choice for the model, as there is some hint of heteroscedasticity for each model. Notice that only one of the heteroscedasticity indicators has an unacceptable value for most of the models, however, which suggests that there might be a problem with influential points.

Table 6.13 Modified Box–Cox Results for Stack Loss Data

$r^2_{Y\hat{Y}_{raw}}$	R^2_{raw}	α_3	$r_{e'e'_n}$	λ	log-likelihood	r_{H_1}	r_{H_2}	SPREAD-RATIO
0.908	0.486	0.995	0.911	-2.0	-23.37	0.376	0.060	[a]
0.913	0.537	0.949	0.914	-1.8	-22.08	0.373	-0.005	[a]
0.917	0.588	0.900	0.918	-1.6	-20.77	0.370	-0.091	2.34
0.920	0.636	0.849	0.921	-1.4	-19.43	0.366	-0.086	14.43
0.923	0.682	0.793	0.926	-1.2	-18.04	0.361	-0.017	12.36
0.925	0.726	0.731	0.932	-1.0	-16.59	0.355	0.019	10.60
0.728	-16.350	0.658	0.938	-0.8	-15.04	-0.827	-0.399	20.20
0.802	-1.363	0.569	0.945	-0.6	-13.38	-0.767	-0.407	55.74
0.859	0.292	0.458	0.954	-0.4	-11.56	-0.710	-0.392	65.90
0.903	0.727	0.313	0.966	-0.2	-9.54	-0.646	-0.355	20.33
0.935	0.881	-0.125	0.981	0.0	-7.28	0.495	0.268	23.18
0.958	0.944	0.110	0.992	0.2	-4.80	0.419	0.238	7.78
0.972	0.969	0.333	0.989	0.4	-2.27	0.159	0.036	6.18
0.979	0.979	0.332	0.987	0.6	-0.29	-0.058	-0.076	5.65
0.979	0.979	-0.041	0.981	0.8	-0.05	0.038	0.351	5.51
0.973	0.973	-0.272	0.986	1.0	-2.29	0.430	0.446	5.72
0.960	0.959	-0.079	0.974	1.2	-6.22	0.550	0.227	6.44
0.939	0.933	0.215	0.946	1.4	-10.66	0.495	0.147	8.49
0.904	0.876	0.440	0.923	1.6	-15.03	0.409	0.003	15.56
0.850	0.803	0.587	0.912	1.8	-19.17	0.363	0.104	107.34
0.922	0.853	0.680	0.908	2.0	-23.07	0.789	0.828	21.58

[a]Unable to compute.

Table 6.14 Modified Box–Cox and Box–Tidwell Results for Stack Loss Data

$r^2_{Y\hat{Y}_{\text{raw}}}$	$r^2_{Y\hat{Y}_{\text{raw}}(BT)}$	λ	α_1	α_2	$r_{H_1(BT)}$	$r_{H_2(BT)}$	SPREAD-RATIO(BT)	$\alpha_{3(BT)}$	$r'e'_{e_n(BT)}$
0.958	0.979	0.2	−0.90	−0.39	−0.112	−0.234	2.899	0.009	0.988
0.966	0.979	0.3	−0.51	−0.18	−0.097	−0.234	2.725	0.036	0.988
0.972	0.979	0.4	−0.13	0.03	−0.083	−0.232	2.748	0.061	0.987
0.976	0.980	0.5	0.24	0.27	−0.070	−0.220	2.919	0.087	0.986
0.979	0.980	0.6	0.61	0.51	−0.059	−0.224	3.026	0.112	0.986
0.980	0.980	0.7	0.98	0.77	−0.049	−0.219	5.803	0.137	0.985
0.979	0.980	0.8	1.34	1.04	−0.039	−0.212	5.014	0.162	0.985
0.977	0.980	0.9	1.70	1.31	−0.030	−0.205	2.014	0.186	0.983
0.973	0.980	1.0	2.05	1.60	−0.022	−0.198	1.939	0.210	0.982

Since the best model is not apparent, we will examine more than one model. One such model would be obtained by using $\lambda = 0.4$ in conjunction with logarithmic transformations of X_1 and X_2. Doing so produces the following results: $r^2_{Y\hat{Y}_{\text{raw}}(\text{BT})} = .979$, $r_{e'e'_n(\text{BT})} = .986$, $r_{H_1(\text{BT})} = .093$, $r_{H_2(\text{BT})} = .376$, SPREAD-RATIO $= 2.76$, and $\alpha_{3(\text{BT})} = 0.114$, where the additional subscript BT is used to indicate that these are the statistics obtained after the Box–Tidwell transformation approach has been applied, as previously noted. When one heteroscedasticity indicator is out of line with the other two, it is desirable to investigate why. A scatter plot of the two components of $r_{H_2(\text{BT})}$ reveals a highly influential point, and when that point is deleted, the value changes to .061.

It is worthwhile to note that the type of influence-symbol scatter plot shown in Figure 6.2 can be useful in determining if there are any points that are strongly influencing the values of any of these indicators. Furthermore, if we use such a scatter plot for $\hat{Y}$ plotted against Y, we can see, in particular, if there are any data points that are making a major contribution to $r^2_{Y\hat{Y}_{\text{raw}}}$. Such a plot might be used in conjunction with an added variable plot for each transformed regressor. The reader is asked in Exercise 6.9 to construct this modified scatter plot for the stack loss data, both for the quadratic model advocated by Daniel and Wood (1980), and for the model suggested later in this section.

The standardized residuals should, of course, also be plotted against the regressors. When that is done here there is some evidence that σ^2_ϵ may decrease over X_2, but that signal is basically due to a single point. There is somewhat stronger evidence that the absolute value of the residuals increases over X_2, however.

Thus, although there is some abnormality in the standardized residual plots, the heteroscedasticity is not severe, so we may wish to use this model (assuming that it passes other diagnostic checks that will be discussed later). The model finally selected by Daniel and Wood (1980) has $\hat{\sigma}^2 = 1.26$, and they indicate that searching for a better model would not likely lead to much improvement since the lowest possible value is 1.05. Here we obtain $\hat{\sigma}^2 = 1.197$ (on the original scale) on the first try. The model selected by Daniel and Wood, which has terms in X_1, X_2, and X^2_1, has small values for the heteroscedasticity indicators, and there is no clear evidence of nonnormality. For their model $R^2 = .9799$, whereas $r^2_{Y\hat{Y}_{\text{raw}}(\text{BT})} = .9795$ for the model obtained using the two-stage procedure that has only two regressors.

Knowing that there is the potential for obtaining a slightly better fit, assume that we now consider the model where Y is not transformed, and X^2_1 and $X^{1.6}_2$ are used—another model that is suggested by Table 6.14. This produces a slightly better model than the one obtained using $Y^{0.4}$. Specifically, we obtain $\hat{\sigma}^2 = 1.17$, and the indicators, of course, have virtually the same values as those given in Table 6.14.

We need to perform some additional checks before settling upon this model, however. Influential data diagnostics are not part of this two-stage procedure, so it would be desirable to construct added variable plots before accepting a particular model. As mentioned in Section 6.1, Cook (1979) discovered that the need for a quadratic term in X_1 (in addition to the linear term) is signaled by a single point (point 2 in Table 6.1), and this becomes apparent when the added variable plot is constructed for X_1^2.

We might expect the same thing to happen with our model since we are also using X_1^2, but the added variable plots for X_1^2 and $X_2^{1.6}$ have no abnormalities, and the t-statistics for both terms are still large after the point is deleted.

Before using either of our two models, it would be a good idea to construct a simulated envelope for the residuals (see Section 2.1.2.1) as a further check on normality. Twenty simulated envelopes were constructed for each of the two models, and there was no evidence of nonnormality.

6.7 OTHER TRANSFORMATION METHODS

In a much-referenced paper, Breiman and Friedman (1985) presented the *alternating conditional expectations* procedure that is best known by its acronym, ACE. Although it can be used when the joint distribution of $Y, X_1, X_2, \ldots, X_p$ is known, it is essentially a nonparametric procedure. When the joint distribution is unknown (the usual case), the procedure minimizes the regression mean-squared error over all possible transformations of Y and the regressors. Thus, the emphasis is solely on fitting, with no guarantee that normality or homoscedasticity will exist. Other less-than-desirable features of ACE include the fact that the procedure can be undermined by outliers and may also be affected by the distributions of the regressors. Other anomalies of ACE are discussed by Hastie and Tibshirani (1990, pp. 184–186). See also Tibshirani (1988) and the comments of the discussants of Breiman and Friedman (1985). Problems can also ensue when the error term has a nonnormal distribution, as is illustrated by Statistical Sciences, Inc. (1991, p. 16–10).

Tibshirani (1988) presents an alternative approach that has the acronym AVAS (for *additivity and variance-stabilizing transformation*) and conjectures that ACE may be more suitable for correlation analysis than for regression. See also Chapter 7 of Hastie and Tibshirani (1990) for additional reading on AVAS, ACE, and transformations in general.

Both procedures seek to obtain a good-fitting model with additive regressors and an additive error term. The essential difference between ACE and AVAS is that the latter uses a variance-stabilizing transformation for the response variable. Approximate normality is not assured, however, so normality would have to be assessed for the selected transformation.

It should be noted that these two procedures are almost totally graphical, as no regression summary statistics are available for the transformed model that would result from using the transformation that is suggested by each graph. The user of either procedure must determine the transformation of Y by looking at the scatter plot of the transformed Y plotted against Y, and the transformation of each regressor would similarly be determined from the plot of each transformed regressor plotted against the regressor. Thus, the determination of each transformation is made subjectively, and a user would do well to have a set of graphs of nonlinear functions to use as a guide in matching the ACE and AVAS graphs with known nonlinear functions.

Also, we cannot easily determine the quality of the fit when there is more than one regressor, since the only R^2 value that is produced is the one that results from using the transformed values, and in order to (approximately) capture that R^2 value we would have to determine the transformations that produce approximately the transformed values. (In simple regression, a plot of the transformed Y against the transformed X will give us an indication of whether or not the fit is adequate, but it will be helpful to have one or more numerical measures.) It would be desirable to have $r^2_{Y\hat{Y}_{\text{raw}}}$ to view for each transformation, and also to have some indication as to whether the assumptions of homoscedasticity and normality are at least approximately met.

Thus, the advantages of the two-stage procedure relative to ACE and AVAS are due in part to the additional information that is available, such as being able to see the worth of a transformation in terms of $r^2_{Y\hat{Y}_{\text{raw}}}$, in addition to being able to see if *both* assumptions appear to be met. Other advantages include the ability to see how various transformations compare in terms of fit and in regard to the assumptions. This can be helpful if an experimenter observes that a transformation that does have physical meaning produces results that are almost as good as transformations that do not have physical meaning. A disadvantage of the proposed procedure is that only power transformations are considered for both Y and the regressors. The procedure could be easily modified to test other possible transformations of Y, but Box–Tidwell would have to be replaced by some other method if other types of transformations of the regressors are to be considered. It can be shown that both ACE and AVAS have trouble with the Minitab tree data and the stack loss data, in that a good-fitting model with acceptable values for the nonnormality and heteroscedasticity indicators is not produced for either data set.

Neither ACE nor AVAS is widely available in statistical software packages; see Section 6.9 for details.

6.7.1. Transform Both Sides

As mentioned in Section 2.3.4, another transformation approach is the transform both sides (TBS) approach advocated by Carroll and Ruppert (1984, 1988). With that approach it is assumed that a model that provides a satisfactory fit has already been identified, but there is a problem with nonnormality and/or heteroscedasticity. Then Y is transformed as $Y^{(\lambda)}$ so that the assumptions of normality and homoscedasticity are approximately met (if possible), and the same transformation is then applied to the right side of the model, thus creating a nonlinear regression model. Specifically, the model is

$$Y^{(\lambda)} = (\beta_0 + \beta_1 X_1 + \beta_2 X_2 + \cdots + \beta_p X_p)^{(\lambda)} + \epsilon$$

Nonlinear regression models are covered in detail in Chapter 13. Consequently, we will not discuss the mechanics of parameter estimation at this time.

The reader may wonder why we cannot simply transform each regressor in the same manner in which we transform Y. Certainly this is one way to transform each side, and Carroll and Ruppert (1988, p. 119) give conditions under which this approach has been used when there is a single regressor. In general, transforming each regressor individually cannot be relied upon to retrieve the R^2 value that existed before the transformation.

The two-stage procedure discussed in Section 6.6 has some advantages over the TBS approach. First, there is no assumption with the former that a good-fitting model has been found, so the two-stage procedure can be used to identify a good-fitting model if one has not been discovered. The two-stage procedure also produces a *linear* regression model, so there is the advantage in terms of simplicity that a linear model has over a nonlinear model.

But what if the two-stage procedure does not identify transformations of Y and/or the regressors such that the assumptions seem to be approximately met? Recall that this happened with the Minitab tree data. One possibility would be to start with a model for which the assumptions appear not to be seriously violated, and apply a Box–Cox transformation. Assume that a transformation of Y is identified so that the assumptions appear to be met. If there is a considerable drop in the value of $r^2_{Y\hat{Y}_{\text{raw}}}$, then the TBS approach might be applied, with the same transformation applied to the right side of the equation (excluding, of course, the error term).

6.8 TRANSFORMATION DIAGNOSTICS

In Section 6.1 an added variable plot for the constructed variable $X \log(X)$, using symbols that represent the amount of (linear) influence of each point, was presented as one method of determining whether the signal for a transformation is due to only one or two observations. In this section we briefly survey other developments and also illustrate the importance of using diagnostic procedures *after* a transformation has been made.

The diagnostics that have been proposed for assessing influence on a transformation parameter can be classified as (1) deletion diagnostics and (2) data perturbation. The former is self-explanatory and has been used in preceding chapters. The latter, which has been referred to as *local influence* in the literature, involves making small changes in Y and/or the regressors and may be viewed as a follow-up to (1). That is, if a particular observation is found to be influential, an experimenter might then wish to determine the coordinate(s) that make the observation influential. Lawrance (1988) considers only the perturbation of Y, while indicating that the regressors might also be perturbed, and emphasizes that one advantage of the local influence approach is that it allows the assessment of influence for all observations considered together, as opposed to the one-at-a-time assessment of influential observations with some of the earlier deletion methods. An alternative local influence approach has been proposed by Tsai and Wu (1992).

Various deletion methods have been proposed, including those of Cook and Wang (1983) and Hinkley and Wang (1988). Tsai and Wu (1990) claim that these methods may not provide satisfactory results in the presence of outliers or influential data, however, and provide an alternative approach. The interested reader is referred to these papers for details.

We have a somewhat delicate problem regarding transformation diagnostics since the extreme observations will frequently be the ones that rightfully signal the need for a transformation, and by definition we will have only a few of them. Thus, if we decide that we do not want to allow only one or two points to signal the need for a transformation, we will frequently miss a necessary transformation.

6.8.1 Diagnostics after a Transformation

In addition to diagnostics for determining the influence of points in suggesting a transformation, we should also examine the influence of those points *after* a transformation has been used. We would expect that points that were influential before a transformation might not be influential after the transformation, however, because as noted by Cook and Wang (1983), the transformation is largely determined so as to accommodate the influential points.

In was suggested in Section 6.6.5 that one possible approach would be to construct a plot of the type used in Figure 6.2, with $\hat{Y}$ plotted against Y. This would show whether there are only a few points that have considerable influence on the value of R^2. Similarly, the size of the symbol in the $\hat{Y}-Y$ scatter plot might reflect the value of Cook's-D, DFBETAS, or some other statistic.

6.9 SOFTWARE

Capability for performing the types of transformations discussed in this chapter is available in differing degrees in the widely available statistical software packages.

PROC TRANSREG in SAS Software can be used for producing optimal and nonoptimal transformations. The former are obtained iteratively using the *method of alternating least squares* (Young, 1981). The search for optimal transformations is performed using a squared-error loss function, and, depending on what is requested, can produce transformations that are similar to those produced by ACE (SAS Institute, Inc., 1990, p. 1513).

Desired transformations may be produced in BMDP by using the TRANS-FORM paragraph in conjunction with one of the regression programs (such as 2R). A search for an optimal Box–Cox transformation can be performed.

Transformations may also be produced in a general way in SPSS, but there is no provision to allow a search for an optimal transformation, or for an optimal parameter value in a particular type of transformation.

Similarly, there is no transformation command in MINITAB, but desired transformations can be performed by writing macros, as was done to produce some of the tables in this chapter.

S-Plus does have separate procedures for ACE and AVAS, as described and illustrated in Statistical Sciences, Inc. (1991). As was stated in Section 6.7, ACE and AVAS are graphical procedures, so the graphs would have to be interpreted in order to identify the manner in which each variable should be transformed. For example, if the plot of the transformed Y against Y has a logarithmic configuration, then a log transform of Y would be used, and similarly for the regressors.

The Minitab macros BOXCOXA and BOXCOXAM can be used for the Box–Cox and combined Box–Cox and Box–Tidwell transformation approaches, respectively, and were used in producing Tables 6.5–6.9 and 6.11–6.14.

Dynamic graphics (Cook and Weisberg, 1989) are useful for determining transformations and for the other components of a regression analysis. The software that comes with Cook and Weisberg (1994b) can be used efficiently and effectively for determining transformations.

SUMMARY

We have examined some of the commonly used methods for deciding upon a transformation of Y and/or the regressors. The Box–Tidwell procedure should be viewed as an alternative to the regression plots that were given in Chapter 5. We saw in that chapter that all of the plots touted as being able to identify the appropriate functional form of a regressor can fail under various scenarios, and although the Box–Tidwell approach can also fail under certain conditions, it should generally be more reliable.

The modified Box–Cox approach, which consists of using auxiliary information combined with the standard Box–Cox approach, was seen as a useful first-stage procedure for assessing simultaneously normality and homoscedasticity, while examining the quality of the model in terms of $r^2_{Y\hat{Y}_{\mathrm{raw}}}$. Following this with the use of a Box–Tidwell approach in the second stage was seen to provide useful information, and this two-stage procedure will frequently identify a good model.

It is worth noting that none of the transformation procedures presented in this chapter may be used to identify the need for interaction terms. One way to identify the need for such terms is to use projection pursuit regression (Friedman and Stuetzle, 1981). Another approach is to allow the inclusion of interaction terms in a variable selection scheme, as is discussed in Section 7.6.

REFERENCES

Atkinson, A. C. (1985). *Plots, Transformations, and Regression.* New York: Oxford University Press.

Box, G. E. P. (1980). Sampling and Bayes inference in scientific modeling and robustness. *Journal of the Royal Statistical Society, Series A,* **143,** 383–404 (discussion: 404–430).

Box, G. E. P. and D. R. Cox (1964). An analysis of transformations. *Journal of the Royal Statistical Society, Series B,* **26,** 211–243 (discussion: 244–252).

Box, G. E. P. and G. S. Watson (1962). Robustness to non-normality of regression tests. *Biometrika,* **49,** 93–106.

Breiman, L. and J. H. Friedman (1985). Estimating optimal transformations for multiple regression and correlation. *Journal of the American Statistical Association,* **80,** 580–598 (discussion: 598–619).

Brownlee, K. A. (1965). *Statistical Theory and Methodology in Science and Engineering,* 2nd edition. New York: Wiley.

Carroll, R. J. and D. Ruppert (1984). Power transformations when fitting theoretical models to data. *Journal of the American Statistical Association,* **79,** 321–328.

Carroll, R. J. and D. Ruppert (1988). *Transformation and Weighting in Regression.* New York: Chapman and Hall.

Cook, R. D. (1979). Influential observations in regression. *Journal of the American Statistical Association,* **74,** 169–174.

Cook, R. D. and P. C. Wang (1983). Transformations and influential cases in regression. *Technometrics,* **25,** 337–343.

Cook, R. D. and S. Weisberg (1982). *Residuals and Influence in Regression.* New York: Chapman and Hall.

Cook, R. D. and S. Weisberg (1989). Regression diagnostics with dynamic graphics. *Technometrics,* **31,** 277–291 (discussion: 293–311).

Cook, R. D. and S. Weisberg (1994a). Transforming a response variable for linearity *Biometrika,* **81,** 731–737.

Cook, R. D. and S. Weisberg (1994b). *An Introduction to Regression Graphics.* New York: Wiley.

Daniel, C. and F. S. Wood (1980). *Fitting Equations to Data,* 2nd edition. New York: Wiley.

Draper, N. R. and H. Smith (1981). *Applied Regression Analysis,* 2nd edition. New York: Wiley.

Friedman, J. H. and W. Stuetzle (1981). Projection pursuit regression. *Journal of the American Statistical Association,* **76,** 817–823.

Hastie, T. J. and R. J. Tibshirani (1990). *Generalized Additive Models.* New York: Chapman and Hall.

Hernandez, F. and R. A. Johnson (1980). The large sample behavior of transformation to normality. *Journal of the American Statistical Association,* **75,** 855–861.

Hinkley, D. V. and S. Wang (1988). More about transformations and influential cases in regression. *Technometrics,* **30,** 435–440.

Kvålseth, T. O. (1985). Cautionary note about R^2. *The American Statistician,* **39,** 279–285.

Lawrance, A. J. (1988). Regression transformation diagnostics using local influence. *Journal of the American Statistical Association,* **83,** 1067–1072.

Montgomery, D. C. and E. A. Peck (1992). *Introduction to Linear Regression Analysis,* 2nd edition. New York: Wiley.

Ruppert D. and B. Aldershof (1989). Transformations to symmetry and homoscedasticity. *Journal of the American Statistical Association*, **84**, 437–446.

Ryan, B. F., B. L. Joiner, and T. A. Ryan, Jr. (1985). *Minitab Handbook*, 2nd edition. North Scituate, MA: Duxbury Press.

SAS Institute, Inc. (1990). *SAS/STAT User's Guide, Version 6*, 4th edition, Volume 2. Cary, NC: SAS Institute, Inc.

Scott, A. and C. Wild (1991). Transformations and R^2. *The American Statistician*, **45**, 127–129.

Seber, G. A. F. (1977). *Linear Regression Analysis*. New York: Wiley.

Statistical Sciences, Inc. (1991). *S-Plus for DOS, User's Manual*, Vol. 2, Version 2. Seattle, WA: Statistical Sciences, Inc.

Tibshirani, R. (1988). Estimating transformations for regression via additivity and variance stabilization. *Journal of the American Statistical Association*, **83**, 394–405.

Tsai, C.-L. and X. Wu (1990). Diagnostics in transformation and weighted regression. *Technometrics*, **32**, 315–322.

Tsai, C.-L. and X. Wu (1992). Transformation-model diagnostics. *Technometrics*, **34**, 197–202.

Tukey, J. W. (1957). On the comparative anatomy of transformations. *Annals of Mathematical Statistics*, **28**, 602–632.

Wilson, E. B. and M. M. Hilferty (1931). The distribution of chi-square. *Proceedings of the National Academy of Sciences, Washington*, **17**, 684–688.

Young, F. W. (1981). Quantitative analysis of qualitative data. *Psychometrika*, **46**, 357–388.

EXERCISES

6.1. Explain why it is necessary to compute R^2_{raw} (or $r^2_{Y\hat{Y}_{\text{raw}}}$) instead of using R^2 for the transformed model whenever Y is transformed.

6.2. Assume that a model with transformed regressors has been identified, and the normality and homoscedasticity assumptions appear to be met. What plot(s) should be constructed before the model is used?

6.3. The Y values (given on the following page) are not the values that have been simulated using a particular model. Can you determine how Y must be transformed so as to give the true model?

6.4. Construct the plot of the standardized residuals against X_2 for the tree data in Table 6.9 to show that the plot shows evidence of heteroscedasticity when the transformation $Y^{1/3}$ is used.

Y	X_1	X_2
6.46	28	33
5.65	21	25
6.02	22	35
5.60	17	35
5.47	16	35
6.32	28	24
5.80	22	29
5.52	22	22
6.05	28	26
6.10	29	24
5.31	17	25
5.21	15	33
5.19	19	21
5.80	17	39
5.84	22	31
6.06	25	38
5.75	24	20
4.91	15	25

6.5. Assume that a practitioner decides to use the two-stage procedure discussed in Section 6.6. When will it not be necessary to use the second stage?

6.6. Explain the consequences that could result if the algorithm ACE is used to obtain a transformation of Y and/or the regressors, and confidence intervals, prediction intervals, and hypothesis tests are subsequently constructed.

6.7. It was mentioned in Section 6.7.1 that the combined Box–Cox and Box–Tidwell approach might be followed by the use of the modified Box–Cox approach for a specific set of regressor transformations in an effort to see if nonnormality and/or heteroscedasticity that seems to be present in all models identified by the two-stage procedure might be removed. Examine this possibility for the Minitab tree data by using the regressor transformations suggested in Table 6.11, searching for a different value of λ in increments of .05 from $\log(Y)$. Is an acceptable model apparent? Would the results likely have been different if you had used logarithmic terms in both regressors?

6.8. Sketch a plot of the standardized residuals graphed against $\hat{Y}$ that would produce a value of SPREAD-RATIO that is very close to 1, and values for r_{H_1} and r_{H_2} that far exceed 0.1. Then produce a sketch where r_{H_1}

is approximately 0 and SPREAD-RATIO is much greater than 1. What does this suggest about the need to use more than one numerical measure of heteroscedasticity?

6.9. Consider the stack loss data discussed in Section 6.6.5 and given in Table 6.1. If available, use SYSTAT (or other software) to construct an influence-symbol plot of $\hat{Y}$ against Y for the quadratic model suggested by Daniel and Wood, and the model using the terms X_1^2 and $X_2^{1.6}$ that was discussed in Section 6.6.5. Does the plot for the quadratic model corroborate the observation made in Section 6.1 that there is a single point signaling the need for a quadratic term? (How can that point be identified from the plot?) Do any points exert a major contribution to the value of $r_{Y\hat{Y}_{raw}}^2$ for the other model? If so, what action, if any, should be taken? Construct the added variable plots for each of the transformed regressors in this model and determine if the message from the scatter plot essentially agrees with the messages from the two added variable plots.

6.10. Obtain the inverse response plot for the data in Table 6.2. Make the appropriate transformation (from Section 6.2.1), then transform the predicted values back to the original scale and compute $r_{Y\hat{Y}_{raw}}^2$. Compare this value with the R^2 value obtained before the transformation and explain why the appropriate transformation has very little effect on the quality of the fit.

CHAPTER 7

Selection of Regressors

In preceding chapters we have considered graphical and numerical techniques for determining transformations of regressors. This presupposes that we have already identified all of the regressors that should be used in the model, in some form. This raises the question as to how we first determine this set of regressors. In particular, should we use all of the regressors that seem relevant or should we use some subset of this larger set?

Intuitively, it would seem desirable to use as regressors all of the variables that appear to be relevant. Actually, we would *not* want to automatically use all of the available variables because the magnitude of $Var(\hat{Y})$ is influenced by the number of regressors that are used. Specifically, Walls and Weeks (1969) showed that $Var(\hat{Y})$ cannot decrease, and will almost certainly increase, when new regressors are added to a model. Similarly, it can be shown that the $Var(\hat{\beta_i})$ will also almost certainly increase as the number of regressors is increased (see Seber, 1977, p. 139).

Consequently, some procedure (or a combination of procedures) needs to be used to select a set of regressors. We will assume in Sections 7.1, 7.2, and 7.3 that only linear terms are needed in the model, and then we shall relax that assumption in Section 7.4. The latter is important because the linear form of a potential regressor might seem unimportant, whereas a nonlinear form of the regressor could make a substantial contribution to the model.

Partial-F tests and (equivalently) *t*-tests were illustrated in Section 4.1.2.1 for the case of two available regression variables. If, however, there were, say, 10 available variables, the variables would have to be ordered so that the best variable at each stage was tested for inclusion. The first variable to be tested for inclusion is determined by $\max_i r_{YX_i}$. In subsequent stages (assuming that the process does not terminate at the first stage—a somewhat unlikely scenario), the regressor to be tested for inclusion would be determined by $\max_i r_{YX_i \cdot X}$, where X denotes the variables presently in the model, X_i denotes a regressor not in the model, and $r_{YX_i \cdot X}$ is the partial correlation between Y and X_i, with the relationship of each adjusted for their respective relationships with X.

7.1   FORWARD SELECTION

This approach has been termed *forward selection* by Draper and Smith (1981) and others, and it has several shortcomings. First, if there are k available variables, there are only (at most) k possible models that are examined with this approach, whereas there are $2^k - 1$ possible subsets, in general. (Each variable can be either in or out of the model, and we exclude the case where they are all out, thus producing $2^k - 1$ possible subsets.)

Accordingly, when p is large, only a small percentage of possible subset models are examined, and there is no guarantee that the subset that is used will satisfy certain optimality criteria (such as maximum R^2 for a given subset size). See also the discussion in the chapter summary.

Furthermore, the partial-F (or t) tests are correlated, so the significance level associated with the set of hypothesis tests used in determining the model is unknown. The significance levels are also unknown because the best variable at each stage is tested for model inclusion, whereas one of the remaining candidate variables would have to be picked arbitrarily in order to have a true significance level for any one particular test. This uncertainty should not preclude the use of hypothesis tests in subset selection procedures, however. It simply means that some thought should be given to the selection of the significance level for each test. There has been much discussion of this problem in the literature regarding subset selection (and not restricted to just forward selection), but definitive guidelines are lacking, especially for the case of correlated regressors.

Papers in which this problem has been discussed include Pope and Webster (1972), Freedman (1983), Butler (1984), Pinault (1988), and Freedman and Pee (1989). The message that emerges is that there can be a considerable difference between the actual significance levels and the nominal levels when the ratio of the sample size to the number of regressors is small. Therefore, it is desirable to try to keep this ratio reasonably large, if possible. (Recall the rough rule-of-thumb given in Section 1.5.4 that the ratio should be at least $10:1$, whenever possible.)

The difference between the actual and nominal levels will also increase as the number of variables that have entered the model increases, as discussed by Butler (1984). Because of this result, Butler (1984) suggests that the nominal significance level be increased as more variables enter the model, as the nominal significance level is closer to the true level when the number of variables in the model is small, with $p = k/2$ given as somewhat of a line of demarcation.

Butler (1984) supports the use of forward selection, but the procedure has been found to be inferior to other selection procedures [see, e.g., Draper and Smith (1981, p. 308) or Seber (1977, p. 217)]. One condition under which forward selection will fail is when variables are jointly significant, but not individually significant. In particular, recall the data in Table 5.1, which had neither of the two variables highly correlated with Y, although R^2 was virtually 1.0 when both were used in the model. When forward selection is applied to that data set, the better of the two variables, X_2, barely enters the model, using $t = 2$ as a rough cutoff, as the value of the t-statistic is only 2.142. (If we had arbitrarily selected X_2, we would then have a valid hypothesis test with the nominal level the same as the true level, and the calculated value of 2.142 could then be compared directly to the .05 significance level t-value of 2.101.)

With minor modification of the data in Table 5.1, we may obtain a two-regressor model with $R^2 = .949$, with neither variable entering the model with forward selection, using $t = 2$ as the cutoff. With a larger number of candidate variables we might also encounter subsets for which "the whole is greater than the sum of the parts" (see Section 5.8), which could similarly have deleterious effects upon the performance of the forward selection approach.

7.2 BACKWARD ELIMINATION

A related procedure is *backward elimination*, in which the starting point is the model with all of the available variables. The partial-F tests are then used to determine if the *worst* variable at each testing stage can be deleted, with "worst" determined by $\min_i r_{YX_i \cdot \mathbf{x}}$, where the latter is as defined in Section 7.1. Although many data analysts prefer backward elimination to forward selection, the former has the same general shortcomings as the latter: Not all subset models are examined, and the hypothesis tests are correlated.

7.3 STEPWISE REGRESSION

Stepwise regression is a combination of these two procedures. The procedure starts in "forward mode," but unlike forward selection, testing is performed after each regressor has been entered to see if any of the other regressors in the model can be deleted. The procedure terminates when no regressor can be added or deleted. Although this procedure also has the aforementioned shortcomings of correlated hypothesis tests and the failure to examine all possible models, stepwise regression can be used where there are so many variables (say, more than 50) that it is impractical to consider implicitly or explicitly all possible subsets.

The results that it produces can be misleading, however, since all possible subsets are not considered, as will be seen in Section 7.6.

7.3.1 Significance Levels

One determination that must be made when stepwise regression is used is the choice of (nominal) significance levels for the hypothesis tests. As in forward selection, the true significance levels are unknown, since the tests are not independent. The actual significance levels are determined by the specification of F_{IN} and F_{OUT} when statistical software is used for the computations. The general idea is to choose these two F-values in such a way that the nominal significance levels will translate into desired actual significance levels, while recognizing that this cannot be done exactly.

Some statistical software have default values for which $F_{IN} = F_{OUT}$, but it isn't mandatory that they be equal. It is necessary, however, to have $F_{OUT} \leq F_{IN}$; otherwise, a variable that was added in a particular step could then be removed in the next step.

7.4   ALL POSSIBLE REGRESSIONS

The consideration of all possible subsets is generally referred to as *all possible regressions*. When the number of available variables is small, it is practical to compute all possible regressions explicitly. When k is larger than, say, 5 or 6, however, it is desirable to have all possible regressions computed *implicitly*.

To illustrate the latter, let's assume that we wish to determine the subset of each size that has the largest R^2 value, and that we have 10 available variables. Of the 1023 subset models, we would expect that many of these would have low R^2 values simply because not all of the 10 regressors are apt to be highly correlated with Y. If these inferior subsets can be identified, then the computations need not be performed for them.

For example, if the model with only X_1 has a higher R^2 value than the model with both X_4 and X_5, then we know that the latter subset will be inferior, in terms of R^2, to all subsets (X_1, X_i), $i = 2, 3, \ldots, 10$. This is the general idea behind the *branch-and-bound algorithm* of Furnival and Wilson (1974), which can be used to efficiently (and implicitly) perform all possible regressions so as to identify the best subset(s) of each size (and/or overall) according to some criterion.

Obviously, the latter provides the user with more information than is available when forward selection, backward elimination, or stepwise regression is used. Another disadvantage of these last three is that they select a single subset, whereas it is unlikely that one subset will stand out as being clearly superior to all other subsets. It is preferable to provide the user with a list of good subsets and then allow the user to select a subset, possibly using nonstatistical considerations (e.g., costs, ease of measurement, etc.) in reaching a decision.

It should also be noted that our objective should be to obtain a good model for the *population* (univariate or multivariate depending on whether the regressors are fixed or random), not just for the sample data that we have. The all-possible-regressions approach utilizes only sample statistics, whereas the other three methods involve hypothesis tests relative to a population. As pointed out by Berk (1978), it is important that the four procedures be compared relative to known population values, as a comparison based solely on the sample data will show an inflated advantage for all possible regressions. Berk's comparison study showed the latter to have a very slight advantage over forward selection and backward elimination (stepwise regression was not included in the study).

7.4.1 Criteria

We may use one or more criteria in conjunction with all possible regressions. We might choose to look at the best subset(s) of each size in terms of R^2, or perhaps in terms of adjusted R^2. [Recall that the latter is a modification of R^2. It can be written in terms of R^2 as $R^2_{\text{adjusted}} = 1 - (1 - R^2)(n - 1)/(n - p - 1)$, with p denoting the number of parameters. Unlike R^2, R^2_{adjusted} can decrease as the number of regressors increases.] With the former we would have to determine the "point of diminishing returns," which might suggest (roughly) the number of regressors that should be used, whereas the latter would determine that point directly.

7.4.1.1 Mallows's C_p

Alternatively, one might use Mallow's C_p statistic. The statistic was developed by Colin Mallows (and named for another prominent statistician, Cuthbert Daniel) as a vehicle for determining a reasonable range for the number of regressors to use. It was not intended to be used to select a particular subset, although it is often used for that purpose. Originally presented at a regional statistics conference in 1964, the C_p statistic is described extensively in Mallows (1973) and Seber (1977, pp. 364–369).

The statistic is defined here as

$$C_p = \frac{\text{SSE}_p}{\hat{\sigma}^2_{\text{FULL}}} - (n - 2p) \tag{7.1}$$

where p is defined here, as in previous chapters, as the number of parameters in the model, $\hat{\sigma}^2_{\text{FULL}}$ is the estimate of σ^2 obtained when all of the available variables are used, SSE_p is the residual sum of squares for a particular subset model with $m = p - 1$ regressors, and n is the number of observations. [Note: We are tacitly assuming that the full model is the correct model so that $\hat{\sigma}^2$ is an unbiased estimator of σ^2. Prior information on σ^2 might also be used in obtaining an estimate, but we will assume in this section that σ^2 is to be estimated using all of the available regressors. Mallows (1973) did not define C_p with σ^2 estimated in any specific manner, but the possible use of $\hat{\sigma}^2_{\text{FULL}}$ was implied, and in fact this is the customary approach.]

The reason that the C_p statistic can be used for determining the approximate number of regressors to use can be explained as follows. If we use a subset model and assume that the full model is the correct model, then $E(Y_s) \neq \xi$, where $E(Y) = \xi$ for the full (true) model, and Y_s denotes the left side of a subset model. Thus, there is bias in the subset model—the least squares estimators are biased so that $E(\hat{Y}_s) \neq \xi$, in addition to $E(Y_s) \neq \xi$. This bias might be more than offset, however, by a reduction in the variance of the predicted response resulting from use of the subset model.

When two estimators are to be compared and at least one of the two estimators is biased, it is reasonable to use *mean-squared error* (variance plus squared bias) to compare the two estimators.

Here we are comparing two models over n data points rather than making a comparison of estimators, however, so a slightly different approach is required. The *total squared error* is defined as $E[\sum_{i=1}^{n} (\hat{Y}_{s,i} - \xi_i)^2]$, and we may easily show that

$$E\left[\sum_{i=1}^{n} (\hat{Y}_{s,i} - \xi_i)^2\right] \equiv E\left[\sum_{i=1}^{n} (\hat{Y}_{s,i} - \eta_{s,i} + \eta_{s,i} - \xi_i)^2\right]$$

$$= E\left[\sum_{i=1}^{n} (\hat{Y}_{s,i} - \eta_{s,i})^2\right] + \sum_{i=1}^{n} (\eta_{s,i} - \xi_i)^2$$

since $\eta_{s,i}$ here denotes $E(\hat{Y}_{s,i})$, so the middle term vanishes, $E(\eta_{s,i} - \xi_i)^2 = (\eta_{s,i} - \xi_i)^2$, and we will use SSB (sum of squares due to bias) to denote $\sum_{i=1}^{n} (\eta_{s,i} - \xi_i)^2$. By definition,

$$E\left[\sum_{i=1}^{n} (\hat{Y}_{s,i} - \eta_{s,i})^2\right] = \sum_{i=1}^{n} \text{Var}(\hat{Y}_{s,i})$$

It is desirable to divide by σ^2 so as to have a unit-free quantity, so putting these simplified components together and then dividing by σ^2 produces the *total standardized squared error* given by

$$T = \frac{\text{SSB} + \sum_{i=1}^{n} \text{Var}(\hat{Y}_{s,i})}{\sigma^2}$$

$$= \frac{\text{SSB}}{\sigma^2} + p$$

since it can be shown (see the chapter appendix) that $\sum_{i=1}^{n} \text{Var}(\hat{Y}_{s,i}) = p\sigma^2$.

When the model is correct, $E(\hat{\sigma}^2) = E[\sum (Y - \hat{Y})^2/(n - p)] = \sigma^2$, so that $E[\sum (Y - \hat{Y})^2] \equiv E(\text{SSE}) = (n-p)\sigma^2$. When a subset model is used, however, $\hat{\sigma}^2$ will be a biased estimator of σ^2, just as the least squares estimators are biased. Specifically, $E(\hat{\sigma}^2) = \text{SSB}/(n - p) + \sigma^2$, as is shown in the chapter appendix.

Thus, SSB $= E(\text{SSE}) - (n - p)\sigma^2$, so SSB could be estimated as $\widehat{\text{SSB}} = \text{SSE} - (n - p)\hat{\sigma}^2$. It follows that we would estimate T as

$$\hat{T} = \frac{\text{SSE} - (n - p)\hat{\sigma}^2}{\hat{\sigma}^2} + p$$

$$= \frac{\text{SSE}}{\hat{\sigma}^2} - (n - p) + p$$

$$= \frac{\text{SSE}}{\hat{\sigma}^2} - (n - 2p) \qquad (7.2)$$

As indicated previously, one approach would be to estimate σ^2 using the full model, so that $\hat{\sigma}^2 = \hat{\sigma}^2_{\text{FULL}}$. With such an estimator we then have Eq. (7.1), so C_p is an estimator of the total standardized squared error.

If a subset model were the true model (and $\hat{\sigma}^2 = \sigma^2$), we would have

$$E(C_p) = \frac{E(\text{SSE}_p)}{\sigma^2} - (n - 2p)$$

$$= \frac{(n - p)\sigma^2}{\sigma^2} - (n - 2p)$$

$$= p$$

Thus, a reasonable starting point would be to examine subsets for which C_p is close to p, and in particular, less than p.

One approach would be to plot C_p against p and draw the line $C_p = p$ on the chart, with subsets that have small C_p values then plotted on the chart. This type of display can be found in many regression texts including Draper and Smith (1981, p. 302), Neter et al. (1989, p. 449), Daniel and Wood (1980, p. 88), and Montgomery and Peck (1992, p. 274). Here the emphasis is on looking at those subsets that have C_p close to p. Should we look at subsets that have $C_p \doteq p$, or should we look at those subsets that have the smallest C_p values? Draper and Smith (1981, p. 300) suggest that this is largely a matter of preference, whereas Mallows (1973) argues against automatically selecting the subset with the smallest C_p value.

If, however, we have two subsets of different sizes and $C_p \doteq p$ for each subset, it will frequently be desirable to use the smaller of the two subsets. The logic behind such a decision is as follows. If we select a (larger) subset with a larger C_p value (where C_p has increased by more than the increase in p), we have in essence proceeded past the point of diminishing returns in terms of R^2. Recall from Section 7.4.1 that R^2_{adjusted} can be used to determine this point, and Kennard (1971) has noted the relationship between R^2_{adjusted} and C_p. [See also Seber (1977, p. 368).] In particular, it can be shown (as the reader is asked to do in Exercise 7.12) that if the relationship $C_p = p + a$ holds for two values of p (where a is a constant), the smaller value of R^2_{adjusted} will occur at the larger of the two values of p.

When we use all of the available variables, we have $C_p = p$ (as can be easily shown). This will often be the smallest value of C_p, but that doesn't necessarily mean that we should use all of the variables. In particular, this can happen when the full model has a very high R^2 value (such as $R^2 > .99$), and σ^2 is estimated from the full model.

Although it would seem to be highly desirable to use the full model when it has a very high R^2 value, such an R^2 value will cause $\hat{\sigma}^2$ to be quite small, which will in turn magnify the differences between subsets in terms of SSE. This follows because SSE is divided by $\hat{\sigma}^2$ in the computation of C_p.

Specifically, C_{p+1} will be smaller than C_p if $(\text{SSE}_p - \text{SSE}_{p+1})/\hat{\sigma}^2 > 2$, which could happen solely because of a very small value of $\hat{\sigma}^2$. This underscores the fact that C_p should be used only as a guide in selecting the best subset size.

The difference between the two SSE values divided by $\hat{\sigma}^2$ is the usual F-statistic for testing the addition of a new regressor, with the minor difference that σ^2 is typically estimated using the larger of the two models, whereas here it is estimated from the model using all of the available variables. Thus, we may say that seeking the minimum value of C_p is roughly equivalent to using an F_{IN} value of 2.

7.4.1.1.1 C_p and Influential Data

Since C_p can be greatly affected by influential data (as can the other criteria), it can be worthwhile to determine the influence of individual observations on C_p. A method for doing so is given by Weisberg (1981).

Specifically, Weisberg defines C_{pi} as

$$C_{pi} = \frac{(\hat{y}_i - \hat{Y}_i)^2}{\hat{\sigma}^2} + v_{ii} - (u_{ii} - v_{ii})$$

where $v_{ii} = \text{Var}(\hat{y}_i)/\sigma^2$, with $\hat{y}_i$ denoting the predicted value of Y from the subset model, $\hat{Y}_i$ is the predicted value of Y from the full model, and u_{ii} is the ith diagonal element of

$$U = [\mathbf{X_P} \quad \mathbf{X_2}] \begin{bmatrix} \mathbf{X_P'X_P} & \mathbf{X_P'X_2} \\ \mathbf{X_2'X_P} & \mathbf{X_2'X_2} \end{bmatrix}^{-1} \begin{bmatrix} \mathbf{X_P'} \\ \mathbf{X_2'} \end{bmatrix}$$

where $\mathbf{X_2}$ is a matrix that contains the candidate regressors that are in the full model but not in the subset model, and $\mathbf{X_p}$ is the matrix for the subset model.

We may compare the C_{pi} with the v_{ii} (as suggested by Mallows, 1973), just as C_p can be compared with p. It can be shown that the C_{pi} and the v_{ii} sum to C_p and p, respectively. Since it is desirable to have $C_p \doteq p$, it is natural to compare each C_{pi} against its corresponding v_{ii}. As indicated by Weisberg (1981), we may also use the C_{pi} as influential data statistics. If use of the C_{pi} leads to identification of influential data points, what then? One possible approach would be to use bounded influence regression (see Section 11.8) to produce the values of SSE_p. But Ronchetti and Staudte (1994), in presenting a robust version of C_p, contend that unmodified selection procedures such as C_p should not be used when regression parameters are estimated robustly.

7.4.1.2 Minimum $\hat{\sigma}^2$

What if we wished to select the subset that minimizes $\hat{\sigma}^2$? This would be the same subset that maximizes R^2_{adjusted}, as can be seen by substituting $\hat{\sigma}^2 = \sum (Y - \hat{Y})^2 / (n-p)$ in the expression for R^2_{adjusted} that was given in Section 7.4.1, simplifying, and recognizing the components of the result that do not vary with p (as the reader is asked to do in Exercise 7.13).

7.4.1.3 t-Statistics

Another approach is to select regressors based on the magnitude of the t-statistics. This approach can fail completely in the presence of correlated variables, however, as is shown in Section 7.5.

7.4.1.4 Other Criteria

There are several other criteria that have been proposed for subset selection. Perhaps the best known of these is Akaike's information criterion (AIC) (Akaike, 1974), but as indicated by Young (1982), it is equivalent to Mallows's C_p when σ^2 is known. See Young (1982) for information on AIC and other criteria.

7.5 EXAMPLES

It is important to recognize that variable selection is not necessarily a good strategy when the available variables are (highly) correlated, as small data perturbations can cause the results to differ greatly. This is not to imply that we should use all of the available regressors. Rather, we should recognize the limitations in trying to select a subset when the data are weak (multicollinear).

In a designed experiment we can generally select the values of the regressors in such a way that the regressors are orthogonal or near orthogonal, but as has been stated previously, regression is most frequently used when the regressors are random, and often highly correlated.

An alternative to either using variable selection or applying least squares to all of the available variables is discussed briefly in Section 7.7 and covered in detail in Chapter 12.

In this section we consider variable selection under the conditions of both orthogonal regressors and multicollinearity, with the latter being addressed since variable selection is frequently performed in practice in the presence of multicollinearity.

Assume first that we have four orthogonal regressors and we are trying to decide if all four should be used in the model. Since they are orthogonal, their respective contributions to the prediction of Y are the same whether they are used individually or in some combination. Hence, t-statistics have meaning in assessing the merits of each variable in the model, and we could initially include all of the available variables in the model and decide which ones to retain based upon the t-statistics.

When multicollinearity exists, however, the marginal contributions of the individual regressors are not easily assessed, and small perturbations in the data can greatly affect the subset that is obtained, as stated previously. [An attempt at determining the marginal contributions has been made by Chevan and Sutherland (1991), however.]

We first consider the data in Table 4.3, which were used in Section 4.1.2. When both X_1 and X_2 are used, we obtain $R^2 = .9948$, whereas using only X_1 produces $R^2 = .9881$. Thus, there is very little difference, but the fact that R^2 is so close to 1.0 when both X_1 and X_2 are used will cause the values of the statistics that have been presented in preceding sections to give "overwhelming" evidence that both regressors should be used. For example, the value of C_p drops from 23.2 to 3.0 when X_2 is added to the model, and the t-statistic for determining if X_2 should be added has a value of -4.72. Such a large C_p value when only X_1 is in the model suggests that this model is a relatively poor one, but it would be difficult to make such an argument when $R^2 = .9881$. Common sense suggests that we use only X_1 and ignore the C_p values and the value of the t-statistic.

When there is at least a moderate number of available variables, the graph of the smallest value of C_p for each subset size plotted against p will frequently be U-shaped. For example, for the stack loss data of Table 6.1, $C_p = 2.9$ for the subset (X_1, X_2), which is less than the C_p values for the best subsets for the other two sizes: $C_p = 13.3$ for (X_1) and, by definition, $C_p = 4$ for (X_1, X_2, X_3). (A U shape will be shown also to exist for the stack loss data when nonlinear terms are considered in the next section.)

7.6 VARIABLE SELECTION FOR NONLINEAR TERMS

In the preceding sections the emphasis was on obtaining a good subset of regressors *in linear form*. If the true model contains a nonlinear term in one of the available variables in addition to the linear term, we may be omitting an important variable if the nonlinear term is not highly correlated with the linear term. How can we protect against this possibility? One approach would be to use $X_i \ln(X_i)$ as a proxy for a potentially important nonlinear term in each regressor. This, of course, could create some serious multicollinearity problems (and remember that variable selection is risky when multicollinearity exists), but the approach will frequently be useful, especially if there is only a small number of available variables. If $X_i \ln(X_i)$ is selected by some selection procedure, then the Box–Tidwell procedure might be applied to X_i to determine if a power transformation of X_i would be suitable.

It would also be a good idea to include interaction (i.e., cross-product) terms as candidate regressors, provided that the number of available regressors is only moderate. [The number of such cross-product terms is $k(k-1)/2$, for k available regressors.]

To illustrate, we will apply stepwise regression to the Minitab tree data that were given in Table 6.10. Since $k = 2$, there are five candidate regressors: X_1, X_2, $X_1 \ln(X_1)$, $X_2 \ln(X_2)$, and $X_1 X_2$. Using $F_{\text{IN}} = F_{\text{OUT}} = 4.0$, we obtain the subset $(X_1 X_2, X_2, X_1)$, with the three selected variables entered in the order in which they are listed, and $R^2 = .976$. (This subset is also identified as best when all possible regressions are performed.)

We should not be surprised that the cross-product term would enter first, since the volume of a tree is proportional to $X_1^2 X_2$. Having a term that is similar to the latter in the model, however, should cause us to suspect that there may be a problem with heteroscedasticity, since it was observed in Section 6.6.4.1 that there was a severe heteroscedasticity problem when $X_1^2 X_2$ was the sole regressor.

And in fact this is the case because $r_{H_1} = .305$, $r_{H_2} = .354$, and SPREAD-RATIO $= 6.44$ (see Section 6.4 for the definition of these statistics). The existence of heteroscedasticity is confirmed when the standardized residuals are plotted against X_2.

This illustrates why it is important to check the model assumptions when automated procedures such as stepwise regression are used to select a subset, since stepwise regression programs do not generally provide a check on the assumptions for the selected model.

It is also important to remember that influential observations may affect the subset that is selected, so it is a good idea to construct an added variable plot for each regressor that is selected by some automated procedure, whether the regressor is in nonlinear form or not. Another possibility would be to compute the C_{pi} values.

As a final example we consider the stack loss data in Table 6.1. Recall that the need for X_1^2 rested solely on a single observation, so we would hope that this would be apparent if we looked at the added variable plot and/or the C_{pi} values. A plot of the C_{pi} values against v_{ii} is given in Figure 7.1. Here the subset model is (X_1, X_2), the full model is (X_1, X_2, X_1^2), and observations 1, 3, 4, and 21 are not used. Since C_{p2} is much greater than v_{22}, it is of interest to determine why C_{p2} is so large. It can be shown that the difference is due primarily to the difference between $\hat{y}_2$ and $\hat{Y}_2$. Thus, the plot shows that the signal for a quadratic term in X_1 results from a single point, as has been stated previously.

Remember that it was also tacitly assumed in the analysis in Section 6.1 that X_3 was not of value, so we will see if influential data and the weakness of X_3

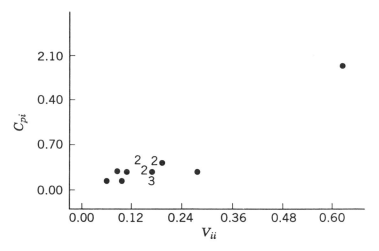

Figure 7.1 C_p plot for stack loss data, nonlinear terms included.

as a potential regressor show up in the analysis. In addition to all three linear terms, we will also consider all cross-product terms and terms in $X_i \ln(X_i)$ as potential regressor terms.

When stepwise regression is applied, using all nine potential regressors with $F_{IN} = F_{OUT} = 4$, the selected subset is $(X_1 \ln(X_1), X_1X_2)$. This subset has $R^2 = .978$ and $\hat{\sigma}^2 = 1.28$. The Box–Tidwell procedure could then be used to determine the best nonlinear transformation, fixing X_1X_2 as the other regressor. (Note that, as in Chapter 6, we are using only the 17 good data points.)

In general, however, if one or more interaction terms were selected, then it would be desirable to see if Y could be transformed so as to render the interaction term(s) nonsignificant. It is generally preferable to have an additive model (i.e., one without interaction terms), and the presence of interaction terms may mean that the wrong metric (of Y) is being used [see, e.g., Box (1988)].

There is a problem in trying to apply all possible regressions in this example because the correlation between each linear term and the corresponding $X_i \ln(X_i)$ term is virtually 1.0. Excluding the term $X_i \ln(X_i)$ from entering the model for $i = 2, 3$ permits the best subsets to be determined (using Minitab). The results for the best two subsets of each size in terms of R^2 are given in Table 7.1. Notice that X_3 does not show up in a subset model until there are at least five variables in the subset, thus confirming the belief that X_3 is not of value. Notice also the two negative values of C_p. This will rarely occur, and its occurrence means that the subset model is far superior to the full model. Notice the constant change in the C_p values for the best subset of sizes 4, 5, 6, and 7. This is the largest possible increase that can occur for consecutive subset sizes, and will occur only when adding a variable to a subset causes no reduction in SSE. If we compare this benchmark figure of 2.0 to the changes in going from

Table 7.1   All Possible Regressions Applied to Stack Loss Data

Number of Variables	R^2	C_p	$\hat{\sigma}$	Subset[a]
1	95.4	8.1	1.5897	{4}
1	95.0	9.7	1.6483	{1}
2	97.9	−1.5	1.1064	{2, 12}
2	97.8	−1.0	1.1327	{4, 12}
3	98.0	0.1	1.1214	{12, 13, 23}
3	98.0	0.2	1.1265	{1, 4, 12}
4	98.0	2.0	1.1617	{1, 2, 12, 13}
4	98.0	2.0	1.1618	{2, 4, 12, 13}
5	98.0	4.0	1.2122	{1, 2, 3, 12, 13}
5	98.0	4.0	1.2123	{2, 3, 4, 12, 13}
6	98.0	6.0	1.2713	{1, 2, 3, 12, 13, 23}
6	98.0	6.0	1.2713	{1, 2, 3, 4, 12, 13}
7	98.0	8.0	1.3399	{1, 2, 3, 4, 12, 13, 23}

[a] 1, 2, 3, represent X_1, X_2, and X_3, respectively, 12, 13, and 23 denote the three cross-product terms, and 4 represents $X_1 \ln(X_1)$.

two variables to three and from three variables to four, we see that using more than two regressors would not be productive. (And, of course, we receive the same message from the R^2 values.)

7.6.1   Negative C_p Values

To illustrate how a negative value of C_p can occur, assume that there are six available regressors but four of them are orthogonal to Y. For the subset consisting solely of the two regressors that are not orthogonal to Y, it can be shown using Eq. (7.1) that C_p equals -1. (In general, the lower bound on C_p is $2p - k - 1$, for k available regressors.) Notice that this says nothing about how good the other two regressors are; a very small or negative value of C_p could simply mean that there are sizable differences in the worth of the individual regressors. Thus, the value of C_p cannot be used as an indicator of the worth of a regression model.

For the stack loss data $R^2 = .979$ and $\hat{\sigma}^2 = 1.22$ for the subset $(X_2, X_1 X_2)$, so this subset is clearly of value. But, as stated earlier, it would be preferable not to have interaction terms in the model. How can we avoid that here since the best subsets are clearly those that have at least one interaction term? Notice that the smallest value of $\hat{\sigma}^2$ (obtainable) from Table 7.1 is 1.22, but smaller values were obtained for this data set in Section 6.6.5 *without* using an interaction term. Therefore, we might look at more subsets of each size and see how the subsets without interaction terms compare to the best subsets with interaction terms.

If we looked at the five best subsets of each size here, we would observe that the fifth best subset of size 2 is $\{X_1, X_2\}$, which has $R^2 = .973$. Since this is only slightly less than the best subset of size 2, which has an interaction term, we might consider applying the Box–Tidwell approach to each of these regressors, as was done in Section 6.6.5.

The point to be made is that we should not necessarily use what appears to be the best subset obtained from use of a subset selection program. A better model may be obtainable, and of course we should also check to make sure that the assumptions appear to be approximately met.

There is an upper bound on the number of variables that can be used in all possible regressions algorithms, so we would generally need fewer than 10 available variables if the algorithm is to be applied to the X_i, X_iX_j, and $X_i \ln(X_i)$. Stepwise regression programs generally allow many more variables, but then there is the distinct possibility that a misleading message may be emitted.

As mentioned in the summary of Chapter 6, another approach to detecting the need for interaction terms is to use projection pursuit regression.

7.7 MUST WE SELECT A SUBSET?

When multicollinearity exists, we do not want to use redundant variables, but we also do not want to accidentally delete any important variables. Since we should seek the smallest number of useful regressors so as to not unnecessarily inflate $\text{Var}(\hat{Y})$, variable selection is thus *seemingly* necessary, but difficult.

But this conclusion is based on the assumption that we will be using ordinary least squares. One alternative is to use *ridge regression* instead of least squares and to use (possibly) all of the available variables. This allows us to reduce $\text{Var}(\hat{Y})$ without having to reduce the number of variables. Ridge regression is covered in detail in Chapter 12.

Another reason to be wary of selecting a subset of regression variables is that noise variables can frequently be included in the chosen subset, especially when n is not large relative to the number of available variables, as discussed by Flack and Chang (1987).

7.8 MODEL VALIDATION

As discussed by Roecker (1991) and Miller (1984), the theory that supports statistics such as C_p is based on the assumption that model selection and parameter estimation are not performed using the same data. Since the same data are generally used for both purposes, relative performances of the various subset selection procedures must be determined numerically. As stated earlier, one simulation study was performed by Berk (1978); another study was performed by Roecker (1991).

One way to have the model determination and parameter estimation functions performed independently would be to split a data set in half and use each of the two halves for each function. This is what has often been done for model validation, but if the data set has been split into homogeneous parts, we would expect the model to fit each half approximately the same. Roecker (1991) also points out that there is a considerable loss in prediction efficiency when the splitting is performed and suggests that *not* splitting the data appears to be preferable.

Therefore, the best approach would be to use new data for model validation, but acquiring additional data is not always practical. If this were done, it would be necessary to check to make sure that the new data fall within the region occupied by the first set of data, so that extrapolation does not occur.

Picard and Berk (1990) provide a way to quantify the degree of extrapolation and point out that at least moderate extrapolation can be expected when the sample size is small and the regressors are random variables. This is intuitively apparent when we recognize that the original data can lie in a rather small region when multicollinearity exists. Nevertheless, it would be improper to attempt to validate a model by using data that are (far) outside the region occupied by the original data.

7.9 SOFTWARE

It is virtually imperative that we use specialized software for variable selection, rather than attempting to obtain the necessary statistics from software that has only general regression capabilities. Fortunately, the major statistical software packages have excellent capabilities for variable selection.

All possible regressions can be performed in MINITAB by using the BREG command, but only 20 "free predictors" (i.e., those not forced into the model) can be specified. The output gives the best subsets (up to five) of each size in terms of R^2 and also prints the corresponding values of R^2_{adjusted} and C_p. Stepwise regression can also be performed, using the STEPWISE command, and as many as 100 variables may be used. Forward selection and backward elimination may also be used by setting the appropriate values of FENTER and FREMOVE when STEPWISE is used.

Stepwise regression is performed in BMDP using program 2R (which is also the general all-purpose regression program in BMDP), and all possible regressions can be performed using program 9R. The latter can be used to select the best subsets in terms of R^2, R^2_{adjusted}, or C_p. A maximum of 10 subsets of each subset size can be identified for each of the three criteria, and the best 5 subsets overall in terms of C_p may also be identified.

The REGRESSION command in SPSS can be used for stepwise regression (and also forward selection and backward elimination), but all possible regressions cannot be performed.

PROC REG in SAS permits the use of forward, backward, and stepwise regression, in addition to all possible regressions. Also, a "maximum R^2 improvement" technique can be used that requires much less computer time than the method (originally PROC RSQUARE) that is used for computing all possible regressions. There is no guarantee, however, that the best subsets are selected in terms of the three criteria that have been discussed in conjunction with all possible regressions.

SYSTAT has the capability for stepwise regression, as well as forward selection and backward elimination, but all possible regressions cannot be performed.

SUMMARY

Regression equations will frequently contain at most five or six regressors, so variable selection methods are employed to select a reasonably small subset from a larger set of available regressors. The use of such methods can be very helpful when there are many potential regressors, but selecting regressors when there is multicollinearity can be difficult.

The various methods used for selecting a subset of regression variables have been discussed, in addition to the software that can be used to implement them. The all-possible-regressions approach provides the most information, but the subset selected using this approach is not guaranteed to be superior to the (possibly different) subsets selected using the other approaches, relative to the population(s) from which the data came.

But methods such as forward selection and backward elimination can perform very poorly relative to criteria such as R^2. Extreme examples of this are given by Berk (1978) and Miller (1984). In the former there is an example in which a subset of size 2 that is selected by both forward selection and backward elimination has $R^2 = .015$, whereas the best subset of size 2 in terms of R^2 has $R^2 = .99$. Another extreme example is given by Miller (1984), who states that in the physical sciences the difference of two variables is often a proxy for a rate of change in position or time. Thus, it is important to recognize the limitations of techniques such as forward selection and backward elimination and also to think about the possible existence of lurking variables (as discussed in Section 5.7).

Variable selection techniques are usually discussed relative to the selection of linear terms, but they may also be used to detect the need for nonlinear and interaction terms, as was discussed herein.

There is (apparently) no variable selection program that will reject a particular subset if the assumptions appear not to be met, so it is important for the user of such programs to remember to perform the appropriate checks. The latter should include checks on the assumptions of homoscedasticity and normality, as well as a check for influential observations.

Once a subset model (or any model) has been selected, it is advisable to collect new data, if possible, to validate the model before using it. (Model validation has often been performed by splitting a data set into two homogeneous halves, using one for model determination and the other for model validation. But since the data have been split into halves using a computer program that makes the halves very similar, it should not be surprising that the data half used for validation has supported the model that was obtained from the other half.)

Readers seeking additional information on selection of regressors are referred to Miller (1990) and to review papers such as Hocking (1976) and Miller (1984).

APPENDIX

7.A Derivation of $\sum \text{Var}(\hat{Y}_{s,i})$

As given in Eq. (3.8), $\text{Var}(\hat{\mathbf{Y}}) = \sigma^2 \mathbf{X}(\mathbf{X}'\mathbf{X})^{-1}\mathbf{X}'$. Using this result, it follows that $\sum \text{Var}(\hat{Y}_i) = \sigma^2 \text{tr}[\mathbf{X}(\mathbf{X}'\mathbf{X})^{-1}\mathbf{X}']$, where tr denotes *trace* of a matrix, which is defined to be the sum of the diagonal elements. For two matrices $\mathbf{A}$ and $\mathbf{B}$, $\text{tr}(\mathbf{AB}) = \text{tr}(\mathbf{BA})$. Using this result, it follows that $\text{tr}[\mathbf{X}(\mathbf{X}'\mathbf{X})^{-1}\mathbf{X}'] = \text{tr}[(\mathbf{X}'\mathbf{X})^{-1}\mathbf{X}'\mathbf{X}] = \text{tr}(\mathbf{I_P}) = p$, where $\mathbf{I_p}$ denotes the identity matrix with p diagonal elements. It then follows that $\sum \text{Var}(\hat{Y}) = p\sigma^2$.

7.B Derivation of $E(\hat{\sigma}^2)$ When the Model is Incorrect

We let $\hat{\sigma}_s^2$ denote the estimate of σ^2 obtained from the subset model. Since $\hat{\sigma}_s^2 = \sum (Y - \hat{Y}_s)^2/(n - p)$, where p denotes the number of parameters, we need to obtain $E\{\sum (Y - \hat{Y}_s)^2\}$. We proceed as follows.

$$E\left\{ \sum (Y - \hat{Y}_s)^2 \right\} = \sum \{E(Y - \hat{Y}_s)^2\}$$

and

$$
\begin{aligned}
E(Y - \hat{Y}_s)^2 &\equiv E\{(Y - \xi) - (\hat{Y}_s - \xi)\}^2 \\
&= E\{(Y - \xi)^2 + (\hat{Y}_s - \eta + \eta - \xi)^2\} - 2E(Y - \xi)(\hat{Y}_s - \eta) \\
&= E(Y - \xi)^2 + E(\hat{Y}_s - \eta)^2 + (\eta - \xi)^2 - 2\sigma^2 h_{ii} \\
&= \sigma^2 + \text{Var}(\hat{Y}_s) + (\eta - \xi)^2 - 2\sigma^2 h_{ii}
\end{aligned}
$$

where η denotes $E(\hat{Y}_s)$, ξ denotes $E(Y)$, and h_{ii} is the ith diagonal element of $\mathbf{H} = \mathbf{X}(\mathbf{X}'\mathbf{X})^{-1}\mathbf{X}'$, with the third line resulting from the fact that $E(Y - \xi)(\hat{Y}_s - \eta) = \text{Cov}(Y, \hat{Y}_s) = \text{Cov}[Y_i, \mathbf{x}'(\mathbf{X}'\mathbf{X})^{-1}\mathbf{X}'\mathbf{Y}] = \sigma^2 h_{ii}$, where $\mathbf{x}'$ denotes one of the rows of $\mathbf{X}$.

It then follows that

$$
\begin{aligned}
\sum E(Y - \hat{Y}_s)^2 &= n\sigma^2 + p\sigma^2 + \text{SSB} - 2\sigma^2 p \\
&= (n - p)\sigma^2 + \text{SSB}
\end{aligned}
$$

using the fact that $\sum h_{ii} = p$, and the result for $\sum \text{Var}(\hat{Y}_s)$ that was given in Section 7.A. It then follows that

$$E(\hat{\sigma}^2) = \sigma^2 + \frac{\text{SSB}}{n-p} \tag{A.1}$$

It should be noted that this result applies when $\hat{\sigma}^2$ is computed when *underfitting* occurs; that is, when the subset model is assumed to be part of the true model. If *overfitting* has occurred so that the true model is a subset of the fitted model, Eq. (A.1) simplifies to $E(\hat{\sigma}^2) = \sigma^2$ because SSB = 0. Specifically, the least squares estimators can be shown to be unbiased when overfitting occurs, so $\eta = \epsilon$. See also Myers (1990, p. 184) for a different approach to establishing this result.

To put this result in the perspective of variable selection, this would mean that a subset model would contain the true model, which would thus not be the model with all of the available regressors.

REFERENCES

Akaike, H. (1974). A new look at the statistical model identification. *IEEE Transactions on Automatic Control*, AC-19, 716–723.

Berk, K. N. (1978). Comparing subset regression procedures. *Technometrics*, **20**, 1–6.

Box, G. E. P. (1988). Response. *Technometrics*, **30**, 38–40.

Butler, R. W. (1984). The significance attained by the best-fitting regressor variable. *Journal of the American Statistical Association*, **79**, 341–348.

Chevan, A. and M. Sutherland (1991). Hierarchical partitioning. *The American Statistician*, **45**, 90–96.

Daniel, C. and F. S. Wood (1980). *Fitting Equations to Data*, 2nd edition. New York: Wiley.

Draper, N. R. and H. S. Smith (1981). *Applied Regression Analysis*, 2nd edition. New York: Wiley.

Flack, V. F. and P. C. Chang (1987). Frequency of selecting noise variables in subset regression analysis: a simulation study. *The American Statistician*, **41**, 84–86.

Freedman, D. A. (1983). A note on screening regression equations. *The American Statistician*, **37**, 152–155.

Freedman, L. S. and D. Pee (1989). Return to a note on screening regression equations. *The American Statistician*, **43**, 279–282.

Furnival, G. M. and R. W. Wilson (1974). Regression by leaps and bounds. *Technometrics*, **16**, 499–511.

Hocking, R. R. (1976). The analysis and selection of variables in linear regression *Biometrics*, **32**, 1–49.

Kennard, R. W. (1971). A note on the C_p statistic. *Technometrics*, **13**, 899–900.

Mallows, C. L. (1973). Some comments on C_p. *Technometrics*, **15**, 661–675.

Miller, A. J. (1984). Selection of subsets of regression variables. *Journal of the Royal Statistical Society, Series A,* **147,** 389–410 (discussion: 410–425).

Miller, A. J. (1990). *Subset Selection in Regression.* New York: Chapman and Hall.

Montgomery, D. C. and E. A. Peck (1992). *Introduction to Linear Regression Analysis,* 2nd edition. New York: Wiley.

Myers, R. H. (1990). *Classical and Modern Regression with Applications,* 2nd edition. Boston: Duxbury.

Neter, J., W. Wasserman, and M. H. Kutner (1989). *Applied Linear Regression Models,* 2nd edition. Homewood, IL: Richard D. Irwin, Inc.

Picard, R. R. and K. N. Berk (1990). Data splitting. *The American Statistician,* **44,** 140–147.

Pinault, S. C. (1988). An analysis of subset regression for orthogonal designs. *The American Statistician,* **42,** 275–277.

Pope, P. T. and J. T. Webster (1972). The use of an *F*-statistic in stepwise regression procedures. *Technometrics,* **14,** 327–340.

Roecker, E. B. (1991). Prediction error and its estimation for subset-selected models. *Technometrics,* **33,** 459–468.

Ronchetti, E. and R. G. Staudte (1994). A robust version of Mallows' C_p. *Journal of the American Statistical Association,* **89,** 550–559.

Seber, G. A. F. (1977). *Linear Regression Analysis.* New York: Wiley.

Walls, R. C. and D. L. Weeks (1969). A note on the variance of a predicted response in regression. *The American Statistician,* **23,** 24–26.

Weisberg, S. (1981). A statistic for allocating C_p to individual cases. *Technometrics,* **23,** 27–31.

Young, A. S. (1982). The Bivar criterion for selecting regressors. *Technometrics,* **24,** 181–189.

EXERCISES

7.1. Can the values of the correlation coefficients, r_{YX_i}, suggest what subset is likely to be chosen when any of the subset selection techniques are applied to a set of orthogonal regressors? If so, explain how.

7.2. Apply each of the four techniques (forward, backward, stepwise, and all possible regressions using C_p) to the stack loss data in Table 6.1 (omitting points 1, 3, 4, and 21) and compare the results. (Use $F = 4$ as the threshold value for entering and removing variables.) Then check the assumptions and the possible presence of influential data for the selected model(s).

7.3. How many possible subset models are there if there are seven available variables?

7.4. Assume that there are five available variables, and the best subset of size

3, in terms of C_p, is a proper subset of the best subset of size 4. Determine a lower bound on the value of the F-statistic (with σ^2 estimated using all of the variables) for testing that the coefficient of the added (fourth) regressor is zero if $C_3 - C_4 > 1$ and both C_p values are less than p (i.e., the C_p values suggest that the four-variable subset is better than the three-variable subset). While recognizing that n and the significance level are not given, does your lower bound on the F-value suggest that a subset can be selected from the amount by which C_p is less than p? Or should the relationship between C_p and p simply be used as a rough guideline?

7.5. Construct a data set with two regressors for which both t-statistics are less than one even though each regressor is highly correlated with Y. Explain what this means. In particular, would the use of a stepwise regression procedure lead to the selection of neither regressor?

7.6. Would it be possible to obtain true significance levels for the hypothesis tests used in forward selection if the regressors are orthogonal? Explain.

7.7. Assume that there is evidence to suggest that only linear terms will be needed in a regression model, and there are 10 available regressors. What is the smallest possible value of C_p? Can the largest possible value be determined?

7.8. Assume that an experimenter wishes to use forward selection. If the (nominal) significance level for the first hypothesis test is .05, should the nominal levels for subsequent tests be greater than, less than, or equal to .05? Explain.

7.9. Apply forward selection, backward elimination, stepwise regression, and all possible regressions with C_p to the data in Table 5.1, using $F = 4$ as the threshold value for entering and removing data, and compare the results.

7.10. Critique the following statement: When the number of candidate regressors is small, we might as well just use all possible regressions.

7.11. Show that the largest possible increase in C_p for the best subset in consecutive subset sizes is 2.0.

7.12. Show that if $C_p = p + a$ holds for two values of p (where a is a constant), the smaller of the two corresponding values of R^2_{adjusted} occurs at the larger value of p.

7.13. Show that the subset that minimizes $\hat{\sigma}^2$ is also the subset that minimizes R^2_{adjusted}.

CHAPTER 8

Polynomial and Trigonometric Terms

In Chapter 6 we observed how the use of nonlinear regressor terms can produce a parsimonious model. The need for one or more polynomial terms can (frequently) be identified by using the Box–Tidwell procedure, but the need for trigonometric terms cannot be determined from use of this procedure. In this chapter we consider various issues in the use of polynomial terms and also indicate when trigonometric terms are likely to be of value. The combination of polynomial and trigonometric terms in a regression model will often be the best choice.

8.1 POLYNOMIAL TERMS

There are many situations in which polynomial terms will be needed. Some care must be exercised when such terms are considered, however, as there are some pitfalls that should be recognized.

We start with some basic definitions. In particular, the *degree* of a polynomial regression model is defined as the order of the highest order term, with the order of a term $X_1^{\alpha_1} X_2^{\alpha_2} \cdots X_p^{\alpha_p}$ defined as $\sum_{i=1}^{n} \alpha_i$. (In this section we consider only one regressor, so the order will be the highest exponent used.)

As with regression models with only linear terms, we may always obtain an exact fit when polynomial models are used. Specifically, if we have n data points in the form (x, y), we could obtain an exact fit to the data by using a polynomial model with terms of order 1 through $n - 1$. Although being able to obtain a model with $R^2 = 1.0$ may seem tempting, we should quickly realize that this would generally require a very large number of terms, and parsimony should always be one of the objectives in developing a regression model. [Recall from Chapter 7 that $\text{Var}(\hat{Y})$ will almost certainly increase, and can never decrease, when regressors are added one at a time to a model, so it would be unwise to use very many polynomial terms.]

Centering regression terms so as to remove nonessential ill-conditioning was recommended in Section 4.2.1, provided that the intercept is not of interest.

240

Centering is also important in polynomial regression. [Again, however, the intercept must not be of interest. See Belsley (1984) and the discussions of that paper for details. We will assume in this chapter that the intercept is not of intrinsic interest.]

Bradley and Srivastava (1979) advocated that the polynomial terms not only be centered, but that when the regressor values can be selected, they should be chosen to be symmetrical and equally spaced about the center. This relates to the comment of Wood (1984) that just subtracting the mean (i.e., centering) may not be sufficient when the data are skewed.

Daniel and Wood (1980, p. 125) present a slight variation of a method given by Robson (1959) that can be used to make a (transformed) polynomial term orthogonal to the corresponding (centered) linear term. Specifically, instead of using X and X^2 as the terms in the model, the experimenter would use $X - \overline{X}$ and $(X - d)^2$, where

$$d = \frac{\sum X_j^2 (X_j - \overline{X})}{2 \sum (X_j - \overline{X})^2} \tag{8.1}$$

The utility of this approach can be illustrated with the following simple example. Assume that we have data given as

Y	3	5	7	10	14	15	11	6	5	2
X	1	2	3	4	5	6	7	8	9	10

We can see by inspecting the data that the relationship between Y and X is not linear and appears to be roughly quadratic. If we were to construct the X–Y scatter plot and draw a parabola through (or close to) the points, the vertex of the parabola would obviously not be at $X = 0$, so a linear term in X will be needed. Thus, we would expect that a model with X and X^2 would fit the data reasonably well.

We will see what happens when we first do not center the data, and then when we do perform the centering. For the uncentered model we obtain $\hat{Y} = -3.7 + 5.8561X - 0.53788X^2$, and $R^2 = .843$. The t-statistics for the linear and quadratic terms are 5.90 and -6.12, respectively, which suggest that both terms should be in the model. (The reader is asked to show in Exercise 8.9 that a different two-regressor model can be found that fits the data somewhat better.)

When we then center the data by using $X - \overline{X}$ as the linear term and $(X - d)^2$ as the quadratic term, we obtain $\hat{Y} = 12.2375 - 0.0606(X - \overline{X}) - 0.53788(X - d)^2$. This also produces $R^2 = .843$ (and thus the same predicted values and residuals), but the t-statistics are different. In particular, $t = -0.27$ for the centered linear term, which suggests that this term is not needed. This should not be surprising, however, because $(X - d)^2$ contains a linear term in X, and d is chosen so as to make the two terms orthogonal, thus essentially rendering the term $X - \overline{X}$ unnecessary. If we delete that term, however, we would obviously obtain a prediction equation that is slightly different from the equation that results when centering is not used. In general, t-statistics for centered terms should be used only to identify the order of the polynomial to use, rather than trying to determine if any lower-order terms should be omitted.

As in the case when only linear terms are used, the use of centering for polynomial terms is potentially helpful only in terms of improving numerical accuracy. Piexoto (1987, 1990) discusses the fact that polynomial regression models are invariant under linear transformation if and only if the polynomial is "hierarchically well formulated," and contends that polynomial regression models should have this property. (This term is used to describe a polynomial model that includes all of the polynomial terms up through the highest order term that is used.)

In our example we are using an uncentered model that has this property, and hence the centered and uncentered models are equivalent.

Despite this equivalence, however, completely different results can be obtained when variable selection methods are applied to the two different forms. If we apply either forward selection (with $F_{IN} = 4$) or stepwise regression (with $F_{IN} = F_{OUT} = 4$) to the terms X and X^2, the two procedures will terminate with neither term being selected since each term has a low correlation with Y. But if the two centered terms are used, then $(X - d)^2$ is selected. Thus, even though the two full models are equivalent, the application of variable selection methods to centered and uncentered terms can produce conflicting results. (The results for this example are in general agreement if all possible regressions are used, however.) Therefore, if we use centered polynomial terms for computational reasons, we should generally follow the advice of Freund and Littell (1981), Hocking (1984), McCullagh and Nelder (1989, p. 69), and Piexoto (1987, 1990) and include the lower-order terms (and of course also do so if the terms are not centered).

If the regressor values are symmetric about their mean, as was the case with our example, we may simply subtract $\overline{X}$, as this will be equivalent to using d as defined in Eq. (8.1).

Snee and Marquardt (1984) recommend centering the data in such a way so as to produce the model

$$Y = \overline{Y} + \alpha_2(X - \overline{X}) + \beta_3(Z - \overline{Z}) + \epsilon$$

where $Z = (X - \overline{X})^2$. They motivate this approach by stating that this reduces the variance inflation factors for the constant and the squared term. (We should keep in mind, however, that this does not reduce multicollinearity; it simply reduces the likelihood of having numerical problems.)

8.1.1 Orthogonal Polynomial Regression

Practitioners who wish to have orthogonal polynomial terms may use orthogonal polynomial regression, which can be used whether the regressor values are equispaced or unequally spaced. Polynomial regression and orthogonal polynomial regression are compared by Narula (1979), who illustrates the computation of the necessary constants when the independent variables are unequally spaced and the values also occur with unequal frequency. See also Narula (1978), Dutka and Ewens (1971), and Hahn (1977) for information on orthogonal polynomial regression.

Users of this approach should keep in mind that we are not simply "transforming away" the correlation between polynomial terms, as it is not possible to do that. Rather, each polynomial term additionally contains a function of lower-order terms, with the function determined in such a way as to provide the orthogonality.

8.1.1.1 When to Stop?

Although most authors recommend that polynomial terms of successively higher orders be fit until the first nonsignificant result occurs, Graybill (1961, p. 182) recommends that two nonsignificant results be obtained before deciding upon the degree of the polynomial to use. The rationale is as follows. If a pth-order polynomial is the appropriate model to fit, the $(p - 1)$st term may be nonsignificant, so the correct model might be missed if the experimenter stops after obtaining the first nonsignificant result.

Graybill assumes the use of orthogonal polynomial regression, but we may also apply this strategy to uncentered regressor terms. When centered terms are used we would expect certain lower-ordered terms to not be significant, simply because the lower-order terms are represented in part by the higher-order terms. For example, when a quadratic term in the centered regressors is significant, the linear term may not be significant because the centered quadratic term contains a function of the linear term, as was seen in the example in Section 8.1. When uncentered regressors are used, however, it is possible for all of the polynomial terms to be nonsignificant because of multicollinearity, as will be seen in the example in the next section.

8.1.2 An Example

In order to effectively examine and critique the proposed methods for selecting a polynomial model, we need data from a known model. Accordingly, we will generate values for Y as

$$Y = 6 + 2X - 2X^2 + 3X^3 + \epsilon \qquad (8.2)$$

where $\epsilon \sim N(0, \sigma_\epsilon = 100)$. The data are given in Table 8.1. When the known model is fit (without centering), all three of the regressor terms have t-statistics less than 2 in absolute value, and two of the three are much less than 1. This is obviously due to extreme multicollinearity, which will frequently exist when polynomial terms are used. The prediction equation is

Table 8.1   Simulated Data

Y	X
−125.35	1.7
218.83	2.9
205.78	3.0
244.82	3.2
92.31	3.6
192.02	3.8
220.39	4.2
152.93	4.3
115.18	4.8
291.37	5.4
502.64	6.0
836.76	6.8
1128.26	7.4
1710.14	8.5
2434.10	9.3
2996.33	10.2
4825.61	12.0

$\hat{Y} = 192 - 100X + 13.7X^2 + 2.26X^3$. This obviously differs greatly from the known equation, despite the fact that $R^2 = .994$, so the data have very little error. This considerable discrepancy between the known equation and the fitted model is also caused by the extreme multicollinearity. The extent of the latter is indicated by the three variance inflation factors: 230.2, 1098, and 359.8. (Recall from Section 4.3.2 that one variance inflation factor in excess of 10 is regarded as signaling multicollinearity.)

When stepwise regression and all possible regressions (see Chapter 7) are applied, using the three polynomial terms in the known model and a quartic term as candidates, the model using only X^3 is identified as the best model. In particular, $R^2 = .993$ and $\hat{\sigma}^2 = 12,940$ for that model, whereas $R^2 = .994$ and $\hat{\sigma}^2 = 14,290$ using the correct model. Thus, we see evidence of something that has been suggested in previous chapters: *the true model is not necessarily the model that best fits the data*. In this instance it is not even the second best model (that being the model with X and X^3).

We would expect that the true model would contain the linear and quadratic terms (as in this case we know that it does), and we would subsequently include them in the model in accordance with the recommendation given in Section 8.1.1, although in this case that would not give us the best fitting model in terms of minimizing $\hat{\sigma}^2$.

Subset selection can be expected to have considerable sampling variability in the presence of multicollinearity, however, so results such as this one should not be viewed very rigidly. To illustrate, 30 additional error vectors were randomly generated, with Y then computed from Eq. (8.2), and the results varied greatly. Specifically, although $\{X^3\}$ was the subset selected most frequently, several other subsets were selected more than once, including the subset that corresponds to the correct model.

We will compare this result (including subset selection variability) with what we obtain using orthogonal polynomial regression. The transformed linear term is defined as $\xi_{1i} = X_i - \overline{X}$; the quadratic term is given by

$$\xi_{2i} = X_i^2 - \xi_{1i} \frac{\sum \xi_{1j} X_j^2}{\sum \xi_{1j} X_j} - \frac{\sum X_j^2}{n}$$

and the cubic term is given by

$$\xi_{3i} = X_i^3 - \xi_{2i} \frac{\sum \xi_{2j} X_j^3}{\sum \xi_{2j} X_j^2} - \xi_{1i} \frac{\sum \xi_{ij} X_j^3}{\sum \xi_{1j} Y_j} - \frac{\sum X_j^3}{n}$$

Applying these formulas to the regressor values in the current example, we obtain $\hat{Y} = 944 + 426\xi_1 + 59.9\xi_2 + 2.26\xi_3$. Unlike when the uncentered-form regressors were used, the t-statistics for ξ_1 and ξ_2 are quite large (41.45 and 16.30, respectively), and $t = 1.88$ for ξ_3. Thus, the cubic term is somewhat borderline. When 30 additional sets of Y-values were generated, the model with all three polynomial terms was clearly superior in the majority of cases, and for the other sets of simulated values a model with linear and quadratic terms only was selected.

Thus, the results are much more satisfactory, in terms of subset selection variability, when variable selection is performed with orthogonal polynomial regression rather than with polynomial regression, and this should hold in general. Consequently, subset selection should be performed using orthogonal polynomial regression.

When there is more than one independent variable, the need to use polynomial terms may not be quite so apparent. When there are only two variables, however, we can graph the response surface (such as using a two-dimensional display that gives contours of constant values of $\hat{Y}$, or perhaps using a three-dimensional display), and this should provide some insight.

8.2 POLYNOMIAL-TRIGONOMETRIC REGRESSION

We can frequently combine polynomial terms with other nonlinear terms and obtain a better model than would be obtained if we used only polynomial terms. In particular, we will see how we may combine polynomial and trigonometric terms to produce a parsimonious model. This is analogous to what is often done in time series analysis when an autoregressive moving-average (ARMA) model is fit with fewer terms than would have been used if either an autoregressive or moving-average model had been fit.

If we have a single regressor and the scatter plot shows some evidence of periodicity (i.e., a cyclical-type configuration), the useof trigonometric terms may be beneficial. Eubank and Speckman (1990) discuss polynomial-trigonometric regression and point out that the use of only trigonometric terms can produce a poor fit at the end points of the interval on X when the true regression function is not periodic. (Regression data that exhibit curvature cannot be expected to have exact periodicity.) Accordingly, they advance the idea of using both polynomial and trigonometric terms as a means of, in particular, removing this potential problem. [The potential benefit of using both polynomial and trigonometric terms has also been noted by Graybill (1976, Chapter 8).]

The model is thus

$$Y = \beta_0 + \sum_{i=1}^{d} \beta_i X^i + \sum_{j=1}^{\lambda} [\gamma_j \cos(jX) + \delta_j \sin(jX)] + \epsilon \qquad (8.3)$$

with the values of d and γ to be determined. Eubank and Speckman (1990) indicate a preference for $d = 2$, rather than letting d be determined from the data, and suggest two ways in which λ might be determined. (If we always use $d = 2$, then the use of orthogonal polynomial terms is not so important, although we still might prefer to use the orthogonal terms for numerical accuracy. Eubank and Speckman do not address this issue.)

One approach to determining λ is to use the value of λ that minimizes the risk function

$$R(\lambda) = \frac{\text{SSE}(\lambda) + 2\sigma^2(2\lambda + d + 1)}{n}$$

where, if σ^2 were known, this would be equivalent to using Mallows's C_p [see Eubank (1988, p. 25)]. With Mallows's C_p, which was discussed in Section 7.4.1.1, one typically estimates σ^2 from the full model, but there is obviously no such thing as a full model when we are trying to determine λ. Eubank and Speckman (1990) suggest that σ^2 might be estimated using the estimator proposed by Gasser et al. (1986).

8.2.1 Orthogonality of Trigonometric Terms

If the X_i are equally spaced (as, for example, when X represents time), then $\sin(X)$ and $\cos(X)$ are orthogonal, as are $\sin(jX)$ and $\cos(jX)$, in general, $j = 2, 3, \ldots$. In the examples given by Eubank and Speckman (1990), the X_i are equally spaced. They additionally mention possible applications where the X_i are order statistics or are the n quantiles of the distribution of X—applications for which the X_i will be equally spaced.

There would likely be many applications for which we would want to consider the use of *sine* and *cosine* terms, but would not have equal spacing of the regressor, however. Without equal spacing, $\sin(jX)$ and $\cos(jX)$ will not be orthogonal, nor will $\sin(jX)$ and $\sin(j'X)$, $j \neq j'$, or $\cos(jX)$ and $\cos(j'X)$ be orthogonal. Nevertheless, the correlations between these terms will usually be small unless the unequal spacing is very extreme.

8.2.2 Practical Considerations

Obviously, we may view polynomial-trigonometric regression as an alternative to polynomial regression or trigonometric regression. In either case, we would want to plot the data first if there is a single independent variable, as will be the case in the examples that follow. Since the fitting of trigonometric terms will induce some periodicity in the predicted values, we would want to see if there is any hint of periodicity in the observed values. If there is, then the trigonometric terms may be helpful; if not, then polynomial terms may be sufficient. In general, the configuration of plotted points at the extreme values of X may be suggestive of the type of terms to consider.

8.2.3 Examples

In illustrating the utility of their approach, Eubank and Speckman (1990) use an example of voltage drop data that is in Montgomery and Peck (1992, p. 213). (They chose this data set because the scatter plot indicated that the true regression function might be periodic.) The independent variable is time, in equal intervals, and Montgomery and Peck fit a cubic spline with five terms (splines are discussed in Chapter 10). After rescaling X so that all values are in the interval $(0, 2\pi)$, Eubank and Speckman (1990) obtain a good fit to the data with four terms—two polynomial and two trigonometric.

Specifically, their model, which contains a linear and quadratic term in addition to a sine and a cosine term, produces $R^2 = .9895$ and $\hat{\sigma}^2 = 0.0767$, as compared with $R^2 = .9904$ and $\hat{\sigma}^2 = 0.0717$ for the model obtained by Montgomery and Peck. Since there is very little difference in the R^2 values, we might prefer the model with the fewer terms, although the five-term model does have a slightly smaller value of $\hat{\sigma}^2$.

While it is desirable to orthogonalize the polynomial terms before using a variable selection procedure, there still may be considerable correlation between certain candidate terms. Specifically, in addition to the possibility of having considerable correlation between sine terms (assuming the use of two or more) and between cosine terms, there can also be high correlations between the orthogonal polynomial terms and the trigonometric terms. For this data set the correlation between ξ_1 and $\sin(\xi_1)$ is .750, and the correlation between ξ_2 and $\cos(\xi_1)$ is $-.956$. Therefore, some high correlations may still appear, despite our efforts to avoid them.

It is of interest to compare the four-term model for this example with the models that result from using only trigonometric terms and only polynomial terms. Montgomery and Peck (1992) first fit a cubic polynomial to the data but rejected that model because the plot of the residuals against the predicted values showed evidence of curvature. When the standardized residuals are plotted against $\hat{Y}$, the configuration of points virtually appears to be a sine curve and a cosine curve plotted on the same graph. We have $R^2 = .877$ for this model, and it thus appears as though we need to add the sine and cosine terms to gain much improvement. In particular, the fit is poorest at the first and last observations, and the trigonometric terms might substantially improve the fit at those points.

For this example, however, if we add a quartic polynomial model rather than adding sine and cosine terms to a quadratic polynomial model, we obtain $R^2 = .9909$, which is higher than the R^2 value for any of the other models discussed. This model appears to be satisfactory in terms of the assumptions because $r_{e'e'_n} = .991$, $r_{H_1} = .106$, $r_{H_2} = -.011$, and SPREAD-RATIO = 1.85. (The definitions of these nonnormality and heteroscedasticity indicators were given in Chapter 6.) Thus, the quartic model appears to be satisfactory in terms of the assumptions. If we use only the sine and cosine terms, we have $R^2 = .966$ using one set, and $R^2 = .979$ using two sine and cosine terms. And there are distinct curved lines in each plot of the standardized residuals against $\hat{Y}$, indicating that polynomial terms are needed.

Although a user would almost certainly be satisfied with the fit obtained with either the model given by Montgomery and Peck (1992) or the one suggested by Eubank and Speckman (1990), we might proceed somewhat differently if our objective were to minimize $\hat{\sigma}^2$. There are several possibilities. If an X–Y scatter plot exhibits periodicity, one might use a variable selection approach to determine the number of sine and cosine terms to use, and investigate the fit at the extremes of $\hat{Y}$. If the fit is poor for the selected number of terms, the Box–Tidwell approach might be applied only to X, fixing the number of sine and cosine terms. Recall from the analysis of the stack loss data in Section 6.6.5 that a term in X_1^α, with α a noninteger, produced a better model than the model with both X_1 and X_1^2.

If this still does not provide a satisfactory fit at the extremes, then X might be forced into the model, and the Box–Tidwell approach applied to a separate term in X^γ (where γ would have to be at least slightly different from 1.0 in order to start the Box–Tidwell procedure). This should affirm or disaffirm that X^2 would be the appropriate term to use in addition to the trigonometric terms and the linear term. In general, we would not expect the Box–Tidwell procedure to produce X^2, but the exponent might be close to 2.

This does not happen with the Montgomery and Peck example, however, as the optimal value of the exponent appears to be in the vicinity of 1.05. Using $X^{1.05}$ in addition to the linear term and trigonometric terms produces $R^2 = .9905$ and $\hat{\sigma}^2 = 0.0695$. The latter is only slightly lower than the corresponding values for the models given by Montgomery and Peck (1992) and Eubank and Speckman (1990), but we might as well seek the best-fitting model. (Notice that the quartic polynomial model is very slightly superior to the model with the term in $X^{1.05}$, however.)

Table 8.2   Simulated Data

Y	X
54.3849	1
57.9038	2
41.9740	3
39.6582	4
20.6492	5
7.2486	6
−5.2081	7
−20.2649	8
−17.5781	9
−11.0819	10
−6.4355	11
26.3571	12
61.5750	13
78.0212	14
98.3823	15

There are obvious possible modifications to this approach, such as applying a variable selection approach to determine the number of trigonometric terms to use *after* the terms that are a function of X are selected through use of the Box–Tidwell procedure.

As a second example, consider the data in Table 8.2 and the accompanying scatterplot in Figure 8.1. Obviously, we will need at least a quadratic term in

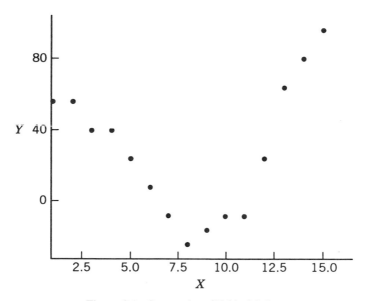

Figure 8.1   Scatter plot of Table 8.2 data.

the model. When we fit the quadratic model we obtain $R^2 = .864$, but the plot of the standardized residuals against $\hat{Y}$, shown in Figure 8.2, obviously has some curvature. In particular, note that all but six points can be connected by the curved dotted line. Curvature such as this indicates the need for either higher-order polynomial terms or perhaps sine and cosine terms.

When sine and cosine terms are added to the model, R^2 increases to .984, and this model is clearly superior to the cubic polynomial model and the quartic polynomial model ($R^2 = .979$ for the latter). All possible regressions applied to the four terms suggests that the polynomial-trigonometric model is appropriate. The data were generated as $Y = 6 + 0.2X + X^2 + 14\sin(X) + 45\cos(X) + \epsilon$, with $\epsilon \sim N(\mu = 0, \sigma = 10)$, so the correct model has been identified.

It is interesting to note that the addition of the trigonometric terms causes the worst fit to be at the middle of the curve rather than at the end points, as was the case before the terms were added. (This also happened with the preceding example.) This is not surprising because, as Eubank and Speckman (1990) indicate, polynomial-trigonometric regression is not good for "bump hunting" (i.e., fitting peaks in a curved configuration of the plotted data).

8.2.4 Multiple Independent Variables

The need to substitute trigonometric terms for higher-order polynomial terms may not be so easy to detect in multiple regression, since individual scatter plots in multiple regression will frequently be misleading (as was illustrated in Section 5.2). Consequently, we need to rely more heavily on the plot

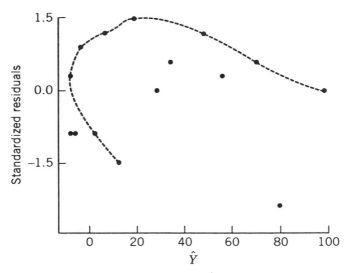

Figure 8.2 Plot of the standardized residuals against $\hat{Y}$ for a quadratic model fit to the data of Table 8.1.

of the standardized residuals against $\hat{Y}$, and see if the model without the trigonometric terms provides a relatively poor fit at the largest and smallest values
of $\hat{Y}$.

8.3 SOFTWARE

None of the statistical software packages whose capabilities were discussed in
previous chapters have a program for polynomial-trigonometric regression, but
the latter can be easily performed by any statistical software that allows variables to be created. All of the statistical software discussed in previous chapters
allow the user to do so.

Some care must be exercised in selecting software for polynomial regression
if variable selection is to be used, as orthogonal polynomial regression is then
preferable from the standpoint of stable results, as was discussed in Section
8.1.2.

SUMMARY

Polynomial regression has been contrasted with polynomial-trigonometric
regression, and the latter will often be preferable, especially if the data exhibit
at least a trace of periodicity. Polynomial-trigonometric regression may also
be viewed as an alternative to nonparametric regression techniques (covered in
Chapter 10), such as splines, and as shown by Eubank and Speckman (1990),
polynomial-trigonometric regression performs well relative to nonparametric
regression.

Whether used alone or in conjunction with trigonometric terms, the polynomial terms should be orthogonalized so that variable selection may be used
more effectively, and there will not be any nonessential ill-conditioning. (Once
the analysis has been performed, the user can always convert the model back
to the usual, raw form.)

Regression models with polynomial and/or trigonometric terms should be
used in conjunction with the same checks for influential data and possible model
violations that were illustrated in Chapters 2 and 5. A plot of the standardized
residuals against $\hat{Y}$ is also important to see if the use of the selected polynomial
terms (in addition to the trigonometric terms) provide a satisfactory fit at the
extremes of the $\hat{Y}$ values.

It is also worth noting that we might combine the methods of this chapter with those of earlier chapters. In particular, there is no need to restrict our
attention to terms in $X_i^{\alpha_i}$ for which α_i is an integer. One approach would be
to determine the number of sine and cosine terms to use, and then apply the
Box–Tidwell procedure to determine the α_i, either instead of or in addition to
a linear term in X_i. Alternatively, terms in $X_i^{\alpha_i}$ might be determined first, with
the possible need for sine and/or cosine terms then investigated.

REFERENCES

Belsley, D. A. (1984). Demeaning regression diagnostics through centering. *The American Statistician*, **38**, 73–77 (with discussion).

Bradley, R. A. and S. S. Srivastava (1979). Correlation in polynomial regression. *The American Statistician*, **33**, 11–14.

Daniel, C. and F. S. Wood (1980). *Fitting Equations to Data*, 2nd edition. New York: Wiley.

Dutka, A. F. and F. J. Ewens (1971). A method of improving the accuracy of polynomial regression analysis. *Journal of Quality Technology*, **3**, 149–155.

Eubank, R. L. (1988). *Spline Smoothing and Nonparametric Regression*. New York: Dekker.

Eubank, R. L. and P. Speckman (1990). Curve fitting by polynomial-trigonometric regression. *Biometrika*, **77**, 1–9.

Freund, R. J. and R. C. Littell (1981). *SAS for Linear Models*. Cary, NC: SAS Institute, Inc.

Gasser, T., L. Sroka, and C. Jennen-Steinmetz (1986). Residual variance and residual pattern in nonlinear regression. *Biometrika*, **73**, 625–633.

Graybill, F. A. (1961). *An Introduction to Linear Statistical Models*, Vol. 1. New York: McGraw-Hill.

Graybill, F. A. (1976). *Theory and Application of the Linear Model*. North Scituate, MA: Duxbury.

Hahn, G. J. (1977). The hazards of extrapolation in regression analysis. *Journal of Quality Technology*, **9**, 159–165.

Hocking, R. (1984). Response to David A. Belsley. *Technometrics*, **26**, 299–301.

McCullagh, P. and J. A. Nelder (1989). *Generalized Linear Models*, 2nd edition. New York: Chapman and Hall.

Montgomery, D. C. and E. A. Peck (1992). *Introduction to Linear Regression*, 2nd edition. New York: Wiley.

Narula, S. C. (1978). Orthogonal polynomial regression for unequal spacing and frequencies. *Journal of Quality Technology*, **9**, 170–179.

Narula, S. C. (1979). Orthogonal polynomial regression. *International Statistical Review*, **47**, 31–36.

Piexoto, J. L. (1987). Hierarchical variable selection in polynomial regression models. *The American Statistican*, **41**, 311–313.

Piexoto, J. L. (1990). A property of well-formulated polynomial regression models. *The American Statistician*, **44**, 26–30.

Robson, D. S. (1959). A simple method for constructing orthogonal polynomials when the independent variable is unequally spaced. *Biometrics*, **15**, 187–191.

Snee, R. D. and D. Marquardt (1984). Comment on "Demeaning conditioning diagnostics through centering" by D. A. Belsley. *The American Statistician*, **38**, 83–87.

Wood, F. S. (1984). Comment on "Demeaning conditioning diagnostics through centering" by D. A. Belsley. *The Amercan Statistician*, **38**, 88–90.

EXERCISES

8.1. Explain the advantage(s) of using the Box–Tidwell procedure to determine the exponent of a regressor, rather than automatically using a linear term and a quadratic term.

8.2. Consider the following data:

Y	X
2.609	0.5
3.740	1.0
15.006	1.5
12.538	2.0
17.627	2.5
21.034	3.0
18.853	3.5
29.378	4.0
42.869	4.5
53.442	5.0
47.103	5.5
61.606	6.0
70.930	6.5
68.985	7.0
68.313	7.5
88.458	8.0
95.781	8.5
97.350	9.0
129.926	9.5
132.511	10.0

First plot the data, and then try to determine what nonlinear term(s) are needed. If you elect to use trigonometric terms, remember to scale the X values so that they all lie within the interval $(0, 2\pi)$. *Hint:* The true model differs slightly from any of the models given in this chapter, but there are other models, with polynomial and/or trigonometric terms, that actually fit the data slightly better than the true model.

8.3. Let the values of X consist of the integers 1–20. Scale these values so that they all lie within $(0, 2\pi)$, and then graph the cosine of the scaled values against the sine of the scaled values. What is the correlation between the sine and cosine values that is suggested by your graph?

8.4. Assume that a model with only trigonometric terms has been used, but the fit is poor at the extreme values of $\hat{Y}$. Would it probably be better to use additional trigonometric terms or should the user try polynomial terms? Explain.

8.5. For what type of data configurations in an X–Y scatter plot would you suggest using a model with polynomial terms, but not trigonometric terms?

8.6. Given the following data

Y	11	13	14	10	11	12	15	10	9	15
X	1	2	3	4	5	6	7	8	9	10

What would happen if a ninth-degree polynomial model were fit to the data? Use a statistical software program to do this after first setting the tolerance at the smallest possible value. Now construct a scatter plot of the data (which, of course, would usually be done first). What model does your plot suggest should be used? What is the moral that is to be learned from this exercise?

8.7. What is likely to happen when an experimenter attempts to use two or more polynomial terms in a regression model when the range of X is small?

8.8. Assume that an experimenter has used a variable selection approach (only) and has determined a good-fitting model with polynomial and/or trigonometric terms. What should the experimenter do before using the model?

8.9. Consider the example that was given in Section 8.1, in which a reasonably good fit was obtained by using a linear and a quadratic term. First graph the data. Does the quadratic model look reasonable based on the graph? Fit the quadratic model, as was done in Section 8.1, and then plot the standardized residuals against $\hat{Y}$. Does this plot suggest adding a quadratic term, which is already in the model, or is there another type of term discussed in this chapter that might be expected to remove the trend in the plot? Add that term to the model and again plot the standardized residuals against $\hat{Y}$. Does this plot still exhibit a trend? Use one or more variable selection methods (such as all possible regressions) to select a subset from these three candidate terms. What model do you recommend?

CHAPTER 9

Logistic Regression

Unlike previous chapters in which there were no restrictions placed on Y, in this chapter there is a restriction on the possible values of Y, and this precludes the use of the linear regression model that was first described in Chapter 1

9.1 INTRODUCTION

Assume that we have a set of independent variables (regressors) that are to be used for prediction in each of the following situations: (1) predicting whether a company's dealer(s) will soon be mired in dire financial straits, (2) predicting if a person is likely to develop heart disease, or (3) predicting whether a hospital patient will survive until being discharged. These are obviously important applications. This writer has been involved in developing a model for scenario 1, and data illustrating scenario 3 are analyzed by Hosmer and Lemeshow (1989).

Notice that with each of these scenarios the response variable has only two possible outcomes. Specifically, the hospital patient will either survive or not survive, a person will either develop heart disease or not develop it, and a dealer will either have or not have financial problems.

9.2 ONE REGRESSOR

Assuming that we have a single regressor, let's try to write the model as

$$Y = \beta_0 + \beta_1 X + \epsilon \tag{9.1}$$

We would logically let $Y_i = 0$ if the ith unit does not have the characteristic that Y represents (e.g., heart disease), and $Y_i = 1$ if the unit does possess that characteristic.

It thus follows that ϵ_i can also take on only two values: $1 - \beta_0 - \beta_1 X_i$ if $Y_i = 1$ and $-\beta_0 - \beta_1 X_i$ if $Y_i = 0$. Therefore, ϵ_i cannot be even approximately normally

255

distributed. Consequently, the model given by Eq. (9.1) is inapplicable for a binary dependent variable.

In simple linear regression the starting point in determining a model is a scatter plot of Y versus X, but this is of limited value when there are only two possible values of Y. Consequently, it is necessary to consider other plots. One such plot results from smoothing the Y values, and then plotting the smoothed values against X. (One method of smoothing that has been suggested is discussed and illustrated in Section 9.7.5.) We might also consider a plot of $E(Y|X)$ against X. But now we must postulate a relationship between the two, rather than plotting points, since the ordinate of the plot is not related to data.

It is customary to let $E(Y_i|X_i) = \pi_i$, which is $P(Y_i = 1)$. [This relates to the general result: $E(W) = p$, where W is a binomial random variable.] Here π_i represents the probability of, for example, someone dying within a stated time period who has a cholesterol level given by X_i. So what relationship should there be between π_i and X_i?

We certainly wouldn't expect the relationship to be linear over a wide range of X_i, as we would expect π_i to be almost 1.0 for any extremely large value of X_i (assuming that the time interval is not short and there are no other risk factors), and to be very close to zero for any low value of X_i that is within the normal range.

What about the relationship between π_i and X_i for intermediate values of X_i? Since we would expect to observe an asymptotic-type graph segment at very large and very small values of X_i, we would expect somewhat of a curvilinear relationship in the neighborhood of the extreme values, whereas the rest of the graph might be closer to linear. Thus, the graph might resemble Figure 9.1. This S-curve configuration has been found to be appropriate in many applications for which Y is a binary random variable.

The function

$$\pi(X) = \frac{\exp(\beta_0 + \beta_1 X)}{1 + \exp(\beta_0 + \beta_1 X)} \tag{9.2}$$

will produce a graph similar to Figure 9.1 when $\beta_0 < 0$ and $\beta_1 > 0$. (Here exp denotes exponentiation.) The model given by Eq. (9.2) satisfies the important requirement that $0 \leq \pi_i \leq 1$ and will be a satisfactory model in many applications. The model in terms of Y would be written as $Y = \pi(X) + \epsilon$.

The value of X at which $P(Y = 1) = 0.5$, which is given by $-\beta_0/\beta_1$, provides useful information. For example, in the cholesterol example this would be the cholesterol level such that there is a 50–50 chance that a person with this level will die from heart disease within a stated time interval.

It follows from Eq. (9.2) that $\pi/(1 - \pi) = \exp(\beta_0 + \beta_1 X)$, so

$$\log\left(\frac{\pi}{1 - \pi}\right) = \beta_0 + \beta_1 X \tag{9.3}$$

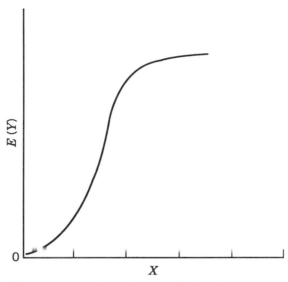

Figure 9.1 Typical function graph for logistic regression.

Since Eq. (9.3) results from using a *logistic transform* (also called a *logit transform*), the model is called a *logistic regression model*. (Throughout this chapter "log" shall designate the base e logarithm. The left side of Eq. (9.3) is also called the *log odds ratio*, and this can be explained as follows. Since $\pi = P(Y = 1)$, it follows that $1 - \pi = P(Y = 0)$, and so $\pi/(1 - \pi)$ is the ratio of the two probabilities, which, when stated in the form of odds, gives the odds of having $Y = 1$, for a given value of X. For example, if $\pi = \frac{3}{4}$ the odds of having $Y = 1$ are $3:1$. Odds are frequently stated in terms of "against" rather than "for," so the odds against having $Y = 1$ would be obtained from $(1 - \pi)/\pi$.

Notice that there is no error term on the right side of Eq. (9.3). This is because the left side is a function of $E(Y|X)$, instead of Y, which serves to remove the error term.

The interpretation of β_1 is naturally somewhat different from the interpretation in linear regression. In Eq. (9.3), β_1 obviously represents the amount by which the log odds change per unit change in X. It is somewhat more meaningful, however, to state that a one-unit increase in X increases the odds by the multiplicative factor e^{β_1}. This is apparent if we exponentiate both sides of Eq. (9.3).

As discussed by McCullagh and Nelder (1989, p. 110), the interpretation of the parameters on a *probability scale* is more complicated, since the derivative of π with respect to X depends on π. For example, from Eq. (9.2) we have $\partial\pi/\partial X = \beta_1(\pi)(1 - \pi)$.

9.2.1   Estimating β_0 and β_1

Why can't we simply let, say, $L_i = \pi_i/(1 - \pi_i)$ for the model in Eq. (9.3), and then use simple linear regression to estimate β_0 and β_1? With simple linear regression we have values of the response variable from the sample, but here we would not have values for the L_i because we would not know the π_i. If each X_i were repeated a large number of times, we could estimate π_i as $\hat{\pi}_i = \Sigma_{i=1}^{n_i} Y_i/n_i$ and use these estimates in obtaining values for L_i.

Even if there were sufficient data to enable us to obtain good estimates of the π_i, and thus obtain $\hat{L}_i$ as the dependent variable, we would still have the problem of unequal variances of the $\hat{L}_i$, as $\text{Var}(\hat{L}_i) \doteq 1/[n_i\pi_i(1 - \pi_i)]$ when the n_i are large (see Neter et al. 1989, p. 585, or Montgomery and Peck, 1992, p. 239). Consequently, weighted least squares would have to be used (which of course also requires large n_i unless the variance is to be modeled).

Even in simple logistic regression we would not expect to have large n_i very often, and in multiple logistic regression it is extremely improbable that we would have each combination of regressor values repeated a large number of times. Therefore, the method of (noniterative) weighted least squares will not be illustrated for logistic regression. The interested reader is referred to Neter et al. (1989, pp. 584–589).

9.2.1.1   Method of Maximum Likelihood

The most commonly used method of estimating the parameters of a logistic regression model is the *method of maximum likelihood*. The (sample) likelihood function is, in general, defined as the joint probability function of the random variables whose realizations constitute the sample. Specifically, for a sample of size n whose observations are $(y_1, y_2, \ldots, y_n)$, the corresponding random variables are $(Y_1, Y_2, \ldots, Y_n)$.

Since the Y_i are assumed to be independent, the joint probability density function is

$$g(Y_1, Y_2, \ldots, Y_n) = \prod_{i=1}^{n} f_i(Y_i)$$

$$= \prod_{i=1}^{n} \pi_i^{Y_i}(1 - \pi_i)^{1 - Y_i} \qquad (9.4)$$

since $f_i(Y_i) = \pi_i^{Y_i}(1 - \pi_i)^{1 - Y_i}$, $Y_i = 0, 1$; $i = 1, 2, \ldots, n$. (The latter is simply the probability that Y_i equals 0 or 1 and is thus a Bernoulli random variable relative to π_i.) In words, Eq. (9.4) gives the probability of a particular sequence of 0's and 1's.

It should be noted that the assumption of independent Bernoulli random variables may not always be plausible (see McCullagh and Nelder, 1989, p. 124). Nevertheless, we will assume that the assumption holds.

Maximum likelihood estimators are generally obtained by maximizing the logarithm of the likelihood function, as this is easier to solve than the likelihood function (and is a suitable substitute since the logarithm is a monotonic function). Taking the logarithm of each side of Eq. (9.4) produces

$$\log(g(Y_1, \ldots, Y_n)) = \log\left[\prod_{i=1}^{n} \pi_i^{Y_i}(1 - \pi_i)^{1 - Y_i}\right]$$

$$= \sum_{i=1}^{n} Y_i \log(\pi_i) + \sum_{i=1}^{n} (1 - Y_i)\log(1 - \pi_i) \quad (9.5)$$

$$= \sum_{i=1}^{n} Y_i \log\left(\frac{\pi_i}{1 - \pi_i}\right) + \sum_{i=1}^{n} \log(1 - \pi_i)$$

Using Eq. (9.2) and the expression for $1 - \pi$ that can be obtained from Eq. (9.2), we may write

$$\log(g(Y_1, \ldots, Y_n)) = \sum_{i=1}^{n} Y_i(\beta_0 + \beta_1 X_i) - \sum_{i=1}^{n} \log[1 + \exp(\beta_0 + \beta_1 X_i)] \quad (9.6)$$

The logarithm of the likelihood function (i.e., the *log likelihood*) is thus given by Eq. (9.6), and differentiating Eq. (9.6) with respect to β_0 and then with respect to β_1 produces the two likelihood equations. Specifically, letting $\log(L(\beta_0, \beta_1))$ denote the logarithm of the likelihood function, we obtain

$$\frac{\partial \log(L(\beta_0, \beta_1))}{\partial \beta_0} = \sum Y_i - \sum \frac{\exp(\beta_0 + \beta_1 X_i)}{1 + \exp(\beta_0 + \beta_1 X_i)}$$

$$\frac{\partial \log(L(\beta_0, \beta_1))}{\partial \beta_1} = \sum X_i Y_i - \sum \frac{X_i \exp(\beta_0 + \beta_1 X_i)}{1 + \exp(\beta_0 + \beta_1 X_i)}$$

The maximum likelihood estimators of β_0 and β_1 are obtained by setting the right side of each of these equations equal to zero, and then solving the equations simultaneously (and iteratively) so as to produce $\hat{\beta}_0$ and $\hat{\beta}_1$. Iteration would continue until certain convergence criteria are met. These are discussed later in this section.

For convenience we will write the equations in matrix notation, a form that will also aid in the transition to multiple logistic regression. If we let **X** denote an $n \times 2$ matrix with each row given by $(1, x_i)$, **Y** denote the vector of response values, and π denote the vector $E(\mathbf{Y})$, the likelihood equations can be written as

$$\frac{\partial l(\boldsymbol{\beta})}{\partial \beta} = \mathbf{X}'(\mathbf{Y} - \boldsymbol{\pi}) \tag{9.7}$$

where $l(\boldsymbol{\beta}) = \log(L(\beta_0, \beta_1))$. From Eq. (9.7) it follows that

$$\mathbf{X}'\boldsymbol{\pi} = \mathbf{X}'\mathbf{Y} \tag{9.8}$$

since $\partial l\boldsymbol{\beta}/\partial \boldsymbol{\beta}$ is set equal to zero. Since $\hat{\mathbf{Y}} = \hat{\boldsymbol{\pi}}$ (i.e., the predicted value of Y_i is the estimated probability that $Y = 1$), the solution to Eq. (9.8) will satisfy

$$\mathbf{X}'(\mathbf{Y} - \hat{\mathbf{Y}}) = 0 \tag{9.9}$$

(Recall that this result also holds for simple and multiple linear regression.)

Equation (9.9) is generally solved using the Newton–Raphson method. This entails first determining $(\partial/\partial \boldsymbol{\beta})\mathbf{X}'(\mathbf{Y} - \boldsymbol{\pi})$, which is equivalent to computing $-(\partial/\partial \boldsymbol{\beta})\mathbf{X}'\boldsymbol{\pi}$, which equals $-[(\partial/\partial \boldsymbol{\beta})\boldsymbol{\pi}]\mathbf{X}$. From Eq. (9.2) we may obtain $\partial \pi/\partial \beta_0 = \pi(1 - \pi)$ and $\partial \pi/\partial \beta_1 = X\pi(1 - \pi)$. Thus, $-(\partial \pi/\partial \boldsymbol{\beta}) = \mathbf{X}'\mathbf{W}$, so $-(\partial/\partial \boldsymbol{\beta})\mathbf{X}'\boldsymbol{\pi} = \mathbf{X}'\mathbf{W}\mathbf{X}$, with $\mathbf{W}$ a diagonal matrix with elements $\pi_i(1 - \pi_i)$ that would have to be estimated, and $\mathbf{X}$ is as previously defined. (See the appendix to Chapter 3 for results regarding matrix derivatives.)

Iterative estimates of $\boldsymbol{\beta}$ are then obtained as

$$\hat{\boldsymbol{\beta}} = (\mathbf{X}'\mathbf{W}\mathbf{X})^{-1}\mathbf{X}'\mathbf{W}\mathbf{Z} \tag{9.10}$$

with $\mathbf{Z}$ playing the role of $\mathbf{Y}$ in this *iteratively reweighted least squares* approach. Specifically,

$$Z_i = \hat{\eta}_i + \frac{Y_i - \hat{\pi}_i}{\hat{\pi}_i(1 - \hat{\pi}_i)} \tag{9.11}$$

with $\eta_i = \log(\pi_i/(1-\pi_i))$. Notice that $\hat{\eta}_i$ plays the role of $\hat{Y}_i$, and $(Y_i-\hat{\pi}_i)/[\hat{\pi}_i(1-\hat{\pi}_i)]$ is the residual corresponding to Y_i divided by the estimated variance of Y_i.

If we wish to write Eq. (9.10) in an equivalent form that shows the updating of $\hat{\boldsymbol{\beta}}$, we may write $\mathbf{Z}$ as

$$\mathbf{Z} = \mathbf{X}\hat{\boldsymbol{\beta}}^{(i)} + \mathbf{W}^{-1}\mathbf{e}$$

with $\mathbf{e} = \mathbf{Y} - \hat{\boldsymbol{\pi}}$. We then obtain

$$\begin{aligned} \hat{\boldsymbol{\beta}}^{(i+1)} &= (\mathbf{X}'\mathbf{W}\mathbf{X})^{-1}\mathbf{X}'\mathbf{W}(\mathbf{X}\hat{\boldsymbol{\beta}}^{(i)} + \mathbf{W}^{-1}\mathbf{e}) \\ &= \hat{\boldsymbol{\beta}}^{(i)} + (\mathbf{X}'\mathbf{W}\mathbf{X})^{-1}\mathbf{X}'\mathbf{e} \end{aligned} \tag{9.12}$$

The updating formula given by Eq. (9.12) is used until the estimates converge. Unlike nonlinear regression (see Chapter 13), there is no one convergence criterion that appears to be superior to the others. Common criteria include a change of less than 10^{-9} in the parameter estimates, and a percentage change of less than 0.01% in the log likelihood. The latter is more in line with the recommendation of McCullagh and Nelder (1989, p. 117), who suggest that the primary criterion should be based on changes in the $\hat{\pi}_i$. A supplementary test could be based on the change in $\hat{\boldsymbol{\beta}}$ or $\hat{\boldsymbol{\eta}}$.

The first step is to obtain the initial estimates, $\hat{\boldsymbol{\beta}}^{(0)}$. Various approaches are used to obtain these, including the use of discriminant analysis. It should be noted that discriminant analysis has often been used in the past when classification is the objective of the study. It is now well known, however, that discriminant analysis is quite sensitive to the assumption of normality of the regressors, so logistic regression will generally be the preferred alternative. Only if we knew that we had normally distributed regressors (an unlikely scenario) would we want to use discriminant analysis. See Huberty (1994) for information on discriminant analysis. As originally derived by Lachenbruch (1975) and displayed by Hosmer and Lemeshow (1989, p. 19), the initial estimates obtained using the discriminant function are given by

$$
\hat{\boldsymbol{\beta}}^{(0)} = \begin{bmatrix} \hat{\beta}_0^{(0)} \\ \hat{\beta}_1^{(0)} \end{bmatrix} = \begin{bmatrix} \ln(\hat{\theta}_1/\hat{\theta}_0) - 0.5(\hat{\mu}_1^2 - \hat{\mu}_0^2)/\hat{\sigma}^2 \\ (\hat{\mu}_1 - \hat{\mu}_0)/\hat{\sigma}^2 \end{bmatrix} \tag{9.13}
$$

The estimators in Eq. (9.13) that are used in computing $\hat{\beta}_0^{(0)}$ and $\hat{\beta}_1^{(0)}$ can be explained as follows. The conditional distributions of X given $Y = 0$ and X given $Y = 1$ are assumed to be normal with means of μ_0 and μ_1, respectively. Logical estimators of these two parameters are $\hat{\mu}_0 = \overline{X}_0$ and $\hat{\mu}_1 = \overline{X}_1$, where $\overline{X}_0$ and $\overline{X}_1$ are the average of the x-values when $Y = 0$ and $Y = 1$, respectively. The estimators $\hat{\theta}_0$ and $\hat{\theta}_1$ estimate $P(Y = 0)$ and $P(Y = 1)$, respectively, and are defined as $\hat{\theta}_1 = \overline{Y}$ and $\hat{\theta}_0 = 1 - \hat{\theta}_1$. Thus, $\hat{\theta}_1$ and $\hat{\theta}_0$ are the percentages of 1's and 0's, respectively, in the data set. [Alternatively, we might let $\hat{\theta}_1 = (\sum Y + 0.5)/(n+1)$, as suggested by McCullagh and Nelder (1989, p. 117).] The difference between the two estimators of θ_1 is obviously slight, and little if anything would be gained in using one over the other. Although good starting estimates can minimize the number of iterations that are required, minute differences in estimators may have no effect. At the other extreme, we should keep in mind that poor initial estimates can keep the iterative process from converging.

The estimate of σ^2 is obtained using the usual pooled estimator. That is, $\hat{\sigma}^2 = [(n_0 - 1)s_0^2 + (n_1 - 1)s_1^2]/(n_0 + n_1 - 2)$, where s_0^2 and s_1^2 are the usual sample variances computed using $Y = 0$ and $Y = 1$, respectively, and n_0 and n_1 are the corresponding sample sizes. (The pooled estimator is used because in discriminant analysis it is assumed that $\sigma_1^2 = \sigma_0^2$.) Studies have shown that these discriminant function estimators are quite sensitive to the assumption of normality, but they should generally be sufficient in providing starting estimates.

9.2.1.2   Exact Logistic Regression

The method of maximum likelihood that was discussed in Section 9.2.1.1 is the most commonly used approach for logistic regression, and will generally perform well for large sample sizes. But for small data sets or data sets in which the average value of Y is close to zero or one, the method of maximum likelihood can produce poor results, or even fail to converge. Cox (1970, pp. 44–48) outlined the general theory, for a single parameter, that is necessary for obtaining exact p values and confidence intervals, but until recently the method was considered to be impractical because of the considerable computation that is required. Hirji et al. (1987) present a recursive algorithm that allows the computations to be performed efficiently, and the algorithm is implemented in LogXact, a software program that is discussed in Section 9.8. The mechanics involved in computing the exact p-values are discussed in detail in Section 9.5.4.

In discussing its LogXact program, Cytel Software Corporation (1993) states:

> We recommend using the asymptotic capabilities of LogXact routinely, and calling on its exact capabilities only for small, sparse, or imbalanced data sets where there is real doubt concerning the validity of the usual asymptotic likelihood based inference.

Accordingly, we shall emphasize the use of the method of maximum likelihood in this chapter since (1) this is what is generally used and (2) it should be adequate for many, if not most, data sets. We shall, however, often give both the exact results and the results based on the method of maximum likelihood.

9.3   SIMULATED EXAMPLE

Before looking at an actual data set, we will simulate a data set for the purpose of illustrating the basic concepts and some problems that could be encountered. We simulate 50 values for X that are discrete and uniform on the interval $(10,30)$ and will then use $\beta_0 = -78.8$ and $\beta_1 = 4.0$ to produce

$$\pi(X) = \frac{\exp(\beta_0 + \beta_1 X)}{1 + \exp(\beta_0 + \beta_1 X)}$$

The value of Y_i, $i = 1, 2, \ldots, 50$, is then produced by randomly generating the value of a Bernoulli variable, using $p_i = \pi(X_i)$ to generate each random value. The simulated data are given in Table 9.1.

Table 9.1 Simulated Data

X	Y	X	Y	X	Y	X	Y	X	Y
29	1	25	1	12	0	24	1	12	0
26	1	15	0	15	0	30	1	20	1
29	1	16	0	24	1	25	1	14	0
26	1	18	0	24	1	15	0	27	1
21	1	15	0	14	0	29	1	28	1
24	1	29	1	26	1	26	1	23	1
18	0	22	1	27	1	19	0	16	0
17	0	22	1	16	0	17	0	24	1
26	1	10	0	19	0	11	0	19	0
22	1	10	0	18	0	13	0	21	1

9.3.1 Complete and Quasicomplete Separation

We will use Table 9.1 and a modification of the table to illustrate problems that could be encountered in using the maximum likelihood approach. If the data in Table 9.1 were ordered by their X-value, it could be observed that the X-values that correspond to $Y = 1$ exceed all of the X-values that correspond to $Y = 0$. When this occurs, there is *complete separation* of the data, and the maximum likelihood estimators do not exist, as was demonstrated by Albert and Anderson (1984) and discussed by Santner and Duffy (1986). [See also Hosmer and Lemeshow (1989, pp. 129–131).] One or more X-values corresponding to both $Y = 0$ and $Y = 1$ constitutes *quasicomplete separation* (Albert and Anderson, 1984), but the maximum likelihood estimators still do not exist. Only when there is *overlap*, such as an X-value corresponding to $Y = 1$ exceeding some but not all of the X-values corresponding to $Y = 0$, will the maximum likelihood estimators exist.

From Eq. (9.5) and remembering that $\hat{Y} = \hat{\pi}_i$, we may write

$$\text{log likelihood} = \sum [Y_i \log(\hat{Y}_i) + (1 - Y_i)\log(1 - \hat{Y}_i)] \qquad (9.14)$$

When $Y_i = 1$, we would, of course, ideally want to have $\hat{Y}_i = 1$, and this might seem likely with complete separation. But having $Y_i = \hat{Y}_i$, $i = 1, 2, \ldots, n$ would cause the log likelihood to be undefined because $\log(1 - \hat{Y}_i) = \log(0)$ would be undefined. Thus, unlike linear regression, in logistic regression we cannot solve for the maximum likelihood parameter estimates that would produce perfect prediction, but perfect prediction is what we might expect when there is complete separation. Furthermore, we may note from the expression for Z_i in Section 9.2.1.1 that Z_i is undefined if $\hat{\pi}_i$ is either one or zero, which would prevent the iteratively reweighted least squares approach from proceeding.

Fortunately, we would not expect to encounter complete separation very often with real data (except perhaps for very small samples), since this would mean that every subject that possessed the characteristic of interest (e.g., heart disease) would have an X-value (e.g., cholesterol level) greater than the X-value of every subject that did not have the attribute. Quasicomplete separation will also cause the iterative procedure to "blow up," and a very small amount of overlap can cause at least some minor convergence problems.

When there is a single regressor, we need only order the values of the regressor and the accompanying Y-values in order to detect complete or quasicomplete separation, but more sophisticated approaches are required when there is more than one regressor. Some such methods are given by Albert and Anderson (1984) and Santner and Duffy (1986).

What should be done when complete or quasicomplete separation is encountered in practice? The literature is not helpful here: Albert and Anderson (1984) indicate that there is no simple answer to this question, and Hosmer and Lemeshow (1989, p. 131) state that it is a problem that one "will have to work around." Of course, we could still use the discriminant analysis estimators as starting values and iterate toward the (nonexistent) maximum likelihood estimators until trouble is encountered. The starting values obtained in this manner are highly dependent on the assumption of normality, and may be poor if the conditional distributions $f(X|Y_j)$, $j = 0, 1$ are not normal.

For the data in Table 9.1 the initial estimates are $\hat{\beta}_0 = -25.3270$ and $\hat{\beta}_1 = 1.2642$. Obviously, these differ considerably from the known parameter values of -78.8 and 4.0, respectively. If we iterate toward the nonexistent maximum likelihood solutions, the procedure terminates immediately because there are several $\hat{\pi}_i$ that are only infinitesimally different from either one or zero.

So what should be done when this type of data is encountered? Although the initial parameter estimates for the Table 9.1 data differ considerably from the known parameter values, the fitted values are virtually identical to the observed values for all but a few of the observed values (as the reader is asked to show in Exercise 9.1). This is what we might expect for complete separation. If the separation had been considerably greater than that seen in Table 9.1, then all of the fitted values would have been virtually one or zero. Although this makes fairly easy our job of determining if a subject to which the logistic regression model is applied in the future has (or will soon have) the attribute of interest, it also renders the analysis rather trivial, because we essentially know what the predicted values will be before we determine those values.

The exact parameter estimates are $\hat{\beta}_{0(\text{exact})} = -12.4801$ and $\hat{\beta}_{1(\text{exact})} = 1.7414$, which differ noticeably from the discriminant analysis estimates given earlier in this section. Neither could be judged superior to the other in this case, however, because whereas the exact estimate is closer to the value of β_1 than is the discriminant analysis estimate, the reverse is true for β_0.

One advantage of the exact approach, however, is that exact confidence intervals and exact p-values can be obtained in addition to the exact parameter estimates, whereas neither exact nor approximate results can be obtained using the discriminant analysis approach. Here the exact confidence intervals are

$(.70, \infty)$ for β_1 and $(-\infty, -8.88)$ for β_0. The exact p-value for β_1 is less than .0001. (Certainly we would expect the p-value to be small and the confidence interval to not cover zero in the case of complete separation, however, since complete separation means that there is a strong relationship between X and Y.)

The exact estimation approach was not developed for prediction problems, and in this example the $\hat{\pi}_i$ values produced using the exact estimates are all virtually equal to 1.0. Thus, for classification purposes the fitted values obtained using the discriminant analysis estimates are much superior in this case, and this may generally occur in the case of complete separation with continuous covariates.

This does not necessarily hold true for binary covariates, however. In particular, a data set with grouped data is given by Cytel Software Corporation (1993), in which data were obtained on 2493 patients in a clinical study. Only five patients were given a particular antibiotic, Cephalexin, and all of them developed diarrhea. If the latter is used as the response variable, and this antibiotic is used as the single covariate (the full data set actually contained four covariates), the maximum likelihood estimates will not exist because of the absence of overlap. Of the other 2488 patients who were not given this antibiotic, only 55 got diarrhea. When the exact estimation approach is used, the exact estimates are $\hat{\beta}_0 = -3.7895$ and $\hat{\beta}_1 = 5.6729$. The predicted probabilities are then $\hat{\pi}(X = 0) = .022$ and $\hat{\pi}(X = 1) = .868$. Thus, the predicted probabilities are appropriately close to 0 and 1, respectively. We will see another example in Section 9.9 in which the exact approach is applied when the data preclude computation of the maximum likelihood estimates.

The nonexistence of the maximum likelihood estimators becomes more likely as regressors are added to a model, since complete separation for a single regressor will cause the maximum likelihood approach to fail. We should not think of complete separation as being bad, however, because it roughly corresponds to having an almost perfect correlation between Y and one or more regressors in linear regression.

9.3.2 Overlap: Modifying Table 9.1

We will change the twelfth value of X in Table 9.1 from 15 to 22 and not change the corresponding Y-value, thus creating an overlap. The sequence of iterations toward the maximum likelihood estimator is shown in Table 9.2.

The next step would be to compute the predicted values, the $\hat{\pi}_i$, and compare these with a cutoff value, which might be 0.5. As in linear regression, our sole interest in computing the predicted (fitted) values for observed values of Y is to see how well the observed values are fitted, reasoning that if the observed values are well fit, then future, unobserved Y values should be well predicted, assuming that the X values that are used are within the interval on X in the data set used to obtain the logistic regression equation.

Notice that the parameter estimates do not differ greatly from the estimates that were obtained before the one change was made in the data. Thus, although

**Table 9.2   Iterative Solution of Logistic Regression
Coefficients for Modified Table 9.1 Data**

Iteration Number	$\hat{\beta}_0$	$\hat{\beta}_1$
i^a	-22.1215	1.0969
1	-27.0198	1.3325
2	-29.2740	1.4415
3	-29.5984	1.4572
4	-29.6036	1.4575
5	-29.6031	1.4574
6	-29.6033	1.4574
7	-29.6036	1.4575
8	-29.6036	1.4575

[a] i denotes the discriminant analysis estimators. Converegence is achieved after eight iterations.

we do not know the parameter values for the Table 9.1 data, we may still safely state that the estimates are quite poor. Nevertheless, if we use a cutoff value of 0.5 for $\hat{\pi}$, only two of the 50 Y values are misfitted. This parallels what can frequently happen in linear regression: The parameter estimates might differ greatly from known parameter values, but the model may still have excellent predictive ability.

9.4   MEASURING THE WORTH OF THE MODEL

There are various statistics that have been proposed for assessing the worth of a logistic regression model, analogous to those that are used in linear regression. We examine some of the proposed statistics in the following sections, and it will be seen that most of the suggested statistics are inadequate.

9.4.1 R^2 in Logistic Regression

The worth of a linear regression model was determined by using R^2, but R^2 computed as in linear regression should *not* be used in logistic regression, at least not when the possible values of Y are zero and one. To see this, again consider Table 9.1 with the single change. We could hardly expect to do a better job of fitting Y than we do here, but R^2 is only .88, despite the fact that the vast majority of the fitted values are extremely close to the observed values. When $\hat{Y}$ is regressed against Y, the two points at which the misfitting occurs have extremely large standardized residuals: 5.39 for point 12 and -3.49 for point 42. When point 12 is deleted from the computation of R^2 (but still used for the parameter estimates), R^2 increases to .95, with the additional deletion of point 42 increasing R^2 further to .98.

This shows that R^2 can drop considerably for every misfitted point, so R^2 can be less than .9 even for near-perfect fitting. Cox and Wermuth (1992) also conclude that R^2 should not be used when Y has only two possible values, and show that frequently $R^2 \doteq .1$ when good models are used.

Various alternative forms of R^2 have been proposed for the binomial logit model. Maddala (1983) and Magee (1990) proposed using

$$R^2 = 1 - \{L(0)/L(\hat{\beta})\}^{2/n} \tag{9.15}$$

with $L(0)$ denoting the likelihood for the null model (i.e., with no regressors) and $L(\hat{\beta})$ representing the likelihood function that would result when $\hat{\pi}_i$ replaces π_i in Eq. (9.4). Essentially the same expression, except that $2/n$ was misprinted as $1/n$, was given by Cox and Snell (1989). [Equation (9.15) is motivated by the form of the likelihood ratio test for testing the fitted model against the null model. It can be shown that R^2 as defined in linear regression is equivalent to the right-hand side of Eq. (9.15). Hence, this is a natural form for R^2 in logistic regression.] Since the likelihood function $L(\hat{\beta})$ is a product of probabilities, it follows that the value of the function must be less than 1. Thus, the maximum possible value for R^2 defined by Eq. (9.15) is $\max(R^2) = 1 - (L(0))^{2/n}$. In linear regression $\hat{Y} = \overline{Y}$ for the null model. Similarly, in logistic regression we would have $\hat{\pi} = \gamma_1$ for the null model, with γ_1 denoting the percentage of 1's in the data set. It follows that $\max(R^2) = 1 - \{(\gamma_1)^{\gamma_1 n}(1 - \gamma_1)^{n - \gamma_1 n}\}^{2/n}$. For example, if $\gamma_1 = 0.5$, then $\max(R^2) = .75$. This is the largest possible value of R^2 defined by Eq. (9.15). When the data are quite sparse, the maximum possible value will be close to zero. Therefore, Nagelkerke (1991) suggests that $\overline{R}^2$ be used, with $\overline{R}^2 = R^2/\max(R^2)$.

9.4.2 Deviance

The *deviance* in logistic regression corresponds to SSE in linear regression, with the deviance, D, defined as

$$D = -2 \sum_{i=1}^{n} \left\{ y_i \log\left(\frac{\hat{\pi}_i}{y_i}\right) + (1 - y_i)\log\left(\frac{1 - \hat{\pi}_i}{1 - y_i}\right) \right\} \tag{9.16}$$

Although this statistic has been suggested for use in determining if a variable should be added to the model [Hosmer and Lemeshow (1989, p. 14), McCullagh and Nelder (1989, p. 119)], it should not necessarily be used as an absolute measure of the quality of the fit of the model.

There are two reasons why the deviance can be inadequate when applied to the full model. Although we have written the deviance in Eq. (9.16) in the same form that it is given in Hosmer and Lemeshow (1989, p. 14), potential inadequacy of the deviance can be seen better if we replace each 1 in the second term in Eq. (9.16) with m_i. The latter would then represent the number of repeats of the ith combination of regressor values.

The value of D is generally compared with the value of χ^2 with $n-p$ degrees of freedom, perhaps using a significance level of .05. But such a comparison is based on the assumption that $m_i \rightarrow \infty$ for each i. That is, each m_i would have to be large before D could be assumed to have approximately a chi-square distribution. Clearly D should not be used in this manner when, in particular, there are no repeats of the combinations of regressor values. When the asymptotic result is clearly inappropriate, D will not be even approximately independent of $\hat{\pi}$, the vector of predicted responses (McCullagh and Nelder, 1989, p. 119).

Many logistic regression data sets do have large m_i, however, and for such data sets the deviance can be useful. It is important, though, that the data be grouped by covariate patterns before the deviance is computed, but most of the software that is used for logistic regression does not automatically perform this grouping. Consequently, with such software the user would have to input the data in grouped form for the deviance to be computed properly. The deleterious effects of not fully grouping the data by covariate patterns is illustrated in detail in Appendix B of Cytel Software Corporation (1993). This issue is also discussed further in Section 9.7.3.1.

9.4.3 Other Measures of Model Fit

Other statistics that have been proposed for assessing the quality of fit of a logistic regression model include the Hosmer–Lemeshow (1980) goodness-of-fit statistic, which entails the calculation of the Pearson chi-square statistic after the data have been grouped in a prescribed manner. This statistic is discussed in detail in Section 9.6.1. Landwehr et al. (1984) proposed a procedure for partitioning the deviance into a lack-of-fit component and a pure error component, but the procedure has been criticized by Jennings (1986b), who, similar to McCullagh and Nelder (1989), noted that the expected value of each component of the deviance depends on the π_i. Another method for assessing the adequacy of the model, proposed by Landwehr et al. (1984), is to construct a probability plot of the residuals, with confidence bounds on the set of residuals obtained by simulation. (Recall that confidence bounds were also employed, and obtained by simulation, in the method discussed in Section 2.1.2.3 for assessing normality of errors in linear regression.)

The value of any goodness-of-fit statistic is highly dependent on the number of classes that are used, and the statistical significance of the statistic is also dependent to some extent on the number of classes. This clearly mitigates somewhat against their use as a measure of the quality of fit. We might also criticize any statistic that is a function of the $\hat{\pi}_i$ when Y is binary. That is, why directly use, say, $\hat{\pi}_i = .39$ when Y must be 0 or 1, and when each $\hat{\pi}_i$ is influenced by the percentage of 1's in the data set? That is, each $\hat{\pi}_i$ and its closeness to Y_i depends on more than the worth of the model. If our objective is to predict whether a subject will or will not have the attribute of interest, a more meaningful measure of the worth of the model would be the percentage of subjects in the data set that are classified correctly. Accordingly, we will use the *correct classification rate (CCR)* as a measure of the fit of the model. For the modification of the Table 9.1 data given in Section 9.3.2, CCR = $\frac{51}{53}$ = .96 and $\overline{R}^2$ = .926. Such statistics are potentially more informative than the p-value of a goodness-of-fit statistic.

Like virtually any summary statistic, CCR is not perfect, however. As discussed by Hosmer and Lemeshow (1989, p. 147), in simple logisic regression the CCR is a function of the slope, β_1. That is not necessarily a sufficient reason for eschewing CCR, however, since R^2 in simple *linear* regression also depends on the value of the slope (see Section 1.5.2). Other criticisms of the CCR are given by Van Houwelingen and Le Cessie (1990). Since the predicted values are rounded when the CCR is computed, two models could produce the same value for CCR but have $\overline{R}^2$ values that differed considerably. Therefore, when comparing models it would be advisable to use $\overline{R}^2$ as a supplementary statistic (only). Additional measures of model fit are described in SAS Institute, Inc. (1990, p. 1091).

9.5 DETERMINING THE WORTH OF THE INDIVIDUAL REGRESSORS

In subsequent sections we discuss various statistics that have been suggested for assessing the worth of each individual regressor. (The results that are given easily generalize to the situation in which more than one regressor is added to an existing model.)

9.5.1 Wald Test

In linear regression, t-statistics are used in assessing the value of individual regressors when other regressors are in the model. In logistic regression, $W = \hat{\beta}_i/s_{\hat{\beta}_i}$ is called a *Wald statistic*. First, it should be noted that W does not have a t-distribution, even though it does have the same form as a t-statistic. Rather, W is asymptotically normally distributed, with a large sample size, n, being required, instead of large m_i.

It should be noted that there is no agreement as to the general form of what is being called a Wald statistic. The definition given herein is that given by Hosmer and Lemeshow (1989, p. 17) and Hauck and Donner (1977). But $\hat{\beta}_i^2/s_{\hat{\beta}_i}^2$, written in a different but equivalent form, is termed a Wald statistic by Rao (1973, p. 417), Cytel Software Corporation (1993, p. A-5), and also by Wald (1943). If the latter definition is used, the statistic would be regarded as (approximately) a chi-square random variable with one degree of freedom. As discussed by Hosmer and Lemeshow (1989, p. 17), Hauck and Donner (1977) found that the Wald statistic performed poorly, and Jennings (1986a) also questioned the use of $\hat{\beta}_i/s_{\hat{\beta}_i}$ (without labeling it as a Wald statistic).

9.5.2 Likelihood Ratio Test

A superior alternative is the *likelihood ratio test* for each regressor. The likelihood ratio test statistic for a particular regressor is the difference between two deviance statistics: the deviance *without* the regressor in the model minus the deviance *with* the regressor in the model.

The problem with using the deviance as a measure of model adequacy was noted in Section 9.4.2, so we might expect that a statistic that is the difference of two deviances would be similarly inadequate. McCullagh and Nelder (1989, p. 119) state: "The χ^2 approximation is usually quite accurate for differences of deviances even though it is inaccurate for the deviances themselves." As with the second definition of a Wald statistic given in Section 9.5.1, the likelihood ratio test statistic is assumed to be approximately a chi-square random variable with one degree of freedom.

Nevertheless, Moulton et al. (1993) give correction factors for improving the likelihood ratio test. They conclude that it is better to improve this test than to try to improve the Wald test.

9.5.3 Scores Test

Recall from Section 9.2.1.1 that the matrix of partial derivatives of the log-likelihood function is used in obtaining the parameter estimates. Both the matrix of first derivatives and the matrix of second derivatives—each matrix evaluated at the maximum likelihood estimates—are used in computing the score statistic.

These two matrices are used in the iterative solution of the maximum likelihood estimators and are given (implicitly) in Eq. (9.12). Specifically, it can be seen by considering Eq. (9.7) that the matrix of first derivatives can be written as $\mathbf{X'e}$, and the reader is asked to show in Exercise 9.13 that -1 times the matrix of second derivatives is given by $\mathbf{X'WX}$.

Let $\boldsymbol{\beta} = (\boldsymbol{\beta}^*, \beta_i)'$ denote the full parameter vector that is to be estimated, with H_0: $\beta_i = 0$ the hypothesis that is to be tested with the scores test. (The test can be applied to more than one parameter, but we will use just one parameter for simplicity.) Let $\tilde{\boldsymbol{\beta}} = (\tilde{\boldsymbol{\beta}}^*, 0)'$ denote the maximum likelihood estimators of the parameters other than β_i when X_i is not in the model. Similarly, let $\mathbf{A}(\tilde{\boldsymbol{\beta}})$ denote $\mathbf{X'WX}$ evaluated at $\tilde{\boldsymbol{\beta}}$, and let $\mathbf{C}(\tilde{\boldsymbol{\beta}})$ denote $\mathbf{X'e}$ evaluated at $\tilde{\boldsymbol{\beta}}$.

The scores statistic for testing H_0 is then given by $S = \mathbf{C'}(\tilde{\boldsymbol{\beta}})\mathbf{A}^{-1}(\tilde{\boldsymbol{\beta}})\mathbf{C}(\tilde{\boldsymbol{\beta}})$ and is assumed to be approximately a chi-square random variable with one degree of freedom.

9.5.4 Exact Conditional Scores Test

Generally speaking, exact tests involving binary or categorical data are performed using all possible outcome combinations, and defining the *p*-value as the proportion of possible outcomes that are at least as extreme, relative to the null hypothesis, as the outcome that was observed. This concept is used in Fisher's exact test for a 2×2 contingency table, in addition to computing exact p-values in logistic regression. As discussed by Hirji et al. (1987), $\mathbf{X'Y}$ is a vector of sufficient statistics for $\boldsymbol{\beta}$, and they let $\mathbf{T} = \mathbf{X'Y}$. (A *sufficient statistic* is one that contains all of the sample information necessary to estimate one or more parameters.)

As in Section 9.5.3, we will partition $\boldsymbol{\beta}$ as $(\boldsymbol{\beta}^*, \beta_i)'$, with the intention of testing $H_0: \beta_i = 0$. We will similarly partition $\mathbf{T}$ as $(\mathbf{T}^*, T_i)$ and will let $\mathbf{t} = (\mathbf{t}^*, t_i)$ represent the sample values of the random variable $\mathbf{T}$. It can be shown that T_i is a sufficient statistic for β_i, with the sample value t_i computed as $t_i = \sum_{j=1}^{n} x_{ij} y_j$.

Computing the exact p-value for the hypothesis test entails determining the probability of observing a value for T_i that is more extreme, under H_0, than t_i. The task of computing the probability for each such T_i is complicated by the fact that different y-vectors can produce the same value of t_i. Consequently, the number of such y-vectors must be computed for each t_i. The recursive algorithm of Hirji et al. (1987) reduces this computational burden considerably.

Once these computations have been performed, the p-value is computed as $p = \sum_{u_i \in R} f(u_i | \mathbf{t}^*)$, with $R \subset \Omega_2$ denoting a region of the conditional sample space of T_i given $\mathbf{t}^*$ (given by Ω_2) in which the values $T_i = u_i$ are all considered to be more extreme under H_0 than the observed value $T_i = t_i$, and $f_0(u_i | \mathbf{t}^*)$ is the exact conditional probability that $T_i = t_i$, given $\mathbf{t}^*$, under H_0.

The latter is computed from the exact conditional probability that $T_i = t_i$, given $\mathbf{t}^*$, which is

$$f_0(t_i | \mathbf{t}*) = \frac{C(t_i, \mathbf{t}^*)}{\sum_{u_i \in \Omega_2} C(u_i, \mathbf{t}^*)} \tag{9.17}$$

with $C(t_i, \mathbf{t}^*)$ denoting the number of y-vectors that will produce the combination $(t_i, \mathbf{t}^*)$, and similarly for $C(u_i, \mathbf{t}*)$, with u_i as previously defined.

When the method of scoring is used, the region R is defined as follows. Let μ_i and σ_i^2 denote the mean and variance, respectively, of the random variable T_i, based on its conditional distribution given in Eq. (9.16). The exact conditional scores statistic is given by $S_{\text{exact}} = (T_i - \mu_i)^2 / \sigma_i^2$, with the corresponding sample statistic given by $s_{\text{exact}} = (t_i - \mu_i)^2 / \sigma_i^2$. The region R is then defined as $R = \{u_i \in \Omega_2 : (u_i - \mu_i)^2 / \sigma_i^2 \geq s_{\text{exact}}\}$, with the p-value given by $p =$ the probability that $S_{\text{exact}} \geq s_{\text{exact}}$.

9.5.5 Exact p-value

The p-value given in Section 9.5.4 was based on the exact conditional scores test. The p-value that corresponds to the exact parameter estimates is defined somewhat differently. Let

$$F_0(t_i) = \sum_{u_i = t_{i,\min}}^{t_i} f_0(u_i|\mathbf{t}^*)$$

and

$$G_0(t_i) = \sum_{u_i = t_i}^{t_{i,\max}} f_0(u_i|\mathbf{t}^*)$$

with $t_{1,\min}$ and $t_{1,\max}$ denoting the minimum and maximum values, respectively, in the range of values assumed by the random variable T_i, conditional on $\mathbf{t}^*$. The exact (two-sided) p-value is given by $p_2 = 2p_1$, with $p_1 = \min\{F_0(t_i), G_0(t_i)\}$.

Additional information concerning exact p-values and exact conditional scores p-values, in particular, can be found in Cytel Software Corporation (1993, Appendix A).

9.6 CONFIDENCE INTERVALS

In this section we discuss (approximate) confidence intervals based on the maximum likelihood approach, followed by the corresponding exact confidence intervals. Specifically, we present (1) a confidence interval for β_1, (2) a confidence interval for the change in the odds ratio, and (3) a confidence interval for π_j. The latter corresponds to a prediction interval for Y in linear regression. There is technically not a prediction interval in logistic regression since Y is not the dependent variable that is used in the model.

9.6.1 Confidence Interval for β_1

A confidence interval for β_1 is a confidence interval for the change in the "log odds." The interval is obtained as $\hat{\beta}_1 \pm z_{\alpha/2}s_{\hat{\beta}_1}$, where $z_{\alpha/2}$ denotes the standard normal deviate with a tail area of $\alpha/2$. Notice that z is used here rather than t, which is used for confidence intervals in linear regression. This is because there is no normality in logistic regression, as is assumed in linear regression. Rather, in logistic regression we must rely upon asymptotics to argue that $\hat{\beta}_1$ will be approximately normally distributed if the sample size is sufficiently large. Similarly, $\hat{\beta}_1$ and $s_{\hat{\beta}_1}$ are only asymptotically unbiased (McCullagh and Nelder, 1989, p. 119). Consequently, a confidence interval for $\hat{\beta}_1$ should be used cautiously, especially if the sample size is not large. We may obtain $s_{\hat{\beta}_1}$ directly from the output of the weighted least squares analysis [given by Eq. (9.10)] at the point of convergence. Then, if n is large, we may obtain the (approximate) confidence interval for a desired degree of confidence.

As discussed by Hosmer and Lemeshow (1989, p. 56), it will usually be more meaningful to have a confidence interval for more than a one-unit change in X. For a change of cX the form of the interval is $c\hat{\beta}_1 \pm z_{\alpha/2}(cs_{\hat{\beta}_1})$.

9.6.2 Confidence Interval for Change in Odds Ratio

A confidence interval for the change in the log odds is similar to a confidence interval for the change in the *odds ratio*. Since for a single regressor, $\ln[\pi/(1 - \pi)] = \beta_0 + \beta_1 X$, we have $\pi/(1 - \pi) = \exp(\beta_0 + \beta_1 X)$. Thus, $\exp(\beta_1)$ represents the *change* in the odds ratio per unit change in X. [If X is binary, $\exp(\beta_1)$ then represents the ratio of the odds for $x = 1$ to the odds for $x = 0$, and is then more appropriately viewed as an odds ratio rather than a change in an odds ratio.]

A confidence interval for $\exp(\beta_1)$ is thus a confidence interval for the change in the odds ratio. *One* way to obtain a confidence interval is to use $\exp[(c\hat{\beta}_1 \pm z_{\alpha/2}(cs_{\hat{\beta}_1})]$, where c represents the increase in X for which the interval is desired. Thus, if $c = 1$, we would simply exponentiate the interval end points that are obtained for the confidence interval for β_1. Hosmer and Lemeshow (1989, p. 57) suggest using $c = 5$ or $c = 10$, however. Whatever the choice, we may write the interval as $[\exp(\hat{\beta}_1 \pm z_{\alpha/2}s_{\hat{\beta}_1})]^c$. If we had first obtained a confidence interval for β_1, we would simply exponentiate each end point, and then raise each of those numbers to the cth power. As discussed by Hosmer and Lemeshow (1989, p. 44), this approach is preferable to working directly with the estimate of the odds ratio, as that estimate will not be approximately normally distributed for the sample sizes usually encountered in practice. We should note that the confidence interval will not be symmetric about the estimate of the log odds, $\exp(\hat{\beta}_1)$, with the amount of the asymmetry influenced by the value of c.

Notice also that our interval is *not* a function of X, which might seem illogical. That is, this result is based on the assumption that the increase in the risk of what Y represents is independent of the level of X. But this assumption results from use of the logistic regression model, which is linear in X. (Notice the analogy with simple *linear* regression, as there the expected increase in Y for a given increase in X does not depend on the starting value of X.)

If a constant risk increase seems illogical, one remedy would be to use a nonlinear term in X. See Hosmer and Lemeshow (1989, p. 57) for additional discussion.

9.6.3 Confidence Interval for π

Although a confidence interval for the change in the odds ratio is obviously important, it is also important to know what the risk is that is being increased. Specifically, if we say that the risk increases 2.3 times for every 10-unit increase in X, we need to know what the risk is at a given value of X. Clearly there is a considerable difference between a risk that doubles starting at 1% and a risk that doubles starting at 20%.

Thus, we need to know $\hat{\pi}_i$, and we need to know how well π_i is estimated. Therefore, we need a confidence interval for π_i. There are at least two ways in which we might proceed, but if we are going to assume approximate normality and use the standard normal deviate as was done in Sections 9.6.1 and 9.6.2, we need to be careful that what we are using will be approximately normally distributed for large n. The most obvious approach would be to use $\hat{\pi} \pm zs_{\hat{\pi}}$, but since $0 \leq \hat{\pi} \leq 1$, $\hat{\pi}$ will not be even approximately normally distributed for most values of π. Therefore, this method of obtaining the confidence interval would not be defensible.

What about first constructing a confidence interval for $\pi/(1 - \pi)$, and then solving for the end points of the interval for π? Unfortunately, the estimate of $\pi/(1 - \pi)$ will not be normally distributed for sample sizes usually encountered in practice. Consequently, this approach would be meaningful only if we had a very large sample.

If we had a very large sample, we could proceed as follows. Assume that there is a single regressor. Since $\log[\pi/(1 - \pi)] = \beta_0 + \beta_1 X$, a confidence interval for $\log[\pi/(1 - \pi)]$ would be given by $\hat{\beta}_0 + \hat{\beta}_1 X \pm zs_{\hat{\beta}_0 + \hat{\beta}_1 X}$. Therefore, a confidence interval for $\pi/(1 - \pi)$ would be obtained by exponentiating the end points of the interval for $\log[\pi/1 - \pi)]$. The confidence interval for $\pi|x_j$ would then be obtained by performing the necessary algebra. Doing so produces the following results:

Lower limit	$\exp(\hat{\beta}_0 + \hat{\beta}_1 X_i - zs\sqrt{h_{ii}})/(1 + \exp(\hat{\beta}_0 + \hat{\beta}_1 X_i - zs\sqrt{h_{ii}}))$
Upper limit	$\exp(\hat{\beta}_0 + \hat{\beta}_1 X_i + zs\sqrt{h_{ii}})/(1 + \exp(\hat{\beta}_0 + \hat{\beta}_1 X_i + zs\sqrt{h_{ii}}))$

where $\text{Var}(\hat{Y}|X_i) = \sigma^2 h_{ii}$, h_{ii} is as previously defined, and s would be obtained from the weighted least squares output.

9.6.4 Exact Confidence Intervals

The confidence intervals covered in Sections 9.6.1–9.6.3 are "inexact" in two ways: the parameter estimates are obtained using maximum likelihood, and normality is assumed. Consequently, we might expect the exact confidence intervals to differ considerably from the approximate intervals in some cases.

9.6.4.1 *Exact Confidence Interval for* β_1

In general, an exact confidence interval for β_i would be obtained using the left and right tails of the conditional distribution of T_i given $\mathbf{T}^*$, with T_i as defined in Section 9.5.4. For β_1 we are concerned with the distribution of t_1, with the latter denoting the sample value of T_1. Let

$$F_{\beta_1}(t_1) = \sum_{u_1 = t_{1,\min}}^{t_1} f_{\beta_1}(u_1 | \mathbf{t}^*)$$

and

$$G_{\beta_1}(t_1) = \sum_{u_1 = t_1}^{t_{1,\max}} f_{\beta_1}(u_1 | \mathbf{t}^*)$$

with $f_{\beta_1}(u_1 | \mathbf{t}^*)$ defined analogous to $f_{\beta_1}(t_i | \mathbf{t}^*)$ in Eq. (9.16).

For a $(1 - \alpha)\%$ confidence interval on β_1, the lower limit, say $\beta_{1(LL)}$, is determined from $G_{\beta_{1(LL)}}(t_1) = \alpha/2$ if $t_{\min} < t_1 \leq t_{1,\max}$, with $\beta_{1(LL)} = -\infty$ if $t_1 = t_{1,\min}$. Similarly, the upper confidence limit, $\beta_{1(UL)}$, is determined from $F_{\beta_{1(UL)}}(t_1) = \alpha/2$ if $t_{\min} \leq t_1 < t_{1,\max}$, with $\beta_{1(UL)} = \infty$ if $t_1 = t_{1,\max}$.

9.6.4.2 Exact Confidence Interval for Change in Odds Ratio

Analogous to Section 9.6.2, we can obtain an exact confidence interval for a change in the odds ratio due to a c-unit change in X_i by first exponentiating the end points of the exact confidence interval for β_i and then raising that result to the cth power.

9.6.4.3 Exact Confidence Interval for π

There is not at present a published approach for obtaining an exact confidence for π, so the approximate confidence interval provided in Section 9.6.3 would have to be used.

9.7 AN EXAMPLE WITH REAL DATA

We will further illustrate logistic regression by using part of a data set in Brown (1980). The original objective was to see whether an elevated level of acid phosphatase in the blood serum would be of value as an additional regressor for predicting whether or not prostate cancer patients also had lymph node involvement. The data set in Brown et al. (1980) additionally contains data on the four more commonly used regressors, but we will use only acid phosphatase in illustrating simple logistic regression. The data on the 53 patients are given in Table 9.3. The dependent variable is nodal involvement, with 1 denoting the presence of nodal involvement and 0 indicating the absence of such involvement.

It was stated in Section 9.2 that an X–Y scatter plot of logistic regression data is of limited value in terms of suggesting a possible model. Such a plot can be helpful in other ways, however, and the scatter plot of the Table 9.3 data is shown in Figure 9.2.

There is clearly an outlier (187) among the patients without nodal involve-

Table 9.3   Data from Brown (1980) for Illustrating Simple Logistic Regression[a]

Y	A.P.	Y	A.P.	Y	A.P.	Y	A.P.
0	48	0	47	0	50	1	81
0	56	0	49	0	40	1	76
0	50	0	50	0	55	1	70
0	52	0	78	0	59	1	78
0	50	0	83	1	48	1	70
0	49	0	98	1	51	1	67
0	46	0	52	1	49	1	82
0	62	0	75	0	48	1	67
1	56	1	99	0	63	1	72
0	55	0	187	0	102	1	89
0	62	1	136	0	76	1	126
0	71	1	82	0	95		
0	65	0	40	0	66		
1	67	0	50	1	84		

[a]Y denotes nodal (cancer) involvement [presence (1) or absence (0)]. A.P. denotes acid phosphatase level.

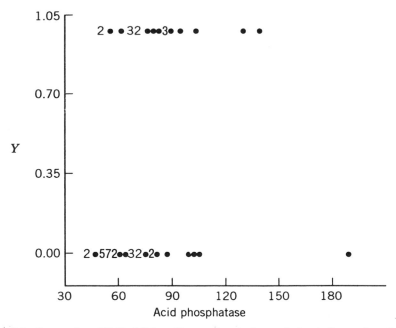

Figure 9.2   Scatter plot of Table 9.3 data. The numbers in the graph denote the number of points at each location.

ment. For the other patients, those without nodal involvement have, as a group, lower acid phosphatase levels than those patients who do have nodal involvement. But the data clearly overlap, without the outlier, so the maximum likelihood estimates will exist. Of course, we do not want to see too much overlap, as that would suggest that acid phosphatase may not be an important predictor. At the extreme case, if the configuration of points at $Y = 1$ could be superimposed on the points at $Y = 0$, then acid phosphatase would be of no value whatsoever as a single predictor.

Following Hosmer and Lemeshow (1989), we might group acid phosphatase into intervals and compute the proportion of individuals in each interval who have nodal involvement. Those proportions would then be plotted against acid phosphatase, using the midpoint of each interval, the objective being to see if the basic logistic regression model appears to be a plausible model.

This raises the question, "How many intervals should we use?" We obviously need to use enough intervals so that there will be a sufficient number of points to construct a meaningful plot. In particular, the plot should be curvilinear if the logistic regression model is appropriate, so there should be enough points to show the nature of the nonlinearity. We should also have enough points so that the estimate of the proportion within each interval is meaningful.

The sample size in this data set is smaller than usual for a medical application, so we need to use a relatively small number of classes. We will use six classes, and these are shown in Table 9.4. We will initially use the outlier in the computations and will later see how it affects the results. The points are plotted in Figure 9.3. The point for the last class is somewhat bothersome, as it does not conform to the pattern that we would expect. (The point would also be of concern if the outlier were not used, since the value of $\hat{\pi}_6$ without the outlier, .5, would still be out of line with the other points.).

When the simple logistic regression model is fit to these data using the method of maximum likelihood, one or more convergence criteria must be used, since the equations represented by Eq. (9.12) are solved iteratively. As indicated in Section 9.2.1.1, we might use changes in the deviance (to reflect changes in

Table 9.4 Grouped Data from Table 9.3[a]

Interval	Midpoint	n	$\hat{\pi}_i$
40–48	44	7	.143
49–56	52.5	15	.200
57–67	62	9	.333
68–78	73	9	.556
79–89	84	6	.833
90–187	102 (median)	7	.429

[a]The median was used for the last class rather than the midpoint because of the outlier, 187.

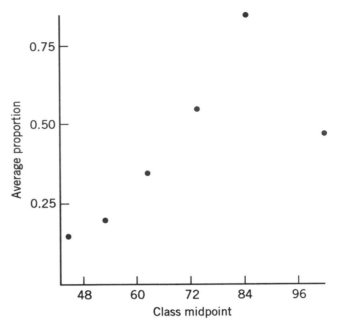

Figure 9.3 Scatter plot of Table 9.4 data, grouped: average proportion per class with characteristic plotted against class midpoint.

$\hat{\pi}_i$) as the primary criterion, with changes in the $\hat{\beta}_i$ as a secondary criterion. Using 0.0001 as the convergence requirement for both the deviance and the $\hat{\beta}_i$, convergence is achieved after four iterations, and the prediction equation is

$$\hat{\pi}(X) = \hat{Y} = \frac{\exp(-1.927 + 0.0204X)}{1 + \exp(-1.927 + 0.0204X)}$$

9.7.1 Hosmer–Lemeshow Goodness-of-Fit Tests

The Hosmer–Lemeshow goodness-of-fit statistic is, using their notation, given by

$$\hat{C} = \sum_{k=1}^{g} \frac{(o_k - n'_k \overline{\pi}_k)^2}{n'_k \overline{\pi}_k (1 - \overline{\pi}_k)} \tag{9.18}$$

where g denotes the number of groups, n'_k is the number of observations in the kth group, o_k is the sum of the Y values for the kth group, and $\overline{\pi}_k$ is the average of the $\hat{\pi}$ for the kth group. Notice that this differs slightly from the usual chi-squared goodness-of-fit test, as the denominator in Eq. (9.18) is not the expected frequency. Rather, it is the expected frequency for the kth group multiplied times one minus the average of the estimated probabilities for the kth group. Thus, each of the g denominators will be less than the g expected frequencies, and there will be a considerable difference when $\bar{\pi}_k$ is close to 1. Small denominators in a goodness-of-fit test can create the type of problems that are discussed and illustrated later in this section.

As discussed by Hosmer and Lemeshow (1989, p. 144), this statistic will almost always indicate an adequate fit for the model when g is less than 6. But if we use many groups with a small-to-moderate sample size, we run the risk of having cell sizes that are too small for the chi-squared goodness-of-fit test. The methods of Hosmer and Lemeshow (1980) and Lemeshow and Hosmer (1982) specify that 10 groups be used. Following that suggestion and the option of having class widths of 0.1 for the $\hat{\pi}_j$ produces $\hat{C} = 12.127$ and an associated p-value of .033 for the Table 9.3 data.

Since we are using a *goodness-of-fit* test, we want $\hat{C}$ to be small, and the p-value suggests that it is not sufficiently small. But we can expect to have some empty cells when $g = 10$ and $n = 53$. Here there are three cells that are empty, and three others have one observation each. Consequently, the use of this goodness-of-fit test with the fixed class widths would be inadvisable without considerable combining of cells.

Hosmer and Lemeshow (1989, p. 142) actually indicate a preference for determining the class intervals from the percentiles of the estimated probabilities, rather than using fixed cutoff points. This, of course, precludes the possibility of empty cells. For the present example we obtain $\hat{C} = 11.149$ if no classes are combined, and this has a p-value of .1933. Therefore, we now receive the message that the fit is adequate. But the grouping performed here is somewhat arbitrary because of the replications of regressor values. This prevents us from obtaining classes with approximately $n/10$ observations per class, and the allocation for the last class strongly influences the value of $\hat{C}$, since $\bar{\pi}_{10}$ will be close to 1.0.

Due to the fact that the data are sparse relative to the number of classes that are being used, we would expect that the outlier could have a considerable effect on the goodness-of-fit test, and may also cause the parameter estimates to change considerably. The estimates for β_0 and β_1 change from -1.927 and 0.0204 to -3.6621 and 0.0471, respectively, when the outlier is deleted, and the point also has a considerable effect on $\hat{C}$. Specifically, for the option of equal class widths $\hat{C}$ drops from 12.127 to 4.491, and the associated p-value increases from .033 to .611. Thus, the message changes from an inadequate fit of the model to a good fit.

It is disturbing that the deletion of 1 out of 53 observations should have such a dramatic effect on the results of a goodness-of-fit test. Hosmer and Lemeshow (1989, p. 143) have, as they admit, a liberal view regarding the use of their goodness-of-fit tests when the expected frequencies are small. The value of the test statistic for any goodness-of-fit test can certainly become inflated when the denominator of any of the summed components is less than 1, in particular.

Various rules-of-thumb have been given regarding the magnitude of the denominators in a goodness-of-fit test, the most extreme being that all of the expected frequencies must exceed 5, whereas other rules stipulate the maximum percentage of expected frequencies that can be less than 5.

In this example the use of 10 groups (*without* the outlier) causes 2 of the groups to be empty, and causes some of the other groups to have only two or three observations, and similarly, small expected frequencies. In particular, since there are considerably more Y values equal to zero than equal to one, some of the intervals on $\hat{\pi}_j$ that are greater than 0.5 may have a very small number of observations, since the magnitudes of the $\hat{\pi}_j$ are influenced by the relative percentage of 0's and 1's (Hosmer and Lemeshow, 1989, p. 146).

For all of these reasons, a goodness-of-fit test such as the Hosmer–Lemeshow test should be used cautiously, if at all. If it is used, the grouping method based on percentiles of the estimated probabilities seems preferable, which is also the conclusion drawn by Hosmer et al. (1988).

What other options do we have? Landwehr et al. (1984) proposed a procedure for partitioning the (total) deviance into a lack-of-fit component and a pure error component, but Jennings (1986b) argues that the test is undermined by the fact that the test depends on $E(d_i^2)$, which strongly depends upon π_i. This is discussed further in Section 9.7.2. [The components of the deviance in Eq. (9.16) are given by d_i.] As indicated previously, the deviance is best used for determining if a variable should be added to the model, not for testing the adequacy of the model.

The correct classification rate (CCR) was suggested in Section 9.4.3. This is certainly the easiest statistic to relate to of those that have been discussed and is an important statistic when classification is one of the objectives of the study [as it was with Brown (1980)]. The value of CCR is 60.4 in this example, as 32 of the 53 patients are properly classified, using 0.5 as the cutoff value. The value of $\overline{R}^2$, defined in Section 9.4.1, is .078, which contrasts sharply with the value of CCR. When the outlier (187) is removed, CCR increases to 69.8, and the value of $\overline{R}^2$ increases to .212. As indicated by Hosmer and Lemeshow (1989, p. 147), such a classification percentage is fairly typical. Nevertheless, there is obviously considerable misclassification, which is caused by the considerable overlap in the data. It is also apparent that $\overline{R}^2$ cannot be easily interpreted, and it is disturbing that very small values of $\overline{R}^2$ can result for typical classification percentages. This is because $\overline{R}^2$ will be close to zero whenever at least one value for Y is poorly predicted, since $L(\hat{\boldsymbol{\beta}})$ is the *product* of components which reflect the quality of the fit at each point. So a single poorly predicted point is sufficient to produce a very small value of $\overline{R}^2$. Consequently, CCR, even with its obvious shortcomings, seems to be a better indicator of the worth of the model, with $\overline{R}^2$ perhaps used only as a supplementary statistic.

We might be tempted to try to increase the CCR by seeing what happens when we use a different cutoff percentage. Although this might seem as though we would be trying to play games with numbers, Hosmer and Lemeshow (1989, p. 147) indicate that "classification ... will always favor classification into the larger group." That is, if there are, say considerably more 1's than 0's among the Y values, we would expect most of the $\hat{\pi}_j$ to be closer to 1 than to 0. This seems apparent, as we would expect the greater proportion of 1's to cause there to be more $\hat{\pi}_j$ above 0.5 than below 0.5.

Neter et al. (1989, p. 610) suggest that the proportion of 1's serve as the initial cutoff value in searching for an optimal value. Implementing this suggestion for the current example produces an initial cutoff of $\frac{20}{53} = .3774$. This does not change the number of misclassifications, however, but reducing the cutoff to .37 reduces the number of misclassifications from 16 to 13. (Here we assume the exclusion of the outlier.)

This raises an important question. Should we use a cutoff value that minimizes the number of mislcassifications? This was done by Brown (1980) and is also illustrated by Neter et al. (1989, p. 611). This seems quite reasonable, provided that the costs of each type of misclassification are either approximately equal or else are inversely related to the number of misclassifications of each type. For example, if the cost of not detecting cancer is greater than the cost of erroneously concluding that a person has cancer, then we would want the number of misclassifications of the first type to be smaller than the number of misclassifications of the second type. Although exact costs could obviously not be determined when the "unit" is human suffering and there is possible loss of life, a rough ratio of costs could seemingly be determined.

We will henceforth use the maximum correct classification rate, maximized over the cutoff value, as a global measure of goodness of fit. (This will sometimes be designated as MCCR.) For the current example, a cutoff value of .37 causes nine false positives and four subjects with cancer who are not detected. These numbers seem reasonable relative to the type of error that we would prefer most to avoid.

It would be reasonable to use diagnostics to see if the model can be improved when MCCR is less than 100, just as one does in linear regression even when R^2 exceeds 0.9.

9.7.2 Which Residuals?

In our study of the adequacy of the fitted model, we must decide which of the proposed residuals we are to use. Specifically, we can choose between the *Pearson residual* and the *deviance residual*.

The former is defined as

$$r_i = r(y_i, \hat{\pi}_i) = \frac{y_i - m_i \hat{\pi}_i}{\sqrt{m_i \hat{\pi}_i (1 - \hat{\pi}_i)}} \tag{9.19}$$

where, if the response values are aggregated, y_i denotes the number of times that $y = 1$ among the m_i repeats of X_i. (Of course, all m_i could conceivably equal 1.) If the response values are *not* aggregated, then the y_i are the actual response values, with each m_i equal to 1. (See Section 9.7.3 for a discussion of data aggregation.) This is called the Pearson residual because $\sum r(y_i, \hat{\pi}_i)^2$ is the Pearson chi-square statistic.

The deviance residual is defined as

$$
d_i = d(y_i, \hat{\pi}_i) = \text{sign}\,(y_i - m_i\hat{\pi}_i)\left\{2\left[y_i \log\left(\frac{y_i}{m_i\hat{\pi}_i}\right)\right.\right.
$$
$$
\left.\left. + (m_i - y_i)\log\left(\frac{m_i - y_i}{m_i(1 - \hat{\pi}_i)}\right)\right]^{1/2}\right\} \tag{9.20}
$$

where the components of Eq. (9.20) are defined the same as they are for Eq. (9.19), depending on whether the response values are aggregated. We may simplify Eq. (9.20) by removing undefined and zero components. If we rewrite the expression within the curly brackets, excluding $2^{1/2}$, we obtain (assuming each $m_i = 1$)

$$
y_i \log\,(y_i) - y_i \log\,(\hat{\pi}_i) + (1 - y_i)\log\,(1 - y_i) - (1 - y_i)\log\,(1 - \hat{\pi}_i)
$$

Clearly only the second and fourth terms are needed, since the first and third terms will be either zero or undefined, depending on whether y_i is 0 or 1.

Thus, if each $m_i = 1$, we may simplify Eq. (9.20) to the equivalent expression

$$
d_i = \text{sign}\,(y_i - \hat{\pi}_i)\{2[-y_i \log\,(\hat{\pi}_i) - (1 - y_i)\log\,(1 - \hat{\pi}_i)]\}^{1/2} \tag{9.21}
$$

where one of the terms in the radicand must be zero. The connection between the deviance residual and the log likelihood should now be apparent, since the parenthetical expression in the radicand in Eq. (9.21) is the same as the expression in Eq. (9.5) that is being summed.

We may simplify Eq. (9.21) further to obtain separate expressions for d_i for $y_i = 0$ and $y_i = 1$. For $y_i = 0$, $d_i = -\{2[-\log\,(1 - \hat{\pi}_i)]\}^{1/2}$, and for $y_i = 1$, $d_i = \{2[-(\log\,(\hat{\pi}_i)]\}^{1/2}$.

As discussed by Pregibon (1981), d_i measures the disagreement between the ith component of the log likelihood of the fitted model and the corresponding component of the log likelihood that would result if each point were fitted exactly.

McCullagh and Nelder (1989, p. 398) express a preference for the deviance residuals because they are closer to being normally distributed than are the Pearson residuals, as demonstrated by Pierce and Schafer (1986). The latter do admit, however, that their evidence that supports the deviance residual is rather vague. Furthermore, they were looking primarily at distributional properties and not at the possible need to modify the existing model. Approximate normality is certainly a desirable property of residuals, but it is also desirable to use some type of residual that will detect the need for necessary modifications to a logistic regression model so as to, in particular, improve CCR or MCCR.

A perhaps more compelling reason for preferring the deviance residual was given by Pregibon (1981), who noted that the Pearson residuals are unstable when $\hat{\pi}_i$ is close to either 0 or 1.

A related issue is the implied normal approximation assumption. When viewed from a different perspective, the Pearson residual in Eq. (9.19) is in the general form of the normal approximation to the binomial, which will not work well when p (or in this case, $\hat{\pi}_i$) is close to 0 or 1. Since a continuity correction factor is used when the normal distribution is used to approximate binomial probabilities, we should ask if the same correction factor (i.e., $\pm\frac{1}{2}$) should be used in computing the residuals, since they are assumed to be approximately normally distributed. Pierce and Schafer (1986) did recommend that this correction factor be used for deviance, Pearson, and other types of residuals, but Duffy (1990) examined their suggestion and concluded that "continuity correction is a poor idea for residuals in logistic regression."

Although Pregibon (1981), Landwehr et al. (1984), and McCullagh and Nelder (1989) prefer the deviance residual over the Pearson residual, Hosmer and Lemeshow (1989) advocate plotting a function of the square of each against $\hat{\pi}_i$. Specifically, the latter recommend plotting $r_{is}^2 - r_i^2/(1-h_{ii})$ against $\hat{\pi}_i$, where h_{ii} is the ith leverage value and r_{is}^2 is the square of the *standardized Pearson residual*. (See Section 9.7.6 for the form of the matrix **H** in logistic regression). The motivation for this plot is that $r_i^2/(1 - h_{ii})$ is the approximate change in the Pearson chi-square statistic that would result from deleting X_i. (Here we assume that X_i is not repeated.) Similarly, $d_i^2/(1 - h_{ii})$ estimates the change in the deviance that would result from the deletion of X_i, so the value of this statistic would also be plotted against $\hat{\pi}_i$.

It should be noted that these estimates are not very precise (Pregibon, 1981), although they may be adequate for detecting outliers. Hosmer and Lemeshow (1989) use 4 as the cutoff value for each statistic, relative to identifying outliers. They motivate the use of this cutoff value by pointing out that the 95th percentile of a chi-square distribution with one degree of freedom is 3.84. But the distribution of r_i approaches a normal distribution only as $m_i \to \infty$. Hardly ever will we encounter data where at least some of the m_i are quite large, and even a large m_i may be insufficient if $\hat{\pi}_i$ is close to 0 or 1. Therefore, using their suggested cutoff value seems unwise, and the lack of an obvious threshold value for r_{is}^2 mitigates against its use.

There are also problems with both d_i and d_i^2. As Jennings (1986b) has shown, both $E(d)$ and $E(d^2)$ depend on π, and $E(d^2)$ varies greatly over π. In particular, $E(d^2|\pi = .01) = E(d^2|\pi = .99) = 0.112$, as compared to $E(d^2|\pi = .50) = 1.386$. Thus, departures from the model at certain points could be confounded with $E(d_i^2|\pi_i)$. Jennings suggests that one possible approach to remedying this problem is to use the standardized statistic $d^2/E(d^2)$, although the properties of such a statistic when π_i is replaced by $\hat{\pi}_i$ are apparently unknown.

Following McCullagh and Nelder (1989, p. 105) and Pierce and Schafer

(1986), we might replace d_i by a modified deviance statistic given by

$$d_i^* = d_i + \frac{1 - 2\hat{\pi}_i}{\{m_i\hat{\pi}_i(1 - \hat{\pi}_i)\}^{1/2}} \qquad (9.22)$$

Pierce and Schafer (1986) show that d_i^* is approximately normally distributed, even when the m_i are very small. Thus, provided that we have at least a few repeats of the X_i, we might consider using d_i^* as a statistic for detecting outliers. But d_i^* will be inflated when $\hat{\pi}_i$ is close to zero and m_i is small, however, with the possible consequence that a point is spotlighted at which the fit is quite good. (This will be seen in Exercise 15.4 of Chapter 15.) Consequently, the modified deviance statistic should be used cautiously, if at all.

Another possibility would be to use $d_i' = d_i/\hat{E}(d_i)$, where $\hat{E}(d_i)$ is the estimate of the expected value. [See Myers (1990) for more information on this approach.]

9.7.3 Application to Table 9.3 Data

Before the statistics given in Section 9.7.2 can be applied, a decision must be made as to whether the Y_i that have a common X_i are to be aggregated before the statistics are computed, assuming that at least some of the X_i are repeated. This issue is discussed by Hosmer and Lemeshow (1989, p. 152), who recommend that aggregation be used if the number of different X_i is much less than n, or if some of the m_i are greater than 5.

9.7.3.1 Pearson Residuals

If the Y_i are aggregated, the Pearson residuals are then defined in accordance with the first definition for r_i given in Eq. (9.19), and we would similarly use the form for the deviance residuals given by Eq. (9.20). Why is aggregation desirable before computing diagnostics? With both the Pearson and deviance residuals we are essentially looking at the difference between y_i and $m_i\hat{\pi}_i$, and if there is a considerable discrepancy between $\hat{\pi}_i$ and the response value (if the response values are the same) at a particular X_i, then this discrepancy is being magnified by the value of m_i.

For example, assume that each response value is zero at each of the m_i values of X_i. We can see from Eq. (9.19) that aggregating the response values would cause r_i to be $\sqrt{m_i}$ times what it would be without aggregation. Consequently, what might be only a moderate value of r_i could become fairly large if m_i is not small, and thus perhaps spotlight a poor fit at X_i. (This magnification could also produce false signals, however, as fits that are not particularly bad could also be magnified.) The aggregation issue does not arise in linear regression, although one could argue that some form of aggregation should also be used there for the same reason that it is suggested for logistic regression.

For the data in Table 9.3 there are 33 different X_j (excluding the obvious outlier, 187), and $n = 52$; this difference suggests that aggregation be used. The standardized Pearson residuals are given in Table 9.5. If, following Hosmer and Lemeshow (1989), we flag for scrutiny any X_i (and the corresponding response values) for which $r_{is}^2 \geq 4$, the only value that is flagged is $X_i = 67$. Although this is one of the lowest acid phosphatase levels associated with nodal involvement, the three people who had this level all had nodal involvement. Since there are patients with lower acid phosphatase levels who had nodal involvement, $(X, Y) = (67, 1)$ does not seem all that unusual for these three patients.

Table 9.5 Standardized Pearson Residuals for Table 9.3 Data

r_{is}	$\hat{\pi}_i$ (ordered)	X_i (ordered)
−0.608	0.145	40
−0.482	0.183	46
−0.493	0.190	47
0.626	0.198	48
0.582	0.205	49
−1.273	0.213	50
1.911	0.221	51
−0.796	0.229	52
−0.851	0.255	55
0.780	0.264	56
−0.650	0.293	59
−0.998	0.323	62
−0.714	0.333	63
−0.748	0.354	65
−0.766	0.365	66
2.315	0.376	67
1.742	0.410	70
−0.862	0.421	71
1.162	0.433	72
−0.950	0.468	75
0.062	0.479	76
−0.006	0.503	78
0.946	0.538	81
1.338	0.550	82
−1.155	0.561	83
0.885	0.573	84
0.792	0.629	89
−1.557	0.693	95
−1.677	0.722	98
0.634	0.731	99
−1.852	0.758	102
0.340	0.907	126
0.267	0.940	136

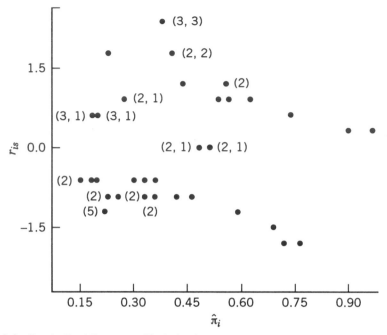

Figure 9.4 Standardized Pearson residual plot for Table 9.3 data. A single number beside (or under) a plotted point signifies the number of repeats of that point. When there are two numbers, the first is the number of repeats and the second is the number of misclassified points at that value of $\hat{\pi}_i$, using a cutoff value of 0.5.

Furthermore, $\hat{\pi}$ = .3760, which is virtually equal to the cutoff value of .37 that was used in Section 9.4. Thus, the value of $\hat{\pi}$ is certainly not unreasonable, whether the response value is 0 or 1. Recall from Section 9.4 that the values of $\hat{\pi}_i$ will be affected by the relationship between the number of 1's and the number of 0's, but this is not taken into acocunt when r_{is} is computed. Consequently, here we receive a signal that the response value at X_i = 67 should be investigated.

This example illustrates why the numerical explanation of a large value of r_{is} should be sought before regarding the response value(s) at X_i as being somewhat suspicious.

Plotting r_i (or r_{is}) against $\hat{\pi}_i$ can provide useful information. For example, a plot of r_{is} against $\hat{\pi}_i$ gives a graphical representation of the correct classification rate if 0.5 were used as a cutoff value, as can be seen from Figure 9.4, which is the plot of r_{is} against $\hat{\pi}_i$ for the current example. A single number beside a point indicates the number of repeats of a value of acid phosphatase that produces the indicated value of $\hat{\pi}_i$. Two numbers are given for certain points for which there are repeat values for X_i (and hence for $\hat{\pi}_i$). The first is the value of m_i, and the second is the number of misclassified points for this $\hat{\pi}_i$. Of the quadrants that result from displaying the zero line for the r_{is} and the cutoff value for $\hat{\pi}_i$ as a horizontal line, the first and third quadrants show the $\hat{\pi}_i$ (and thus, indirectly, identify the X_i) at which at least one misclassification occurs. (These would denote the misclassified points if all $m_i = 1$.)

Displaying these two numbers provides additional insight. In particular, a moderately large value of r_{is} that occurs with the two numbers being equal may mean that the value of r_{is} is due more to the aggregation of the data than to a bad fit at that value of X_i. The second number in each pair can be used to determine the total number of misclassified points, as well as the number that correspond to $Y = 0$ and to $Y = 1$.

We may also see from this plot how close each point is to the cutoff value. Recall that a cutoff value of .37 was found to maximize the correct classification rate. If we think about sliding the vertical line at $\hat{\pi} = .50$ to the left, we will be shifting some points into the second quadrant, and some into the third quadrant. Since the second and fourth quadrants contain the points that have been correctly classified, the CCR will increase only if there are more misclassified points in the first quadrant that are close to the line than there are properly classified points that are close to the line in the fourth quadrant.

We can see from Figure 9.4 that the CCR will increase if we lower the cutoff value since the multiple misclassifications in the first quadrant that are close to the $\hat{\pi} = .50$ line exceed the number of correctly classified points that are in the fourth quadrant and are close to the line. Thus, a modified plot of r_{is} versus $\hat{\pi}_i$ such as is given here can be of value, although we should remember not to view the value of r_{is} too rigidly.

9.7.3.2 Deviance Residuals

Although a preference for the deviance residuals has been demonstrated in the literature, in this example the deviance residuals are very highly correlated with the (unstandardized) Pearson residuals, as the (Pearson) correlation coefficient is .993. The reader is asked to construct the plot of d_i versus $\hat{\pi}_i$ in Exercise 9.2. Because of this high correlation, we would expect the plot to look very similar to the plot in Figure 9.4. Although the distribution of the deviance residuals will, in general, be closer to a normal distribution than will be the distribution of the Pearson residuals, that is not true for the empirical distributions of the two types of residuals for this example.

9.7.4 Other Diagnostics

Following Cook and Weisberg (1982, p. 192), we might adapt Cook's-D statistic to logistic regression. Doing so for one regressor produces the statistic

$$D_i^* = \frac{1}{2}\, r_{is}^2 \left(\frac{h_{ii}}{1 - h_{ii}} \right) \tag{9.23}$$

but D_i^* will obviously not contribute anything beyond what is obtained using r_{is} or r_{is}^2, unless there are considerable differences in the leverages. The leverages may differ very little when there are many repeats, since the upper bound on each leverage value is $1/m_i$. For the current example the values of the parenthetical expression in Eq. (9.23) are approximately 0.10 for each i.

Another possible diagnostic is the logistic regression adaptation of Atkinson's (1981) modification of Cook's-D statistic, which is given by

$$C_i^* = \left\{ \frac{n-p}{p} \cdot \frac{h_{ii}}{1 - h_{ii}} \right\}^{1/2} |r_{is}^*| \tag{9.24}$$

where

$$r_{is}^* = \frac{y_i - m_i \hat{\pi}_{(i)}}{\sqrt{m_i \hat{\pi}_{(i)}(1 - \hat{\pi}_{(i)})} \sqrt{1 - h_{ii}}} \tag{9.25}$$

with (i) signifying that the ith observation is not used in the computations, p is the number of parameters, and h_{ii} is the ith diagonal element of the $\mathbf{H}$ matrix for logistic regression that is discussed in Section 9.7.6. As in linear regression, the user of this statistic would be faced with the problem of determining what constitutes a large value of the statistic. Therefore, it might be necessary to use simulated envelopes for the modified Cook's-D, just as Atkinson (1981) advocated for linear regression.

One possible problem in using C_i^* is that r_{is}^* will not be easily obtainable unless software is being used that computes deletion residuals for logistic regression. Unfortunately, these cannot be obtained with the major statistical packages. Therefore, it might be necessary to use a one-step approximation to the deletion (standardized) Pearson residual in (9.24). [One-step approximations for this and other statistics in logistic regression are discussed by Pregibon (1981).] Alternatively, we could use the deviance residual rather than the Pearson residual and use a one-step approximation to the deletion deviance residual, as suggested by McCullagh and Nelder (1989, p. 407).

9.7.5 Partial Residual Plot

A partial residual plot was shown in Chapter 5 to be sometimes useful in detecting the need to modify a linear regression model. This plot has also been suggested for use in logistic regression by Landwehr et al. (1984), and a method based on smoothing is given by Fowlkes (1987).

As emphasized by Fowlkes (1987), smoothing of binary data is necessary if certain diagnostic plots are to be meaningful. We will illustrate the method of Fowlkes (1987) for a single regressor (only), as the smoothing is considerably more involved for multiple regressors (see Fowlkes, 1987). Each observed value of Y is replaced by a smoothed value, Y_s, which is defined as $Y_s = \sum (w_k Y_k / \sum w_k)$ with the weights w_k computed when there is a single regressor as $w_k = \{1 - (|X_k - X_s|/d_s)^3\}$. The X_k are the $(n/10)\%$ of the values that are closest to each X_s (including X_s), and d_s is the largest difference between X_k and X_s, for a given value of X_s. It is highly desirable to use a computer program to perform this smoothing, even when there is only a single regressor.

The partial residuals are then computed as

$$r_i^* = x_i \hat{\beta}_1 + \frac{Y_{s,i} - \hat{\pi}_i}{\hat{\pi}_i (1 - \hat{\pi}_i)} \tag{9.26}$$

where $\hat{\beta}_1$ and $\hat{\pi}_i$ are obtained from the logistic regression model that is obtained using the unsmoothed Y values. The plot is then produced by plotting the r_i^* against the X_i.

The partial residual plot for the Table 9.3 data (using acid phosphatase only) is given in Figure 9.5. Notice that there is strong evidence of a linear trend, except at large values of acid phosphatase. The latter results from the almost equal number of patients with nodal involvement and with no nodal involvement at the largest acid levels. Thus, acid phosphatase is not a good predictor at its highest levels, so these smoothed values do not fit with the other smoothed values. Although this causes the plot to exhibit some nonlinearity, this appears to be more of a lack of a linear relationship at large acid levels rather than being

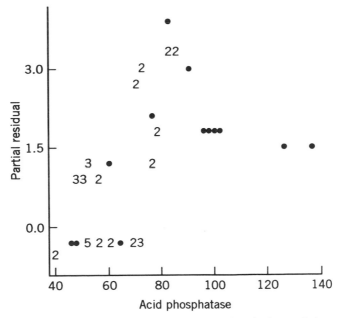

Figure 9.5 Partial residual plot for Table 9.3 data. The numbers in the graph denote the number of points at each location.

suggestive of a nonlinear relationship at those levels. When a quadratic term is added to the model, the value of CCR declines slightly, thus suggesting that only the linear term is needed.

It could be shown that the scatter plot of Y_s against X is very similar to the partial residual plot, as would be expected. The reason for this is explained in the chapter appendix.

The partial residual plot suggested by Landwehr et al. (1984) differs from the plot that results from using the expression given in Eq. (9.26) only in that Y_i is used instead of $Y_{s,i}$. They do suggest that the plot be smoothed, however.

9.7.6 Added Variable Plot

An added variable plot for logistic regression has been given by Pregibon (1985) and Wang (1985), and a different version was given by O'Hara Hines and Carter (1993).

Since, as indicated in Section 9.2.1.1, Z as used herein plays the role of Y in linear regression, the natural extension of an added variable plot to logistic regression entails plotting $(\mathbf{I} - \mathbf{H_X})\mathbf{W}^{1/2}\mathbf{Z}$ against $(\mathbf{I} - \mathbf{H_X})\mathbf{W}^{1/2}\mathbf{A}$, with $\mathbf{H_X} = \mathbf{W}^{1/2}\mathbf{X}(\mathbf{X'WX})^{-1}\mathbf{X'W}^{1/2}$. The matrix $\mathbf{X}$ does not contain the regressor that is represented by $\mathbf{A}$, for which the added variable plot is constructed, and $\mathbf{W}$ is the matrix of weights, defined for Eq. (9.10), that results from *excluding* the regressor represented by $\mathbf{A}$. Notice that the ordinate and abscissa for this plot are of the same general form as those for the added variable plot in linear regression, as described in Section 5.5. Thus, with this plot for logistic regression we are also plotting two sets of residuals: the residuals that result from regressing Y against all of the regressors except the one in $\mathbf{A}$ and the residuals that result from regressing the regressor in $\mathbf{A}$ against all of the other regressors.

The modification suggested by O'Hara Hines and Carter (1993) entails replacing $\mathbf{H_X}$, $\mathbf{W}^{1/2}$, and $\mathbf{Z}$ by their counterparts that result when $\mathbf{A}$ is *included* in the regression, rather than being excluded. They argue that this produces a more reliable plot for detecting influential observations and for estimating their effect.

9.7.7 Confidence Intervals for Table 9.3 Data

Once it has been determined that the model cannot be improved upon, and there are no extreme observations that have a deleterious effect, inferences can then be made.

As stated previously, $\hat{\beta}_1 = 0.0471$ for the Table 9.3 data when the outlier is excluded. It can be shown that $s_{\hat{\beta}_1} = \sqrt{0.00033}$, so an approximate 95% confidence interval for β_1 is obtained as

$$\hat{\beta}_1 \pm z_{\alpha/2} s_{\hat{\beta}_1} = 0.0471 \pm 1.96\sqrt{0.00033} = 0.0471 \pm 0.0356$$

so the interval is given by (0.0115, 0.0827). By comparison, the exact parameter estimate is 0.0461, and the exact 95% confidence interval is (0.0133, 0.0844). Thus, in this case the exact parameter estimate and exact confidence interval differ only slightly from the approximate values.

Consequently, there will also be only slight differences when the approximate and exact confidence intervals for the change in the log odds ratio are obtained. Assume that we wish to obtain the latter for a change of 10 units in X. We obtain (1.1219, 2.2864) for the approximate interval, with $\exp(10\hat{\beta}_1) = 1.6016$. How do we interpret these numbers? We would estimate that for every 10-unit increase in acid phosphatase level, the risk of nodal involvement increases 1.6 times, with the 95% bounds on our estimate being 1.1219 and 2.2864. The exact confidence bounds are 1.1422 and 2.3256. Clearly the approximate values are adequate in this example.

9.8 AN EXAMPLE OF MULTIPLE LOGISTIC REGRESSION

We will now consider the full data set given by Brown (1980), which is given here in Table 9.6. We will employ primarily the maximum likelihood estimation approach, although some p-values from the use of exact logistic regression are also given in Section 9.8.3.

Table 9.6 Full Data Set from Brown (1980)

Y	Acid Phosphatase	X-ray	Stage	Grade	Age
0	48	0	0	0	66
0	56	0	0	0	68
0	50	0	0	0	66
0	52	0	0	0	56
0	50	0	0	0	58
0	49	0	0	0	60
0	46	1	0	0	65
0	62	1	0	0	60
1	56	0	0	1	50
0	55	1	0	0	49
0	62	0	0	0	61
0	71	0	0	0	58
0	65	0	0	0	51
1	67	1	0	1	67
0	47	0	0	1	67
0	49	0	0	0	51
0	50	0	0	1	56
0	78	0	0	0	60
0	83	0	0	0	52

Table 9.6 (*Continued*)

Y	Acid Phosphatase	X-ray	Stage	Grade	Age
0	98	0	0	0	56
0	52	0	0	0	67
0	75	0	0	0	63
1	99	0	0	1	59
0	187	0	0	0	64
1	136	1	0	0	61
1	82	0	0	0	56
0	40	0	1	1	64
0	50	0	1	0	61
0	50	0	1	1	64
0	40	0	1	0	63
0	55	0	1	1	52
0	59	0	1	1	66
1	48	1	1	0	58
1	51	1	1	1	57
1	49	0	1	0	65
0	48	0	1	1	65
0	63	1	1	1	59
0	102	0	1	0	61
0	76	0	1	0	53
0	95	0	1	0	67
0	66	0	1	1	53
1	84	1	1	1	65
1	81	1	1	1	50
1	76	1	1	1	60
1	70	0	1	1	45
1	78	1	1	1	56
1	70	0	1	0	46
1	67	0	1	0	67
1	82	0	1	0	63
1	67	0	1	1	57
1	72	1	1	0	51
1	89	1	1	0	64
1	126	1	1	1	68

One of the objectives of the investigator was to determine if an elevated level of acid phosphatase has any diagnostic value in predicting lymph node involvement beyond the contribution of the other four, customarily used, variables. We can obviously answer this question by fitting both models and determining if the marginal contribution of acid phosphatase is significant. We will do this, but we will also proceed as if we did not know that the "true model" was one of these two models.

When the full model is used we obtain the prediction equation

$$\hat{\pi} = \frac{\exp[0.0618 + 0.0243\text{A.P.} + 2.0453\text{X-ray} + 1.5641\text{Stage} + 0.7614\text{Grade} - 0.0693\text{Age}]}{1 + \exp[N]}$$

where N denotes the bracketed expression in the numerator of the fraction. Here we are initially using all 53 observations, including the observation that was eventually deleted in the univariate analysis. We will initially include it in our analysis, however, primarily for the purpose of seeing if it can be detected by using the diagnostics that have been discussed in preceding sections, assuming that it turns out to be influential. We should also bear in mind that a data point that has only one extreme coordinate should not be automatically deleted, because the corresponding variable might not be included in the eventual model.

It was mentioned in Section 9.5.1 that some research suggests that the Wald statistics are unreliable, and that the likelihood ratio statistics are a better choice. For this data set the two sets of p-values differ very little, however. (If the two sets of values did differ considerably, this might mean that the maximum likelihood approach could produce misleading results.) Only X-ray and Stage have p-values less than .05, with the (likelihood ratio) p-value for Acid Phosphatase (A.P.) being .0634. (It is interesting to note that the standard error of the constant term is 56 times the value of the constant.)

It is of interest to compare these results with the results given by Brown (1980). The latter dichotomized the continuous variables Acid Phosphatase and Age and analyzed the data using contingency tables and multiple logistic regression. The contingency table analysis revealed that X-ray was the best single predictor, and this was also the result when stepwise logistic regression was used. (The latter is discussed in Section 9.8.3.1.) X-ray has the smallest p-value in our analysis, so the results are consistent.

It appears that Age is of essentially no value as a predictor. Its correlation with Acid Phosphatase is .054, and it seems to be essentially unrelated to the other three dichotomized variables. Thus, its contribution, which seems to be close to nil, is almost independent of the contributions of the other predictors.

The dichotomization performed by Brown was obviously done so that a contingency table analysis could be performed. (This is a common occurrence in the medical literature.) Although such dichotomization is unnecessary for multiple logistic regression, Brown indicates a preference for it since "the regression coefficients for the five variables are all in the same units." Nevertheless, we are discarding potentially useful information whenever we dichotomize a continuous random variable. Brown does not select a model, but simply concludes that acid phosphatase is "a useful prognostic indicator" that adds to the predictive ability of the model with the other four variables.

Although measurements on the other four variables might continue to be routinely made, there is certainly a question of whether they should be included in the model. As in linear regression, we must realize that investigators will often insist on using a variable that they believe to be important, even if it shows up nonsignificant in the analysis. Indeed Hosmer and Lemeshow (1989, p. 88) state that excluded variables should be "biologically or statistically unimportant," thus leaving open the possibility that biologically important variables might be included. Nevertheless, the inclusion of variables that are not statistically significant could unnecessarily inflate $Var(\hat{\pi})$. (Recall the discussion at the beginning of Chapter 7 regarding the increase in $Var(\hat{Y})$ in multiple linear regression when variables are added to a model.)

This is of particular importance relative to the present example, since Brown (1980) states that "in certain cases the logistic model may provide a probability estimate so close to 0 or 1 as to obviate the need for surgery, especially in the patient considered a poor risk with regard to the surgical procedure." Therefore, good estimation of the π_i values is obviously of paramount importance for this application, especially if a decision to operate or not operate depends on how close $\hat{\pi}_i$ is to 0 or 1, rather than whether $\hat{\pi}_i$ exceeds the chosen threshold value.

In the next section we investigate the classification rate and the extent to which it is improved by using all five variables.

9.8.1 Correct Classification Rate for Full Data Set

If we use 0.5 as the threshold value, we have a CCR of $\frac{41}{53} = .77$. As stated in Section 9.7, we are initially using all 53 observations. If the point that was deleted in the one-variable analysis were also deleted here, the CCR would increase to .83, but the Hosmer–Lemeshow statistic, using fixed cutoff points, increases from 5.757 to 8.803. This again underscores problems in using this option, since a model that affords better classification should be judged superior by a goodness-of-fit test that indirectly assesses the classification performance of the model.

When we compare the CCR for the full model with the CCR for the model containing only acid phosphatase, we see that the rate has improved from 69.8 to 77.4%. Again, following Neter et al. (1989), we will use an initial cutoff of $\frac{20}{53} = .377$ and see if the CCR improves. Unfortunately, that increases the number of misclassifications by one.

Figure 9.6 is the plot of r_i against $\hat{\pi}_i$ that corresponds to Figure 9.4 for the one-variable analysis, except that here r_i is plotted instead of r_{is}, and no covariance patterns are repeated. In addition to allowing us to detect aberrant points, it also allows us to see if the CCR will increase or decrease for a change in the cutoff value. Here we see that we should *increase* the cutoff from .50 in order to lower the number of misclassifications, and raising the cutoff to .60 lowers the number of misclassifications to 10. We observe that we cannot reduce the number of misclassifications further, so the best CCR is $\frac{43}{53} = .81$.

Interestingly, when Brown used the model with all dichotomized variables, the optimal cutoff value was .35, which produced only nine misclassifications. We might guess that this could signal the presence of an influential outlier. When observation 24 is deleted, the number of misclassifications is nine using

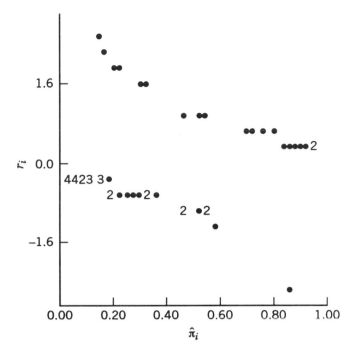

Figure 9.6 Pearson residual plot for Table 9.6 data. The numbers in the graph denote the number of points at each location.

.5 as the cutoff, and that number can be reduced to eight if a cutoff of .513 is used. Thus, the MCCR is $\frac{44}{52} = .85$.

9.8.2 Influential Observations

We saw in the one-variable case that observation 24 had a sizable effect on the fitted model. Here with all five variables the leverage at that point is 0.56. Since $p/n = .2264$, this is roughly $2.5(p/n)$ and is thus potentially influential. The value of Cook's-D for this point is 0.58. The corresponding values for the other points are much smaller, so this point should be scrutinized.

Unfortunately, the point is not indicated as being unusual by the residuals, however, as the (standardized) Pearson residual is -1.643, and the deviance residual is -1.251. But the point is clearly influential, as $\hat{\pi}_{24}$ changes from 0.54 to 0.97 when the point is deleted. Thus, the point is not declared extreme by the Pearson residual or the deviance residual because it exerts considerable influence on the fitting of itself.

Although most of the $\hat{\pi}_j$ do not change greatly when the point is deleted, many of them do change by more than 0.1, and 21 of the 53 change by more than 0.05, so the point is clearly influential.

This shows the importance of using deletion residuals such as r_{is}^*, given in

Eq. (9.25). The value of r_{is}^* is obtained as

$$r_{24s}^* = \frac{0 - .9652}{\sqrt{(.9652)(.0348)} \sqrt{1 - .5607}}$$

$$= -7.946$$

We thus have an extremely large value for the Pearson deletion residual, and, if we computed the value of C_{24}^*, we would obtain $C_{24}^* = 25.127$. (Certainly we should not have to use a simulated envelope in order to declare that this is a very large value.)

Some of the regression coefficients do change considerably when the observation is deleted, as $\hat{\beta}_0$ changes from 0.0618 to -2.0643, and the coefficient of acid phosphatase doubles from 0.0243 to 0.0490. There is the obvious question of what to do about observation 24. Two possible approaches would be to (1) delete the point or (2) downweight it in some manner. [See Copas (1988), Pregibon (1982), and Carroll and Pederson (1993) for robust logistic regression methodology.]

Bedrick and Hill (1990) give a computationally intensive approach for detecting outliers that is based on exact conditional methods, and they apply their approach to this data set. They use all of the variables in dichotomized form, however, and fail to detect an outlier. Their result is an example of why continuous regressors should not be dichotomized for the sake of convenience, as is often done.

9.8.3   Which Variables?

Although the investigator was primarily interested in seeing if it is desirable to add Acid Phosphatase to the model, the statistical evidence suggests that Age and Grade should not be included in the model. There is apparently no biological reason for including Age in the model, and a plot of Y against Age shows no apparent dependence upon Age. When simple logistic regression is used with Age as the sole regressor, the exact p-value for the Age parameter is .35. (The asymptotic value is .34.)

When Grade is used as the sole regressor, however, the exact p-value is .1006 (which differs considerably from the asymptotic value of .0565). Since Grade is binary, we could see the relationship between Grade and Y by constructing a two-way table. Doing so, with the influential point deleted, produces

$$\text{GRADE} \quad \begin{matrix} & & Y \\ & & 0 \quad 1 \\ 0 & \begin{pmatrix} 23 & 9 \\ 9 & 11 \end{pmatrix} \\ 1 & \end{matrix}$$

which shows some relationship between the two. We may use *Fisher's exact test* for a 2 × 2 table (see Cox, 1970, p. 50), and doing so produces a *p*-value of .079. We observe that this differs from the logistic regression exact *p*-value that was given, but it will always equal the *p*-value from the exact conditional probability test. The latter is not covered in this chapter, but is described in Cytel Software Corporation (1993).

Then why isn't Grade a significant regressor in the multiple logistic regression equation? Obviously, this is because there is at least one other regressor that has a stronger relationship with Y and that is also related to Grade. In particular, the relationship between X-ray and Y is considerably stronger than the relationship between Grade and Y, as is apparent from comparing the following table with the one for Y and Grade.

$$
\begin{array}{c}
Y \\
\begin{array}{cc}
0 & 1
\end{array}
\end{array}
$$

$$
\text{X-RAY} \quad \begin{array}{c} 0 \\ 1 \end{array} \begin{pmatrix} 28 & 9 \\ 4 & 11 \end{pmatrix}
$$

This table has a *p*-value (from Fisher's exact test) of .002. There is some relationship between X-ray and Grade, as is apparent from the following table,

$$
\begin{array}{c}
\text{GRADE} \\
\begin{array}{cc}
0 & 1
\end{array}
\end{array}
$$

$$
\text{X-RAY} \quad \begin{array}{c} 0 \\ 1 \end{array} \begin{pmatrix} 25 & 12 \\ 7 & 8 \end{pmatrix}
$$

although the *p*-value (.213) is not particularly small.

It can be similarly shown that Stage is more strongly related to Y than is Grade. The exact *p*-value for the 2 × 2 table of Y and Stage is .011. There is also a strong relationship between Stage and Grade ($p = .011$). Thus, Grade is inferior to Stage as a predictor, and the two predictors are strongly related. Therefore, we can see why Grade would not show up as being significant in the multiple logistic regression model. Notice that the use of these two-way tables is analogous to the use of pairwise correlations in multiple linear regression.

Technically, we should simultaneously select the regressors and the observations that are to be retained. For example, if we had deleted observation 24 and then decided not to use Acid Phosphatase in the model, we would have deleted the point needlessly, since it was the reading on Acid Phosphatase that was causing the point to be influential. On the other hand, if we decide not to use a particular regressor, that decision could have been strongly influenced by one or more extreme values of that regressor.

9.8.3.1 Algorithmic Approaches to Variable Selection

To this point we have tentatively decided that X-ray, Stage, and Acid Phosphatase should be in the model, although the marginal contribution of the latter is indeed "marginal" after observation 24 is deleted. This decision was reached using the p values in the multiple logistic regression model, supplemented by the two-way tables and the scatter plot of Y against Acid Phosphatase.

When we have a small number of potential regressors, as we do here, examining all possible combinations of regressors is feasible. To do so, however, we need either a software package that will do all possible regressions for linear regression or a logistic regression package that will either produce all possible regressions directly or allow them to be easily obtained manually.

As discussed by Hosmer and Lemeshow (1989, p. 121) and Hosmer et al. (1989), we may use a best subsets feature in a linear regression program as a satisfactory, but not equivalent, substitute for a logistic regression program with a best subsets option. To proceed, we would use the same pseudo-variable Z that was given in Eq. (9.11). The best subsets would then be produced from the linear regression program with weights given by $\hat{\pi}_i(1 - \hat{\pi}_i)$. Notice that this requires the linear regression program to be able to select best subsets with weighted least squares. If this is not possible (as with Minitab), then Z, the X_i, and the column of 1's would all be multiplied by $\sqrt{\hat{\pi}_i(1 - \hat{\pi}_i)}$, and an equivalent unweighted analysis would be performed.

Hosmer et al. (1989) show that Mallows's C_p statistic has the same intuitive appeal when best subsets logistic regression is performed in this manner as it does in best subsets linear regression. [Of course, the representation of the C_p statistic is different in logistic regression; see Hosmer and Lemeshow (1989, p. 123).] The C_p values for this example are given in Table 9.7. Notice that the subset {X-ray, Acid Phosphatase, Stage} would be a good choice.

Another possibility would be to use stepwise logistic regression, although the need for a stepwise procedure is lessened when there is only a small number

Table 9.7 C_p Values for Table 9.5 Data[a]

Subset	C_p	p
X-ray	8.0	2
Stage	8.6	2
X-ray, Acid	6.5	3
X-ray, Stage	6.6	3
X-ray, Acid Phos., Stage	4.3	4
X-ray, Acid Phos., Grade	6.2	4
X-ray, Stage, Acid Phos., Grade	5.0	5
X-ray, Stage, Acid Phos., Age	5.2	5
All	6.0	6

[a]The best two subsets of each size are given.

of potential regressors, as in this example. Nevertheless, we still might use a stepwise procedure and compare the results. We need not confine our attention to stepwise logistic regression, as stepwise discriminant analysis might be used for variable selection, even if logistic regression is to be used for estimation. O'Gorman and Woolson (1991) indicate a preference for stepwise discriminant analysis when the sample size is less than 100, based on their simulation study. We will not pursue stepwise discriminant analysis here.

The mechanics of stepwise logistic regression are very similar to those of stepwise linear regression, although the tests have a different form. In testing for the possible inclusion of a regressor, it is a matter of determining if twice the change in the log-likelihood function is statistically significant, as this statistic has asymptotically a chi-square distribution. When testing for possible deletion of a variable, it is a matter of determining if the change in twice the log-likelihood function is *less* than the critical value.

As in stepwise linear regression, a value of α must be chosen for the hypothesis tests. As discussed by Hosmer and Lemeshow (1989, p. 108), no research results have been reported that would suggest a reasonable value of α in stepwise logistic regression, but they suggest that the same range of α that has been recommended for stepwise linear regression (.15 to .20) should also be suitable for stepwise logistic regression. It turns out that the selection of α is not critical for this example, however, because X-ray, Acid Phosphatase, and Stage all enter with p-values less than .05, and the p-value is .2605 for entering the next best variable, Grade.

9.8.3.2 What about Nonlinear Terms?

To this point we have determined the best subset model when only linear terms are considered. Before we declare this model as the one to use, and proceed to examine its predictive properties, we would want some assurance that there is not a better model that contains nonlinear and/or interaction terms.

When the number of potential regressors is not great, we could construct all possible interaction terms and allow them to be potential regressors in a variable selection scheme. We will not do that for this example, however, since Brown (1980) did study the interactions and found that the Stage × Grade interaction was the only one that was significant, and it was declared to be borderline significant.

As far as possible nonlinear terms are concerned, we have only one continuous regressor (Acid Phosphatase) that is of any importance (as a linear term). Perhaps a nonlinear term in Acid Phosphatase would also be important. We have essentially two possible courses of action in this regard: we could use the partial residual plot approach of Fowlkes (1987) or we could use the Box–Tidwell transformation concept, add Acid × log(Acid) to the selected model, and see if the term is significant. If the term were significant, we might use a logistic regression adaptation of the Box–Tidwell approach to determine the power transformation to apply to Acid Phosphatase. Another possible approach is the "marginal distribution of covariates" method of Kay and Little (1987), but that approach requires enough observations to be able to determine the distribution of a predictor at $Y = 0$ and at $Y = 1$.

When the term Acid Phosphatase $\times$ log (Acid Phosphatase) is tested for entry into the model that already contains X-ray, Stage, and Acid Phosphatase, the term is declared not significant, so we select the model that has these three linear terms. [Another transformation approach that might sometimes be useful is the adaptation of the Box–Cox method to logistic regression, as described by Guerrero and Johnson (1982).]

9.9 MULTICOLLINEARITY IN MULTIPLE LOGISTIC REGRESSION

Although multicollinearity has been treated extensively in the linear regression literature, very little attention has been devoted to it in the logistic regression literature. We know from Section 4.1.2.1. that multicollinearity in linear regression can cause the values of the t-statistics to be small when each of two regressors are highly correlated with Y, and we would expect similar problems to occur in logistic regression.

We must select one or more indicators of multicollinearity in logistic regression. If the predictors were all continuous, then pairwise (Pearson) correlations and variance inflation factors might be used. The problem is more complicated when not all of the predictors are continuous, however.

Hosmer and Lemeshow (1989, p. 132) briefly discuss the detection of collinearity in logistic regression and imply that diagnostics similar to those used in linear regression should be helpful. They indicate that large standard errors is a tip-off that collinearity could exist. We will see in this section, however, that harmful collinearity can exist without large standard errors.

Consider the data in Table 9.8 with the two binary predictors. The output for each predictor using both the maximum likelihood and exact estimation approaches is given in Table 9.9. Notice the large differences between the approximate and exact p-values; we might ascribe these differences to the fact that the sample size, $n = 24$, is small. The exact p-values suggest either that both X_1 and X_2 are not needed in the model or that neither is a useful predictor.

From inspection of Table 9.8 we can see that the values of X_1 and Y agree on 18 of the 24 cases, and the same is true for X_2 and Y. Thus, we would suspect that X_1 and X_2 are strongly related. We can see that X_1 and X_2 agree on 16 of the 24 cases, so they seem to be related. It would be helpful to have one or more formal checks for multicollinearity, however. Hosmer and Lemeshow (1989, p. 131) state that most software packages have some type of check for multicollinearity, but no such check is described for the appropriate programs in SAS Software or BMDP.

One possible check for two binary predictors would be to use *Cohen's kappa*, which assesses the extent of agreement between two binary variables relative

Table 9.8   Data for Illustrating Multicollinearity in Multiple Logistic Regression

X_1	X_2	Y
1	1	1
1	1	1
1	0	1
0	1	1
1	1	1
0	0	0
0	0	0
1	0	0
0	0	0
0	0	0
0	0	0
0	0	0
0	1	0
0	0	0
1	1	1
1	0	1
1	1	0
0	1	1
1	1	1
1	1	1
1	1	1
0	0	1
1	0	0
1	0	1

to what would be expected due to chance. The value of this statistic, using X_1 and X_2, is .338. This value can be compared with the discussion in Wilkinson (1990, p. 511), in which it is stated that values below .40 signify poor agreement. Similarly, the p-value for Fisher's exact test is .123, which also suggests a nonsignificant relationship between X_1 and X_2.

Yet when separate models are fit using X_1 and X_2, individually, the exact p-value for X_1 is .0414, and .0341 for X_2. (The corresponding asymptotic p-values are .0207 and .0189, respectively.) Thus, two statistics that can be used for assessing the possible relationship between two binary variables both indicate that there is no relationship between X_1 and X_2, although the data suggest otherwise.

The user of logistic regression is advised to proceed cautiously in the absence of reliable checks for multicollinearity.

Table 9.9 Maximum-Likelihood Estimates and Exact Parameter Estimates for the Table 9.8 Data

Logistic Regression

⟨MODEL: 1⟩

File: TABLE98.CYL

Model: $Y = X1 + X2$
Strat Var: ⟨Unstratified⟩
Liklihd Ratio Stat: 10.4016 on 3 df #Obs: 24

Weight: ⟨Unweighted⟩
#Groups: 4

Parameter Estimation

Variable	Inference Type	Beta	SE (Beta)	95.0% Conf.	Interval	P-Value 2 * 1_Sided
	Asymptotic	1.9044	1.0422	−0.1382	3.9470	0.0677
X1	Exact	1.7111	NA	−0.5269	4.3532	0.1666
	Asymptotic	2.0457	1.0713	−0.0540	4.1454	0.0562
X2	Exact	1.8316	NA	−0.4307	4.5375	0.1369

[↑↓ → ←] Scroll   [TAB] Switch Window   [Esc]Done
                                                   Compute Time 0: 0:23

9.10   OSTEOGENIC SARCOMA DATA SET

To illustrate further how the exact approach can perform quite well when the asymptotic approach fails, we will use a data set given in Hirji et al. (1987). The investigators (Goorin et al. 1987) studied the treatment of non-metastatic osteogenic sarcoma with the objective of determining the predictors of a disease-free interval of at least 3 years. The use of Fisher's exact test had led to the identification of some variables that were statistically significant when considered individually. The investigators then wanted to use these variables in a logistic regression model and estimate the odds ratio for each predictor.

We will study the same part of the full data set that was given in Hirji et al. (1987). The data on the 46 patients involved in the study are given in Table 9.10.

As stated in Cytel Software Corporation (1993, p. 5-23), all of the commonly used software packages fail on this data set, since the maximum likelihood estimates do not exist. This occurs because all of the 10 patients without lymphocytic infiltration were disease-free for at least 3 years.

Table 9.10   Nonmetastatic Osteogenic Sarcoma Data Set[a]

LI	SEX	AOP	DFI	LI	SEX	AOP	DFI	LI	SEX	AOP	DFI
1	0	1	1	0	1	1	1	1	1	0	0
1	1	1	1	0	0	1	1	1	1	1	0
1	1	0	1	1	1	1	1	1	0	1	0
0	0	0	1	1	0	0	1	1	1	1	0
1	1	0	1	0	1	0	1	1	1	1	0
0	1	0	1	1	0	1	1	1	1	1	0
1	1	1	1	1	0	0	1				
1	1	1	1	1	1	0	1				
1	1	0	1	1	0	0	1				
0	0	0	1	1	1	1	0				
1	1	1	1	1	1	1	0				
0	0	1	1	1	1	1	0				
0	1	0	1	1	1	1	0				
1	0	0	1	1	1	1	0				
0	1	0	1	1	1	0	0				
0	0	0	1	1	1	0	0				
1	1	0	1	1	1	1	0				
1	0	0	1	1	0	1	0				
1	1	1	1	1	1	0	0				
1	0	1	1	1	1	1	0				

[a]LI, lymphocytic infiltration (1 = yes, 0 = no); SEX, Male = 1, Female = 0; AOP, any osteoid pathology? (1 = yes, 0 = no); DFI, disease-free interval (1 = ≥ 3 years, 0 = <3 years).

We will see that the exact approach performs well for the Table 9.10 data, but we clearly need a good yardstick that will allow us to make such an assessment. If we construct two-way tables (as in Section 9.8.3), using DFI and each of the three potential predictors, in turn, (thus essentially repeating what the investigators did), we would observe that LI has the strongest relationship with DFI. All three of the p-values are less than .05, however, as the values for LI, SEX, and AOP are .008, .026, and .032, respectively.

The exact parameter estimates are $\hat{\beta}_0 = 3.5352$, $\hat{\beta}_{LI} = -1.8859$, $\hat{\beta}_{SEX} = -1.5479$, and $\hat{\beta}_{AOP} = -1.1561$, and the corresponding confidence intervals are $(1.4773, \infty)$, $(-\infty, 0.1600)$, $(-4.0246, 0.3632)$, and $(-2.9969, 0.5122)$, respectively. The first two estimates could be suspect since the corresponding confidence intervals do not have a finite length. Therefore, it is of interest to compare the estimates with the discriminant analysis estimates. These are 4.0350, -1.9656, -1.5275, and -1.3220, respectively. Obviously these do not differ greatly from the exact estimates, so we would expect the three odds ratios to be comparable for the two sets of estimates. For the exact estimates (using $c = 1$), we obtain 0.152, 0.213, and 0.315, and the corresponding confidence intervals are $(0, 1.1735)$, $(0.0179, 1.4380)$, and $(0.0499, 1.6689)$, respectively. The estimates of the log odds ratios using the discriminant analysis estimates are 0.140, 0.217, and 0.267, respectively. Only for AOP would the difference in the odds ratios likely be considered significant.

Each of these odds ratios indicates that a patient is much more likely to have a 3-year disease-free interval if the patient does not have lymphocytic infiltration (LI), is female (SEX), and is without osteoid pathology (AOP). Stated differently, a patient with lymphocytic infiltration is about one-seventh as likely to be free of disease for 3 years as a patient who does have that characteristic, with the other fractions being approximately one-fifth and one-third for males and patients with osteoid pathology, respectively. [It should be noted that the use of adjusted odds ratios will often be necessary, but we will not pursue that here. The interested reader is referred to Hosmer and Lemeshow (1989, pp. 58–63).]

The p-values for the exact estimates are, in the same order, .0742, .1392, and .2182. (The corresponding p-values using exact conditional scores are .0606, .1169, and .1535.) Thus, none of the variables shows as being a strong, or even statistically significant, predictor. This occurs despite the fact that all three predictors had small p-values when used individually in Fisher's exact test. It can be shown that the use of Fisher's test indicates that the predictors are not strongly related when examined in pairs (as the reader is asked to show in Exercise 9.13).

Because of this, the use of point estimates in the computation of odds ratios is highly questionable, although this is what the investigators wanted to do. A value of zero for a regression parameter corresponds to a value of one for an odds ratio, so if we cannot reject the hypothesis that the value of a logistic regression parameter is zero, then we similarly cannot conclude that an odds ratio differs from one. Consequently, it would be better to use either or both of the end points of confidence intervals when the end points are finite and not trivially zero, so as to give bounds on the odds ratios. (Here the lower bound for LI is trivially zero since there is not a finite lower bound in the confidence interval for β_{LI}.)

Although classification was apparently not one of the objectives of the study, we might still use the value of CCR as a rough indicator of the worth of the model. The optimal cutoff value for $\hat{\pi}$ is actually any value in the interval (.259, .524). (It is helpful to plot Y against $\hat{\pi}$, as this enables us to easily see the optimal cutoff value. The plot is given in Figure 9.7.) For any value in this interval we obtain MCCR = $\frac{34}{46}$ = .739. (We obtain the same value of MCCR in this example whether we use the exact estimates or the discriminant analysis estimates.) This is a fair value, so the model does have some predictive ability.

We would want to conduct a thorough analysis, including the use of diagnostics, before accepting any of these models, however. We will not do so for this data set, however, since diagnostics have been illustrated in previous sections. Our objective here was simply to provide further evidence that the exact estimates can give satisfactory results when the maximum likelihood estimates do not exist.

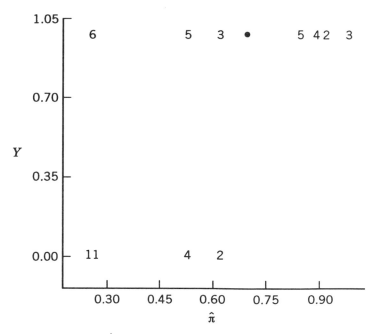

Figure 9.7 Y plotted against $\hat{\pi}$ for Table 9.10 data. The numbers in the graph denote the number of points at each location.

9.11 SAMPLE SIZE DETERMINATION

It was stated in Section 1.5.4 that having at least 10 times as many observations as regressors is desirable in linear regression. Logistic regression presents a somewhat different problem: as if when π is small the data could be sparse, with the result that hypothesis tests could have low power. This was studied by Whittemore (1981) who provided tables of required sample sizes. The necessary sample sizes were found to be strongly influenced by the distribution of the covariate(s). In general, very large sample sizes will generally be required when π is small.

9.12 ALTERNATIVES TO LOGISTIC REGRESSION

It was stated in Section 9.2.1.1 that logistic regression is a logical alternative to discriminant analysis. Similarly, tree-based models are an alternative to logistic regression models for classification problems (and to linear models for regression problems). Tree-based models derive their name from the fact that the model fit is displayed in the form of a binary tree. As discussed by Clark and Pregibon (1992), tree-based models are superior to linear models under certain scenarios, such as when the regressors are a mixture of numeric and categorical variables. Recall that the data in Table 9.6 are of this type. Tree-based models are also considered to be better than linear models in capturing nonadditive behavior (i.e., the need for interaction terms). See also Breiman et al. (1984) for information on tree-based models.

9.13 SOFTWARE FOR LOGISTIC REGRESSION

Logistic regression capability varies considerably among some of the commonly used software packages. PROC LOGISTIC in SAS Software has extensive capabilities, including some diagnostics that have not been discussed in this chapter. There is no option for performing Hosmer–Lemeshow goodness-of-fit tests, however. Rather, fit can be assessed in several other ways. In particular, several types of rank correlation coefficients can be computed between the observed and predicted values, the model can be tested using the likelihood ratio test (i.e., -2 log likelihood), and a 2×2 classification table can be produced for the observed and predicted values.

BMDP also provides extensive capabilities in program LR. Although the program is entitled Stepwise Logistic Regression, it has general logistic regression capabilities. Several goodness-of-fit tests are available, including the Hosmer–Lemeshow test with the cells determined from the percentiles of the observed probabilities. Classification tables can be produced for user-specified cutoff values, and confidence intervals and an influence statistic are also available.

SPSS also has a logistic regression program, with similar capabilities, including diagnostics and variable selection. Logistic regression can also be performed in S-Plus using the generalized linear model capability, and logistic regression is also available in Minitab. (Logistic regression can also be performed using the LOGISTIC Minitab macro described in the chapter appendix.)

Collett (1991) describes and illustrates the use of SAS, BMDP, and SPSS in analyzing logistic regression data, as well as Genstat, GLIM, and EGRET. Furthermore, GLIM macros are given that produce various diagnostics as well as half-normal plots of standardized deviance residuals with simulated envelopes.

When software such as GLIM, SAS Software, and BMDP are used for logistic regression with multiple covariate patterns, the grouping will have to be performed by the user and the data entered as completely grouped data in order for the deviance to be computed correctly.

A shareware program for logistic regression that is also available is described in Dallal (1988).

All of these programs utilize the method of maximum likelihood, with the output providing asymptotic results. There are several software programs that perform exact logistic regression, one of which is LogXact (Cytel Software Corporation). The latter additionally has the capability for logistic regression using the maximum likelihood approach.

STATA also has considerable capabilities for logistic regression, including diagnostics.

SUMMARY

This chapter contains a survey of the state-of-the-art of many topics in logistic regression. Some topics were not covered, however. These include the use of logistic regression in case-control studies, polytomous logistic regression (i.e., more than two values of the dependent variable), and logistic regression with an ordinal dependent variable. Readers are referred to books such as Hosmer and Lemeshow (1989), McCullagh and Nelder (1989), Agresti (1984, 1990), Freeman (1987), Collett (1991), and Kleinbaum (1993) for further reading. Also, a thorough case study is given by Kay and Little (1986), which includes the illustration of some methods not discussed in this chapter, and we analyze an actual logistic regression data set in Chapter 15.

Logistic regression presents various technical problems that are not encountered in linear regression. For example, unless the exact approach is used, the regression coefficients must be estimated iteratively, and one or more convergence criteria must therefore be selected. There are also choices that must be made regarding diagnostics, such as the choice between Pearson residuals and deviance residuals, with most writers expressing a preference for the latter.

Diagnostics are just as important in logistic regression as they are in linear regression. Unfortunately, there are problems with some of the proposed diagnostics, as has been discussed, and more research is needed in this area. There is also the matter of the validity of inferences being based on asymptotic theory, with no guidance available from the literature regarding required sample sizes.

The manner in which the worth of a regression model is assessed must also be determined. In particular, there is the problem of R^2 and $\overline{R}^2$ not being useful in logistic regression, thus necessitating the use of other measures of model worth such as CCR and MCCR.

Some care must be exercised when data are multicollinear. As in linear regression, p-values cannot be relied on to determine which predictors should remain in the model when all of them are initially used. Instead, variable selection techniques should be used.

APPENDIX

9.A Partial Residuals

The relationship between an $X-Y_s$ scatter plot in simple logistic regression and a partial residual plot can be seen by considering Z_i as given in Eq. (9.11). Since $\hat{\eta}_i = \hat{\beta}_0 + \hat{\beta}_1 X_i$, it follows that $Z_i = \hat{\beta}_0 + r_i^*$. Therefore, the pseudo-observations Z_i differ from the partial residuals by $\hat{\beta}_0$. The Z_i are not quite equal to the Y_i (or the $Y_{i,s}$), however, so the partial residual plot is not exactly the same as the $X-Y_s$ scatter plot.

9.B Minitab Macro LOGISTIC

Included among the set of Minitab macros is LOGISTIC, which can be used for simple or multiple logistic regression. The output contains the regression coefficients in addition to the number of 0's and 1's, and the number of predicted 0's and 1's using a nominal cutoff of 0.5.

REFERENCES

Agresti, A. (1984). *Analysis of Ordinal Categorical Data*. New York: Wiley.

Agresti, A. (1990). *Categorical Data Analysis*. New York: Wiley.

Albert, A. and J. A. Anderson (1984). On the existence of maximum likelihood estimates in logistic models. *Biometrika*, **71**, 1–10.

Atkinson, A. C. (1981). Two graphical displays for outlying and influential observations in regression. *Biometrika*, **68**, 13–20.

Bedrick, E. J. and J. R. Hill (1990). Outlier tests for logistic regression: A conditional approach. *Biometrika*, **77**, 815–827.

Breiman, L., J. H. Friedman, R. Olshen, and C. J. Stone (1984). *Classification and Regression Trees*. Belmont, CA: Wadsworth International Group.

Brown, B. W., Jr. (1980). Prediction analyses for binary data, in *Biostatistics Casebook*, R. G. Miller, Jr., B. Efron, B. W. Brown, Jr., and L. E. Moses, eds. New York: Wiley.

Carroll, R. and S. Pederson (1993). On robustness in the logistic regression model. *Journal of the Royal Statistical Society, Series B,* **55,** 693–706.

Clark, L. A. and D. Pregibon (1992). Tree-Based Models, *Statistical Models in S,* J. M. Chambers and T. J. Hastie, eds. Pacific Grove, CA: Wadsworth and Brooks/Cole, Chapter 9.

Collett, D. (1991). *Modelling Binary Data.* London: Chapman and Hall.

Cook, R. D. and S. Weisberg (1982). *Residuals and Influence in Regression.* New York: Chapman and Hall.

Copas, J. B. (1988). Binary regression models for contaminated data (with discussion). *Journal of the Royal Statistical Society, Series B,* **50,** 225–265.

Cox, D. R. (1970). *The Analysis of Binary Data.* London: Chapman and Hall.

Cox, D. R. and Snell, E. J. (1989). *The Analysis of Binary Data,* 2nd edition. London: Chapman and Hall.

Cox, D. R. and N. Wermuth (1992). A comment on the coefficient of determination for binary responses. *The American Statistician,* **46,** 1–4.

Cytel Software Corporation (1993). *LogXact-Turbo User Guide.* Cambridge, MA: Cytel Software Corporation.

Dallal, G. (1988). LOGISTIC: A logistic regression program for the IBM PC. *The American Statistician,* **42,** 272.

Duffy, D. E. (1990). On continuity-corrected residuals in logistic regression. *Biometrika,* **77,** 287–293.

Fowlkes, E. B. (1987). Some diagnostics for binary regression via smoothing. *Biometrika,* **74,** 503–515.

Freeman, D. H., Jr. (1987). *Applied Categorical Data Analysis.* New York: Marcel Dekker, Inc.

Goorin, A. M., A. Perez-Atayde, M. Gebhardt, and J. Andersen (1987). Weekly High-Dose Methotrexate and Doxorubicin for Osteosarcoma: The Dana Farber Cancer Institute/The Children's Hospital—Study III. *Journal of Clinical Oncology,* **5,** 1178–1184.

Guerrero, V. M. and R. A. Johnson (1982). Use of the Box–Cox transformation with binary response models. *Biometrika,* **69,** 309–314.

Hauck, W. W. and A. Donner (1977). Wald's Test as applied to hypotheses in logit analysis. *Journal of the American Statistical Association,* **72,** 851–853.

Hirji, K. F., C. R. Mehta, and N. R. Patel (1987). Computing distributions for exact logistic regression. *Journal of the American Statistical Association,* **82,** 1110–1117.

Hosmer, D. W., Jr. and S. Lemeshow (1980). A goodness-of-fit test for the multiple logistic regression model. *Communications in Statistics,* **A10,** 1043–1069.

Hosmer, D. W., Jr. and S. Lemeshow (1989). *Applied Logistic Regression.* New York: Wiley.

Hosmer, D. W., Jr., B. Jovanovic, and S. Lemeshow (1989). Best subsets logistic regression. *Biometrics,* **45,** 1265–1270.

Hosmer, D. W., Jr., S. Lemeshow, and J. Klar (1988). A goodness-of-fit test for the multiple logistic regression model. *Communications in Statistics,* **A10,** 1043–1069.

Huberty, C. J. (1994). *Applied Discriminant Analysis.* New York: Wiley.

Jennings, D. E. (1986a). Judging inference adequacy in logistic regression. *Journal of the American Statistical Association*, **81**, 471–476.

Jennings, D. E. (1986b). Outliers and residual distributions in logistic regression. *Journal of the American Statistical Association*, **81**, 987–990.

Kay, R. and S. Little (1986). Assessing the fit of the logistic model: A case study of children and the haemolytic uraemic syndrome. *Applied Statistics*, **35**, 16–30.

Kay, R. and S. Little (1987). Transformation of the explanatory variables in the logistic regression model. *Biometrika*, **74**, 495–501.

Kleinbaum, D. G. (1993). *Logistic Regression: A Self-Learning Text*. New York: Springer-Verlag.

Lachenbruch, P. A. (1975). *Discriminant Analysis*. New York: Hafner.

Landwehr, J. M., D. Pregibon, and A. Shoemaker (1984). Graphical methods for assessing logistic regression models. *Journal of the American Statistical Association*, **79**, 61–71 (discussion: 72–83).

Lemeshow, S. and D. W. Hosmer, Jr. (1982). The use of goodness-of-fit statistics in the assessment of logistic regression models. *American Journal of Epidemiology*, **115**, 92–106.

Maddala, G. S. (1983). *Limited-Dependent and Qualitative Variables in Econometrics*. Cambridge: Cambridge University Press.

Magee, L. (1990). R^2 measures based on Wald and likelihood ratio joint significance tests. *The American Statistician*, **44**, 250–253.

McCullagh, P. and J. A. Nelder (1989). *Generalized Linear Models*, 2nd edition. New York: Chapman and Hall.

Montgomery, D. C. and E. A. Peck (1992). *Introduction to Linear Regression Analysis*, 2nd edition. New York: Wiley.

Moulton, L. H., L. A. Weissfeld, and R. T. St. Laurent (1993). Bartlett correction factors in logistic regression models. *Computational Statistics and Data Analysis*, **15**, 1–11.

Myers, R. H. (1990). *Classical and Modern Regression with Applications*, 2nd edition. Boston: Duxbury.

Nagelkerke, N. J. D. (1991). A note on the general definition of the coefficient of determination. *Biometrika*, **78**, 691–692.

Neter, J., W. Wasserman, and M. H. Kutner (1989). *Applied Linear Regression Models*, 2nd edition. Homewood, IL: Irwin.

O'Gorman, T. W. and R. F. Woolson (1991). Variable selection to discriminate between two groups: stepwise logistic regression or stepwise discriminant analysis? *The American Statistician*, **45**, 187–193.

O'Hara Hines, R. J. and E. M. Carter (1993). Improved added variable and partial residual plots for the detection of influential observations in generalized linear models. *Applied Statistics*, **42**, 3–20.

Pierce, D. A. and D. W. Schafer (1986). Residuals in generalized linear models. *Journal of the American Statistical Association*, **81**, 977–986.

Pregibon, D. (1981). Logistic regression diagnostics. *Annals of Statistics*, **9**, 705–724.

Pregibon, D. (1982). Resistant fits for some commonly used logistic models with medical applications. *Biometrics*, **38**, 485–498.

Pregibon, D. (1985). Link Tests, in *Encyclopedia of Statistical Sciences*, Vol. 5, S. Kotz and N. L. Johnson, eds. New York: Wiley, pp. 82–85.

Rao, C. R. (1973). *Linear Statistical Inference and Its Applications*, 2nd edition. New York: Wiley.

Santner, T. J. and D. E. Duffy (1986). A note on A. Albert's and J. A. Anderson's conditions for the existence of maximum likelihood estimates in logistic regression models. *Biometrika*, **73**, 755–758.

SAS Institute, Inc. (1990). *SAS/STAT User's Guide, Version 6*, 4th edition, Volume 2. Cary, NC: SAS Institute, Inc.

Van Houwelingen, J. C. and S. le Cessie (1990). Predictive value of statistical models. *Statistics in Medicine*, **8**, 1301–1325.

Wald, A. (1943). Tests of statistical hypotheses concerning several parameters when the number of observations is large. *Transactions of the American Mathematical Society*, **54**, 426–482.

Wang, P. C. (1985). Adding a variable in generalized linear models. *Technometrics*, **27**, 213–216.

Whittemore, A. S. (1981). Sample size for logistic regression with small response probability. *Journal of the American Statistical Association*, **76**, 27–32.

Wilkinson, L. (1990). *SYSTAT: The System for Statistics (Statistics Volume)*. Evanston, IL: SYSTAT, Inc.

EXERCISES

9.1. Compute the predicted values for the data set in Table 9.1 (after first deleting observation 24) and compare with the observed values.

9.2. Construct the plot of d_i against $\hat{\pi}_i$ for the (modified) data of Table 9.1.

9.3. Assume that there is a single regressor, and the data are given as follows.

Y	0	1	0	1	0	0	1	0	1	1	1	0	0	0	1
X	14	26	18	21	15	13	29	22	20	30	19	20	13	16	31

Determine the logistic regression coefficients using either the Minitab macro LOGISTIC or other software. Also determine the values of CCR and MCCR. Assume that certain inferences are to be made such as confidence intervals for π and $c\beta_1$. Should there be concern about constructing such confidence intervals for this sample size? Explain.

9.4. Assuming that classification is an experimenter's primary objective in using logistic regression, explain why logistic regression is generally preferred over discriminant analysis.

9.5. Consider the data in Table 9.3, which were analyzed using logistic regression. Compute the discriminant analysis estimators given by Eq. (9.13) and compare with the logistic regression estimators. How can we tell whether or not it might have been "safe" to use discriminant analysis for this data set? Perform the necessary analysis.

9.6. Assume that a logistic regression data set has a single regressor. What graph could be used to determine if complete separation exists? Now assume that it does exist. What would you recommend if it is necessary to develop a model for classification purposes, and additional data cannot be easily obtained?

9.7. Interpretation of β_1 in the simple logistic regression model was discussed in Section 9.2. Would it be meaningful to try to interpret β_0 for that model? If so, what would be the interpretation?

9.8. Consider the subset model with regressors X-ray, Acid Phosphatase, and Stage that was selected in Section 9.8.3. Using all 53 observations and appropriate software, produce the three partial residual plots and the three added variable plots using each of the two versions of the latter given in Section 9.7.6. Comment on the two added variable plots, especially in regard to observation 24. Do the partial residual plots suggest that a linear term in each of the three variables is appropriate?

9.9. Determine the regression coefficients and the values of CCR and MCCR for the subset model referred to in Problem 9.8.

9.10. Assume that there is a single regressor. Explain why an X–Y scatter plot will not indicate whether a linear term in X is appropriate. Can a modified version of the scatter plot be used that will be more informative?

9.11. Assume that there are six potential regressors and the experimenter believes that some interaction terms may be needed in the model. How many possible interactions are there, and how might we determine if any interaction terms are needed?

9.12. Explain why the deviance residuals are generally preferred over the Pearson residuals.

9.13. Show that minus one multiplied times the matrix of second partial derivatives produces $\mathbf{X'WX}$, where this matrix is as described in Section 9.2.1.1.

9.14. Construct two-way tables for the predictors in Table 9.8, and use appropriate software to obtain the p-value for Fisher's exact test in an effort to determine the extent to which the predictors are related.

9.15. No attempt was made to determine a model for the Table 9.10 data such that all of the predictors were statistically significant. Since the p-value for each of the predictors in the full model is greater than .05, a subset model would seem to be more appropriate. Determine what would seem to be the best subset model.

CHAPTER 10

Nonparametric Regression

In both linear regression and logistic regression the user starts with a tentative model and then applies diagnostics to see if the model should be modified. With *nonparametric regression* the user employs a strictly adaptive approach with the regression equation determined from the data. Hastie and Tibshirani (1987) contrast the two approaches in stating that "residual and partial residual plots are used to detect departures from linearity and often suggest parametric fixes. An attractive alternative to this indirect approach is to model the regression function nonparametrically and let the data decide on the functional form." Silverman (1985) similarly states that "an initial non-parametric estimate may well suggest a suitable parametric model (such as linear regression), but nevertheless will give the data more of a chance to speak for themselves in choosing the model to be fitted."

Silverman makes an important point. We will not always be able to start with a linear model, then perhaps make a transformation or two and obtain a model that provides a good fit to the data. If the data set is allowed to essentially "model itself," the result may be fitted values determined nonparametrically that are superior in some sense to the fitted values obtained from a parametric model.

As defined by Altman (1992), "nonparametric regression is a collection of techniques for fitting a curve when there is little a priori knowledge about its shape." As with nonparametric procedures in general, however, nonparametric regression methods will not be as efficient as model-based regression techniques when there is an assumed model, and the model is appropriate. Consequently, if subject-matter theory suggests a particular model, then the practitioner should fit that model.

In subsequent sections we will compare nonparametric regression with linear regression and discuss conditions under which the former could be used.

10.1 RELAXING REGRESSION ASSUMPTIONS

We recall the assumptions for the use of linear regression that were given in Section 1.5.3: (1) A linear model is assumed to be appropriate and (2) the errors

314

are independent, have a common normal distribution with a mean of zero, and have a constant variance. What if we relax the second assumption? We then have a problem with the inferential procedures not being strictly applicable, since those are based on the assumption of normal errors. If we knew that the errors were more than slightly nonnormal, we would not want to use the method of least squares, and hence we would not use the linear regression techniques that have been presented in previous chapters.

Consider the case where the error term is assumed to have a nonnormal distribution, the single regressor is assumed to be fixed, and the relationship between X and Y is linear. What if we used the method of least squares, recognizing that it is inappropriate? The least squares estimators are still "best" in the sense of having the smallest variance of any unbiased estimators that are linear functions of the Y_i. This is the *Gauss–Markov theorem*, which is frequently given in books on regression and books on linear models. See, for example, Neter et al. (1989, p. 43).

Since the normal theory inference procedures cannot be used, one possible approach would be to have the subsequent inferences based on nonparametric procedures. As discussed by Iman and Conover (1980), we could use a nonparametric test for the hypothesis that β_1 is zero. The test is to compute the *Spearman rank correlation coefficient*, using the ranks of the pairs (X_i, Y_i), and compare it with the appropriate tabular value. The Spearman rank correlation coefficient is obtained by applying the formula for the Pearson correlation coefficient, given in Eq. (1.20), to the ranks of X and Y. This would indirectly test the hypothesis that $\beta_1 = 0$. Similarly, a confidence interval for the slope could also be obtained (see Iman and Conover, 1980, p. 266).

We may also obtain confidence intervals for regression parameters by using bootstrapping, which is discussed in Section 10.1.1. Another possible alternative when the error term is nonnormal is to use robust regression, which is discussed in Chapter 11.

10.1.1 Bootstrapping

Assume that we wish to construct a confidence interval for β_i, but the normality assumption seems untenable. If we had some estimate of the distribution of $\hat{\beta}_i$, or some function thereof, we could proceed by using the percentage points of the estimated distribution. This can be accomplished by using *bootstrapping*. Simply stated, bootstrapping is a resampling technique in which samples of size n are obtained, *with replacement*, from the original sample of size n.

As discussed by Hamilton (1992, p. 321), one possible approach is to estimate not the percentiles of the distribution of $\hat{\beta}_i$, but rather to compute

$$t_j^* = \frac{\hat{\beta}_{ij}^* - \hat{\beta}_i}{s_{\hat{\beta}_i^*}} \qquad j = 1, 2, \ldots, N$$

where N denotes the number of bootstrap samples that are obtained, $\hat{\beta}_{ij}^*$ is the estimate of β_i obtained from the jth bootstrap sample, $s_{\hat{\beta}_i^*}$ is the estimate of the standard deviation of the bootstrap estimate of β_i, and $\hat{\beta}_i$ is the least squares estimator of β_i from the original sample. A 95% confidence interval for β_i would be computed as: lower limit $= \hat{\beta}_i - t_{.975}^* s_{\hat{\beta}_i^*}$; upper limit $= \hat{\beta}_i + t_{.025}^* s_{\hat{\beta}_i^*}$, where $t_{.025}^*$ and $t_{.975}^*$ denote the .975 and .025 percentage points, respectively, of the empirical distribution of t^*.

A very large number of bootstrap samples must be used so that the percentage points of t^* are stable. Recall from Section 1.5.4 that, when we use Student's-t distribution in computing a confidence interval for β_i, the tabular value of t that is used does not depend on any sample statistics. It is determined by α and the degrees of freedom for the estimate of σ. Similarly, in obtaining a bootstrap confidence interval we want the values of $t_{.025}^*$ and $t_{.975}^*$ to be as stable as possible. Accordingly, N should be quite large, and Hamilton (1992, p. 314) recommends using $N = 2000$.

For additional information on bootstrapping see Léger et al. (1992), Efron (1982), Efron and Gong (1983), Efron and Tibshirani (1986), and Hall (1988). See also LePage and Billard (1992) for information on the limitations of bootstrapping.

10.2 MONOTONE REGRESSION

What if we now additionally relax the first assumption, and thus do not assume that a linear regression model is appropriate? This leads to consideration of *monotone regression*. Assume that we have regression data with a single regressor that would graph as in Figure 10.1. It is clear that the relationship between X and Y is nonlinear. We might attempt to use polynomial and/or other nonlinear terms in a linear regression model, or perhaps use a nonlinear regression model. (Nonlinear regression is covered in Chapter 13.) A simpler approach, however, would be to use some form of monotone regression, as there is clearly almost a strictly monotonic relationship between X and Y.

Iman and Conover (1979) introduced the idea of using *rank regression* as an alternative to formulating a nonlinear model. For a single regressor, the method entails converting X and Y to ranks and then applying simple linear regression to the ranks. The data that are graphed in Figure 10.1 are given in Table 10.1. We will use these data to illustrate rank regression.

When the formulas for $\hat{\beta}_0$ and $\hat{\beta}_1$ are applied to the rank data, the resultant formulas are

$$\hat{\beta}_1 = \frac{\sum R_{x_i} R_{y_i} - n(n+1)^2/4}{\sum R_{x_i}^2 - n(n+1)^2/4} \tag{10.1}$$

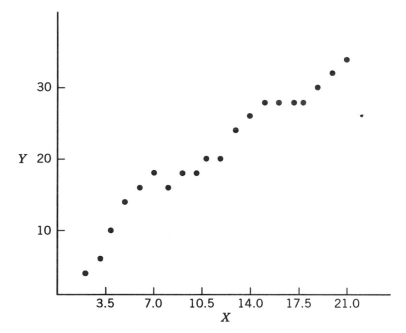

Figure 10.1 Scatter plot of Table 10.1 data.

**Table 10.1   Data for Illustrating
Monotone Regression**

Y	X
3.0	2
6.0	3
10.0	4
13.0	5
16.0	6
17.0	7
16.5	8
17.5	9
18.0	10
19.0	11
20.0	12
23.0	13
25.0	14
28.5	15
27.0	16
27.5	17
28.0	18
30.0	19
31.0	20
33.0	21

and

$$\hat{\beta}_0 = \frac{(1 - \hat{\beta}_1)(n + 1)}{2} \tag{10.2}$$

where R_{x_i} denotes the rank of X_{i} and R_{y_i} denotes the rank of Y_i. Recall that the usual expressions for $\hat{\beta}_0$ and $\hat{\beta}_1$ [as given in Eq. (1.6)] are

$$\hat{\beta}_1 = \frac{\sum XY - \dfrac{(\sum X)(\sum Y)}{n}}{\sum X^2 - \dfrac{(\sum X)^2}{n}} \tag{10.3}$$

and

$$\hat{\beta}_0 = \overline{Y} - \hat{\beta}_1 \overline{X} \tag{10.4}$$

The conversion of (10.3) and (10.4) to (10.1) and (10.2), respectively, results from the fact that $\sum_{i=1}^{n} i = n(n + 1)/2$, so $\bar{i} = (n + 1)/2$.

Regressing Rank(Y) against Rank(X) produces the regression equation $\widehat{R(Y)} = 0.1105 + 0.9895R(X)$. The almost strictly monotonic relationship between X and Y is evidenced by the fact that $R^2 = .979$. Here R^2 can be interpreted as the square of the correlation between the ranks of the Y values and the predicted ranks.

Our objective is, of course, to predict Y rather than the rank of Y, so it is necessary to obtain $\hat{Y}$ from $\widehat{R(Y)}$. Assume that we wish to predict Y when $x_0 = 11$. Since this is one of the observed X values, it follows that $R_{x_0} = 10$ and $\widehat{R(Y)} = 10.0052$. Since this is virtually the same as R_{y_0}, we could let $\hat{Y}_0 = Y_{10} = 19$.

More formally, we would interpolate to obtain $\hat{Y}_0$ as

$$\hat{Y}_0 = 19 + \frac{10.0052 - 10}{11 - 10}(20 - 19)$$
$$= 19.0052$$

where 11 and 10 are the two adjacent ranks that bracket 10.0052, and 20 and 19 are the corresponding values of Y. In this instance it wasn't necessary to interpolate to obtain R_{x_0}, since x_0 was one of the observed X values. If this were not the case, then linear interpretation would be used to obtain R_{x_0}.

If we had erroneously used simple linear regression ("erroneous" relative to the message conveyed by Figure 10.1), we would have obtained $\hat{Y}_0 = 19.757$.

More generally, when simple linear regression is used, we obtain $R^2 = .951$, whereas with monotone (rank) regression we have $(r_{Y,\hat{Y}_0})^2 = .997$. Thus, we obtain better results using a nonparametric result than by using an obviously incorrect parametric approach. Since .997 is so close to 1.0, we could hardly expect to obtain a better fit if we used a linear regression model with polynomial terms, so this illustrates the claim for rank regression that it serves as a substitute for a curvilinear model.

It is analogous to least squares with no assumptions in that we are simply fitting a line through the (ranks of the) points. Since no assumptions are being made, there are essentially no confidence intervals, hypothesis tests, prediction intervals, and so forth, either, nor are the regression coefficients interpretable. Certainly these are limitations of monotone regression, as the user is limited to predicting Y. Obviously, there would also not be any regression plots, residual plots, and so forth, as there would be no checking of the model since there is no model.

Monotone regression may also be used in multiple regression, as is illustrated in Iman and Conover (1979).

10.3 SMOOTHERS

In subsequent sections we present several smoothers that are used in nonparametric regression. A *smoother* is a tool for describing the trend in Y as a function of one or more regressors. Smoothers are useful because the amount of horizontal scatter in data will often make it difficult to see the trend in a data set when there is a trend, and of course there is also the advantage of not having to specify a model.

When there is only a single regressor, the smoothing is called *scatter plot smoothing*, which entails estimating $E(Y|x_0)$, where x_0 denotes the regressor value for which the estimate is desired. The set of such estimates for the n sample x_i values is called a *smooth*. The smoothers that we discuss and illustrate in this chapter are all *linear smoothers*; that is, $\hat{Y}_i$ can be written as a linear combination of the Y_i.

In general, with nonparametric regression for a single regressor we are trying to estimate $\mu(x_j)$, where x_j is an arbitrary value of the regressor, and the "model" is assumed to be

$$Y_i = \mu(X_i) + \epsilon_i \qquad i = 1, 2, \ldots, n \qquad (10.5)$$

for which we would like to be able to assume that $\mu(X)$ is a smooth function. If so, then we would expect to be able to use X-values that are close to x_i in estimating $\mu(x_i)$.

We might ask why we couldn't simply use $\hat{Y}_i = \hat{\mu}(X_i) = Y_i$. This issue is addressed by Eubank (1988), who points out that Y will have an unacceptably large mean-squared error that does not become smaller when the sample size is increased, and who also points out that the use of Y as an estimator would preclude obtaining $\hat{Y}_j$ for X_j not included in the sample.

When we use a smoother, we are using a biased estimator, but we hope that what we lose in having a biased estimator is more than offset by having a small variance. This trade-off between bias and variance is certainly not unique to nonparametric regression since the same thing occurs when ridge regression (see Chapter 12) is used instead of least squares.

We may easily show that $\hat{\mu}(X_i)$ obtained through smoothing will be biased, in general. Let n^* denote the number of observations in a sample of size n that will be used in estimating $\mu(X_j)$, with Y_j being one of the n^* data values. From Eq. (10.5), and using $E(\epsilon_i) = 0$ and $\hat{\mu}(X_j) = (Y_j + \sum_{i \neq j}^{n^*} Y_i)/n^*$, we obtain

$$
E[\hat{\mu}(X_j)] = \frac{1}{n^*} \left\{ \mu(X_j) + \sum_{i \neq j}^{n^*} \mu(X_i) \right\}
$$

$$
= \mu(X_j) + \frac{1}{n^*} \left\{ \sum_{i=1}^{n^*} [\mu(X_i) - \mu(X_j)] \right\}
$$

with the second term in the last equation representing the bias.

One of our goals in selecting a smoother should thus be to use one for which the mean-squared error (i.e., variance plus squared bias) is the smallest, or at least close to the smallest, among all possible smoothers.

Smoothers have long been used in various applications, including outside the field of regression. For example, a moving average is a method for smoothing time series data, and it has also been used extensively in quality control work.

Whether a smoother is to be used only for description or additionally for estimation, two fundamental decisions must be made: (1) how should the Y values be averaged in each "neighborhood" and (2) what should be the size of each neighborhood. That is, if we have a single regressor, how wide should each interval on the regressor be for which the Y values will be averaged. A *neighborhood* is the set of data points (x_i, y_i) that are above and below X_j (except at X_{max} and X_{min}) that is used in computing $\hat{\mu}(X_j)$.

Smoothers must be used with care. To see how a false impression can be created by a smoother, consider Figure 10.2. Notice that there is no evidence of any trend in the Y-values, and this is because these are normal $(0,1)$ random numbers. Figure 10.3 is the corresponding graph of moving averages of size 3 (except that the first two and last two observations are individual values). Notice that smoothing the data in this way provides evidence of a trend when there should be no trend. Similarly, in discussing smoothers in general, McCullagh and Nelder (1989, p. 384) state that "such smoothed curves must be treated with some caution, however, since the algorithm is quite capable of producing convincing-looking curves from entirely random configurations."

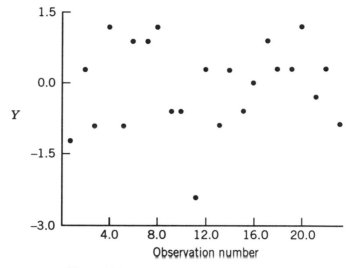

Figure 10.2 Plot of $N(0, 1)$ random numbers.

Therefore, smoothers should be used judiciously. If the smoothed values are plotted on the same graph with the original observations and a time sequence or scatter plot of the latter shows no evidence of a trend, then the smoother could be giving a false signal.

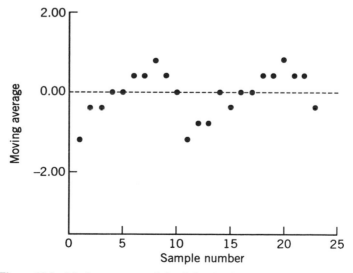

Figure 10.3 Moving averages of size 3 for the data graphed in Figure 10.2.

10.3.1 Running Line

One of the simplest types of smoothers for regression data is a *running line smoother* (Cleveland, 1979). This is similar to a moving average discussed in Section 10.3 in that a moving "window" of points is used. For the moving average, an average is computed each time a point is deleted and another point added, whereas with a running line smoother a simple linear regression line is computed for each window.

As with a moving average, a decision must be made as to how many points should be used in each window when the running line smoother is computed. As discussed by Hastie and Tibshirani (1987), the window size, which is also called the "size of the neighborhood," is typically 10–50% of the data.

If we were writing a computer program for a running line smoother, we could use updating formulas that would take advantage of previous computations. Specifically, recall from Section 1.4 that $\hat{\beta}_1 = S_{xy}/S_{xx}$ and $\hat{\beta}_0 = \overline{Y} - \hat{\beta}_1\overline{X}$. Therefore, we need the updating formulas for S_{xy}, S_{xx}, $\overline{Y}$, and $\overline{X}$. Let k denote the size of the neighborhood, with (x^+, y^+) representing the point that is to be added, and (x^-, y^-) the point that is to be removed. We will let S_{xy}^*, S_{xx}^*, $\overline{Y}_k^*$, and $\overline{X}_k^*$ denote the updated statistics, and the corresponding symbols without the asterisks will denote the statistics from the previous calculation. [The (x, y) data points will be assumed to have been sorted on X, and we will initially assume that all of the X-values are distinct.] It should be apparent that

$$\overline{X}_k^* = \frac{1}{k}(k\overline{X}_k + x^+ - x^-)$$

$$\overline{Y}_k^* = \frac{1}{k}(k\overline{Y}_k + y^+ - y^-)$$

whereas the updating formulas for S_{xx}^* and S_{xy}^* are not so obvious. It is shown in the chapter appendix that

$$S_{xx}^* = S_{xx} + (x^+ - \overline{X}_k)^2 - (x^- - \overline{X}_k)^2 - \frac{1}{k}(x^+ - x^-)^2$$

and

$$S_{xy}^* = S_{xy} + (x^+ - \overline{X}_k)(y^+ - \overline{Y}_k) - (x^- - \overline{X}_k)(y^- - \overline{Y}_k) - \frac{1}{k}(x^+ - x^-)(y^+ - y^-)$$

In addition to selecting the value of k, there are other decisions that must be made. When a moving average is computed, the average is roughly in the center of the neighborhood, and it seems reasonable to similarly predict Y_i for X_i that is in the center of the neighborhood. For example, let $k = 11$, and assume that we wish to obtain $\hat{Y}_{14}$. It would be reasonable to use points (x_9, y_9) through

(x_{19}, y_{19}) to compute the regression line, with $\hat{Y}_{14} = \hat{Y}(x_{14})$. This, of course, will not work at the two extremes of the data set, as we will not be able to construct neighborhoods that are symmetric about X_i. What do we do at the extremes; should we keep the neighborhood size constant, or should we keep constant the number of points that are greater than X_i (at the left extreme) or less than X_i (at the right extreme)?

Assume that the neighborhood size is k. A *nearest neighborhood* is defined as the set of k points that are closest to x_0. A *symmetric nearest neighborhood* has $k/2$ points on each side of x_0. When it is not possible to have $k/2$ points on a given side, as many points as possible are used. While recognizing that *nearest neighborhoods* will have less bias than *symmetric neighborhoods*, Hastie and Tibshirani (1990, p. 31) argue for the use of symmetric neighborhoods for a running line smoother, indicating that a nearest neighborhood would give too much weight to points far away from X_i, since the neighborhood size at the end points would be the same as the neighborhood size in the center of the data. Assume that k is odd, and that $\hat{Y}(x_0)$ is to be obtained. With a symmetric neighborhood, the size of the neighborhood at the end points of the data will be approximately half the size of the nearest neighborhood at the same point, since there will not be any points to the left of $x_{\min}$ nor to the right of $x_{\max}$. Then in obtaining $\hat{Y}(x_{\min})$ for a nearest neighborhood of size k, we could be using as many as $k/2$ ordered x_i that are further from $x_{\min}$ than any of the x_i would be from an x_0 that is near the middle of data set. With a symmetric neighborhood, however, the maximum number of observations that any X_i is from x_0 is $k/2$. Clearly, we would not want $\hat{Y}(x_{\min})$ or $\hat{Y}(x_{\max})$ to be badly biased by observations that are far removed from $x_{\min}$ and $x_{\max}$, respectively.

Regardless of which approach we use, selecting k too small will cause the data not to be smoothed at all; rather, the running line will have considerable variability. For example, for the data plotted in Figure 10.2 it can be shown that using $k = 5$ or $k = 7$ causes a noticeable trend to be apparent, and since there should be no trend at all, this means that many of the observed values are poorly predicted. Using larger values of k such as 11 and 15 produce almost a horizontal line, which, of course, is what we would want this particular line to be. Therefore, for this example we need to use a neighborhood size of roughly 50% or more of the data in order to avoid having a running line smoother that gives a false impression of the data.

To this point in the chapter we have assumed that predictions are to be made at the X_i in the sample. There is obviously a problem in using symmetric nearest neighborhoods if the user decides to obtain a subsequent prediction for an arbitrary x_0. Hastie and Tibshirani (1990, p. 15) suggest interpolating between the closest X_i in the sample.

It would seem as though one possible use of a running line smoother would be in identifying data points that do not fit with the rest. Consider Figure 2.15 for the lifeline data. We would expect a running line smooth of the data to exhibit slopes for the leftmost data points that are considerably different from zero. A combination of such slopes with the majority of the slopes being practically zero would show the influence of the leftmost data points. Since the actual data corresponding to Figure 2.15 were unobtainable, the data points in Table 10.2 were constructed in such a way that they graph virtually identical to Figure 2.15.

The data in Table 10.2 are graphed in Figure 10.4. In addition to showing the regression line, the influence of each point on the correlation coefficient is indicated symbolically. As discussed in Section 2.4.1 for Figure 2.13, an open circle denotes a point that by its presence causes a positive change in the coefficient, with a filled-in circle representing a negative change. We see that the "significant relationship" between X and Y (the p-value for testing that the slope is zero is less than .001) is due to the four points in the lower left quadrant. Deleting those points would apparently cut r_{XY} approximately in half. We should note that Figure 10.4 shows single-point deletion influence, so we can't determine exactly from the graph what changes would occur when more

Table 10.2 Data Similar to Lifeline Data

Y	X	Y	X	Y	X	Y	X
85	30	86	73	75	64	89	70
60	40	86	74	90	64	94	71
90	45	86	75	90	66	94	71
70	46	86	77	90	66	94	72
71	50	86	79	90	67	94	89
64	53	86	80	90	67	89	73
91	52	85	82	90	67	94	73
90	55	85	83	90	68	94	74
93	55	85	84	90	68	94	74
91	57	85	86	90	69	94	75
88	59	85	88	90	69	94	75
68	60	82	65	90	70	94	79
74	60	82	66	90	70	94	80
82	60	82	67	90	70	94	80
95	60	82	69	90	71	94	82
91	62	81	70	90	71	94	82
91	63	81	74	90	71	94	83
91	64	81	66	90	71	94	83
87	65	79	76	90	75	94	86
87	66	79	66	90	77	94	86
87	67	77	67	90	77	93	89
87	68	77	81	90	77	93	89
87	69	77	65	90	81	90	70
87	70	75	79	89	81	90	71
87	71	75	80	89	81	90	72

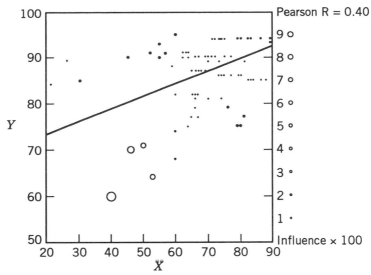

Figure 10.4 Scatter plot of Table 10.2 data, showing the regression line and the influence of each point on the correlation coefficient.

than one point is deleted. We should also note that masking (see Section 2.4.6) could render such statistics almost meaningless. Nevertheless, when the four points are deleted, r_{XY} drops from .401 to .177, so the sum of the individual influences is close to the joint influence in this instance. This also causes the aforementioned p-value to increase to .085, so we now have "no significance."

When there is, for all practical purposes, essentially no relationship between X and Y (as is true for this example), we would want a smoother to show this by giving essentially a horizontal line in the region(s) of the scatter plot where there is clearly no relationship. When we compute the running line smoother for the Table 10.2 data using neighborhood sizes 21, 31, and 51, we obtain a smooth that is very similar to the least squares regression line that is fit to all of the data. Consider Figure 10.5, which shows the smooth for $k = 31$. The plotted points suggest that there is essentially no relationship between X and Y, except perhaps at the smaller values of X. In particular, notice that all but a few data points lie on horizontal lines.

This shows that a running line smoother can be fooled by a few anomalous data points, just as ordinary least squares can similarly be fooled. Therefore, if a smoother is to be used, it should be one that is not influenced by bad data points, or extreme, valid data points that are influential.

10.3.1.1 Modified Running Line
Assume that we want to use a running line smoother that is not affected by influential data points, either bad or valid points. We have many options, one

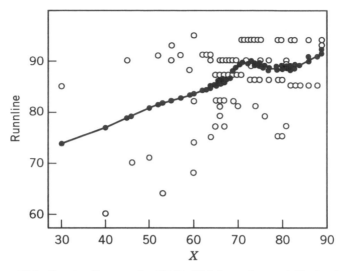

Figure 10.5   Running line smooth of Table 10.2 data using a neighborhood size of 31.

of which would be to adapt one of the robust regression methods discussed in Chapter 11 to a running line smoother (or perhaps to use one of the other smoothers discussed in subsequent sections). Throughout the remainder of this chapter we will use the term *robust* to mean "not influenced by anomalous data points."

We will use a simplified version of one of the methods in Chapter 11 to illustrate the general idea of a robust smoother. Ideally we would want to avoid using influential data points in each neighborhood, but avoiding them could be difficult because of possible problems with masking and swamping. (Swamping is when valid data points are falsely identified as outliers.) One possibility would be to use a certain fraction of the data points in each neighborhood, and to use the data points that seem to be well fit by the regression line. For example, we might use 70% of the data in a given neighborhood and use the subset of the data points that minimized the residual sum of squares. (This technique is discussed in detail in Section 11.5.2.) Such computations can be computer intensive, however, so for illustration we will simple exclude the 30% (or more) of the neighborhood data points (rounded to the nearest integer) that have the largest squared residuals when all of the observations are used, with the subset that is retained used as a substitute for the subset of that size that minimizes the residual sum of squares.

Consider Figures 10.6 and 10.7, each of which is produced using the data in Table 10.2. Trimming as described in the preceding paragraph was used to produce each smooth, with 30% trimming used for Figure 10.6 and 40% trimming used for Figure 10.7. A neighborhood size of 51 was used for each smooth. The graphs differ greatly because with 40% trimming the points in the lower portion of the graph are less likely to be used in producing the fitted values for

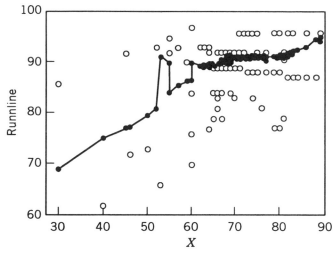

Figure 10.6 Running line smooth of Table 10.2 data with 30% trimming and a neighborhood size of 51.

the smallest X values, whereas Figure 10.6 indicates that at least a few of those points did influence the fitted values for the smallest values of X with 30% trimming. The smooth in Figure 10.6 would seem to be the more reasonable of the two smooths (except for the absence of a smooth curve near $X = 55$), but we want to recognize that the smooths are essentially the same for the interval of X values in the right half of the graphs. This suggests that the smooth is well determined in this region, but perhaps not well determined in the left half. When

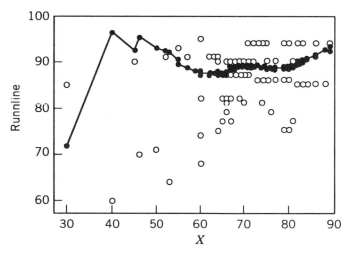

Figure 10.7 Running line smooth of Table 10.2 data with 40% trimming and a neighborhood size of 51.

data are sparse in a particular region of a graph, it is arguable as to whether a smooth should even be attempted. Figure 10.6 suggests that there is a moderate, slightly curvilinear relationship between X and Y. Figure 10.7, however, shows a much different relationship. Notice also that the curve segments are almost identical for the two graphs for $x \geq 65$ and are also almost identical to the same segment in Figure 10.5. This means that the predicted values are well determined for $x \geq 65$, but a single regression line or smooth cannot additionally be used for $x < 65$. Therefore, it isn't very practical to try to fit either to this data set.

Consider the following. Assume a hypothetical example in which all of 20 data points fall exactly on a straight line. It makes no difference which observations (or how many observations) we use to compute the equation for the line; we obtain the same equation for any subset of points. Similarly, if we have a graph that exhibits a very strong linear trend, the equations obtained using different point subsets should not differ greatly, especially if the subsets are not small and are virtually the same size. We have seen for the Table 10.2 data, however, that the equations for the smallest X-values can differ greatly depending on the number of observations that are used in each neighborhood.

10.3.1.2 Inferences for Running Line

Inferences for linear smoothers are considerably more difficult to obtain than inferences for linear regression. This is due in part to the absence of certain exact distributional results.

Assume that we wish to obtain a confidence interval for $\mu(x_0)$ after a running line smooth has been produced. The results that immediately follow apply not just to a running line smooth but to linear smooths in general.

The predicted values from the use of any linear smoother can be written as $\hat{\mathbf{Y}} = \mathbf{SY}$, where $\mathbf{S}$ is an $n \times n$ smoother matrix. It follows that $\text{Var}(\hat{\mathbf{Y}}) = \mathbf{S}(\sigma^2 \mathbf{I})\mathbf{S}' = \sigma^2 \mathbf{SS}'$. Note that for linear regression $\mathbf{S} = \mathbf{H} = \mathbf{X}(\mathbf{X}'\mathbf{X})^{-1}\mathbf{X}'$, which is the hat matrix that was discussed in Section 3.2.6. It follows that the ith row of $\mathbf{S}$ is given by the appropriate row of $\mathbf{H_j}$, where $\mathbf{H_1}, \mathbf{H_2}, \ldots, \mathbf{H_n}$ are the hat matrices for each of the n neighborhoods. The row of $\mathbf{H_j}$ that is used corresponds to the position of x_0 within each neighborhood, here assuming that each x_0 is a sample value.

Hastie and Tibshirani (1990, p. 54) show that the degrees of freedom for the error term is given by $\delta = n - \text{tr}(\mathbf{S} - \mathbf{SS}')$. Therefore, we could estimate σ^2 as $\hat{\sigma}^2 = \sum_{i=1}^{n} (Y_i - \hat{Y}_i)^2/\delta$. An approximate 95% confidence interval for $\mu(x_0)$ is then given by

$$\hat{Y}(x_0) \pm 2\hat{\sigma}_{\hat{Y}(x_0)} = \hat{Y}(x_0) \pm 2\hat{\sigma} \sqrt{\sum_{i=1}^{n} l_i^2(x_0)} \qquad (10.6)$$

where $l_i(x_0)$ is the ith element of the row of $\mathbf{S}$ that is used in computing $\hat{Y}(x_0)$.

Note that here we are interested in a confidence interval rather than a prediction interval, since we are interested in estimating $\mu(x_0)$ rather than obtaining a prediction interval for a future value of Y.

Other inferences for linear smoothers that are based on approximations (because exact distributional results are not known) are discussed by Hastie and Tibshirani (1990, pp. 65–67).

10.3.2   Kernel Regression

As with the running line smoother, in *kernel regression* a neighborhood is selected, and $\hat{\sigma}(x_0)$ is computed. In kernel regression the size of a neighborhood is referred to as the *bandwidth*. As with the running line smoother, in kernel regression the neighborhoods are overlapping.

With a running line smoother the points in a neighborhood are all (ostensibly) equally weighted, as there is no attempt to define weights as a function of their distance from the points x_0 for which predicted values are to be obtained. Of course, influential data points can cause the weighting to be unequal without any attempt at differential weighting.

With a kernel smoother the weights for the X_i depend on their distance from x_0. Specifically, the weight assigned to X_i for obtaining the predicted value at x_0 is

$$w_{0i} = \frac{c_0}{\lambda}\, K\left(\left|\frac{x_0 - x_i}{\lambda}\right|\right) \tag{10.7}$$

where $K(t)$ is an even function decreasing in t, λ is the bandwidth, and c_0 is a constant that forces the weights to sum to 1. The x_i would have to be standardized in some way for $K(t)$ to be meaningful; one simple approach would be to divide the x_i by their standard deviation.

A kernel smoother is frequently written in the general form

$$\hat{\mu}(x_0) = \frac{\sum_{i=1}^{n} K\left(\dfrac{x_0 - x_i}{\lambda}\right) y_i}{\sum_{i=1}^{n} K\left(\dfrac{x_0 - x_i}{\lambda}\right)} \tag{10.8}$$

which might suggest that weights are assigned for all n data points rather than those points in a particular neighborhood. We want to have a zero weight for observations outside a given neighborhood, however. The general idea is to give greatest weight to those y_i for which the corresponding x_i are closest to x_0. It is a matter of selecting a function that will assign the relative weights in a desired manner.

Suggested kernel functions include the standard normal (Gaussian) density and the minimum variance kernel given by

$$K(t) = \begin{cases} \frac{3}{8}(3 - 5t^2) & \text{for} |t| \leq 1 \\ 0 & \text{otherwise} \end{cases}$$

Another frequently used kernel is the Epanechnikov (1969) kernel

$$K(t) = \begin{cases} \frac{3}{4}(1 - t^2) & \text{for} |t| \leq 1 \\ 0 & \text{otherwise} \end{cases}$$

Research results suggest that the selection of the kernel function is relatively unimportant compared to the choice of the bandwidth (Hastie and Tibshirani, 1990, p. 19).

Even though the term *kernel regression* is used, we can see from Eq. (10.7) that $\hat{\mu}(x_0)$ is a weighted average of the y_i. Thus we may speak of kernel smoothing without discussing regression, as in kernel smoothing (regression) there is no functional form that is assumed. [An exception is a kernel-weighted running line fit suggested by Hastie and Tibshirani (1990, p. 38), which would combine the idea of kernel weighting with a running line smoother.]

In discussing the selection of the number of strips (neighborhoods), Altman (1992) draws a parallel between the selection of the number of strips and the choice of the number of regressors when linear regression is used. That is, if we use a very large number of strips, the estimates (predicted values) might be very close to the observed values, but there will be a large variance. At the other extreme, if we use a small number of strips we will have an estimator that has a small variance, but is highly biased. This trade-off also holds for the running line smoother discussed in Section 10.3.1, and for linear smoothers, in general.

See, in particular, Härdle (1990) for additional information on kernel methods, as well as other nonparametric regression techniques.

10.3.2.1 Inferences in Kernel Regression

A confidence interval for $\hat{\mu}(x_0)$ can be obtained using the same general approach as was discussed for a running line. One essential difference is that the elements of the matrix $\mathbf{S}$ are the weights [given generally by Eq. (10.6)] that are used in computing the weighted averages, rather than resulting from a regression computation. Although we could use the approach for estimating σ^2 discussed in Section 10.3.1.2, Altman (1992) presents a different approach. Specifically, σ^2 is estimated by

$$\tilde{\sigma}^2 = \frac{2}{3(n - 2)} \sum_{i=1}^{n-1} r_i^2$$

where $r_i = y_i - (y_{i+1} + y_{i-1})/2$. A confidence interval for $\mu(x_0)$ that utilizes $\tilde{\sigma}$

is then given as

$$\hat{\mu}(x_0) \pm t(h, \alpha)\tilde{\sigma} \left[\sum_{i=1}^{n} \gamma_i^2(x_0) \right]^{1/2}$$

where $h = (n - 2)/2$, and γ_i are the kernel weights such that $\hat{\mu}(x_0) = \sum_{i=1}^{n} \gamma_i(x_0)y_i$.

Although there has been considerable research on kernel regression methods, there is some evidence (Hastie and Loader, 1993) that kernel regression has certain shortcomings and is inferior to local regression, which is covered in the next section. See also Fan and Marron (1993). In particular, kernel estimators can perform poorly at the boundaries of the predictor space and can also perform poorly when the true function is linear (Breiman and Peters, 1992). Hastie and Loader (1993) state that "perhaps the biggest advantage of local regression is when the predictor is two or three dimensional. In this case, a kernel estimate may be influenced by boundary effects over much of the domain, and much structure may be lost by ignoring the effects."

10.3.3 Local Regression

Local regression was introduced by Cleveland (1979) and was called *locally weighted regression*. This term results from the fact that a weight function is used, as in kernel regression.

We may think of local regression as somewhat of a combination of kernel regression and the running line smoother. As in the latter, a least squares fit is produced, but as in kernel regression, weights are used that reflect the distance that each x_i is from x_0. Therefore, the fit in each neighborhood is actually a weighted least squares fit. Local regression thus differs from kernel regression in that with the former a linear regression equation is determined for each segment, as compared to the weighted averages that are computed for the latter.

The objective of local regression is to identify the model (possibly with polynomial terms) that is appropriate for each data segment. As with kernel regression, the general idea of local regression is appealing because points that are well removed from x_0 should be downweighted in order to reduce bias.

Local regression is similar to a running line smoother in that overlapping neighborhoods are used. As with a running line smoother, the fraction of observations to be used in each neighborhood must be specified, and in the literature that fraction is denoted by f. As with other linear smoothers, the determination of f corresponds to determining the number of regressors in linear regression, and Cleveland and Devlin (1988) apply the concept of Mallows's C_p statistic (see Section 7.4.1.1) to local regression for the purpose of determining f.

Once f has been determined, $\hat{\mu}(x_0)$ is produced in four steps, as described in Hastie and Tibshirani (1990, p. 30). The steps consist of (1) identifying the

k nearest neighbors of x_0, (2) determining the distance of the furthest of these points from x_0, which is denoted by Δx_0, (3) assigning weights to each point in the neighborhood by using the *tri-cube* weight function

$$W\left[\frac{|x_0 - x_i|}{\Delta(x_0)}\right]$$

where

$$W(t) = \begin{cases} (1 - t^3)^3 & \text{for } 0 \leq t < 1 \\ 0 & \text{elsewhere} \end{cases}$$

and (4) obtaining the weighted least squares fit using the weights determined in (3).

Notice that nearest neighborhoods are used rather than symmetric nearest neighborhoods. The shortcoming of nearest neighborhoods relative to data end points for the running line smoother that was discussed in Section 10.3.1 does not exist for local regression, since points that are a considerable distance from x_0 are downweighted to practically zero.

What should happen if we apply local regression to the data in Table 10.2? Will local regression be fooled the same way that the running line smoother was fooled? The fit using $f = .43$ is given in Figure 10.8. We see that the local regression smooth is not fooled as badly as the running line smooth. Nevertheless, the smooth does not differ greatly from the least squares regression line and does not perform as well as the robust running line smoother. Here we are using a local regression version (in Systat) for which there is the implicit assumption of normally distributed errors. We would expect a robust version [which is available in S and is described by Cleveland et al. (1992)] to perform better, since that version downweights large residuals. The local regression facility (lowess) in Minitab also has the capability for robust fitting. Although Cleveland (1985) has indicated that two robust steps should adequately smooth outlier effects for most data sets, for the data in Table 10.2 the smooths differ considerably when two robust steps are used and when 10 steps are used. With the former the smooth is almost the same as the smooth that results when no robust steps are used, whereas 10 robust steps flattens out the curve considerably.

The method that was used for the computations for this example, and which is generally used in fitting local regression models, is *loess*, which is short for local regression. This method is implemented in S-Plus and Minitab (see Section 10.4 for additional details). Although locally weighted regression is considered to be a nonparametric regression approach, with loess the errors can be specified to be either normally distributed or to have a symmetric distribution. Robust estimates are produced for the latter, whereas with the former there is the need

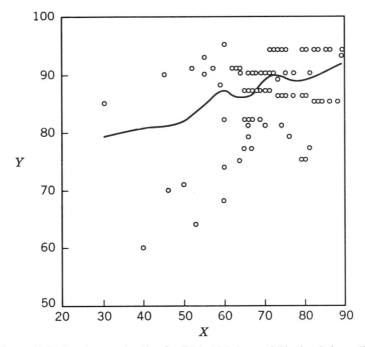

Figure 10.8 Local regression line for Table 10.2 data; neighborhood size = 43.

to perform the same type of assumption checks that are used in linear regression. Computational methods for loess are described in Cleveland and Grosse (1991).

Cleveland and Devlin (1988) provide three illustrative examples of the use of locally weighted regression, including the use of *conditioning plots*, which are generally referred to as *coplots*. These plots are useful in detecting the need for interaction terms in the model. The plot derives its name from the fact that one of the regressors is always "conditioned on."

For example, assume that there are two regressors, X_1 and X_2. One possible coplot would consist of scatter plots of Y against X_1, for each of several intervals on X_2. If the scatter plots differ considerably, this would indicate the need for an $X_1 X_2$ term in the model. (This is the same general idea as interaction graphs that have long been used in design of experiments.)

Coplots are also explained and illustrated in detail in Chambers (1992) and Cleveland (1985).

10.3.3.1 Inferences and Diagnostics

Cleveland et al. (1992) discuss additional capabilities of loess, as well as provide illustrative examples. As in linear regression, a practitioner will generally want to do more than just obtain the locally weighted fit for one or more values of X. That is, we want to check for bad data points and influential data points, construct a prediction interval for $\hat{\mu}(x_0)$, examine residual plots, check for nonnormality, etc.

A confidence interval for $\mu(x_0)$ is discussed by Cleveland et al. (1992). Their approach is quite similar to the method described in Section 10.3.1.2 for the running line smoother, and the appropriate expression differs only slightly from expression (10.6). Using 2σ confidence bounds avoids the problem of not having exact distributional results, but Cleveland et al. (1992) claim that the distribution of $[\hat{\mu}(x_0) - \mu(x_0)]/\hat{\sigma}_{\hat{Y}(x_0)}$, with $\hat{\sigma}_{\hat{Y}(x_0)}$ as defined for expression (10.6), is well approximated by a t distribution with ρ degrees of freedom. The value of ρ is computed as $\rho = \alpha_1^2/\alpha_2$, where $\alpha_1 = n - \text{tr}(\mathbf{S} - \mathbf{SS}')$, and $\alpha_2 = \text{tr}(\mathbf{W}'\mathbf{W})^2$, with $\mathbf{W} = \mathbf{I} - \mathbf{S}$, and $\mathbf{I}$ is an $n \times n$ identity matrix.

Cleveland et al. (1992) also illustrate diagnostic checks for local regression using loess. One distinctive feature of the checks that are discussed is the use of loess to smooth a plot of the residuals against a regressor. See also Cleveland (1993) for information on local regression.

10.3.4 Splines

Splines are generally defined as piecewise polynomials (Eubank, 1988, p. 196; Smith, 1979) in which curve (or line) segments are constructed individually and then pieced together. There are different types of splines. We begin by considering *regression splines* (Wegman and Wright, 1983), which are splines that can be represented by a regression model.

When the equation for each segment contains only linear terms, we then have *linear splines*. This is also referred to as *piecewise linear regression*. We will treat this topic briefly here; readers are referred to Neter et al. (1989, pp. 370–373) for additional details.

10.3.4.1 *Piecewise Linear Regression (Linear Splines)*
We will assume that we have a single regressor. Frequently the assumption of a single slope for a regression model will be untenable. That is, the relationship between X and Y is not constant over the range of X in the data. For example, Teeter (1985) describes the application of piecewise regression to basal body temperature data, under the assumption that there is measurement error in the independent variable.

If we know the value of X at which the relationship changes, then we would apply simple linear regression to each of the two subsets of data. If, however, we do not know the change point, the point would have to be estimated from the data. (This point is called a *knot* in the jargon of splines.)

As discussed by Smith (1979) and others, the use of splines when the position of the knots is not preselected presents special problems. For example, the least squares estimators that minimize the residual sum of squares cannot be obtained directly, and there is the temptation to improve the fit simply by increasing the number of knots and/or changing the positions of the knots, without any theoretical justification for doing so. This relates to the statement made in Section

8.1 that it is possible to obtain an exact fit by simply fitting a polynomial of degree $n - 1$ to n data points. We would similarly have an exact fit with a linear spline if we placed a knot at each data point and thus used $n - 1$ different regression equations.

In general, the use of linear splines with variable knot points is not easily defended unless some stopping rule is used to determine the point of diminishing returns relative to the number of knots. Even then the knots may fall at places for which there is no theoretical justification.

10.3.4.1.1 Model Representation
Still assuming that a linear spline is to be used and that there is a single regressor, the model for the spline with two line segments and a fixed knot position could be written as

$$Y = \beta_0 + \beta_{1a}X + \beta_{1b}(X - k)W + \epsilon \qquad (10.9)$$

where k denotes the position of the knot, and W is an *indicator variable* such that

$$W = \begin{cases} 1 & \text{if } X > k \\ 0 & \text{otherwise} \end{cases}$$

Thus, when $X \leq k$, the slope is β_{1a}, and the slope is $\beta_{1a} + \beta_{1b}$ when $X > k$. [For the general case of d knots there would be d indicator variables, and the generalization from Eq. (10.9) should be apparent.]

We may also use linear splines when there is more than one regressor; see Ertel and Fowlkes (1976) for details.

10.3.4.2 Splines with Polynomial Terms
Consider the data of Table 10.3 that are graphed in Figure 10.9. A close inspection of the latter reveals that quadrature exists up to $X = 5$, with the relationship being essentially linear for larger values of X.

What would happen if we tried to use a regression model with both linear and quadratic terms in an effort to capture the two types of relationships? Obviously, this would not work because the use of such a model would not permit the fitting of the appropriate model for each interval of X. Fitting such a model to these data produces an R^2 value of .91, with the worst fit, a standardized residual of -2.77, occurring at $X = 1$.

We would expect to obtain a better fit if we use a *quadratic spline*, with the first part of the data fit by a quadratic model, and the rest of the data fit by a linear model. This is accomplished as follows. We first need to determine the placement of the knot that "ties" the linear component to the quadratic component. Following guidelines given by Eubank (1988, p. 357), we will place the knot at $X = 5$, as that appears to be roughly an inflection point.

Table 10.3   Data for Illustrating Splines

Y	X
10	1
23	2
34	3
38	4
44	5
45	5
46	6
48	6
47	7
49	8
50	9
51	10
51	11
53	12
54	13
55	14
56	15
57	16

In addition to using X in the model, we need to define a new variable as

$$(X - 5)^2_+ = \begin{cases} (X - 5)^2 & \text{if } X < 5 \\ 0 & \text{if } X \geq 5 \end{cases}$$

Notice that our definition of $(X - 5)^2_+$ causes a model with only a linear term

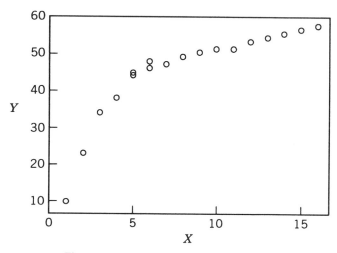

Figure 10.9   Scatter plot of Table 10.3 data.

to be fit to the data for $X \geq 5$, whereas both a linear and quadratic term are used for $X < 5$. Fitting the model with these two terms produces the prediction equation $\hat{Y} = 38.4 + 1.20X - 1.90(X - 5)^2$. The fit is quite good, as evidenced by the fact that $R^2 = .991$.

The worst fit occurs at $X = 4$ and at $(X, Y) = (6, 48)$, at which points the standardized residuals are -2.99 and 2.11, respectively. The magnitude of the first one is of some concern, and we can see from either Table 10.3 or Figure 10.9 why such a large standardized residual occurs at that point. From the latter we can see that the point at $X = 4$ would lie under a smooth curve that could be drawn through points 1, 2, 3, and 5. Thus, the fourth point deviates noticeably from the quadrature that is exhibited by the other four points.

With splines we are fitting portions of the data with different models. We may view this as a type of *local fitting*; that is, we fit segments of the data set individually. If we compare polynomial regression and splines, with the former we are trying to obtain a good fit to the data using a single regression equation and using as many polynomial terms as are needed to obtain a satisfactory fit over the entire range of the regressor values. If for a given scatter plot it is difficult to see how any particular curve with a known functional form could be expected to provide a good fit over the entire range of X, we would certainly expect that we could do a better job of fitting the data by fitting a model to separate pieces of the data.

Practical advice on fitting regression splines can be found in Wold (1974), Smith (1979), and Eubank (1988).

10.3.4.3 Smoothing Splines

A *smoothing spline* differs considerably from a regression spline. In particular, the smooth results from solving an optimization problem, rather than fitting lines and/or curve segments to the data.

A smoothing spline smooth results, generally, from minimizing

$$S(f) = \sum_{i=1}^{n} [y_i - f(x_i)]^2 + \lambda \int_a^b \left[\frac{d f^m(x)}{dx^m} \right] dx \qquad (10.10)$$

where $a \leq x_1 \leq x_2 \leq \cdots \leq x_n \leq b$. The minimization of $S(f)$ produces the function $\hat{f}$ that determines the smooth. Before that minimization can be obtained, however, the smoothing parameter λ must be selected, as well as m, which determines the order of the smoothing spline. A typical choice is $m = 2$, which produces a *cubic smoothing spline*.

The value of λ determines the amount of smoothing. If $\lambda = 0$, we would then be attempting to minimize the residual sum of squares [the first term in Eq. (10.10)], as in the method of least squares. The minimization of $S(f)$ would then obviously occur if the smooth passed through each data point, so that $S(f) = 0$. This would produce an *interpolating spline*, which is of limited use in statistics. At the other extreme, if $\lambda \to \infty$, the smooth approaches the least squares line. Thus, the choice of λ is quite important. The value of λ is often selected by cross validation, which is a model validation approach in which one data point at a time is left out. As discussed by Seber and Wild (1989, p. 488), the cross-validation (CV) choice of λ results from minimizing

$$\text{CV}(\lambda) = \frac{1}{n} \sum_{i=1}^{n} [y_i - \hat{f}_{\lambda}^{(-i)}(x_i)]^2$$

where $\hat{f}_{\lambda}^{(-i)}$ is the estimator of f that results from leaving out the ith data point and using a particular value of λ.

Smoothing splines have some distinct advantages over linear splines. In particular, a smoothing spline is a smooth curve so the question of knot placement that must be addressed with a linear spline does not arise with a smoothing spline. The smooth curve of a smoothing spline is also more esthetically appealing than the general appearance of a linear spline. Smoothing splines have been considered to be much faster computationally than kernel methods, but such an advantage may no longer exist (Fan and Marron, 1993).

For additional information, see Chapter 5 of Eubank (1988) for an introduction to smoothing splines and Wahba (1990) for a detailed account of the theory and applications of smoothing splines. Also of interest is Silverman (1985) and the discussions of that paper. In particular, Silverman (1985) uses the Old Faithful geyser data analyzed by Cook and Weisberg (1982, p. 40) to show how the use of a smoothing spline can identify two distinct segments of a data set that were not apparent when a regression model was fit. [The data set is given in Azzalini and Bowman (1990) and is one of the sample data sets included with the S-Plus statistical software package. See Statistical Sciences, Inc. (1991).]

Diagnostics for smoothing splines that are adaptations of linear regression diagnostics are discussed by Eubank (1985), who also shows how influential data can affect the fit of a smoothing spline. Thus, as with the other linear smoothers discussed in this chapter, a robust smoothing spline approach will frequently be needed. One way to accomplish this would obviously be to replace the first term in Eq. (10.10) by a function that downweights points at which large residuals occur [see, e.g., Hastie and Tibshirani (1990, p. 75)]. The reader is also referred to the first three chapters of Hastie and Tibshirani (1991), in which smoothing splines and the other linear smoothers discussed in this chapter are presented and compared.

10.3.4.4 Splines Compared to Local Regression

It might appear as though local regression and splines are very similar, but there are some major differences. In particular, with splines the data segments that are fit are disjoint, whereas in local regression they are overlapping.

As discussed by Hastie (1992), local regression extends naturally to the case of multiple regressors, whereas the computational complexity of (smoothing) splines increases greatly as the dimension increases. In particular, as discussed by Cleveland and Devlin (1988), when a spline is used with two or more regressors, the number of computations is of the order of n^3, whereas the order is n for the local regression fitting method that was discussed in Section 10.3.3.

10.3.5 Other Smoothers

Another type of smoother is one in which the neighborhood size depends on how well determined the curve segments are that will be fit. Consider the following. When would we want a smoother to do very little, if any, smoothing? Consider Figure 10.10. How much smoothing should be done for points with x-coordinates between 8 and 11? Since there is very little variability in this interval, the curve segment that should be fit is well determined. Therefore, we would not want to use a neighborhood size that would extend beyond this interval and cause unnecessary smoothing.

In general, we would want a smoother to do very little, if any, smoothing in regions where there is obvious curvature or small variability, and more smoothing when there is very little, if any, curvature or considerable variability, as discussed, for example, in Statistical Sciences (1991, p. 16–41).

10.3.6 Which Smoother?

Kafadar (1994) discusses methods for assessing the performance of two-dimensional smoothers on data. Her study included some smoothers not discussed in this chapter. The general finding was that weighted averages, with the weight of

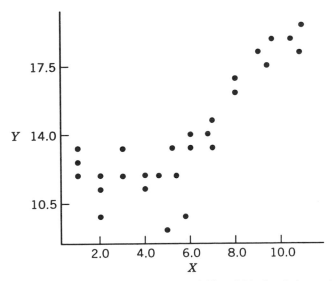

Figure 10.10 Configuration of points for which a variable neighborhood size would be desired.

a point being inversely proportional to its distance from the point that is being smoothed, performed well.

In another study, Breiman and Peters (1992) considered only automatic smoothers, in which the smoothing algorithm selects the window size from the structure of the data. They compared several different types of smoothers in a simulation study and concluded that no smoother in their study dominated the other smoothers in performance.

10.3.7 Smoothers for Multiple Regressors

As discussed in Section 10.3.6, splines cannot be easily applied when there are multiple regressors. Local regression and the running line and kernel smoothers can be used without great difficulty when there is more than one regressor, however. A distance measure must be selected, with Euclidean distance being the most likely choice for most applications. See Hastie and Tibshirani (1990, pp. 32–34) for additional details.

10.4 SOFTWARE

The availability of software for nonparametric regression varies for the methods that have been presented in this chapter. Monotonic (rank) regression can be performed with any statistical software that allows variables to be ranked before performing regression, and this applies to all of the major statistical software packages. Some software allow monotonic regression to be performed directly; this includes PROC TRANSREG in SAS Software.

Bootstrapping is generally unavailable in the software packages that have been discussed at the end of preceding chapters. SPSS, which does not have general capabilities for nonparametric regression, has bootstrapping capability only in one of its nonlinear regression programs. (Nonlinear regression is discussed in Chapter 13.) Bootstrapping is not available in SAS Software, BMDP, Minitab, Systat, or S-Plus. Whereas we could easily program the necessary calculations for use with virtually any such software, we should remember that bootstrapping is very computer intensive.

A running line smoother is available as an option in Stata but is not available in the other software that have been discussed in this and preceding chapters. The Minitab macro RLINESMO described in Appendix A produces the type of running line smooth described in Section 10.3.1, and RLINSMOL produces the modified running line smooth discussed in Section 10.3.1.1.

Local regression is available in S-Plus, Stata, Minitab, and Systat. The latter has only the basic LOWESS capability described by Cleveland (1981), whereas S-Plus provides additional options, including an iterative robust option for nonnormal, but symmetric, errors. The function **scatter.smooth** may be used for a single regressor, and **loess** may be used for one or more regressors. A linear or quadratic model may be specified with either. Minitab also provides (in

Release 10) more than the basic LOWESS capability, with an option to adjust for the influence of outliers by using the NSTEPS subcommand. The larger the specified value of NSTEPS, the less influential should be any outliers that are in the data. Local regression is not available in SPSS, BMDP, or SAS Software.

Kernel regression is available in S-Plus, with four kernels from which to select, including the Gaussian kernel. Although kernel smoothing is available in Systat, the output is in the form of a three-dimensional graph that gives the bivariate density function rather than a smoothed scatter plot. The kernel that is used is the Epanechnikov kernel.

Splines can be fit using PROC TRANSREG in SAS Software, and splines may also be fit in S-Plus. Cubic splines may be fit using Stata. Although a cubic spline may be fit with Systat, the spline that is fit is an interpolating spline. Smoothing splines may be fit in IMSL (IMSL, Inc., 1984) and in JMP (SAS Software, 1989). Splines are not fit by Minitab, BMDP, or SPSS.

SUMMARY

We have presented several nonparametric regression techniques, ranging from monotonic regression to local regression. The former was shown to work well for the example that was given, but its use is obviously limited to situations where the relationship between Y and a single regressor are (practically) monotonic. Splines, kernel regression, local regression, and the running line smoother are all data smoothing techniques that offer the regression user considerably more flexibility than does monotonic regression. With each of these, however, there is the potential danger of overfitting the data, so methods similar to those employed in linear regression should be used to avoid overfitting.

Although splines are frequently used, they do have some drawbacks, especially when there is more than one regressor. It is difficult to justify the use of an ordinary spline in the absence of prior information that would suggest the placement of the knot position(s).

We should also realize that smoothers in general will not necessarily provide a smooth that is insensitive to bad or influential data points, as was seen when each of the smoothers described herein was applied to the data that were graphed in Figure 10.2. The advantage of a smoother is that it allows us to obtain $\hat{\mu}(x_0)$ and a confidence interval for $\mu(x_0)$ without having to specify a model. The predicted values will often be affected by influential observations, so if a smoother is to be used, it is desirable that a robust version be employed.

It is of interest to compare the results obtained from varying the neighborhood size, and also to compare the smooths produced by linear smoothers and robust linear smoothers, as the reader is asked to do in Exercises 10.7–10.11 for a subset of a data set given by Sockett et al. (1987).

Unlike an ordinary smoother, the results produced by using a robust smoother can vary considerably when factors such as the neighborhood size are changed, as was seen in Section 10.3.1.1. Such large changes can reveal instability in one or more regions of the data.

An alternative to an ordinary smoother or to the robust version of loess is to use one of the robust regression methods described in Chapter 11, which require that a model be specified, or adapt one of the better robust regression techniques to a smoother.

APPENDIX

10.A   Updating Formulas for Running Line Smoother

S_{xx}^* and S_{xy}^* can be obtained from S_{xx} and S_{xy}, respectively, as follows. We will let N_k^* denote the new neighborhood of size k for which the updating formulas are to be obtained, and N_k shall denote the previous neighborhood.

We seek a simplification of $S_{xx}^* = \sum (X - \overline{X}_k^*)^2$, so that we may obtain S_{xx}^* recursively in terms of S_{xx}. Since $\overline{X}_k^* = (1/k)(k\overline{X}_k + x^+ - x^-)$, we obtain

$$
S_{xx}^* = \sum \left[X - \frac{1}{k}(k\overline{X}_k + x^+ - x^-) \right]^2
$$

$$
= \sum \left\{ (X - \overline{X}_k) - \frac{1}{k}(x^+ - x^-) \right\}^2
$$

Squaring the bracketed expression and remembering that the summation is still over N_k^* rather than N_k, even though we are using $\overline{X}_k$, we have

$$
S_{xx}^* = S_{xx} + (x^+ - \overline{X}_k)^2 - (x^- - \overline{X}_k)^2 - \frac{2}{k}(x^+ - x^-)\sum (X - \overline{X}_k)
$$

$$
+ \frac{1}{k}(x^+ - x^-)^2
$$

$$
= S_{xx} + (x^+ - \overline{X}_k)^2 - (x^- - \overline{X}_k)^2 - \frac{1}{k}(x^+ - x^-)^2
$$

with the last term resulting from the fact that $\sum (X - \overline{X}_k) = 0 + (x^+ - \overline{X}_k) - (x^- - \overline{X}_k) = x^+ - x^-$.

The derivation for S_{xy}^* proceeds similarly. Using the initial substitution for the means produces

$$S_{xy}^* = \sum \left[X - \frac{1}{k}(k\overline{X}_k + x^+ - x^-) \right] \left[Y - \frac{1}{k}(k\overline{Y}_k + y^+ - y^-) \right]$$

$$= \sum (X - \overline{X}_k)(Y - \overline{Y}_k) - \frac{1}{k}(x^+ - x^-) \sum (Y - \overline{Y}_k)$$

$$- \frac{1}{k}(y^+ - y^-) \sum (X - \overline{X}_k) + \frac{1}{k}(x^+ - x^-)(y^+ - y^-)$$

$$= S_{xy} + (x^+ - \overline{X}_k)(y^+ - \overline{Y}_k) - (x^- - \overline{X}_k)(y^+ - \overline{Y}_k) - \frac{1}{k}(x^+ - x^-)(y^+ - y^-)$$

(The last line follows from using substitutions that are of the same type as the substitution used in obtaining the expression for S_{xx}^*.) Notice that if we replaced each Y by X in the expression for S_{xy}^*, we would obtain the expression for S_{xx}^*, as we would expect.

REFERENCES

Altman, N. S. (1992). An introduction to kernel and nearest-neighbor nonparametric regression. *The American Statistician*, **46**, 175–185.

Azzalini, A. and A. W. Bowman (1990). A look at some data on the Old Faithful geyser. *Applied Statistics*, **39**, 357–365.

Breiman, L. and S. Peters (1992). Comparing automatic smoothers (a public service enterprise). *International Statistical Review*, **60**, 271–290.

Chambers, J. M. (1992). Data for Models, in *Statistical Models in S*, J. M. Chambers and T. J. Hastie, eds. Pacific Grove, CA: Wadsworth and Brooks/Cole, Chapter 3.

Cleveland, W. S. (1979). Robust locally weighted regression and smoothing scatterplots. *Journal of the American Statistical Association*, **74**, 829–836.

Cleveland, W. S. (1981). LOWESS: A program for smoothing scatterplots by robust locally weighted regression. *The American Statistician*, **35**, 54.

Cleveland, W. S. (1985). *The Elements of Graphing Data*. Pacific Grove, CA: Wadsworth.

Cleveland, W. S. (1993). *Visualizing Data*. Summit, NJ: Hobart Press.

Cleveland, W. S. and S. J. Devlin (1988). Locally weighted regression: an approach to regression analysis by local fitting. *Journal of the American Statistical Association*, **83**, 596–610.

Cleveland, W. S. and E. Grosse (1991). Computational methods for local regression. *Statistics and Computing*, **1**, 47–62.

Cleveland, W. S., E. Grosse, and W. M. Shyu (1992). Local Regression Models, in *Statistical Models in S*, J. M. Chambers and T. J. Hastie, eds. Pacific Grove, CA: Wadsworth and Brooks/Cole, Chapter 8.

Cook, R. D. and S. Weisberg (1982). *Residuals and Influence in Regression*. New York: Chapman and Hall.

Efron, B. (1982). *The Jackknife, Bootstrap, and Other Resampling Plans*. Monograph #38. Philadelphia: Society for Industrial and Applied Mathematics.

Efron, B. and G. Gong (1983). A leisurely look at the bootstrap, the jackknife, and cross-validation. *The American Statistican*, **37**, 36–48.

Efron, B. and R. Tibshirani (1986). Bootstrap methods for standard errors, confidence intervals, and other measures of statistical accuracy. *Statistical Science*, **1**, 54–77.

Epanechnikov, V. A. (1969). Nonparametric estimation of a multivariate probability density. *Theory of Probability and its Applications*, **14**, 153–158.

Ertel, J. E. and E. B. Fowlkes (1976). Some algorithms for linear spline and piecewise multiple linear regression. *Journal of the American Statistical Association*, **71**, 640–648.

Eubank, R. L. (1985). Diagnostics for smoothing splines. *Journal of the Royal Statistical Society*, **47**, 332–341.

Eubank, R. L. (1988). *Spline Smoothing and Nonparametric Regression*. New York: Marcel Dekker.

Fan, J. and J. S. Marron (1993). Comment (discussion of Hastie and Loader). *Statistical Science*, **8**, 129–134.

Hall, P. (1988). Theoretical comparison of bootstrap confidence intervals. *Annals of Statistics*, **16**, 927–953.

Hamilton, L. C. (1992). *Regression with Graphics*. Pacific Grove, CA: Brooks/Cole.

Härdle, W. (1990). *Applied Nonparametric Regression*. London: Cambridge University Press.

Hastie, T. J. (1992). Generalized Additive Models, in *Statistical Models in S*, J. M. Chambers and T. J. Hastie, eds. Pacific Grove, CA: Wadsworth and Brooks/Cole, Chapter 7.

Hastie, T. and C. Loader (1993). Local regression: automatic kernel carpentry. *Statistical Science*, **8**, 120–129 (discussion: 129–143).

Hastie, T. and R. Tibshirani (1987). Generalized additive models: some applications. *Journal of the American Statistical Association*, **82**, 371–386.

Hastie, T. and R. Tibshirani (1990). *Generalized Additive Models*. New York: Chapman and Hall.

Iman, R. L. and W. J. Conover (1979). The use of the rank transform in regression. *Technometrics*, **21**, 499–506.

Iman, R. L. and W. J. Conover (1980). *Practical Nonparametric Statistics*, 2nd edition. New York: Wiley.

IMSL, Inc. (1984). *IMSL Library Reference Manual*. Houston: IMSL, Inc.

Kafadar, K. (1994). Choosing among two-dimensional smoothers in practice. *Computational Statistics and Data Analysis*, **18**, 419–439.

Léger, C., D. N. Politis, and J. P. Romano (1992). Bootstrap technology and applications. *Technometrics*, **34**, 378–398.

LePage, R. and L. Billard, eds. (1992). *Exploring the Limits of Bootstrap*. New York: Wiley.

McCullagh, P. and J. A. Nelder (1989). *Generalized Linear Models*, 2nd edition. New York: Chapman and Hall.

Neter, J., W. Wasserman, and M. H. Kutner (1989). *Applied Linear Regression Models*, 2nd edition. Homewood, IL: Irwin.

SAS Institute, Inc. (1989). *JMP User's Guide*. Cary, NC: SAS Institute, Inc.

Seber, G. A. F. and C. J. Wild (1989). *Nonlinear Regression*. New York: Wiley.

Silverman, B. W. (1985). Some aspects of the spline smoothing approach to non-parametric regression curve fitting. *Journal of the Royal Statistical Society, Series B*, **47**, 1–21 (discussion: 21–52).

Smith, P. L. (1979). Splines as a useful and convenient statistical tool. *The American Statistician*, **33**, 57–62.

Sockett, E. B., D. Daneman, C. Clarson, and R. M. Ehrich (1987). Factors affecting and patterns of residual insulin secretion during the first year of type I (insulin dependent) diabetes mellitus in children. *Diabetes*, **30**, 453–459.

Statistical Sciences, Inc. (1991). *S-Plus for DOS, Reference Manual*. Seattle, WA: Statistical Sciences, Inc.

Teeter, R. A. (1985). The application of linear piecewise regression to basal body temperature data. *Biometrical Journal*, **27**, 139–113.

Wahba, G. (1990). *Spline Functions for Observational Data*. CBMS-NSF Regional Conference series. Philadelphia: Society for Industrial and Applied Mathematics.

Wegman, E. J. and I. W. Wright (1983). Splines in statistics. *Journal of the American Statistical Association*, **78**, 351–365.

Wold, S. (1974). Spline functions in data analysis. *Technometrics*, **16**, 1–11.

EXERCISES

10.1. Consider the following data:

Y	2	3	4	5	6	8	11	15	18	23	29	37	47	59	70
X	1	2	3	4	5	6	7	8	9	10	11	12	13	14	15

Which approach do you believe would provide the best fit to these data: monotonic regression or a quadratic spline with the knot determined from inspection of the graph? Use both approaches and compare the results.

10.2. If local regression were applied to the data in Problem 10.1, should f be chosen to be large (close to 1), or small (close to 0)?

10.3. Under what general conditions will the use of monotonic regression produce good results?

10.4. Since locally weighted regression can be used with the usual assumptions for the error term, what distinguishes it from linear regression?

10.5. Would it be appropriate to apply a running line smoother to the data in Table 10.1? Why or why not?

10.6. Consider again the data in Table 10.1. Would it be meaningful to apply *any* linear smoother to these data? In general, when should we consider using a smoother?

Hastie and Tibshirani (1990) use part of a data set given in Sockett et al. (1987) to illustrate and compare various linear smoothers, but their analysis did not consider either robust smoothers, diagnostics, or multiple neighborhood sizes. We will consider each of these in the problems that follow, and the interested reader may wish to compare the results with the results given by Hastie and Tibshirani (1990).

The objective of the experimenters was to determine the dependence of the level of *serum C-peptide* on various other factors. We will use only *age* as an independent variable. The data on the 43 patients are given below.

C-peptide	4.8 4.1 5.2 5.5 5.0 3.4 3.4 4.9 5.6 3.7 3.9 4.5 4.8 4.9 3.0 4.6 4.8 5.5
Age	5.2 8.8 10.5 10.6 10.4 1.8 12.7 15.6 5.8 1.9 2.2 4.8 7.9 5.2 0.9 11.8 7.9 11.5

C-peptide	4.5 5.3 4.7 6.6 5.1 3.9 5.7 5.1 5.2 3.7 4.9 4.8 4.4 5.2
Age	10.6 8.5 11.1 12.8 11.3 1.0 14.5 11.9 8.1 13.8 15.5 9.8 11.0 12.4

C-peptide	5.1 4.6 3.9 5.1 5.1 6.0 4.9 4.1 4.6 4.9 5.1
Age	11.1 5.1 4.8 4.2 6.9 13.2 9.9 12.5 13.2 8.9 10.8

10.7. First construct a scatter plot of these data. Compare your scatter plot with Figure 10.2. Would you expect the "proper" fit to be a horizontal line as in Figure 10.2, or is there evidence for these data that there is a true linear relationship with a nonzero slope for at least part of the data?

10.8. Use either the Minitab macro RLINESMO or other software to produce the running line smooth for $k = 15$, 25, and 35, and compare the results.

10.9. Remembering that a running line smooth can be fooled by anomalous data points, use the Minitab macro RLINSMOL with 25% trimming to produce a robust running line smooth, again using $k = 15$, 25, and 35. Compare the results with the smooths obtained for Problem 10.8. Is there evidence of instability, as in the example in Section 10.3.1.1, or does the smooth appear to be well determined?

10.10. If a regression spline were fit to these data, what order of polynomial might be used? Similarly, where should the knot(s) be placed? Fit the regression spline and compare the smooth with the smooths produced in Problems 10.8 and 10.9.

10.11. What would be the dimensions of the $\mathbf{S}$ (smoother) matrix for the running line smooth in Problem 10.8 with $k = 15$? Remember that the neighborhood size is not constant.

10.12. Determine $\hat{\mu}(x = 5.2)$ for the running line smooth in Problem 10.8 with $k = 15$.

CHAPTER 11

Robust Regression

Robust regression is an alternative to ordinary least squares that can be appropriately used when there is evidence that the distribution of the error term is (considerably) nonnormal, and/or there are outliers that affect the equation.

This chapter contains a survey of robust regression techniques. Most of the proposed methods are covered in detail, and Minitab programs are provided as both an instructional and computational aid, especially the latter since no statistical software package contains routines for all of the methods that are discussed herein. Some techniques that are discussed in detail will later be seen to be inferior to other techniques. The reason for devoting an apparently disproportionate amount of space to them is that they have received considerable attention in the literature.

There is some emphasis on robust regression with a single regressor. The reason for this is that it is much easier to see the need for a robust regression estimator when there are only two dimensions.

11.1 NEED FOR ROBUST REGRESSION

Much of the work in robust regression has been motivated by the Princeton Robustness Study (Andrews et al., 1972), in which it was learned that the ordinary least squares (OLS) estimator can be inferior to other estimation approaches when the distribution of the error term has heavier tails than the tails of the normal distribution. The least squares approach can also produce unstable estimates when a data set contains anomalous data, as was seen in Chapter 6. Consequently, we should require our regression approach to produce resistant estimates.

As introduced by Tukey (1977) and discussed by various authors, including Mosteller and Tukey (1977, p. 203) and Staudte and Sheather (1990, p. 92), a *resistant estimate* is one that is relatively unaffected by large changes in a small part of the data or small changes in a much larger part of the data. (The median is one example of a resistant estimator.) Intuitively, it might seem that we should only use that part of a data set that "fits together." This would mean, however, that we would in essence be rejecting outliers for reasons of statistical expediency, a practice that would be frowned upon by Huber (1977, p. 3), Carroll and Ruppert (1988, p. 5), and many others.

A reasonable compromise would be to judiciously reject uninformative outliers and to limit the influence of influential observations. (An uninformative outlier is an extreme value that does not suggest a need to modify the model.) Consider Figure 11.1 and the point labeled A. Clearly the point is not extreme in terms of its x-coordinate, and no reasonable outlier rule would declare it an outlier in terms of its y-coordinate. But the point will have moderate influence on the regression equation. Since its x-coordinate is close to the average x-coordinate, we would expect the intercept to be primarily affected. That this is the case can be seen from the fact that *with* the point the regression equation is $\hat{Y} = 1.34 + 1.89X$, and *without* the point the equation is $\hat{Y} = 0.45 + 1.82X$. Since this point does not fit the general pattern of the data, nor does it provide any evidence of the need to modify the model, we need a robust regression approach that will essentially assign a weight of zero to that point. If we use a *high breakdown point* (HBP) estimator, the weight will probably be zero. The weight should be close to zero if a *bounded influence* estimator is used. These estimators are discussed later, and the possibility of combining them in a two-stage approach is also discussed.

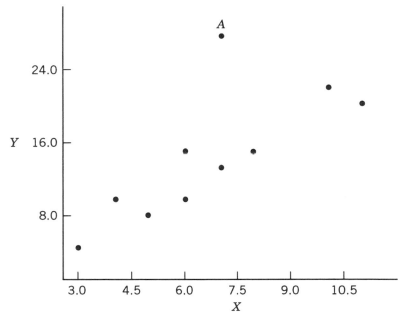

Figure 11.1   Scatter plot showing influential data point.

11.2 TYPES OF OUTLIERS

In the preceding example the extreme point was actually a *regression outlier*, although this was not specifically illustrated. This and other types of outliers are discussed in this section in the context of regression. We should keep in mind, however, that while we may easily categorize outliers, it is not so easy to give a mathematically precise definition of the term. Quoting Rousseeuw and van Zomeren (1990b), "outliers are an empirical reality, but their exact definition is as elusive as the exact definition of a cluster (or, for that matter, an exact definition of data analysis)."

1. **Regression Outlier.** As defined, for example, by Rousseeuw and Leroy (1987), a regression outlier is a point that deviates from the linear relationship determined from the other $n - 1$ points, or at least from the majority of those points.

2. **Residual Outlier.** This is a point that has a large standardized (or standardized deletion) residual when it is used in the calculations. We may distinguish between a regression outlier and a residual outlier by noting that a point can be a regression outlier without being a residual outlier (if the point is influential), and a point can be a residual outlier without there being strong evidence that the point is also a regression outlier. This will be illustrated later in this section.

3. **X-outlier.** This is a point that is outlying only in regard to the x-coordinate. Such a point can cause some robust regression estimators to perform poorly, but the more modern robust estimators are not undermined by X-outliers. This will be discussed later. An X-outlier could also be a regression and/or residual outlier.

4. **Y-outlier.** This is a point that is outlying only because its y-coordinate is extreme. The manner and extent to which such an outlier will affect the parameter estimates will depend on both its x-coordinate and the general configuration of the other points. Thus, the point might also be a regression and/or residual outlier.

5. **X- and Y-outliers.** A point that is outlying in both coordinates may be a regression outlier, or a residual outlier (or both), or it may have a very small effect or even no effect on the regression equation. The determining factor is the general configuration of the other points.

Figure 11.2(a)–(c) illustrates several outlier types. As stated previously, Figure 11.1 illustrates a regression outlier, which is due to the y-coordinate. In Figure 11.2(a) point A illustrates a point that is a regression outlier, but not a (standardized) residual outlier, as the absolute value of the standardized residual is less than 2. The point clearly deviates from the pattern formed by the other points and has a considerable effect upon R^2 and the regression equation. Specifically, $\hat{Y} = 3.22 + 0.942X$ and $R^2 = .54$ when all 12 points are used, and

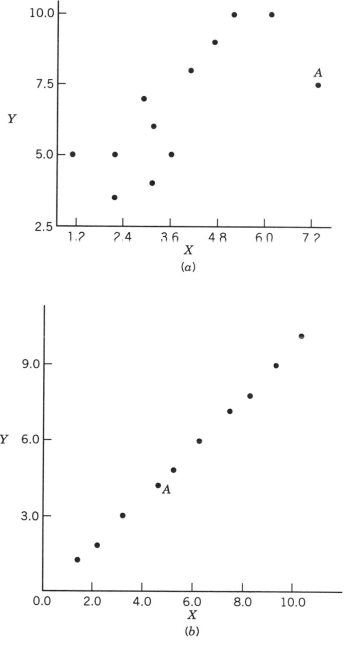

Figure 11.2 Three X–Y scatter plots that show the different types of outliers that can occur in regression. (*a*) Point A is a regression outlier but is neither a (standardized) residual outlier nor an X- or Y-outlier. (*b*) Point A is a (standardized) residual outlier but is neither a regression outlier nor an X- or Y-outlier. (*c*) Point A is both an X- and Y-outlier but is neither a residual outlier nor a regression outlier.

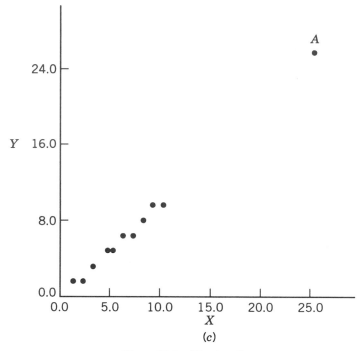

Figure 11.2 (*Continued*)

$\hat{Y} = 2.01 + 1.37X$ with $R^2 = .72$ when the regression outlier is excluded from the computations. Thus, the outlier has considerable influence.

From the graph we would not be inclined to call point A an X-outlier since the x-coordinate of the point is not much larger than the next largest x-coordinate. An X-outlier is generally considered to be the same as a leverage point (see Rousseeuw and Leroy, 1987, p. 6), however, and here the leverage value is .43. Since $2p/n = .33$, we would label the point a leverage point, and hence an X-outlier, if we use this threshold value. It was indicated in Section 2.5, however, that $3p/n$ is a more appropriate choice, at least when p is small, even though $2p/n$ is more frequently used. Here $3p/n = .50$, so the point would not be labeled an X-outlier if we use this cutoff value, and that seems to be consistent with the message given by the graph. This illustrates that a regression outlier need not be either a residual outlier or an X-outlier—a fact that is not well understood.

Figure 11.2(*b*) illustrates that an observation can be a residual outlier without being a regression outlier. The standardized residual at point A is -2.37 (a moderately large value since the corresponding normal tail area is .0089), despite the fact that the point clearly fits with the others. Deleting the point has virtually no effect on the regression equation, as the equation changes from $\hat{Y} = -0.13 + 1.01X$ to $\hat{Y} = -0.07 + 1.00X$. This means that we should be careful in interpreting residual outliers (as the reader is asked to do for this example in Exercise 11.9).

Figure 11.2(c) illustrates a point that is both an X- and Y-outlier, as might occur, for example, if decimal points in both X and Y were lost. The point has no effect on the regression equation, however, which is the same both with and without the point. This does not mean that we can ignore the apparent error, however, because some of the other summary statistics can be considerably affected. If we change the x-coordinate (only) of point A in Figure 11.2(c), setting it equal to the average of the other x-coordinates, we would then have a point that we would label a Y-outlier, a regression outlier, and a standardized residual outlier. (We would have similar results if we were to change only the y-coordinate by setting it equal to the average of the other y-coordinates.)

In much of the robust regression literature there is the tacit assumption that the regressors are random variables, and we will make the same assumption in this chapter. Certainly robust regression can be used when the regressor values are fixed, but then the discussion must take on a different flavor. For example, with fixed regressor values there would be no X-outliers, and this would obviate the use of a certain class of robust regression estimators that were developed to protect against X-outliers. (The elimination of X-outliers is certainly desirable, though, and this can be accomplished through the use of equileverage designs, which are discussed in Chapter 14.) We may thus view the fixed regressor case as a special case in robust regression.

11.3 HISTORICAL DEVELOPMENT OF ROBUST REGRESSION

The historical development of most of the different classes of robust regression estimators is traced by Rousseeuw and Leroy (1987), who indicate that such estimators date from Edgeworth (1887). The latter proposed the idea of *minimizing the sum of the absolute value of the residuals*, an approach that has been discussed extensively in the literature during the past two decades. Although the estimators obtained from using this criterion are not unduly influenced by Y-outliers, they can be seriously affected by X-outliers. (In fact, under certain conditions this robust estimator can be more seriously undermined by X-outliers than will be the OLS estimator.)

Thus, this L_1 estimator, as it is called, should not be used when there are X-outliers. Since only a single outlier can materially affect the parameter estimates, the *breakdown point* of the estimator is only $1/n$. The concept of the breakdown point of an estimator is central to the understanding of robust regression; it is discussed in the next section.

11.3.1 Breakdown Point

The concept of a finite sample breakdown point was first suggested by Hodges (1967) in the context of estimating a population mean, and a definition that is not restricted to the one-dimensional case was given by Donoho and Huber (1983). This definition is generally used in the robust regression literature. Simply stated, the *breakdown point* of a regression estimator is the smallest fraction of anomalous data that can render the estimator useless. [See, e.g., Rousseeuw and Leroy (1987, p. 9) for a more formal definition.] The smallest possible breakdown point is $1/n$, a value that is, unfortunately, attained by many estimators. For example, the sample mean has this breakdown point since a very extreme outlier can cause the sample mean to assume an implausible value.

Hampel et al. (1986) indicate that data generally contain 1–10% gross errors, so we would want the breakdown point of an estimator to exceed .10. During the past 10 years there has been emphasis on developing regression estimators that have a breakdown point of .50. Although the use of such an estimator would seem to be safe, we will see later that such estimators can produce very poor results under certain conditions. A better approach would be to use different breakdown points for a selected estimator and then compare the results. This will be illustrated later.

11.3.2   Efficiency

Another important concept in robust regression is that of *efficiency*. Assume that a data set has no recording or measurement errors (admittedly this may occur infrequently), approximate normality is suggested, and there are no influential observations. If we apply robust regression to such a data set, we would want the regression equation to be virtually identical to the OLS equation, since OLS is the appropriate method of estimation for this scenario. The efficiency of the selected robust regression technique could then be defined as its mean square error obtained using this technique divided by the OLS mean square error. Obviously, we would want this ratio to be close to 1.

In the literature the emphasis is on *asymptotic efficiency*. While such efficiency figures can be helpful in comparing robust estimators, the finite sample efficiencies [which would be data-set specific) can be expected to differ considerably from the asymptotic values (see, e.g., Ruppert (1992)]. When a data set *does* contain outliers, we may again speak of the (finite sample) efficiency of a robust estimator as its mean square error divided by the OLS mean square error, while noting that OLS is applied to only the good data points.

11.3.3   Classes of Estimators

In this section we survey the different classes of robust regression estimators. Those classes that have been discussed extensively in the literature and/or have considerable merit are subsequently discussed in detail starting with Section 11.4.

11.3.3.1   M-estimators
Huber (1973) introduced a class of estimators known as *M-estimators*, whose objective function is

$$\text{Minimize} \sum_{i=1}^{n} \rho(e_i)$$
$$\hat{\theta}$$

where $\rho(e_i)$ is some symmetric function of the residuals. [A symmetric function is one where $\rho(e_i) = \rho(-e_i)$.] Several ρ functions have been proposed; these are covered in Rousseeuw and Leroy (1987).

The M-estimators should properly be viewed as being of only historical significance, however, since they are undermined by X-outliers and thus have a breakdown point of $1/n$.

11.3.3.2 Bounded Influence Estimators

These estimators were developed to overcome the X-outlier deficiency of M-estimators. They are more frequently referred to as *GM-estimators* in the literature, but that label will not be used exclusively in this chapter, since the words "bounded influence" better explain the function of the estimators.

The intention is to have a stable estimator that is not influenced by Y-outliers that occur at X-outliers. These points are used in computing the value of the estimators, but the points are downweighted. This is why they are called *bounded influence estimators*, a name attributed to, in particular, the estimators proposed by Mallows (1975), Welsch (1980), Krasker (1980), and Krasker and Welsch (1982).

One shortcoming of these estimators is that they can have a low breakdown point. Maronna et al. (1979) showed that the breakdown point of a GM-estimator is at most $1/p$. Thus, when p is large, the breakdown point will be small and may not be much larger than the lowest possible breakdown point of $1/n$. When p is small, however, a GM-estimator can, under certain conditions, produce very good results. This will be illustrated later. Even when p is large a bounded influence estimator might be used in tandem with a high breakdown point estimator in a two-stage procedure. This will also be illustrated later.

It should be noted that although the terms GM-estimator and bounded influence estimator have been used synonymously in the literature (see, e.g., Rousseeuw and Leroy, 1987, p. 13), some of the recently developed bounded influence estimators are not also GM-estimators. That is, their breakdown point does not have an upper bound of $1/p$. This includes the two-stage bounded influence estimators of Simpson et al. (1992) that are discussed in Section 11.9.

11.3.3.3 High Breakdown Point Estimators

Most of these estimators have been developed by Peter Rousseeuw and his colleagues. These include the least median of squares estimator, the least trimmed squares estimator, and S-estimators. These estimators have been discussed extensively in the literature, and they will also be discussed extensively in this chapter. Although these estimators have generally been presented as stand-alone procedures, they can also be used in conjunction with a bounded influence estimator, as is discussed in Section 11.9. As stated earlier, it is also preferable to compute a selected HBP estimator using more than one breakdown point.

11.3.3.4 Two-Stage Procedures

The general idea is to combine different classes of estimators in a two-stage procedure so as to capture the benefits of each while avoiding the shortcomings that each one has when used individually. One possibility would be to use an HBP estimator in the first stage and a bounded influence estimator in the second stage. There are many possible combinations; the objective is to find the best one. Although the optimal combination is not known at the present time, some insight is given in a latter section.

11.4 GOALS OF ROBUST REGRESSION

Before examining the proposed estimators and combinations of those estimators, it is desirable to consider the properties that a robust regression estimator (or estimation scheme) should possess. In very general terms, we would want an estimator

1. to perform almost as well as OLS when the latter is the appropriate choice (i.e., when the errors are normally distributed with the data being free of mistakes and influential data points),
2. to perform much better than OLS when the conditions in (1) are not satisfied, and
3. not be overly difficult to compute or understand.

Some additional specific goals are given by Staudte and Sheather (1990, p. 206). In particular, they note that robust regression schemes should provide methods for constructing confidence intervals and hypothesis tests. The area of robust inference in regression is relatively unexplored, however, and is an important area for future work.

11.5 PROPOSED HIGH BREAKDOWN POINT ESTIMATORS

In this section we discuss most of the robust regression estimators that have been presented in the literature, starting with HBP estimators. The advantages and shortcomings of each are discussed, and recommendations are made regarding their use (or disuse).

11.5.1 Least Median of Squares

The *least median of squares* (LMS) estimator is obtained by minimizing the hth-ordered squared residual, where $h = [n/2] + [(p + 1)/2]$, with, as usual, n and p denoting the sample size and number of parameters, respectively. The symbol [·] means "integer portion of." The word "median" is used loosely, as for example, the 12th-ordered residual will be minimized when $n = 21$ and $p = 3$.

Another way to view the LMS estimator is as follows. Assume that we have a single regressor. The LMS regression equation is the equation of the line that is in the center of the narrowest strip that will cover the majority of the data, with distance measured vertically. (In multiple regression it is obtained from the smallest plane or hyperplane that covers the majority of the data with distance measured along the Y-axis.)

Hawkins (1993) states that "the least median of squares criterion is a current standard method of analysis of data when the possibility of severe badly-placed outliers makes an estimate with high breakdown point desirable." This is certainly the most frequently discussed, and probably also the most frequently used, robust regression estimator. Introduced by Rousseeuw (1984) and discussed extensively by Rousseeuw and Leroy (1987) and Rousseeuw (1990), it has since been disfavored by its developer [see Rousseeuw and van Zomeren (1990a)]. Although it should properly be viewed as having only historical significance, it will be discussed extensively in this section because of the attention that it has received.

The LMS estimator is motivated by Rousseeuw (1990, p. 16.14) in the following way. Since OLS minimizes the sum of the squared residuals, it also minimizes the *mean* of the squared residuals, and the mean is, of course, not an appropriate statistic when outliers are present. Minimizing the *median* of the squared residuals would seemingly be a logical alternative. [See also Rousseeuw and Leroy (1987, p. 14).]

At first this seems intuitively appealing, but upon reflection it should be apparent that this estimator should perform poorly relative to OLS when the latter is the appropriate choice. In fact, it is well known that LMS has an asymptotic efficiency of zero.

It should be apparent why this happens. The LMS objective function dictates that a function of the residuals be minimized for a particular point, thus ignoring the fit at the other $n - 1$ points. As $n \to \infty$ we would expect the fit at the "neglected" points to become worse, as a group, relative to OLS. This is one reason why LMS should not be used as a stand-alone procedure. In an effort to improve the efficiency of LMS, Rousseeuw and Leroy (1987, p. 129) suggested using the LMS estimates as starting values for computing a one-step M-estimator. We will see later why LMS should not be used in the first stage of any two-stage procedure, however.

11.5.1.1 Computational Aspects

Another criticism that has been made of LMS involves algorithmic instability. The LMS estimates have usually been obtained using PROGRESS, a computer program written by Annick Leroy under the guidance of Peter Rousseeuw. Assume that we have a single regressor. The slope estimate would be obtained by evaluating all $\binom{n}{2}$ point subsets, and using the slope from the equation that produces the smallest median squared residual. [The intercept is obtained through an intercept adjustment; see Rousseeuw and Leroy (1987, p. 201).] Clearly the number of point subsets that must be used will be large when n is large, and the problem is more acute when there is more than one regressor, since the "2" in the combinatoric would be replaced by the number of parameters, in general. Therefore, a "complete search" of all possible point subsets was not built into the initial version of PROGRESS.

A complete search can be performed using a more recent version of PROGRESS (Dallal, 1991), however, so algorithmic variability due to the evaluation of a fraction of all of the point subsets is no longer a valid criticism. This assumes that n is not so large that it is computationally impractical to evaluate all of the point subsets. To illustrate, 5985 point subsets must be evaluated for the stack loss data that were examined in Chapter 6. Doing so took 2 min and 18 sec on this author's PC, using the program due to Dallal (1991). But if, say, $n = 75$ and $p = 4$, a total of 1,215,450 point subsets must be evaluated, and running the program until completion might require almost a full day! Therefore, a person who wishes to use, or at least study, LMS might wish to use some type of approximation algorithm. Two such algorithms are discussed in Section 11.5.1.3.

11.5.1.2 Illustrative Examples

Perhaps the best way to illustrate how things can go very wrong when the LMS approach is used would be to slightly modify an example given by Stefanski (1991), as will be done later in this section. In the Stefanski example, with $n = 9$, there were three good data points that fell on a line and two outliers that fell on the same line. If the seven good data points were used, the slope would be practically zero, but the collinearity of the three good points and the two bad points caused the LMS equation to be the equation of that line. In that example the LMS equation was considerably different from the correct equation.

The phenomenon illustrated by that example can be explained generally as follows. Since $[n/2] + [(p + 1)/2] = 5$, the fifth-ordered squared residual will be minimized, but since five data points lie on a line, the fifth-ordered squared residual must be zero. Therefore, the LMS equation must be the equation of the line that passes through these five points.

This illustrates the *exact fit property* of LMS, a property that it shares with other high breakdown point estimators. With real data we would certainly not expect slightly more than half of the data to be collinear, or in the case of multiple regression, coplanar. We might, however, expect to encounter a slight variation of this scenario. Since LMS fits just over half of the data, bad data that are aligned with good data could draw the LMS equation away from the appropriate equation for the good data. The data do not have to be collinear for this to happen; we illustrate this with the following example.

Consider the data in Table 11.1 and the graph of these data in Figure 11.3. Notice that there are obviously two outliers, and a regression equation fit to the seven good data points should have a slope of approximately zero, as in

Table 11.1 Data for Illustrating a Shortcoming of LMS

X	Y
2.5	5.0
7.5	2.6
14.0	2.6
15.0	3.9
16.0	5.1
22.0	13.0
23.0	14.2
23.0	5.1
29.0	2.7

the Stefanski example. The OLS equation fit to all nine data points is $\hat{Y} = 2.94 + 0.18X$, so the outliers have very little effect. The LMS equation, however, is $\hat{Y} = -15.41 + 1.29X$, so the line is determined by the two bad points in conjunction with the three good points with which they are closely aligned. Thus, in this example LMS performs poorly, whereas OLS performs well.

One of the touted advantages of LMS is its ability to identify outliers. Once identified, these outliers are given a weight of zero, and *reweighted least squares*

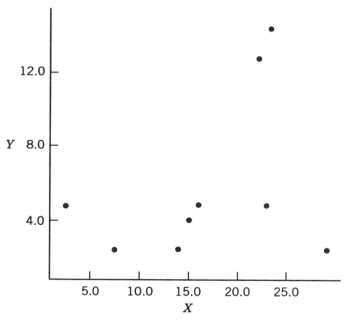

Figure 11.3 Scatter plot of the data in Table 11.1.

is applied to the supposedly good data points. Specifically, a robust estimate of σ is computed (see p. 202 of Rousseeuw and Leroy, 1987), and points corresponding to standardized residuals greater than 2.5 are assigned a weight of zero.

If the LMS solution is influenced by outliers, however, the good data points are apt to be deleted instead of the outliers. For the current example, the other four good data points are given a weight of zero, and the reweighted least squares equation is $\hat{Y} = -15.56 + 1.30X$, which is only slightly different from the LMS equation. Furthermore, the R^2 value is 0.99991, whereas $R^2 = .018$ when the seven good data points are used.

What has happened is the robust estimate of scale is much too small because the LMS equation is almost an exact fit to five of the data points. This magnifies the distance that these four good data points are from the line, thus causing the good data points to appear to be outliers. Thus, LMS performs poorly even though the breakdown point is (ostensibly) 0.5, and we have only 22% bad data. Any high breakdown point estimator will be susceptible to this type of problem, which is why the use of a 50% breakdown point must be questioned if only a single breakdown point is to be used. (An alternative would be to use a high breakdown point estimator with multiple breakdown points, such as .50 and .25, and compare the results.)

With virtually any procedure the identification of multiple outliers will be extremely difficult for certain data configurations, such as a higher-dimensional generalization of what we saw in the preceding example. Rousseeuw and van Zomeren (1990b) introduced a *robust distance* diagnostic for identifying outliers in multiple regression that was shown to perform well for certain data sets. It has the desirable feature of using robust estimates of scale and location, but since those estimates are obtained from the smallest ellipsoid that covers half of the data, it can be expected to have the same failing in multiple regression that we observed that LMS can have in simple regression. Nevertheless, the robust distance concept is a step in the right direction, but the authors's statement that "it turns out that it does not matter so much *which* high-breakdown estimator is used when the purpose is to detect outliers" should be ignored. Not only is the choice of a high breakdown point estimator important, but the breakdown point itself can be even more important. As computing continues to become faster and more efficient, we should probably expect to see the emergence of estimation and outlier detection procedures based on the use of multiple breakdown points.

Another criticism that has been made of LMS is that small perturbations in the data can cause large differences in the regression coefficients. (Thus, the LMS estimator is not a resistant estimator.) This is illustrated by Hettmansperger and Sheather (1992) using a data set from Mason et al. (1989). That is a multiple regression example; in Exercise 11.1 the reader is asked to show this for the simple regression data set that is given.

As pointed out by Stromberg (1993a), however, the instability in the LMS solution for that example results from the fact that point subsets are used, the customary approach, rather than computing the *exact* LMS solution. (There was no instability when the exact solution was obtained.) The latter is discussed in the next section.

11.5.1.3 Exact LMS

Although the LMS estimators have generally been obtained by using PROGRESS, Stromberg (1993b) gives results that indicate that PROGRESS should not be used. The problem is that the "exact fits" approach used by PROGRESS will not minimize the LMS objective function. The minimization is actually accomplished with a *Chebyshev fit* to part of the data. This requires that *l-point* subsets be used, where *l* equals the number of parameters plus one.

Let $\hat{\boldsymbol{\theta}}_{\text{OLS}}$ be the OLS estimator for an arbitrary *l*-point subset. The Chebyshev fit estimator, $\hat{\boldsymbol{\theta}}_C$, is obtained as

$$\hat{\boldsymbol{\theta}}_C = \hat{\boldsymbol{\theta}}_{\text{OLS}} - (\mathbf{X}'\mathbf{X})^{-1}\mathbf{X}'\delta\mathbf{S}$$

where δ equals the sum of the squares of the OLS residuals divided by the sum of the absolute value of the residuals, $\mathbf{S}$ is a vector that contains the signs of the residuals, and $\mathbf{X}$ is defined as in previous chapters. The Chebyshev fit estimator is computed for each *l*-point subset, and the one that minimizes the LMS objective function is the solution, provided that δ is defined for the *l*-point subset that minimizes the objective function.

To illustrate this last condition, consider an example where $Y = 7, 2, 3, 4$, and 1, and $X = 1, 4, 5, 6$, and 2. There is an exact fit to points 2, 3, and 4, which is given by $\hat{Y} = -2 + X$. Thus, three of the residuals are zero, so the median squared residual is zero. Therefore, this must be the exact LMS equation. But this is not a *feasible* Chebyshev fit because the latter is not defined when there is an exact fit to *l* points since δ is then undefined. (The Minitab macro in the chapter appendix flags exact-fit subsets for further scrutiny.) Thus, some care must be exercised in using the Chebyshev fit approach to find the exact LMS solution.

The exact LMS solution can be obtained in simple linear regression without having to use Chebyshev fits, provided that the intercept adjustment is performed for each two-point subset, not just for the one that produces the minimum value of the LMS objective function. This result follows from the results of Steele and Steiger (1986), which indicate that the slope of the exact LMS solution is the same as the slope of at least one of the lines that joins each pair of points. If PROGRESS is used without this modification, however, the regression equation that it produces will not be the exact LMS solution. To illustrate, if PROGRESS (as in the version given by Dallal, 1991) is applied to the example on page 51 of Rousseeuw and Leroy (1987), the regression equation that is produced in $\hat{Y} = -2.468 + 0.102X$, whereas the exact LMS solution is $\hat{Y} = -3.4225 + 0.125X$. (The slope of the line that connects points 2 and 6 is 0.125; 0.102 is the slope of the line that connects points 1 and 6.) With $n = 9$, the objective function is minimized by minimizing the fifth-ordered squared residual. For the exact solution that value is 0.0053, whereas for the PROGRESS solution the value is 0.0071.

As a technical point we should note that the data for this example are in *general position*, a requirement that must be met for the claimed theoretical results of LMS to hold. General position simply means that the $\hat{\theta}_{OLS}$ solutions are all different (Rousseeuw and Leroy, 1987, p. 117).

In summary, if one wished to study LMS, the computations should be performed by a program that will compute the Chebyshev fits when there is more than one regressor. The problem in doing so is that the latter require considerably more computation than do the exact fits, thus necessitating the use of an algorithm that approximates the exact LMS solution when n is not small. Hawkins (1993) presents a linear programming algorithm for approximating this solution. The algorithm provides the exact solution with probability 1 as the number of iterations increases. Ruppert (1992) also gives an algorithm for approximating the exact LMS solution that can additionally be used for approximating the exact solution for other high breakdown estimators. Both of these approaches are explained in detail in Section 11.6 and illustrated in Sections 11.6.1–11.6.3. The Minitab macros described in the chapter appendix can be used for computing the Chebyshev fits for simple regression with a small n.

11.5.2 Least Trimmed Squares

The *least trimmed* (sum of) *squares* (LTS) estimator is obtained from

$$\underset{\hat{\theta}}{\text{Minimize}} \sum_{i=1}^{h} e_{(i)}^2$$

where $e_{(1)}^2,\ e_{(2)}^2, \ldots, e_{(n)}^2$ are the ordered squared residuals, from smallest to largest, and the value of h must be determined.

We might let h have the same value as for LMS, so that it will be a high breakdown point estimator, but it was shown in Section 11.5.1.2 that a 50% breakdown point can sometimes produce poor results. Consequently, it seems preferable to use a larger value of h and to speak of a trimming percentage α. Rousseeuw and Leroy (1987, p. 134) suggest that h might be selected as $h = [n(1 - \alpha)] + 1$.

LTS was proposed by Rousseeuw in a symposium in 1983. A good source is Rousseeuw (1984). LTS is intuitively more appealing than LMS, due in part to the fact that the objective function is not based on the fit at any particular point. This estimator should not be confused with the estimators proposed by Ruppert and Carroll (1980) and Frees (1991). With the latter, which is only for simple regression, the trimming is done on the slopes, whereas with the former the trimming is done on the residuals. Both are analogous to the trimmed mean concept.

Part of the appeal of LTS is that if we are fortunate to trim the exact number of bad data points without trimming any good data points, we will have the (seemingly) optimal estimator: OLS applied to the good data points. If we apply LTS with 25% trimming to the data in Table 11.1, using h as defined earlier in this section, we will trim two of the nine data points, and we would hope that we trim the two bad data points. [We will use the notation LTS(.25) to represent the LTS estimator with 25% trimming.]

Unfortunately, there is not a practical way to obtain the exact LTS solution, except when n is small. For the Table 11.1 data with 25% trimming we need only compute the OLS equation for 36 point subsets, since $\binom{9}{2} = 36$, or 126 subsets if 50% trimming is used. Performing the 36 computations will provide the exact LTS(.25) solution and also guarantee that we will find the two bad data points if these points are regression outliers.

But regression data sets generally have more than nine points, and the number of required computations can easily be prohibitive for large n, just as we saw for LMS. The LTS estimator can be computed using PROGRESS, but the latter will not provide the exact LTS solution. Rousseeuw and Bassett (1991) do show, however, that the approximate LTS solution given by PROGRESS, which as with LMS is based on exact fits, will have the same breakdown point as the exact LTS solution.

Ruppert (1992) gives an algorithm that can be used for approximating the exact LTS solution. Similarly, Hawkins (1994) provides an algorithm for approximating the exact LTS solution that will give the exact solution with high probability for a sufficient number of random starts. These methods are discussed in Section 11.6, and we will use them with multiple trimming in an effort to identify clusters of outliers. The importance of multiple trimming can be explained as follows. Assume that we have 39 data points, and 11 are gross errors that strongly influence the regression equation. If we use LTS with $\alpha = .25$, the result can be expected to correspond to what will happen if we apply OLS to 30 data points that include two gross errors. Conversely, if we overshoot the percentage of gross errors, we run the risk of LTS following the bad data points instead of the good ones, as was seen with LMS. For example, if we apply LTS with 50% trimming to the Table 11.1 data, we should be able to guess that the solution will be very poor, remembering what we saw previously about that data set. This will be illustrated later.

11.5.3 S-estimators

Introduced by Rousseeuw and Yohai (1984), S-estimators are the last HBP estimators that will be presented. They have not been as widely discussed as the LMS estimator, but a detailed discussion can be found in Rousseeuw and Leroy (1987, pp. 135–143).

This class of estimators derives its name from the fact that the estimator, $\hat{\beta}_s$

in the notation used here, is obtained as

$$\underset{\hat{\beta}}{\text{Min}} \ S(e_1(\beta), \ldots, e_n(\beta))$$

where $e_1(\beta), \ldots, e_n(\beta)$ denote the n residuals for a given candidate β, and $S(e_1(\beta), \ldots, e_n(\beta))$ is given by the solution to

$$\frac{1}{n} \sum_{i=1}^{n} \rho(e_i/s) = k \tag{11.1}$$

The function $\rho(\cdot)$ must be selected, and Rousseeuw and Yohai (1984) suggest

$$\rho(x) = \begin{cases} \dfrac{x^2}{2} - \dfrac{x^4}{2c^2} + \dfrac{x^6}{6c^4} & |x| \le c \\[3mm] \dfrac{c^2}{6} & |x| > c \end{cases}$$

The selection of c, which determines k, involves, as we would expect, a trade-off between breakdown point and efficiency.

Table 19 of Rousseeuw and Leroy (1987, p. 142) provides some guidance, but we should keep in mind that the efficiency figures given therein are *asymptotic* efficiencies. The latter should not be viewed rigidly, however, as an S-estimator can still have low efficiency regardless of the value of k that is used. This follows from the fact that, like LMS, S-estimators also possess the exact-fit property, as discussed by Rousseeuw and Leroy (1987, p. 139). As discussed by Stefanski (1991), estimators that have this property can have low efficiency in finite samples. This should be apparent from the previous discussion regarding the Table 11.1 data.

Since all HBP estimators possess the exact-fit property, all HBP estimators can perform poorly relative to OLS when the latter should be used.

11.5.3.1 Are S-estimators Any Better?
Rousseeuw and Leroy (1987, p. 208) report that "some preliminary experience (both for simple and multiple regression) indicates that S-estimators do not really perform better than the LMS, at least from a practical point of view." Ruppert (1992) examined the performance of S-estimators for both real and simulated data and found that S-estimators performed somewhat better than LMS and LTS.

11.5.3.2 Computing S-estimators
S-estimators may be computed from the exact-fits approach using PROGRESS. Regardless of the algorithm that is used, Eq. (11.1) must be solved in an iterative manner, and the solution can be difficult to obtain. The computations may

be reduced considerably by using a comparison discussed by Ruppert (1992) and Yohai and Zamar (1991). There is not, however, a method for obtaining the exact solution, just as there is not a computationally practical method for obtaining the exact LTS solution. A method that could be used for approximating S-estimators (as well as for LMS and LTS) is discussed in the next section.

11.6 APPROXIMATING HBP ESTIMATOR SOLUTIONS

We have observed that it is possible, but computationally impractical, to obtain the exact LMS and LTS solutions when n is not small, and it is not possible to obtain the exact solution for S-estimators for any value of n. Thus, there is a need for a computationally efficient algorithm for approximating the exact solution for each of these estimators.

One such algorithm is given by Ruppert (1992), which was called RAND-DIR. We will discuss and illustrate the use of RANDDIR for LTS only, and it will be applied to several data sets. It is essentially a four-stage procedure in which a number of exact-fit candidate solutions are initially generated randomly from the set of all possible exact fits, and a search is then conducted in the neighborhood of the best exact-fit solution. An additional search is then conducted in the neighborhood of the best second-stage solution, and a final search is conducted in the neighborhood of the best third-stage solution, with the search being progressively narrowed over the last three stages. These steps are summarized generally in the flowchart that is given in Figure 11.4, with N_{samp} denoting the number of point subsets that are used.

It is important to note that the value of γ_l depends on the value of l relative to N_{samp}. In the first stage, $\gamma_l \equiv 1$ when $l < \lceil N_{\text{samp}}/5 \rceil$, with $\lceil \cdot \rceil$ denoting "rounded to the next largest integer." When $l \geq \lceil N_{\text{samp}}/5 \rceil$, $\gamma_l = U^{k(l)}$ with $U \sim \text{Uniform}(0, 1)$ and $k(l) = 1$ if $\lceil N_{\text{samp}}/5 \rceil \leq l < \lceil N_{\text{samp}}/2 \rceil$, equals 2 if $\lceil N_{\text{samp}}/2 \rceil \leq l < \lceil .8N_{\text{samp}} \rceil$, and equals 4 if $\lceil .8N_{\text{samp}} \rceil \leq l \leq N_{\text{samp}}$.

When $l < \lceil N_{\text{samp}}/5 \rceil$, RANDDIR is the same as PROGRESS applied to LTS for the version of PROGRESS that uses random subsampling (instead of using all possible point subsets). In each successive stage the random search is concentrated closer to the best solution that has been found during each stage, which is intuitively appealing.

In order for this procedure to work well, enough exact fits must be performed so that the second-stage search will start in the "right neighborhood." In the simulations performed by Ruppert (1992), 600 subsamples were generated for LTS, which means that 120 exact-fit candidate solutions were produced for the first stage. This would seemingly be sufficient, except perhaps when n is quite large. (Since this is an approximation procedure that involves random sampling, successive applications of the procedure to the same data set should produce different solutions, but the differences should be slight.) The other 480 candidate solutions will not be exact-fit solutions, but each candidate solution will be in a random direction (hence, the name RANDDIR) from what is the best

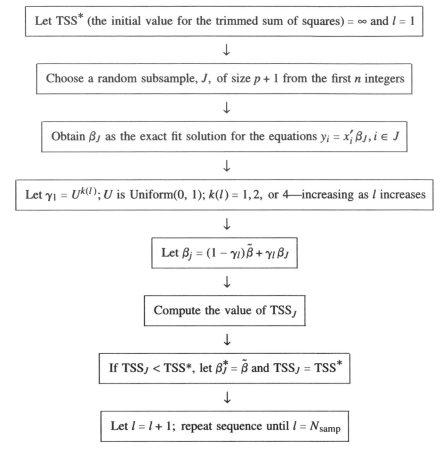

Let TSS* (the initial value for the trimmed sum of squares) = ∞ and $l = 1$

$\downarrow$

Choose a random subsample, J, of size $p + 1$ from the first n integers

$\downarrow$

Obtain β_J as the exact fit solution for the equations $y_i = x_i'\beta_J, i \in J$

$\downarrow$

Let $\gamma_1 = U^{k(l)}$; U is Uniform(0, 1); $k(l) = 1, 2,$ or 4—increasing as l increases

$\downarrow$

Let $\beta_j = (1 - \gamma_l)\tilde{\beta} + \gamma_l\beta_J$

$\downarrow$

Compute the value of TSS$_J$

$\downarrow$

If TSS$_J <$ TSS*, let $\beta_J^* = \tilde{\beta}$ and TSS$_J =$ TSS*

$\downarrow$

Let $l = l + 1$; repeat sequence until $l = N_{\text{samp}}$

Figure 11.4 Flowchart for RANDDIR for LTS.

solution (the one that produces the smallest value of $\hat{\sigma}$), when each candidate solution is generated.

The data in Table 11.1 are at the opposite extreme since $n = 9$. As indicated previously, the use of LTS(.25) requires that only 36 subsamples be computed to obtain the exact LTS solution. Consequently, the Ruppert procedure is unnecessary for this data set, but one might use it anyway since there is apparently no computer program available that will compute the exact LTS solution, especially for user-specified trimming percentages. We would expect the Ruppert procedure to closely approximate this exact LTS solution, even when the number of subsamples generated (N_{samp}) is much less than 600.

The "feasible set algorithm" of Hawkins (1993) was proposed for LMS, but the general idea can also be applied to LTS. When used for LMS, the Chebyshev fit is characterized as the solution to a linear programming problem. The steps that are followed may be summarized as follows. First, a random subset of $p + 1$

observations is obtained, and this forms the starting basis. The dual simplex algorithm is then applied, with pivoting performed in the following manner. With h as defined in Section 11.5.1, if at least h shadow costs are positive, then the current basis satisfies the necessary condition for the LMS optimum. If not, some column with a negative shadow cost is brought into the basis, and the case with the hth smallest shadow cost is removed.

Effective use of the algorithm requires that a sufficiently large number of "random starts" be used (i.e., random subsets of size $p + 1$), as one of the random starts will produce the exact LMS solution. Calculations performed by Hawkins (1993) suggest that 200 may be sufficient for most data sets. The objective is to use enough random starts so that a sufficiently large number of candidate bases will be evaluated. The number of random starts that is used should obviously depend on n and p. If both are large, a correspondingly large number of random starts will be required. For example, if $n = 50$ and $p = 5$, approximately 1000 random starts will be required if the exact LMS solution is to be found with probability of .99 or greater. This many random starts can require considerable time, depending on the computing equipment that is used.

In comparison with RANDDIR, which conducts a random search in the parameter space, the feasible set algorithm uses only Chebyshev fits, moving from one fit to a better one in terms of LMS. The feasible set algorithm can be adapted to LTS so as to approximate the exact LTS solution, as described in Hawkins (1994). The algorithm for LTS will be used to approximate the exact LTS solution for the examples in Sections 11.6.1.–11.6.3.

11.6.1 Application to Hawkins–Bradu–Kass Data Set

A more challenging application is the Hawkins–Bradu–Kass (1984) data set. This is a much-referenced artificial data set consisting of 75 observations and three regressors that is known to be quite troublesome in terms of masking and swamping, and new methods for detecting outliers have been tested on the data set. (As discussed in previous chapters, *masking* refers to bad data points being camouflaged because they are clustered, and *swamping* means that good data points appear to be outliers.) For example, Rousseeuw and van Zomeren (1990b) use it for testing their robust distance diagnostic, Rousseeuw and Leroy (1987) apply LMS combined with an index plot to the data set, and Hawkins et al. (1984), Hadi (1992), and Hadi and Simonoff (1993) use it for illustrating the outlier detection methods that they proposed. (An *index plot*, as defined by Rousseeuw and Leroy (1987), is a plot of the standardized residuals against the index number of the observation, with a robust estimate of scale used to standardize the residuals.) The data set is given in Table 11.2.

As discussed by Rousseeuw and Leroy (1987, p. 93) in using this data set we have the advantage that the bad data points are known. Here the first 10 data points are (regression) outliers and the next four points are X-outliers only. If we apply LTS(.25) to this data set, we would hope that the first 10 points are among the ones that are effectively trimmed.

Table 11.2   Data set from Hawkins et al. (1984)

Y	X_1	X_2	X_3
9.7	10.1	19.6	28.3
10.1	9.5	20.5	28.9
10.3	10.7	20.2	31.0
9.5	9.9	21.5	31.7
10.0	10.3	21.1	31.1
10.0	10.8	20.4	29.2
10.8	10.5	20.9	29.1
10.3	9.9	19.6	28.8
9.6	9.7	20.7	31.0
9.9	9.3	19.7	30.3
−0.2	11.0	24.0	35.0
−0.4	12.0	23.0	37.0
0.7	12.0	26.0	34.0
0.1	11.0	34.0	34.0
−0.4	3.4	2.9	2.1
0.6	3.1	2.2	0.3
−0.2	0.0	1.6	0.2
0.0	2.3	1.6	2.0
0.1	0.8	2.9	1.6
0.4	3.1	3.4	2.2
0.9	2.6	2.2	1.9
0.3	0.4	3.2	1.9
−0.8	2.0	2.3	0.8
0.7	1.3	2.3	0.5
−0.3	1.0	0.0	0.4
−0.8	0.9	3.3	2.5
−0.7	3.3	2.5	2.9
0.3	1.8	0.8	2.0
0.3	1.2	0.9	0.8
−0.3	1.2	0.7	3.4
0.0	3.1	1.4	1.0
−0.4	0.5	2.4	0.3
−0.6	1.5	3.1	1.5
−0.7	0.4	0.0	0.7
0.3	3.1	2.4	3.0
−1.0	1.1	2.2	2.7
−0.6	0.1	3.0	2.6
0.9	1.5	1.2	0.2
−0.7	2.1	0.0	1.2
−0.5	0.5	2.0	1.2
−0.1	3.4	1.6	2.9
−0.7	0.3	1.0	2.7
0.6	0.1	3.3	0.9
−0.7	1.8	0.5	3.2

Table 11.2 (*Continued*)

Y	X_1	X_2	X_3
−0.5	1.9	0.1	0.6
−0.4	1.8	0.5	3.0
−0.9	3.0	0.1	0.8
0.1	3.1	1.6	3.0
0.9	3.1	2.5	1.9
−0.4	2.1	2.8	2.9
0.7	2.3	1.5	0.4
−0.5	3.3	0.6	1.2
0.7	0.3	0.4	3.3
0.7	1.1	3.0	0.3
0.0	0.5	2.4	0.9
0.1	1.8	3.2	0.9
0.7	1.8	0.7	0.7
−0.1	2.4	3.4	1.5
−0.3	1.6	2.1	3.0
−0.9	0.3	1.5	3.3
−0.3	0.4	3.4	3.0
0.6	0.9	0.1	0.3
−0.3	1.1	2.7	0.2
−0.5	2.8	3.0	2.9
0.6	2.0	0.7	2.7
−0.9	0.2	1.8	0.8
−0.7	1.6	2.0	1.2
0.6	0.1	0.0	1.1
0.2	2.0	0.6	0.3
0.7	1.0	2.2	2.9
0.2	2.2	2.5	2.3
−0.2	0.6	2.0	1.5
0.4	0.3	1.7	2.2
−0.9	0.0	2.2	1.6
0.2	0.3	0.4	2.6

The OLS equation obtained after first deleting the 10 outliers is $\hat{Y} = -0.1805 + 0.08138X_1 + 0.03990X_2 - 0.05167X_3$, so this is what we would (seemingly) want an approximation technique to produce. When $N_{\text{samp}} = 600$ is used with the Ruppert procedure for LTS(.25), we obtain $\hat{Y} = -1.04 + 0.20X_1 + 0.28X_2 + 0.11X_3$. From the ordered squared residuals we can determine which points were effectively trimmed. Only *one* of the outliers is in this group, and *all four* of the "good leverage points" are deleted because they are "swamped." This is also what happens when the Hawkins feasible set algorithm is used. Thus, LTS *seems* to have the same failing as some other robust regression techniques that have been applied to this data set.

The efficiency relative to OLS applied to the good data points is poor because LTS(.25) is considerably influenced by the (regression) outliers. For the good data points OLS produces a residual sum of squares of 18.939, whereas 673.96 is produced by LTS(.25). But the sum of the squared errors corresponding to the "good leverage points" is 644.66, so LTS(.25) fits the other 61 data points almost as well as OLS fits them.

Thus, LTS is obviously "riding" the regression outliers instead of the "good" outliers. Even though the latter have been labeled good leverage points by Rousseeuw and van Zomeren (1990b, p. 636) and Rousseeuw and Leroy (1987, p. 93), we will see very shortly that there is no such thing as good or bad leverage points for this data set.

Using the 61 data points that are not outlying in either the regressors or the response, $R^2 = 0.047$, and when the 65 good data points are used $R^2 = 0.043$. Therefore, there is not a set of good data points for the model that is fit. Because of this and the fact that the data are artificial, the separation of the data points into good and bad points is questionable. (It is also apparent from inspection of the data set in tabular form that the data do not all fit together.) Consequently, we should not attempt to assess the merits of LTS or any other estimation procedure for this data set.

We used this data set to stress the importance of having a data set for which some model provides a good fit before examining the performance of various outlier detection schemes. We may draw an analogy between what happens when LTS(.25) and other robust regression and outlier detection schemes are applied to this data set, and what happens when LMS (or LTS without complete enumeration of all point subsets) is applied to the data given by Stefanski (1991). There a model, with a nonzero R^2, was not well determined from the good data points, which is why LMS failed, with certain good points there similarly being swamped.

We should note, however, that LMS performs satisfactorily for the Hawkins, et al. (1984) data set, as both the robust distance diagnostic and the index plot do pick out the 10 outliers. This should not be surprising, however, since the good data points are fairly clustered and are far removed from the bad points. (Recall the definition for LMS for multiple regression being determined from the smallest hyperplane that covers the majority of the data.) Thus, LMS can sometimes, if not frequently, produce the desired results; the performance of LMS can be influenced more by the general configuration of points than by the number of bad data points.

11.6.2   Another Application: One Regressor

We will first look at a simple regression example before moving to multiple regression. We can, of course, see the outliers in the two-dimensional case, but it is instructive to examine how a procedure works when the correct answer is apparent.

Consider the graph in Figure 11.5. Here we may truly speak of good leverage points and bad leverage points. The three points that are (somewhat) outlying on X (only) help to define the regression line and thus increase the precision of the parameter estimates, whereas we would want the three points that are also outlying on Y to be identified as regression outliers. When OLS is applied to the data, which are given in Table 11.3, the three points with extreme y-coordinates cause the regression line to tilt toward them, with the result that the rightmost (good) point has a standardized residual of -2.30 and a standardized deletion residual of -2.71, and all of the other standardized residuals are within $(-2, 2)$. (See Chapter 5 for the different types of residuals and their definitions.)

Let's assume that the points with extreme y-coordinates represent recording errors, and that the other points are all valid data points. We thus have an example that illustrates swamping and masking. Clearly the regression equation should be primarily determined by the cluster of points in the lower left quadrant of Figure 11.5, with the three other good data points not greatly influencing the regression line. (Without those points the regression equation is $\hat{Y} = 5.62 + 0.80X$, and with them the equation is $\hat{Y} = 6.82 + 0.60X$.) But the equation using all of the data is $\hat{Y} = 0.68 + 1.77X$, which shows the influence of the bad data points.

It may be quite a challenge for any robust regression procedure to essentially

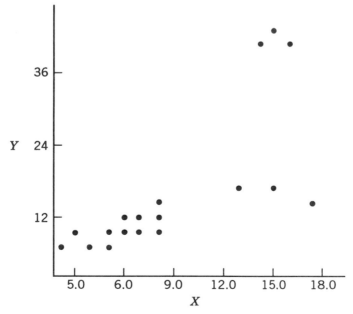

Figure 11.5. Scatter plot of the data in Table 11.3.

**Table 11.3 Data for Illustrating LTS with
Sequential Trimming**

Y	X
8.0	2.0
9.0	3.0
8.3	4.0
8.0	5.0
9.0	5.0
9.5	6.0
11.0	6.0
10.8	7.0
11.8	7.0
12.0	8.0
10.6	8.0
14.8	8.0
17.3	12.9
15.9	15.0
14.6	17.3
41.1	14.0
42.0	15.0
41.0	16.0

separate the bad data points from the extreme good data points, since we can almost visualize a line fit to the bad data points that provides a reasonably good fit to all but the good leverage points. We will apply LTS with multiple trimming percentages. Because n is small, and also because we wish to identify the outliers rather than determine the final equation, we will first try the Ruppert approach with $N_{samp} = 200$ and see if that is sufficient. (We would logically expect that the smallest value of N_{samp} that will produce the desired results should depend on n.)

Consider the data in Table 11.4. Since outlier tests for robust regression have not been developed, we might construct an ad hoc rule for identifying outliers when sequential trimming is used. One possibility is to compute the marginal reduction in the objective function as a multiple of the average squared error, where the average is obtained from OLS applied to all of the data points. We will denote this multiple by k. A large value of k will signify the detection of either a false positive outlier or a true outlier.

The progressive trimming shows the difficulty that LTS has in identifying the real outliers. Notice that the false positive outliers are trimmed before the true outliers. Notice also that when n is small, we can easily obtain the exact solution for each of the first few steps without having to use the Ruppert procedure. For example, to find the best single point to delete we could simply perform the 18 regressions with a different point deleted in each, and determining the best pair to delete would require only 153 regressions. The next two numbers would

Table 11.4 LTS with Sequential Trimming for Table 11.3 Data

Trimming Percentage (number trimmed)	Point(s) Trimmed (Y, X)	Marginal Reduction in TSS[a]	Multiple of Average Squared Error[b]
.10 (1)	(14.6, 17.3)	442.4	7.14
.15 (2)	(14.6, 17.3) (15.9, 15.0)	222.9	3.60
.20 (3)	(41, 16) (41.1, 14) (42, 15)	266.6	4.30
.25 (4)	(41, 16) (41.1, 14) (42, 15) (14.6, 17.3)	112.2	1.81
.30 (5)	(41, 16) (41.1, 14) (42, 15) (14.6, 17.3) (15.9, 15.0)	91.2	1.47

[a]TSS = Trimmed sum of squares; first value is reduction from SSE (OLS).
[b]This is obtained using SSE(OLS)/n as the base, and dividing that into each marginal reduction.

be 816 and 3060, however, so obtaining the exact solution at each step would soon become impractical, especially if n is large. Therefore, one possibility would be to start with the exact solution approach and then switch over to the Ruppert procedure when the combinatorics become excessive. It isn't necessary to obtain the exact solution; we simply need a good approximation.

Ensuring that we will have a good approximation with high probability requires determining N_{samp} relative to n and an upper bound on the expected number of bad data points, a determination that is yet to be made. Even when n is relatively small, as it is here, using a small value for N_{samp} can create problems. For example, the exact solution at step 3 produces a value of 32.383 for the objective function, considerably less than the approximate value. This means that the improvement that results from deleting the three true outliers is much greater than the approximate figure indicates.

Clearly a stopping rule is needed so that the procedure terminates when $k <$ k_0. A seemingly reasonable value would be $k_0 = 3$. We can see that the true outliers are identified if that rule is applied here.

The actual reduction figures are given in Table 11.5. Notice that these differ considerably from those in Table 11.4, indicating that the value of N_{samp} is not sufficient. Nevertheless, the same decision is reached using the stopping rule given above. We should also bear in mind that N_{samp} need only be large enough to allow us to identify the points that should be deleted at each stage.

Table 11.5   Actual Reductions in SSE for Sequential Trimming

Trimming Percentage (number trimmed)	Point(s) Trimmed (Y, X)	Reduction in SSE	Multiple of Average Squared Error
.10 (1)	(14.6, 17.3)	367.2	5.93
.15 (2)	(14.6, 17.3)	296.7	4.79
	(15.9, 15.0)		
.20 (3)	(41, 16)	418.7	6.76
	(41.1, 14)		
	(42, 15)		
.25 (4)	(41, 16)	11.5	0.19
	(41.1, 14)		
	(42, 15)		
	(14.6, 17.3)		
.30 (5)	(41, 16)	3.8	0.06
	(41.1, 14)		
	(42, 15)		
	(14.6, 17.3)		
	(15.9, 15.0)		

If that determination is properly made, then least squares can be applied to the remaining points so as to obtain the actual reduction figures. Here $N_{samp} = 200$ is not sufficient to provide a good approximation to the actual reduction figures, but it does permit the identification of the points that are deleted at each stage.

If RANDDIR is used to identify bad data points, with least squares then applied to the remaining points, analogous to the use of LMS, a smaller value of N_{samp} can be used than would be needed if the algorithm is used simply to produce the regression equation in a one-step operation. When the Hawkins approach is used, the proper identification of the points that should be deleted can often be made with only a very small number of random starts (such as two).

Although a reduction figure is somewhat meaningless at a stage where there is a departure from proper subsets, the number 418.7 in Table 11.5 does indicate the considerable improvement that results from the switch to the true outliers from the false positive outliers. Notice that if we can assume that we will be successful in identifying the bad data points with this approach, then we would consider using a bounded influence approach in a two-stage procedure only if there were influential observations among the good data points.

We will later see an example where the (larger) value of N_{samp} that is used produces reduction figures that are almost exactly equal to the actual figures. One reason that 200 is not sufficient for approximating the regression equation in a one-step operation is that only 40 exact fits were computed for the first trimming stage, which is much less than 153 possible exact fits at that stage, and similarly for the following stages.

11.6.3 Application to Multiple Regression

The data in Table 11.6 were constructed so that there is a moderate, but not extreme, degree of correlation between the three regressors (the three pairwise correlations are close to .75), the value of R^2 is high (.92 when OLS is used with all of the data points), and all of the regressors are significant. Similar to the previous example, most (24) of the 30 data points constitute a homogeneous set of good data, in addition to 3 good data points that are somewhat removed from the first set in terms of the regressor values, and 3 bad data points. The observations were generated using $Y = 30 + 1.5X_1 - 2X_2 + 3X_3 + \epsilon + \delta$, where

Table 11.6 Multiple Regression Example of LTS with Sequential Trimming

Y	X_1	X_2	X_3
192.435	11	32	68
140.125	14	28	50
131.522	16	39	49
167.656	18	35	61
181.396	23	36	63
181.191	22	42	66
131.443	21	40	48
169.829	20	41	64
182.263	16	35	65
137.330	12	33	53
153.592	19	29	50
137.975	17	46	59
200.989	15	28	68
142.088	13	31	52
163.871	12	38	67
162.010	24	41	62
182.666	25	38	65
139.104	20	41	53
137.298	15	45	59
131.868	10	44	61
152.650	13	35	58
138.864	15	26	50
139.007	22	31	52
180.265	24	48	68
232.376	31	42	79
194.843	32	58	78
203.432	34	60	81
239.322	31	63	82
230.985	33	59	80
247.940	35	66	85

$\epsilon \sim N(0, 64)$ with $\delta = 0$ for the first 27 observations, and $\delta = 38$ for the last three observations.

This time we will set $N_{samp} = 600$. This will increase the likelihood of having a good approximation to the value of the objective function for the exact solution, but remember that a good approximation is not absolutely essential, since we could subsequently apply least squares to the points that are not deleted for a given amoung of trimming. The advantage of using a sufficiently large value for N_{samp} is that the identification of the points that are effectively deleted would not have to be made; the experimenter could simply use the regression equation that is produced.

When OLS is applied to the data set, the regression equation is $\hat{Y} = -8.1 + 1.55X_1 - 1.32X_2 + 3.19X_3$, with one of the good (extreme) data points having a standardized residual of -2.13, and one of the bad data points having a standardized residual of 2.28. Thus, these two points appear to be somewhat suspicious, although we would expect one or two points out of 30 to be so spotlighted when there is nothing wrong with the data.

But we know that there are three bad data points, so we would like to see if the suggested procedure is able to detect them. The results are summarized in Table 11.7. Again we recognize that a modified Ruppert procedure would be more efficient, since only 435 regressions are required to obtain the *exact* solution at the second step, but with $N_{samp} = 600$ the approximations should be reasonably good throughout.

This time the procedure has no difficulty in identifying the outliers as the points that are trimmed at each stage form proper subsets. If we use $k_0 = 3$ rigidly, we would additionally delete point 23, but the sharp drop-off in k from step 3 to step 4 suggests that the outliers have already been identified. (Obtaining the exact solution for each of the last two steps would have required 4060 and 27,405 regressions, respectively.)

As a final application we shall apply the procedure to the stack loss data that were analyzed in Chapter 6, and which have been analyzed by many other statisticians. This time we will use all possible point subset regressions for the first two steps, since not very many regressions are required, and then switch to the Ruppert procedure in step 3. The results are given in Table 11.8. Using

Table 11.7 LTS with Sequential Trimming Applied to Table 11.6 Data

Trimming Percentage (number trimmed)	Point(s) Trimmed (Nos.)	Marginal Reduction in TSS	Multiple of Average Squared Error
.03 (1)	28	549.3	5.97
.06 (2)	28, 30	730.8	7.95
.09 (3)	28, 29, 30	588.9	6.41
.12 (4)	28, 29, 30, 23	282.5	3.08

Table 11.8 LTS with Sequential Trimming for Stackloss Data

Trimming Percentage (number trimmed)	Point(s) Trimmed (Nos.)	Marginal Reduction in TSS	Multiple of Average Squared Error
.05 (1)	21	73.2	8.60
.10 (2)	21, 4	45.8	5.38
.14 (3)	21, 4, 3	16.3	1.91
.19 (4)	21, 4, 3, 1	23.1	2.71
.24 (5)	21, 4, 3, 1, 13	7.8	0.92

the suggested value of k_0, we would delete only observations 4 and 21, but the consensus is that points 1 and 3 are also outliers, as the four data points correspond to transitional plant conditions. Here we see an example of masking because the value of k decreases and then increases. This suggests that the procedure should not be stopped when the first value of k below the cutoff is encountered. Since the next value of k is only slightly below the cutoff, we could make an argument for deleting observations 1, 3, 4, and 21 with this procedure. We may also note that with $N_{samp} = 600$ the objective function values given in Table 11.8 are virtually identical to the residual sum of squares when OLS is applied to the points that are not deleted. Thus, as stated earlier, a value of 600 should generally be sufficient.

The results are essentially the same when the Hawkins approach is used, and these results can be produced with only one random start. (The use of only a single random start is certainly not recommended, in general, however, as the use of only one or two random starts will often give suboptimal results.)

This general approach to detecting outliers is simply a variation of the (impractical) brute force method of looking at all possible point subsets of each size, and will not have some of the same shortcomings as the delete-one-point-at-a-time approach.

Although considerable computation may be required for large data sets when substantial trimming is performed, there is no point in trimming more than seems necessary, so when the procedure outlined here (or some modification) is employed, the computations should generally not be excessive, especially if fast computer resources are utilized.

Furthermore, there is no point in using, say, 25% trimming with LTS if the sequential procedure indicates that less than 25% of the data points should be trimmed, and we would hardly ever encounter a data set with as many as 25% bad data points. Remember that our objective should be to apply least squares to all of the good data points, not to a subset of them.

If the sequential strategy is followed, in addition to detecting the outliers we (ideally) should obtain an estimator with 100% efficiency relative to the good data points, so if any of the latter are judged influential, and/or it is desired to subsequently employ a bounded influence approach in a two-stage procedure, the final estimator should have good properties.

The ad hoc procedure illustrated in this section and in Section 11.6.2 (or perhaps some variation of it) will probably be adequate for most data sets. It could be fooled by a substantial fraction of bad data points, but as mentioned in Section 11.3.1, data sets generally contain 1–10% gross errors. When there is very bad swamping, it may be necessary to continue to compute k even after two consecutive small values are observed.

Hadi (1994) gives a procedure for detecting multiple outliers that was shown to work well in the simulations that were performed, and was shown to be considerably superior to the methods given in Hadi and Simonoff (1993). Its performance with real data was not demonstrated, however.

There are difficulties in trying to assess the performance of outlier detection schemes. This is due in part to the fact that there is not a large number of published data sets for which the bad data points are known. Therefore, simulation studies are relied upon to assess the worth of the various proposed outlier detection methods. But if, say, 95% of the data are generated with one equation, and the other 5% are generated using a similar equation, it is arguable whether 5% of the data can be considered bad. The problem stems from the difficulty in trying to provide a quantitative definition of a regression outlier. In the absence of such a definition, the classification of data points in simulated data can be somewhat arbitrary.

11.7   LTS RUNNING LINE SMOOTHER

The LTS concept can also be applied outside linear regression. A running line smoother was discussed in Section 10.3.1, and an LTS-type modification was used in Section 10.3.1.1. Because the LTS concept is intuitively appealing and easily understood, it is desirable to consider adapting it to other techniques such as data smoothers and other nonparametric regression methods.

In applying LTS to a running line smoother, we would simply apply LTS to each neighborhood, thus treating each neighborhood as if it were a separate sample. Since each neighborhood is fit by least squares with a regular running line smoother, bad data points can have a deleterious effect on the fit, as was observed with Figure 10.5. Accordingly, an LTS modification should protect against bad data points in each neighborhood, especially for neighborhoods at the extremes of the data.

11.8   BOUNDED INFLUENCE ESTIMATORS

As noted earlier, *GM*-estimators were developed to overcome the failings of *M*-estimators when bad data occurred at *X*-outliers. The most frequently discussed estimators include those given by Mallows (1975), Krasker (1980),

Welsch (1980), and Krasker and Welsch (1982). We will use the Welsch (1980) estimator for illustration. As implied in Section 11.3.3.2, these estimators will have a low breakdown point when there is a large number of regressors, so a *GM*-estimator should not be routinely used as a stand-alone procedure but might be used efficiently in conjunction with a high breakdown point estimator.

As stated in Section 11.3.3.1, an *M*-estimator is obtained by minimizing the sum of a specified convex function, ρ, of the residuals. When that function is differentiated, the estimating equation is then written (for simple regression) as

$$\sum_{i=1}^{n} \psi \left(\frac{e_i}{\hat{\sigma}_{e_i}} \right) x_i = 0 \tag{11.2}$$

where ψ is the derivative of ρ, and $\hat{\sigma}_i$ is a (preferably) robust estimate of the standard deviation of the *i*th residual. By comparison, with a *GM*-estimator Eq. (11.2) is replaced by

$$\sum_{i=1}^{n} \eta \left(x_i, \frac{e_i}{\hat{\sigma}_{e_i}} \right) x_i = 0 \tag{11.3}$$

The different *GM*-estimators are obtained by using different functions, η.

As discussed by Staudte and Sheather (1990, p. 249), there is a function W such that

$$\eta \left(x, \frac{e}{\sigma} \right) = W \left(x, \frac{e}{\sigma} \right) \frac{e}{\sigma} \tag{11.4}$$

This allows Eq. (11.3) to be written in the form

$$\sum_{i=1}^{n} \left(\frac{e_i}{\hat{\sigma}_{e_i}} \right) W_i x_i = 0 \tag{11.5}$$

Choice of W will determine how an X-outlier will be downweighted when it produces a residual outlier. Staudte and Sheather (1990, p. 258) consider functions of the general form

$$W \left(x, \frac{e}{\sigma} \right) = \left(\frac{\sigma v(x)}{e} \right) \psi_c \left(\frac{e}{\sigma v(x)} \right) \tag{11.6}$$

where $v(x)$ is a function that is to be selected, and ψ_c is Huber's ψ function. The latter is defined as

$$\psi_c(x) = \begin{cases} -c & x < -c \\ x & -c \le x \le c \\ c & x > c \end{cases}$$

where a value of c must be determined. We note that this function may be written equivalently as

$$\psi_c(x) = \min(c, \max(x, -c))$$

as is given by Rousseeuw and Leroy (1987, p. 12). By combining the first expression for the ψ function with a small amount of algebra we can show that Eq. (11.6) may be written equivalently as

$$W\left(x, \frac{e}{\sigma}\right) = \min\left\{\frac{c\sigma v(x)}{|e|}, 1\right\} \tag{11.7}$$

as the reader is asked to show in Exercise 11.2. This is the form given by Staudte and Sheather (1990, p. 258).

Notice that when Eq. (11.6) is substituted into Eq. (11.4) we obtain

$$\eta\left(x, \frac{e}{\sigma}\right) = v(x)\psi_c\left(\frac{e}{\sigma} \cdot \frac{1}{v(x)}\right) \tag{11.8}$$

If we let $v(x) = 1$ and c go to infinity, we obtain the OLS estimator, and if we bound c [and still use $v(x) = 1$], we have the M-estimator given by Huber (1973). The bounded influence modification of an M-estimator is thus obtained by selecting $v(x) \ne 1$.

Welsch (1980) proposed using $v(x) = (1 - h_{ii})/\sqrt{h_{ii}}$ where h_{ii} is the leverage of the ith data point, and estimating σ by $s_{(i)}$, the standard deviation obtained not using the ith data point. Substituting these expressions into Eq. (11.8), we obtain

$$W\left(x, \frac{e}{\sigma}\right) = \min\left\{\frac{cs_{(i)}(1 - h_{ii})}{|e|\sqrt{h_{ii}}}, 1\right\} \tag{11.9}$$

Since the ith value of DFFITS is defined (see Section 2.4.2) as

$$\text{DFFITS}_i = \left(\frac{h_{ii}}{1 - h_{ii}} \right)^{1/2} \frac{e_i}{s_{(i)}\sqrt{1 - h_{ii}}}$$

$$= \frac{\sqrt{h_{ii}}e_i}{s_{(i)}(1 - h_{ii})}$$

it can be easily seen that the absolute value of DFFITS_i is the reciprocal of the first term in the bracketed expression in Eq. (11.9). (We note that DFFITS has also been referred to as DFITS.)

Therefore, Eq. (11.7) may be written as

$$W_i = W\left(x_i, \frac{e_i}{s_{(i)}} \right) = \min \{ c|\text{DFFITS}_i|^{-1}, 1 \} \qquad (11.10)$$

The constant c is defined as $c = k\sqrt{p/n}$. This requires that k be specified. Two proposed values are 1 and 2, with Staudte and Sheather (1990, p. 260) expressing a preference for the former. The choice between the two is a choice between efficiency (relative to OLS when OLS is appropriate) and robustness. The choice of $k = 2$ gives greater efficiency than $k = 1$, but the latter provides more robustness. (Both values could, of course, be used, and the results compared.)

The choice of $k = 2$ can be motivated by the rule-of-thumb given by Belsley et al. (1980), which states that points where $\text{DFFITS} > 2\sqrt{p/n}$ should be declared suspicious observations. Using this critical value, the Welsch procedure would downweight those data points where the critical value is exceeded. Note: Staudte and Sheather (1990, p. 213) suggest replacing 2 by 1.5. If this convention were adopted, it would be appropriate to use $k = 1.5$ as a compromise between $k = 1$ and $k = 2$ for Eq. (11.10). Regardless of the combination of c and k that is used, we will be downweighting points at which the fit is relatively poor and/or an X-outlier occurs.

11.8.1 Shortcomings of Bounded Influence Estimators

As previously noted, the breakdown point of a GM-estimator cannot exceed $1/p$. We reiterate that whereas the estimators that we are calling bounded influence estimators are also GM-estimators, not all bounded influence estimators are GM-estimators, since not all bounded influence estimators have a breakdown point with an upper bound of $1/p$.

We have seen that insistence on a high breakdown point is not necessarily a good idea. Rather, we should seek a breakdown point that is "just sufficient." Unfortunately, the breakdown point of bounded influence estimators is not known. As pointed out by Rousseeuw and Leroy (1987, p. 13), it is not clear if the (GM) bound can be achieved, and if so, which estimator(s) can achieve it. A small numerical study of Rouseeuw and Leroy (1987, p. 68) shows that members of the class of GM-estimators will generally have different breakdown points, even for simple regression.

Thus, when we use a GM-estimator we are "flying blind" somewhat since theoretical results for breakdown points are lacking. This should not rule out their use, however. [See Carroll and Ruppert (1988, p. 188) for a similar view.] We will later see that GM-estimators can perform extremely well under certain conditions, even when used as a stand-alone procedure.

Another criticism that has been voiced of bounded influence estimators is that X-outliers will often provide important information regarding the fit and the model coefficients, and yet these points will be downweighted. Ideally we would want X-outliers to be downweighted if the fit is poor at these points, or if the fit is good but the good fit results from the outlying point drawing the regression line (plane) to it.

The Welsch (1980) estimator may not necessarily downweight points that fall into the latter category because a small residual could more than offset a large leverage value and cause the point to not necessarily be downweighted. (Recall the distinction drawn between a regression outlier and a residual outlier that was made in Section 11.2. Ideally it is regression outliers that we want to downweight, not residual outliers, but we won't necessarily be able to identify the regression outliers when the full data set is used in determining the weights.)

Since leverage indicates only *potential* influence, and either a small or a large residual could occur at points that are truly influential, it seems undesirable to downweight points based on the value of a weight function that is a product of a leverage component and a residual component. The only way to truly measure influence is to use a statistic that is a function of the *change* in the regression coefficients when a point is deleted or when a cluster of similar points is deleted. This suggests the use of DFBETAS (see Section 2.4.2) as a logical alternative.

There is another shortcoming of GM-estimators that has not been emphasized in the literature. Notice that in DFFITS$_i$, σ is estimated by $s_{(i)}$. This will be a poor estimator of σ when there is a cluster of bad data points, as the estimate can be considerably inflated by the cluster. Also, the deviations $\hat{Y}_i - \hat{Y}_{(i)}$ can be deceivingly small in the presence of such clusters. (DFBETAS has the same type of shortcoming.) Consequently, a robust estimator of σ must be used if a bounded influence estimator is to have any chance of producing satisfactory results when such clusters exist. This approach has been taken by Simpson et al. (1992) and will be discussed later.

Some users of robust regression would also criticize the use of the weight function given by Eq. (11.10). Specifically, there are only two possible values: the OLS value ($W_i = 1$) and $c/(|\text{DFFITS}_i|)$. Although the value of $c/(|\text{DFFITS}_i|)$ can downweight a bad data point, the weight may not differ greatly from one.

Many would prefer the *three-part redescending ϕ function* suggested by Hampel (1974) that is given by

$$\psi(t) = \begin{cases} t & |t| < a \\ a\,\mathrm{sgn}(t) & a \le |t| < b \\ \{(c - |t|)/(c - b)\}a\,\mathrm{sgn}(t) & b \le |t| \le c \\ 0 & \text{otherwise} \end{cases}$$

where $\mathrm{sgn}(t) = -1$ if $t < 0$ and $+1$ if $t > 0$. Rousseeuw and Leroy (1987, p. 149) compare the Huber and Hampel ψ functions graphically. The essential difference is that the latter provides considerably more protection against discrepant data points, with badly discrepant data points receiving a weight of zero and less extreme points being smoothly downweighted.

11.8.2 Applications

In this section we will apply the Welsch estimator to data sets that have been used in the literature, as well as those that were used earlier in this chapter. The reader is advised to consult Rousseeuw and Leroy (1987), which lists several of these data sets. The required computation for the Welsch estimator is straightforward since it can be computed using weighted least squares. Staudte and Sheather (1990) provide Minitab macros that can be used for the computations, but the user must decide when the parameter estimates have converged.

Rousseeuw and Leroy (1987) use a data set consisting of international telephone calls (page 26) and a data set from astronomy (page 27), which is referred to in the literature as the Hertzsprung–Russell data, in illustrating LMS. These are both simple regression data sets. The first contains 24 observations, with 6 essentially coming from a different population and 2 from a mixture of the two populations.

When the Welsch estimator is applied to the first data set (using $k = 1$), the initial values of DFFITS do not identify the eight bad data points. In fact, the DFFITS value for the last three data points, all of which are good points, are larger than the DFFITS values for six of the eight bad data points. Three of those bad data points have DFFITS values that are less than the critical value for DFFITS. Consequently, they are given the weight $W_i = 1$ that should be given to good data points.

Since the initial weights are not close to what we would want them to be, we would expect that, when the necessary iterations are performed, the results will not be satisfactory. When 10 iterations are performed, the DFFITS values change very little, and the regression coefficients differ very little from the OLS coefficients. (The regression coefficients differ only slightly when $k = 2$ is used. Then only two of the bad data points are downweighted (slightly), and one of the good data points is also slightly downweighted.) Thus, even though the upper bound on the breakdown point of a GM-estimator is 0.5, here the Welsch estimator performs poorly when one-third of the data are bad.

The (good) data point with the largest x-coordinate is downweighted with both $k = 1$ and $k = 2$. Since its leverage value is very close to $2p/n$, one of the suggested cutoffs for determining (high) leverage points, this example essentially serves as a counterexample to those given by Staudte and Sheather (1990, p. 268). They conclude that the Welsch ($k = 1$) estimator does not downweight "those high leverage points where information is consistent with the bulk of the data" for data sets that they have examined. Here the points with extreme x-coordinates combine with the other good points to form almost an exact fit.

One reason why the Welsch estimator fails for this example is that the bad data points do not occur at leverage points. To see this, we can swap the x-coordinates of the last three (x-ordered) good data points so that they have the x-coordinates of three of the first four bad data points, and the x-coordinates of the last seven bad data points are each increased by three. The Welsch estimator is not fooled as badly for this modified data set, although its performance is short of being satisfactory. It performs well on the astronomy data, however, where the four outliers occur at high leverage values. Those four points are given weights between .08 and .11, and although five of the good data points are also downweighted, the weights are all .89 and higher.

That data set is a good example of why a two-stage procedure will generally be desirable, however, since there are good data points that are slightly outlying and influential. A better approach would be to use a high breakdown estimator such as LTS to trim the four outliers and then use a bounded influence estimator to bound the influence of the good data points. Rousseeuw and Leroy (1987, p. 28) show how well LMS works on this data set. In particular, SSE(LMS) = 11.096 and SSE(Welsch) = 20.144 when the outliers are not used in the computation of SSE, the residual sum of squares. This occurs because the Welsch estimator is influenced somewhat by the outliers. We would want the regression line to be at least slightly influenced by the outlying data points that are valid observations, however, so a desirable regression equation would be one that is slightly different from the LMS equation. Such an equation would result from the two-stage procedure that was suggested above.

It is certainly arguable what the regression line should be for this data set, but if we were to adopt the position that the optimal estimator is OLS applied to the good data points, then we may obtain a bounded influence estimator with 93.9% efficiency simply by using DFBETAS with a cutoff of $1/\sqrt{n}$, instead of using DFFITS. This is not to suggest, however, that DFBETAS should generally be used instead of DFFITS, as DFBETAS can also perform poorly (such as when applied to the stack loss data), nor do we wish to assert that OLS applied to the valid data points should necessarily always be the measuring stick against which robust regression estimators are compared.

11.9 MULTISTAGE PROCEDURES

Although two-stage procedures have been mentioned briefly in the literature, no attempt has been made to determine an optimal two-stage procedure. If we are to use a high breakdown estimator in the first stage and a bounded influence estimator in the second stage, we clearly want to select the best pair.

It is desirable to use either LTS or an S-estimator in the first stage, and we have seen why some care must be exercised in the selection of a bounded influence estimator in the second stage. Simpson et al. (1992) present a two-stage procedure, but they concentrate on the second stage without specifying a choice for the first stage. Their procedure may perform well, however, provided that a good first-stage estimator is used.

The procedure can be explained as follows, where the notation is for multiple regression. Let $\hat{\boldsymbol{\beta}}_*$ denote the estimator of β from the first stage. The one-step GM-estimator is then obtained as $\hat{\boldsymbol{\beta}} = \hat{\boldsymbol{\beta}}_* + \mathbf{H}_0^{-1}\mathbf{g}_0$ with $\mathbf{H}_0 = \sum_{i=1}^{n} w_i \mathbf{z}_i \mathbf{z}_i' \psi^{(1)}(e_i/\hat{\sigma}_0)$ if Newton–Raphson is the method that is used to obtain the estimates. (The authors do indicate a preference for a different method, however.) Here $\mathbf{z}_i' = (1, \mathbf{x}_i)$ is the ith row of the $\mathbf{X}$ matrix, with $\mathbf{g}_0 = \hat{\sigma}_0 \sum_{i=1}^{n} \psi(e_i/\hat{\sigma}_0)w_i \mathbf{z}_i$ and $\psi^{(1)}$ is the first derivative of the selected ψ function. The estimator $\hat{\sigma}_0$ is a high breakdown estimator of σ that will be discussed later. For the ψ function the authors use the three-part redescending Hampel (1974) ψ function that was given in Section 11.5.2.1, with $(a, b, c) = (1.5, 3, 8)$. The weights, w_i, are Mallows (1975) weights given by

$$w_i = \min\left\{\left(\frac{b}{(\mathbf{x_i} - \mathbf{m_x})'\mathbf{C_x^{-1}}(\mathbf{x_i} - \mathbf{m_x})}\right)^{\alpha/2}, 1\right\}$$

where $\mathbf{m_x}$ and $\mathbf{C_x}$ are high breakdown point estimators of $\boldsymbol{\mu_x}$ and Σ_x, the multivariate mean and variance-covariance matrix, respectively. Simpson et al. (1992) use, in their example with three regressors, both $\alpha = 0$ and $\alpha = 2$ and b equal to the 95th percentile of the chi-square distribution with $p - 1$ degrees of freedom. (As in previous chapters, p here denotes the number of parameters.)

An important difference between this and other two-stage procedures that have been advocated is the use of high breakdown estimators in the bounded influence estimator. Although Simpson et al. (1992) suggest that any HPB estimator will suffice for the first stage, in a subsequent paper Ruppert (1992) cautions against the arbitrary selection of an HBP estimator. The reasoning is that a one-step GM-estimator used as a one-step improvement to the LMS estimator, for example, can inherit the shortcomings of the LMS estimator. (We have seen previously that these shortcomings can be quite serious). Simpson et al. (1992) suggest that an S-estimator might be a better choice of a preliminary estimator and also indicate that three-step GM-estimates can be more stable than the one-step estimates. See their paper for details.

We can show that the concern over the selection of the first-stage estimator is well-founded. Consider, for example, the modification to the Stefanski (1991) example given in Table 11.1. Since the LMS regression line is far removed from what would be considered a more appropriate line, we can hardly expect a one-step improvement to alter the LMS line sufficiently.

We will assume that the initial estimates, $\hat{\boldsymbol{\beta}}_*$, are equal to the LMS coefficients. To illustrate the maximum one-step improvement that can be expected, we will proceed optimally in the second stage. That is, we will let $w_i = 1$ and $\psi^{(1)}(e_i/\hat{\sigma}_0) = 1$ for the good data points, and $w_i = 0$ and $\psi^{(1)}(e_i/\hat{\sigma}_0) = 0$ for the bad data points. With this weighting we have

$$\mathbf{H_0} = \sum_{i=1}^{n} w_i \mathbf{z_i z_i'} \psi^{(1)}\left(\frac{e_i}{\hat{\sigma}_0}\right) = \sum_{i=1}^{n*} \mathbf{z_i z_i'}$$

with $n*$ denoting the number of good data points. It follows that $\mathbf{H_0} = \mathbf{X}'_* \mathbf{X}_*$, where $\mathbf{X}_*$ is obtained by deleting the bad data points from $\mathbf{X}$.

For $\hat{\sigma}_0$ we will use the final scale estimate for LMS given by $\hat{\sigma}_0 = \sqrt{(\sum_{i=1}^{n} w_i e_i^2)/(\sum_{i=1}^{n} w_i - p)}$ with $w_i = 0$ or 1. The best estimate of sigma would, of course, be obtained by using only the good data points, knowledge of which would determine the w_i. To be consistent, however, we will use the w_i obtained from LMS.

The expression for $\mathbf{g_0}$ simplifies to $\mathbf{g_0} = \hat{\sigma}_0 \sum_{i=1}^{n*} e_i \mathbf{z_i}$ with $\mathbf{z_i}$ as previously defined, and the summation is over the $n*$ good data points. Thus, $\mathbf{g_0} = \hat{\sigma}_0 \mathbf{p}$, with $\mathbf{p}' = (\sum_{i=1}^{n*} e_i, \sum_{i=1}^{n*} e_i x_i)$. Notice that we would have $\mathbf{g_0} = \mathbf{0}$ if $\hat{\boldsymbol{\beta}}_*$ had been the OLS estimator obtained from using only the same data points that we are now declaring to be separable from the bad ones. Thus, if we had started with the "right" estimator (OLS applied to the good data points), there would not be a one-step improvement, which is obviously the way that it should be.

We may now write the one-step improvement as $\hat{\boldsymbol{\beta}} = \hat{\boldsymbol{\beta}}_* + (\mathbf{X}'_* \mathbf{X}_*)^{-1}\mathbf{p}\hat{\sigma}_0$. Applying these results to the Table 11.1 data with $n = 9$ and $n* = 7$, we obtain $\hat{\sigma}_0 = 0.072$ and $\mathbf{p} = (-2.875, -662.74)'$. (Note: If we had used the estimator $\hat{\sigma}_{01} = 0.6745$ (median $|e_i|$) employed by Simpson et al. (1992), which was not designated for use only with LMS, we would have obtained 0.059 as our scale estimate, which differs only slightly from the one that we are using.) We thus obtain

$$\hat{\sigma}_0(\mathbf{X}'_* \mathbf{X}_*)^{-1}\mathbf{p} = (.072)\begin{bmatrix} 0.6359 & -0.0323 \\ -0.0323 & 0.0021 \end{bmatrix}\begin{bmatrix} -2.875 \\ -662.74 \end{bmatrix}$$

$$= \begin{bmatrix} 1.404 \\ -0.094 \end{bmatrix}$$

Therefore, the one-step improvement is

$$\hat{\beta} = \hat{\beta}_* + \begin{bmatrix} 1.404 \\ -0.094 \end{bmatrix}$$

$$= \begin{bmatrix} -15.412 \\ 1.288 \end{bmatrix} + \begin{bmatrix} 1.404 \\ -0.094 \end{bmatrix}$$

$$= \begin{bmatrix} -14.009 \\ 1.194 \end{bmatrix}$$

Thus, although the one-step improvement has moved the line closer to where it should be (recall that the slope should be approximately 0), the improvement is not sufficient. In general, we cannot expect the one-step improvement to be adequate if the HBP estimator falsely identifies good points as being bad and bad points as being good. Here we are using the proper selection of good and bad points in the second stage, but we cannot expect this to offset the fact that the LMS estimator has made the wrong selections in the first stage.

The example given thus illustrates the statement by Ruppert (1992) that "instability in the cases identified as outliers by an initial estimator can lead to instability of the subsequent one-step estimator."

11.10 SOFTWARE

One reason for the comparatively low usage of robust regression (relative to OLS) is the relative paucity of widely available software. Program 3R in BMDP offers the user some robust regression capability, but only M-estimators can be used, with an option for selecting from among several influence functions, including those given by Huber and Hampel. [Code for program 3R that provides a modified version of PROGRESS, with random subsampling only, was given by Neykov and Neytchev (1991).]

Similarly, SAS offers only (limited) influence function capability that can be used in the NLIN procedure. [It should be noted that Street et al. (1988) report that the NLIN procedure will produce standard errors of the estimators that are not asymptotically correct when NLIN is used with iterative reweighted least squares for computing robust regression estimates.] There is an LMS program available in S-PLUS and Stata also provides limited robust regression capability, but this is essentially all that is available in the well-known general-purpose statistical packages.

Additionally, a shareware program for LMS is described by Dallal (1991). See also Dallal and Rousseeuw (1992). A general survey of algorithms for robust regression is given by Marazzi (1991). One algorithm not given therein is Hawkins and Simonoff (1993).

SUMMARY

It is important to realize that the routine use of least squares in regression analyses will often produce poor results. Even though the field of robust regression is far from being well developed, it has been demonstrated that certain robust estimators can perform well when used alone or in tandem. Practitioners should not wait until research has demonstrated which robust estimation procedure is best before considering the use of robust regression instead of least squares. Since outliers in multiple regression can frequently be very difficult to detect, even ardent believers in least squares should at least use robust regression in addition to least squares and compare the results.

We should also remember, however, that throughout the chapter it was tacitly assumed that the fitted model was the correct model, with attention being focused on obtaining good estimates of the model parameters. Cook et al. (1992) point out that LMS and LTS can perform worse than ordinary least squares when the true model contains nonlinear terms, but a model with only linear terms is fit. Therefore, the user of robust regression methods would be well-advised to try to determine if a linear model with linear terms is appropriate before using a robust regression approach.

APPENDIX

The Minitab macros for robust regression that are included on the accompanying diskette can be used as an aid in understanding the estimators that were discussed in this chapter. These macros will be briefly discussed here; more detailed information can be found at the beginning of each macro and in Appendix A. The macros were written using Release 10.2 and some require that the user have Release 9 or a later release.

Depending on the computing equipment that is used, many users will want to set up a RAM disk for Minitab before running the macros. The reason for this suggestion is that it could take an hour or more to run some of the macros on moderate-sized data sets with a slow PC. This is due in part to the combinatorics that are involved and also because Minitab does not hold macros in memory. Typical running times are given in the documentation for each macro.

The macros are also set up so that the user may, optionally, see how the procedure is seeking the optimal solution while the program is running and/or view other solutions that are close to the optimal solution.

The following macros are provided.

1. **LMSONEX.** This macro computes the ("inexact") LMS solution for simple regression with an intercept using the exact-fits approach of PROGRESS.

2. **EXLMSNB0.** The exact LMS solution for simple regression *without* an intercept is obtained using Chebyshev fits.

3. **EXACTLMS.** This is the same as macro (**2** except it is for use with simple regression *with* an intercept.

4. **RNDLTS1X.** This program implements the RANDDIR algorithm of Ruppert applied to LTS for simple regression.

5. **RNLTSMUL.** This is the same as macro (**4** except that this is used for multiple regression.

6. **LTSONEX.** This macro is provided solely for Exercise 11.7. It is not suitable for general use in obtaining exact LTS solutions.

7. **RWLS1DFB and RWLS2DFB.** RWLS1DFB computes the first iteration of a Welsch-type estimator for simple or multiple regression where the weights are based on DFBETAS. RWLS2DFB is then used for subsequent iterations.

REFERENCES

Andrews, D. F., P. J. Bickel, F. R. Hampel, P. J. Huber, W. H. Rogers, and J. W. Tukey (1972). *Robust Estimates of Location: Survey and Advances*, Princeton, NJ: Princeton University Press.

Belsley, D. A., E. Kuh, and R. E. Welsch (1980). *Regression Diagnostics: Identifying Influential Data and Sources of Collinearity*. New York: Wiley.

Carroll, R. J. and D. Ruppert (1988). *Transformation and Weighting in Regression*. New York: Chapman and Hall.

Cook, R. D., D. M. Hawkins, and S. Weisberg (1992). Comparison of model misspecification diagnostics using residuals from least mean of squares and least median of squares fits. *Journal of the American Statistical Association*, **87**, 419–424.

Dallal, G. E. (1991). LMS: Least median of squares regression. *The American Statistican*, **45**, 74.

Dallal, G. E. and P. J. Rousseeuw (1992). LMS: A program for least median of squares regression. *Computers and Biomedical Research*, **25**, 384–391.

Donoho, D. L., and P. J. Huber (1983). The Notion of Breakdown Point, in *A Festschrift for Erich Lehmann*, P. Bickel, K. Doksum, and J. L. Hodges, Jr., eds. Belmont, CA: Wadsworth.

Edgeworth, F. Y. (1887). On observations relating to several quantities. *Hermathena*, **6**, 279–285.

Frees, E. W. (1991). Trimmed slope estimates for simple linear regression. *Journal of Statistical Planning and Inference*, **27**, 203–221.

Hadi, A. S. (1992). Identifying multiple outliers in multivariate data. *Journal of the Royal Statistical Society, Series B*, **54**, 761–777.

Hadi, A. S. (1994). A modification of a method for the detection of outliers in multivariate samples. *Journal of the Royal Statistical Society, Series B*, **56**, 393–396.

Hadi, A. S. and J. S. Simonoff (1993). Procedures for the identification of multiple outliers in linear models. *Journal of the American Statistical Association*, **88**, 1264–1272.

Hampel, F. R. (1974). The influence curve and its role in estimation. *Journal of the American Statistical Association*, **69**, 383–393.

Hampel, F. R., E. M. Ronchetti, P. J. Rousseeuw, and W. A. Stahel (1986). *Robust Statistics: The Approach Based on Influence Functions*. New York: Wiley.

Hawkins, D. M. (1993). The feasible set algorithm for least median of squares regression. *Computational Statistics and Data Analysis*, **16**, 81–101.

Hawkins, D. M. (1994). The feasible solution algorithm for least trimmed squares regression. *Computational Statistics and Data Analysis*, **17**, 185–196.

Hawkins, D. M. and J. S. Simonoff (1993). Algorithm AS 282: High breakdown regression and multivariate estimation. *Applied Statistics*, **42**, 423–432.

Hawkins, D. M, D. Bradu, and G. V. Kass (1984). Location of several outliers in multiple regression using elemental sets. *Technometrics*, **26**, 197–208.

Hettmansperger, T., and S. Sheather (1992). A cautionary note on the method of least median squares. *The American Statistician*, **46**, 79–83.

Hodges, J. L., Jr. (1967). Efficiency in normal samples and tolerance of extreme values for some estimates of location. *Proceedings of the Fifth Berkeley Symposium on Mathematical Statistics and Probability*. Berkeley, CA: University of California Press.

Huber, P. J. (1973). Robust regression: Asymptotics, conjectures and Monte Carlo. *Annals of Statistics*, **1**, 799–821.

Huber, P. J. (1977). *Robust Statistical Procedures*. Philadelphia: Society for Industrial and Applied Mathematics.

Krasker, W. S. (1980). Estimation in linear regression models with disparate data points. *Econometrica*, **48**, 1333–1346.

Krasker, W. S. and R. E. Welsch (1982). Efficient bounded-influence regression estimation. *Journal of the American Statistical Association*, **77**, 595–604.

Mallows, C. L. (1975). On some topics in robustness. Unpublished memorandum, Bell Telephone Laboratories, Murray Hill, NJ.

Marazzi, A. (1991). Algorithms and Programs for Robust Linear Regression, in *Directions in Robust Statistics and Diagnostics, Part I*, W. Stahel and S. Weisberg, eds., New York: Springer-Verlag, 183–199.

Maronna, R. A., O. Bustos, and V. Yohai (1979). Bias- and Efficiency-Robustness of General M-estimators for Regression with Random Carriers, in *Smoothing Techniques for Curve Estimation*, T. Gasser and M. Rosenblatt, eds., New York: Springer-Verlag.

Mason, R. L., R. F. Gunst, and J. L. Hess (1989). *Statistical Design and Analysis of Experiments with Applications to Engineering and Science*. New York: Wiley.

Mosteller, F. and J. W. Tukey (1977). *Data Analysis and Regression*. Reading, MA: Addison-Wesley.

Neykov, N. M. and P. N. Neytchev (1991). Unmasking Multivariate Outliers and Leverage Points by Means of BMDP3R, in *Directions in Robust Statistics and Diagnostics, Part II*, W. Stahel and S. Weisberg, eds. New York: Springer-Verlag, 115–128.

Rousseeuw, P. J. (1984). Least median of squares. *Journal of the American Statistical Association*, **79**, 871–880.

Rousseeuw, P. J. (1990). Robust Estimation and Identifying Outliers, in *Handbook of Statistical Methods for Engineers and Scientists*, H. M. Wadsworth, ed., New York: McGraw-Hill, Chapter 16.

Rousseeuw, P. J. and G. W. Bassett, Jr. (1991). Robustness of the p-subset Algorithm for Regression with High Breakdown Point, in *Directions in Robust Statistics and Diagnostics: Part II*, W. Stahel and S. Weisberg, eds. New York: Springer-Verlag, 185–194.

Rousseeuw, P. J. and A. Leroy (1987). *Robust Regression and Outlier Detection*. New York: Wiley.

Rousseeuw, P. J. and B. C. van Zomeren (1990a). Rejoinder to: Unmasking multivariate outliers and leverage points. *Journal of the American Statistical Association*, **85**, 648–651.

Rousseeuw, P. J. and B. C. van Zomeren (1990b). Unmasking multivariate outliers and leverage points. *Journal of the American Statistical Association*, **85**, 633–639.

Rousseeuw, P. J. and V. Yohai (1984). Robust Regression by means of S-estimators, in *Robust and Nonlinear Time Series Analysis*, J. Franke, W. Härdle, and R. D. Martin, eds. Lecture Notes in Statistics No. 26. New York: Springer-Verlag, 256–272.

Ruppert, D. (1992). Computing S-estimators for regression and multivariate location/dispersion. *Journal of Computational and Graphical Statistics*, **1**, 253–270.

Ruppert, D. and R. J. Carroll (1980). Trimmed least squares estimation in the linear model. *Journal of the American Statistical Association*, **75**, 828–838.

Simpson, D. G., D. Ruppert, and R. J. Carroll (1992). On one-step GM-estimates and stability of inference in linear regression. *Journal of the American Statistical Association*, **87**, 439–450.

Staudte, R. G. and S. J. Sheather (1990). *Robust Estimation and Testing*. New York: Wiley.

Steele, J. M. and W. L. Steiger (1986). Algorithms and complexity for least median of squares regression. *Discrete Applied Mathematics*, **14**, 93–100.

Stefanski, L. A. (1991). A note on high-breakdown estimators. *Statistics and Probability Letters*, **11**, 353–358.

Street, J. O., R. J. Carroll, and D. Ruppert (1988). A note on computing robust regression estimates via iteratively reweighted least squares. *The American Statistician*, **42**, 152–154.

Stromberg, A. (1993a). Letter to the Editor. *The American Statistician*, **47**, 87.

Stromberg, A. (1993b). Computing the exact least median of squares estimate and stability diagnostics in multiple linear regression. *SIAM Journal on Scientific and Statistical Computing*, **14**, 1289–1299.

Tukey, J. W. (1977). *Exploratory Data Analysis*. Reading, MA: Addison-Wesley.

Welsch, R. E. (1980). Regression Sensitivity Analysis and Bounded-Influence Estimation, in *Evaluation of Econometric Models*, J. Kmenta and J. B. Ramsey, eds. New York: Academic Press, 153–167.

Yohai, V. and R. Zamar (1991). Discussion of: Least median of squares estimation in power systems, by L. Mili, V. Phaniraj, and P. J. Rousseeuw. *IEEE Transactions on Power Systems*, **6**, 520.

EXERCISES

11.1. Consider the following data set, which shows how sensitive the exact fits approach to LMS can be to small perturbations in the data.

X	1	2	3	4	5	1	2	3	4	5
Y	7	8	9	_	11	2.99	3.99	4.99	5.99	7

Fill in the blank with 9.338 and use the Minitab macro LMSONEX to obtain the LMS solution using the PROGRESS approach. Then change 9.338 to 9.339 and repeat. Notice that the LMS coefficients change dramatically. Now use the Minitab macro EXACTLMS to obtain the exact LMS solution using each of the two numbers. Notice that EXACTLMS produces three "best" solutions for each case, and the solutions differ considerably as well as differing considerably from the PROGRESS solution, although all three solutions for each example produce the same value of the median squared residual. What does this exercise suggest regarding the algorithm that should be used to obtain the LMS solution, and what does it also suggest about using an objective function that is based on the median squared residual?

11.2. Derive Eq. (11.6) from Eq. (11.5).

11.3. Explain why a breakdown point of .50 is not necessarily better than a breakdown point of .10.

11.4. Assume that a data set contains $n = 20$ observations with a single regressor whose values have been preselected. Those values are 12(1)22 with each value repeated. The scatter plot reveals Y outliers that are also regression outliers. Which robust regression approach would you recommend?

11.5. Indicate conditions under which the LMS estimator can be expected to provide satisfactory results. How should LMS compare with LTS(.50) under those conditions?

11.6. Explain what is being "bound" when a bounded influence estimator is used?

11.7. a. Use the Minitab macro LMSONEX to compute the LMS estimator, and use LTSONEX to compute the exact LTS(.25) estimator for the following data set.

X	1	2	3	3	3	4	5	5	5	6	7	8	9	10
Y	1	2	2.5	3	3.5	4	4.5	5	5.5	6	7	6	5	4

b. Determine which points were trimmed for the LTS(.25) estimator by looking at the ordered squared residuals.

c. Now construct a scatter plot. Does the set of deleted points seem reasonable relative to your scatter plot? What does your scatter plot suggest should be done first: plot the data or obtain point estimates for an assumed model?

d. Use the Minitab macro EXACTLMS to obtain the exact LMS solution. Notice that there is a dual solution. Can this be determined from inspection of the scatter plot? Notice that many of the Chebyshev fits do not exist because many point triplets constitute exact fits. Could one of these exact fits have produced the exact LMS solution? Explain. Could one of the two triplets for which the regression equation does not exist have produced the exact LMS solution?

e. Assume that the three points with the largest x-coordinates are errors and thus should be deleted. What would be the OLS equation computed from the remaining points? Is this equation apparent from either the scatter plot or the list of the data?

f. What must be the LMS equation computed from these 11 points? What property determines the equation in this instance? Remembering that the LMS estimator is designed to fit the majority of the data, and thus protect against outliers, do either of the exact LMS solutions for the 14 data points seem desirable relative to the configuration of good data points?

11.8. a. Use the Minitab macro RNLTSMUL with $N_{samp} = 600$ to approximate the exact LTS(.25) solution for the following data set. (*Note:* It may take an hour or more for the macro to run on some computers if a RAM disk is not set up for Minitab.)

Y	X_1	X_2
14.3171	6.2	3.8
14.1845	5.8	4.2
13.8039	6.0	3.7
13.5998	5.9	4.0
14.3339	6.1	4.1
12.4540	5.5	3.2
14.1663	5.9	4.1
13.9912	6.3	3.8
14.0843	6.4	3.7
14.7547	6.6	3.9
14.3840	6.8	3.1
13.9815	6.9	3.0
12.9733	5.7	3.4
13.9148	5.5	4.2
14.4456	6.5	4.3
14.1014	6.2	3.8

Y	X_1	X_2
15.7612	6.9	4.5
15.2926	5.1	3.3
13.9870	5.9	4.2
15.4053	6.6	4.4

The Y-values were generated randomly from the regressor values, but one value was changed considerably to represent an error. Can you determine the bad data point? Could this determination have been made using an OLS analysis? Are there any other points that appear to be questionable (from the LTS squared residuals) so that trimming more than one observation seems desirable, or would it be better to bound the influence of those data points? Would it be preferable to use a sequential trimming approach for LTS, even if we didn't know that there was exactly one bad data point?

b. Use OLS and continue trimming each point whose standardized residual is outside $(-2, 2)$. After you have trimmed four points, compare those points with the ones that were trimmed by the LTS(.25) estimator. Comment. Does it make sense to continue trimming points in this manner?

c. What is the value of R^2 with and without the single bad data point? What does the large difference between the two numbers suggest?

11.9. Explain why the fourth point in Figure 11.2(b) has a relatively large standardized residual, as discussed in Section 11.2.

11.10. Explain the difference between a regression outlier and a residual outlier.

11.11. Assume that a two-stage approach is to be used. What would be a good choice for the first-stage estimator, and what would be a reasonable choice for the second-stage estimator.

11.12. Explain why an M-estimator should not be routinely used. In particular, under what conditions would an M-estimator likely produce poor results?

11.13. Consider the data that were graphed in Figure 10.4. What would be the approximate slope of the line if LMS were applied to the data?

11.14. Consider the data in Table 10.2, for which the fitted regression equation should ideally have a slope of zero. For which of the following estimators would we expect the slope to be approximately zero when applied to this data set: (a) LMS or (b) LTS(.25), or both? How many observations will be effectively trimmed if LTS(.25) is used with the definition of h given in Section 11.5.2? When LMS is used with the PROGRESS approach and all exact fits are used, 22 points receive a weight of zero when reweighted least squares is used: 2, 4, 5, 7, 12–14, 19, 22, 23, 25–27, 31, 32, 40, 44, 64, 71, 77, 80, and 83. When LTS(.25) is used, 24 points are deleted, and they are: 2, 4, 5, 6, 12–14, 35–51. When the sequential LTS approach is used with a stopping rule of 3 (as described in Section 11.6.2), the points that are deleted are 2, 4, 5, 6, 12, 13, and 46–51. [The Hawkins approach with 20 random starts was used for the sequential LTS approach, and 100 random starts were used for LTS(.25).] Compute the least-squares regression line using each set of retained points and look at, in particular, the p-value for the slope for each equation. Do all three approaches suggest that the true slope is zero?

11.15. Consider the stack loss data that were given in Table 6.1. Use the Minitab macro RWLS1DFB followed by RWLS2DFB for this data set. Since the known outliers are observations 1, 3, 4, and 21, do the results suggest that the bounded influence approach can be relied upon to identify outliers or should a bounded influence estimator be used in the second stage of a two-stage procedure?

CHAPTER 12

Ridge Regression

Multicollinearity was discussed in Section 4.3, in which it was mentioned that one approach to treating multicollinearity would be discussed in this chapter. Recall that it was also explained in Section 4.3 that the terms *multicollinearity* and *collinearity* are misnomers relative to the way that they are used in the literature since the terms represent an *exact* linear relationship involving two or more variables. The terms are actually used to represent a *near-exact relationship*. Nevertheless, the terms are well-entrenched in the literature, so we will continue to use them. It was similarly mentioned in Chapter 7 that an alternative to *not* deleting regressors would be discussed in this chapter.

12.1 INTRODUCTION

The alternative to not deleting regressors is to use *ridge regression*. Presented by Hoerl (1962) and Hoerl and Kennard (1970a,b), ridge regression is an alternative to ordinary least squares and is one of several biased regression estimators that have been proposed. A *biased regression estimator* (say, $\hat{\beta}_i^*$) is one for which $E(\hat{\beta}_i^*) \neq \beta_i$. The reader should note that, for all practical purposes, the ordinary least squares (OLS) estimator will also generally be biased because we can be certain that it is unbiased only when the model that is being used is the correct model. Since we cannot expect this to be true, we similarly cannot expect the OLS estimator to be unbiased. Therefore, although the choice between OLS and a ridge estimator is often portrayed as a choice between a biased estimator and an unbiased estimator, that really isn't the case. Nevertheless, for the sake of illustration we shall assume that the OLS estimator is unbiased, as this is necessary to show the motivation for using ridge regression.

Ridge regression is sometimes confused with *ridge analysis*. The latter is a technique for determining the optimum point on a response surface (see, e.g., Hoerl, 1985). Although used for a different purpose than ridge regression is used, ridge analysis was used in deriving ridge regression (Hoerl, 1962).

We will first consider an *ordinary ridge estimator*, which has the general

396

form

$$\hat{\boldsymbol{\beta}}^*_{(k)} = (\mathbf{X}'\mathbf{X} + k\mathbf{I})^{-1}\mathbf{X}'\mathbf{Y} \qquad (12.1)$$

where $\mathbf{Y}$ and the columns of $\mathbf{X}$ are such that $\mathbf{X}'\mathbf{X}$ is a correlation matrix, and $\mathbf{X}'\mathbf{Y}$ is a vector of correlations. The $\hat{\boldsymbol{\beta}}^*_{(k)}$ are called *standardized regression coefficients* (see also Section 3.2.5).

Notice that in Eq. (12.1) a constant (the same constant) is added to the diagonal elements (only) of $\mathbf{X}'\mathbf{X}$. Notice also that we obtain the least-squares estimators, for the correlation form, if we let $k = 0$. Just as we convert the least-squares estimators from correlation form back to raw form, we do the same thing for ridge regression. Specifically, $\hat{\beta}_{0,(k)}$, the raw-form ridge estimator of the intercept, must be computed separately. One way to obtain $\hat{\beta}_{0,(k)}$ is $\hat{\beta}_{0,(k)} = \hat{\boldsymbol{\beta}}^{*\prime}_{(k)}\mathbf{Z}$, where $Z_i = \overline{X}^r_i / \sqrt{\sum (X^r_{ij} - \overline{X}^r_i)^2}$, with r denoting the regressor values in raw form. Alternatively, we could convert the other estimators back to raw form first and then use the raw-form ridge estimators to obtain $\hat{\beta}_{0,(k)}$. Thus, ridge regression is not applied directly to the intercept estimate. This is the preferred approach. See Brown (1977).

The other raw-form ridge estimators, $\hat{\beta}_{i(k)}$, would be obtained as $\hat{\beta}_{i(k)} = \hat{\beta}^*_{i(k)}\sqrt{S_{yy}}/\sqrt{S_{x_ix_i}}$, $i \neq 0$, where S_{yy} and $S_{x_ix_i}$ have the same representation as in earlier chapters. If for some reason Y were not also standardized, then the expression for $\hat{\beta}_{i(k)}$ would not have $\sqrt{S_{yy}}$ in the numerator.

Various methods have been proposed for selecting the value of k, and virtually all of these methods result in k being a random variable. The original justification for using ridge regression that was given by Hoerl and Kennard (1970a) was based on the assumption that k was *not* a random variable, however, and this is one reason why ridge regression has been somewhat controversial.

It is customary to use mean square error (MSE) in comparing two biased estimators or when comparing a biased estimator with an unbiased estimator. For any arbitrary estimator β^*_i of β_i, we have the following general result

$$\begin{aligned}\text{MSE}(\beta^*_i) &= E(\beta^*_i - \beta_i)^2 \\ &= \text{Var}(\beta^*_i) + [E(\beta^*_i) - \beta_i]^2 \end{aligned} \qquad (12.2)$$

If β^*_i were unbiased, then $E(\beta^*_i) = \beta_i$, and the second term in Eq. (12.2) is zero. Thus, the MSE of an unbiased estimator is simply the variance of the estimator.

Assume that we wish to compare β^*_i with $\hat{\beta}_i$, the least squares estimator of β_i. The least squares estimator will obviously be superior if $\text{Var}(\beta^*_i) \geq \text{Var}(\hat{\beta}_i)$, if $\hat{\beta}_i$ were unbiased. The motivation for using an alternative to the least squares estimator must therefore be that the alternative estimator has a smaller variance, and the difference in the two variances must more than offset the squared bias term in Eq. (12.2) for the alternative estimator.

Since we are estimating $p + 1$ parameters, we need a counterpart to Eq. (12.2) that is a function of all of the parameters and their corresponding estimators. It is customary to use the *total mean square error (TMSE)*, which is defined as $E[(\boldsymbol{\beta}^* - \boldsymbol{\beta})'(\boldsymbol{\beta}^* - \boldsymbol{\beta})]$. Using the ridge estimator $\hat{\boldsymbol{\beta}}_{(k)}^*$, we have

$$E[(\hat{\boldsymbol{\beta}}_{(k)}^* - \boldsymbol{\beta})'(\hat{\boldsymbol{\beta}}_{(k)}^* - \boldsymbol{\beta})] = \sum_{i=0}^{p} \mathrm{Var}(\beta_{(k),i}^*) + k^2 \boldsymbol{\beta}'(\mathbf{X}'\mathbf{X} + k\mathbf{I})^{-2}\boldsymbol{\beta} \qquad (12.3)$$

where the second term is the bias part. This result is derived in the chapter appendix, where it is also shown that

$$\sum_{i=0}^{p} \mathrm{Var}(\hat{\beta}_{(k),i}^*) = \sigma^2 \sum_{i=1}^{p+1} \frac{\lambda_i}{(\lambda_i + k)^2} \qquad (12.4)$$

where $\lambda_1, \lambda_2, \ldots, \lambda_{p+1}$ are the eigenvalues of $\mathbf{X}'\mathbf{X}$. (See Section 3.1.1 for information on eigenvalues.) Hoerl and Kennard (1970a) proved that there exists a range of values of k such that $\mathrm{TMSE}(\hat{\boldsymbol{\beta}}_{(k)}^*) < \mathrm{TMSE}(\hat{\boldsymbol{\beta}})$. Specifically, a ridge estimator will be superior to the least squares estimator in terms of MSE for $0 < k < \sigma^2/\alpha_{\max}^2$, with $\alpha_{\max}^2$ the largest element in $\boldsymbol{\alpha} = \mathbf{P}\boldsymbol{\beta}$, where $\mathbf{P}$ is the matrix of eigenvectors of $\mathbf{X}'\mathbf{X}$.

Obviously, $\boldsymbol{\beta}$ and σ^2 are unknown, however, so we cannot simply pick a value of k in this interval and know that the ridge estimator will be superior to least squares. This has led some critics, such as Egerton and Laycock (1981), to decry the use of ridge regression because we do not know for sure that we will be better off than we would be by using least squares. This is not a valid criticism, however, as there is always uncertainty involved in the application of statistical methods. We should use ridge regression if we can intelligently select k in such a way that the use of ridge regression is likely to produce superior results, and if it doesn't, we would hope that the results would not be much worse than the results obtained using least squares.

When should we consider using ridge regression, and when would we expect a ridge estimator to produce better results than would be obtained by using least squares? Consider Eq. (12.4), and let $k = 0$. Recall from Section 4.3.2 that multicollinearity will cause at least one of the eigenvalues to be close to zero, and it is apparent from Eq. (12.4) that one or more small eigenvalues can cause the sum of the variances of the estimators to be quite large. Notice also that the closer λ_i is to zero, the smaller k can be and still cause an appreciable reduction in the value of $\lambda_i/(\lambda_i + k)^2$. Thus, the more extreme the multicollinearity, the more room there is for improvement over the least squares estimator. The fact that under extreme multicollinearity we may use a small value of k and still reduce the value of the first term in Eq. (12.3) considerably means that the reduction in the sum of the variances might more than offset the bias that is created.

Hoerl and Kennard (1970a) show a figure that indicates an interval over which ridge would be superior to least squares, and that figure is given here as Figure 12.1. It is worth noting that the variance, in particular, depends on the degree of multicollinearity, and hence the optimal value of k will also depend on the degree of multicollinearity. Consequently, it is not possible to show graphically the apparent optimum value of k. Therefore, Figure 12.1 should be viewed as showing one possible relationship between ridge and least squares in terms of mean square error.

Recall the discussion in Section 4.1.2, in which it was shown that unexpected signs for the regression coefficients are caused in part by multicollinearity. Remember also that having the "right" signs of the regression coefficients is important if regression is being used for estimation or control. Therefore, ridge regression can be an important alternative to least squares when the signs of the least squares estimators differ from the signs of the r_{YX_i} and are consequently viewed as being undesirable.

Another reason for considering ridge regression that has been emphasized in the literature is that the vector $\hat{\beta}$ will be "too long" when there is multicollinearity. Specifically, it follows from Eq. (12.3) that $E(\hat{\beta}'\hat{\beta}) = \beta'\beta + \sigma^2 \sum(1/\lambda_i)$. To see this, let $k = 0$ in Eqs. (12.3) and (12.4), and then observe, by applying the same type of rules to vectors that apply to single random variables, that $E[(\hat{\beta} - \beta)'(\hat{\beta} - \beta)] = E(\hat{\beta}'\hat{\beta}) - \beta'\beta$. Thus, when one or more of the λ_i are

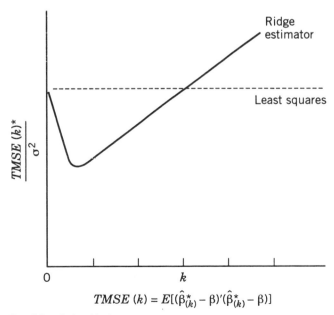

$$TMSE\ (k) = E[(\hat{\beta}^*_{(k)} - \beta)'(\hat{\beta}^*_{(k)} - \beta)]$$

Figure 12.1 Possible relationship between an ordinary ridge estimator and the least-squares estimator in terms of TMSE.

small, the vector $\hat{\boldsymbol{\beta}}$ could be much longer than the vector $\boldsymbol{\beta}$. Notice, however, that $\hat{\boldsymbol{\beta}}$ will be longer than $\boldsymbol{\beta}$ in expected value even when the regressors are uncorrelated, as then $\sum (1/\lambda_i) = p + 1$ since each $\lambda_i = 1$, but the difference will be much greater than $\sigma^2(p + 1)$ when there is multicollinearity.

An instructive way to view ridge regression versus least squares is to recognize that with the latter there is the implied assumption that $\boldsymbol{\beta}'\boldsymbol{\beta}$ is unbounded. Clearly, $\boldsymbol{\beta}'\boldsymbol{\beta}$ must be finite, and as discussed by Draper and Smith (1981, p. 320), if we assume an upper bound on $\boldsymbol{\beta}'\boldsymbol{\beta}$ and apply least squares subject to this restriction, the result has the general form of the ridge regression solution. When viewed in this manner, it is difficult to argue against the use of ridge regression, although $\boldsymbol{\beta}'\boldsymbol{\beta}$ is of course unknown.

12.2   HOW DO WE DETERMINE k?

Many different estimators of k have been proposed, and since theoretical results have been difficult to obtain, these estimators have been compared in a multitude of simulation studies. Simulations performed during the 1970s were criticized for being flawed in various ways. One criticism that is *not* valid is the argument that these simulation studies were biased in favor of ridge estimators, since the value of $\boldsymbol{\beta}'\boldsymbol{\beta}$ was fixed in these simulations, whereas there is no boundedness assumption on $\boldsymbol{\beta}'\boldsymbol{\beta}$ when least squares is used. This is not a valid criticism because, as pointed out in Section 12.1, there is obviously an upper bound on $\boldsymbol{\beta}'\boldsymbol{\beta}$. One valid criticism, though, is that many simulations had low R^2 values (see Peele and Ryan, 1980). Later studies were better conceived. In particular, the simulation study of Hoerl et al. (1986) was conducted in such a way that the F-statistic for the full model was, in almost all cases, large enough to indicate a significant relationship between Y and at least one of the predictors.

One of the early ways of determining k, which is still practiced today, is to construct a *ridge trace*. This is a graph of the values of the parameter estimates against k, an example of which is given in Figure 12.2. The general idea is to use the value of k at which the parameter estimates tend to stabilize. The problem with this approach is that k is being selected very subjectively; two people could look at a ridge trace and reach a different decision. At what value do the parameter estimates tend to stabilize in Figure 12.2? Does this occur at approximately $k = 0.1$, or earlier? Or is there evidence of stability at *any* value of k? Another problem with this approach is that the parameter estimates need not stabilize at all, as discussed by Thisted (1980), although this has generally not been a problem in practice.

Another possible approach is to compute the variance inflation factors (VIF) for each value of k, and use the smallest value of k for which the VIFs are deemed to be sufficiently small. $(\mathbf{X}'\mathbf{X} + k\mathbf{I})$ would replace $\mathbf{X}'\mathbf{X}$ in computing the VIFs.

Of the various estimators that have been proposed, one that has fared well

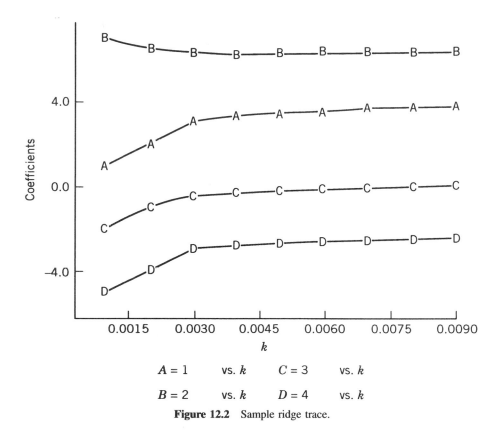

$A = 1$ vs. k $C = 3$ vs. k

$B = 2$ vs. k $D = 4$ vs. k

Figure 12.2 Sample ridge trace.

in simulation studies and has been mentioned prominently in the literature is the estimator proposed by Hoerl et al. (1975). Hoerl and Kennard (1976) indicate that an iterative procedure can be used that produces an estimator that is superior to the Hoerl, Kennard, and Baldwin (1975) estimator, however. The latter is given by $k = p\hat{\sigma}^2/\hat{\beta}'\hat{\beta}$, and will henceforth be referred to as the HKB estimator.

12.3   AN EXAMPLE

Since ridge regression has received considerable (generally unjust) criticism, it seems somewhat appropriate to use for illustration a real data set that was used by Fearn (1983) in attempting to show how ridge regression can produce poor results. The data, which are given in Table 12.1, result from an experiment that was performed to calibrate a near infrared (NIR) reflectance instrument for the measurement of protein content in ground wheat samples. The six potential regressors, here denoted as L1–L6, are measurements of the reflectance of

Table 12.1   Ground Wheat Data

Sample	Protein (%)	L1	L2	L3	L4	L5	L6
1	9.23	468	123	246	374	396	−11
2	8.01	458	112	236	368	383	−15
3	10.95	457	118	240	359	353	−16
4	11.67	450	115	236	352	340	−15
5	10.41	464	119	243	366	371	−16
6	9.51	499	147	273	404	433	5
7	8.67	463	119	242	370	377	−12
8	7.75	462	115	238	370	353	−13
9	8.05	488	134	258	393	377	−5
10	11.39	483	141	264	384	398	−2
11	9.95	463	120	243	367	378	−13
12	8.25	456	111	233	365	365	−15
13	10.57	512	161	288	415	443	12
14	10.23	518	167	293	421	450	19
15	11.87	552	197	324	448	467	32
16	8.09	497	146	271	407	451	11
17	12.55	592	229	360	484	524	51
18	8.38	501	150	274	406	407	11
19	9.64	483	137	260	385	374	−3
20	11.35	491	147	269	389	391	1
21	9.70	463	121	242	366	353	−13
22	10.75	507	159	285	410	445	13
23	10.75	474	132	255	376	383	−7
24	11.47	496	152	276	396	404	6
25	8.66	486	144	266	393	373	26
26	7.90	485	136	260	393	395	6
27	9.27	482	136	260	388	423	−2
28	11.77	443	112	232	346	355	−18
29	9.70	478	134	257	382	390	−5
30	10.46	449	113	233	351	343	−18
31	10.17	461	121	243	366	378	−14
32	11.10	503	155	280	403	414	6
33	12.03	493	146	271	390	378	−3
34	9.43	368	40	158	275	250	−63
35	8.66	462	114	237	367	331	−19
36	14.44	438	109	229	333	326	−28
37	8.50	478	127	252	384	378	−11
38	10.41	405	73	193	311	305	−44
39	9.72	498	146	273	403	415	0
40	11.69	442	106	226	341	303	−28
41	12.19	457	118	240	354	327	−23
42	11.59	439	103	224	339	325	−29
43	8.76	500	146	272	404	398	5
44	8.60	427	85	207	334	319	−36

Table 12.1   (*Continued*)

Sample	Protein (%)	L1	L2	L3	L4	L5	L6
45	8.54	479	128	253	384	382	−10
46	9.34	444	102	224	350	333	−27
47	10.09	458	118	239	362	355	−16
48	8.72	518	162	290	426	464	16
49	10.87	465	124	247	369	386	−13
50	10.89	457	120	242	363	411	−15

NIR radiation by the wheat samples at six different wavelengths in the range 1680–2310, with the measurements being made on a $\log(1/R)$ scale and R representing the reflectance. The objective of the study was to obtain a regression equation that could be used to predict the protein content of future samples.

The first 24 data values in Table 12.1 were used by Fearn (1983) to obtain the regression coefficients. The regression equation was then applied to the next 26 values to see how well the equation performs. Recall from Chapter 7 that one of the methods used for validating a regression model is to use some algorithm for splitting a data set into similar halves (similar in regard to the set of regressor values) and then to see how well the model predicts Y in the half that was not used to estimate the regression coefficients. [The reader is referred to Chapter 10 of Montgomery and Peck (1992) for a detailed discussion of model validation.]

In arguing against the use of ridge regression for this data set, Fearn (1983) points out that any value of k that differs enough from zero to cause a perceptible change in the regression coefficients from the least squares solution will produce a larger value for $\sqrt{\text{SSE}/\text{df}}$ for these additional data points than what is obtained by using least squares. This is of interest if the additional 26 data points are similar to the first 24. Even if this were true, however, one cannot argue that a procedure does not have merit because of what happens with a single example. Expressing doubts about the value of ridge regression because of what happens with one example is clearly inappropriate, and yet this is the same type of criticism made by Egerton and Laycock (1981), whose argument was that ridge will not always be superior to least squares.

Furthermore, the additional points differ considerably from the first set of points. In particular, if a point in the second set has a value for one of the regressors that is outside the range of values for that regressor in the first set, the point must be outside the region covered by the first set of points. We can determine from inspection of Table 12.1 that 11 of the 26 points in the second set have this characteristic, and some points are obviously well outside the region covered by the first set of points. Because of the extreme multicollinearity (as evidenced by the fact that 8 of the 15 pairwise regressor correlations are at least .988), we would expect the experimental region covered by the 24 points to be much smaller than the region obtained by considering only the ranges of the individual regressors. A more detailed analysis would likely reveal additional points that are outside the experimental region.

As mentioned in Section 7.8, Picard and Berk (1990) indicate that more than slight extrapolation is the rule rather than the exception when small sample sizes are used for model development and the regressors are random. Nevertheless, it is inappropriate to compare the performance of ridge against least squares for additional observations that are well outside the initial experimental region. Therefore, the analysis of Fearn (1983) is, strictly speaking, not defensible, since neither ridge nor least squares should be applied to all of the additional data points. Furthermore, Hoerl et al. (1985), in their response to Fearn (1983), showed that if the last 26 data points were used for estimation, and the first 24 points for validation, ridge provides a significant improvement over least squares.

In comparing the two data partitions it is also of interest to compare the correlation matrix (including Y) for the first 24 observations against the correlation matrix for the other 26 observations. The two matrices are given in Table 12.2. Notice that the correlations between the regressors do not differ greatly for the two partitions, but also notice for the second partition that all of the r_{YL_i} are negative, whereas in the first partition they are all positive. The latter might suggest that the data set contains some errors or that the data set is atypical in certain ways. (Recall the claim of Hampel et al. (1986), mentioned in Chapter 11, that data sets generally contain 1–10% gross errors.)

As with regression coefficients, correlation coefficients can be strongly affected by influential observations, however, and the presence of such observa-

Table 12.2   Correlations for Table 12.1 Data

	Protein	L1	L2	L3	L4	L5
(a) First 24 Observations						
L1	0.467					
L2	0.552	0.994				
L3	0.537	0.996	0.999			
L4	0.383	0.995	0.980	0.984		
L5	0.359	0.937	0.925	0.934	0.954	
L6	0.451	0.989	0.988	0.989	0.989	0.949
(b) Last 26 Observations						
L1	−0.247					
L2	−0.127	0.988				
L3	−0.154	0.993	0.999			
L4	−0.340	0.994	0.972	0.981		
L5	−0.271	0.895	0.892	0.900	0.913	
L6	−0.298	0.938	0.948	0.944	0.945	0.857

tions can be seen when the appropriate scatter plots are viewed for the first 24 observations and for the last 26 observations. (Recall from Section 2.4.1 that we can construct scatter plots that show the influence of each observation on the correlation coefficient.) In particular, observation 36 may be anomalous, and the positive values of r_{YL_i} for the first 24 points are caused by two points: 15 and 17. The latter very strongly affect the r_{YL_i} for the first set of points, as there is a large drop in all of the correlations when the two points are deleted. In particular, r_{YL_4} changes from .383 to .00035!

When the two partitions are merged and all points except 15, 17, and 36 are used, a most unusual result is observed involving L_3 and L_4. Specifically, $r_{YL_3} = -.00006$ and $r_{YL_4} = -.19081$, so L_3 and L_4 are of no value as individual regressors, yet $R^2 = .943$ when both are used in the model. Recall that this was the same type of result that was observed for the hypothetical data set given in Table 5.6, which was constructed to illustrate a particular point. Thus, although this data set was used by Fearn (1983) in an attempt to discredit ridge regression, this is clearly not typical regression data.

Fearn (1983) used a ridge trace, given here as Figure 12.3, in selecting $k =$

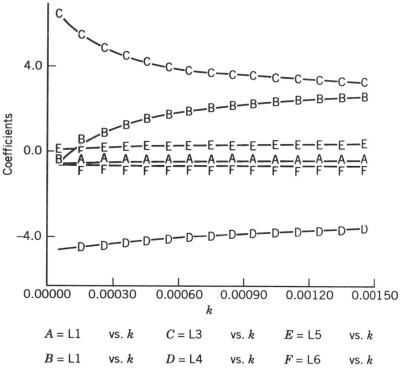

| $A = L1$ | vs. k | $C = L3$ | vs. k | $E = L5$ | vs. k |
| $B = L1$ | vs. k | $D = L4$ | vs. k | $F = L6$ | vs. k |

Figure 12.3 Ridge trace for Table 12.1 data.

0.001. The trace indicates that this is a reasonable choice. By comparison, the HKB estimator is $k = 0.0001$. The trace suggests that the latter value is too small, and this can be attributed to the very extreme multicollinearity, which causes $\hat{\beta}'\hat{\beta}$ to be rather large. This suggests that the HKB estimator should be modified so that it is more sensitive to the degree of multicollinearity, as discussed by Peele and Ryan (1982), especially for those situations in which none of the regressors is deleted before ridge is used.

As implied in Chapter 7, ridge regression permits the use of a set of regressors that might be deemed inappropriate if least squares were used. Specifically, highly correlated variables can be used together, with ridge regression used to reduce the multicollinearity. If, however, the multicollinearity were extreme, such as when regressors are almost perfectly correlated, we would probably prefer to delete one or more regressors before using the ridge approach.

Consider again the pairwise correlations given in Table 12.2. Clearly variables 1–3 are almost perfectly correlated, so even if ridge regression were used, it would seem unwise to attempt to use all six possible regressors. Rather, a better approach would be to delete two of the first three variables (and possibly other variables as well) and then determine if the application of ridge regression to the remaining variables seems necessary.

Hoerl et al. (1985) do delete four of these six variables and then apply ridge regression to the remaining two variables, after first redefining the variables. Specifically, as was also observed by Fearn (1983), the measured values of the potential regressors are related to the size of the particle, but the particle size is unrelated to Y, since the latter is given as a percentage. As discussed by Hoerl et al. (1985), particle size essentially represents unnecessary ill-conditioning, which can be removed by subtracting the average of the regressor coordinates at each data point from each of the coordinates of that point.

12.4 RIDGE REGRESSION FOR PREDICTION?

Ridge regression is useful when regression is used for parameter estimation or control, but what if the objective were prediction, as in the example given by Fearn (1983)? First, we would not use total mean square error, as in Eq. (12.3), as a criterion. Instead, we would prefer a criterion such as $E(\mathbf{Y} - \hat{\mathbf{Y}}^*)'(\mathbf{Y} - \hat{\mathbf{Y}}^*)$. Notice, however, that if we let $\hat{\mathbf{Y}}^* = \hat{\mathbf{Y}}$, the least squares estimator, we then have $E[\sum (Y - \hat{Y})^2] = E(\text{SSE}) = (n - p - 1)\sigma^2$, which does not depend on the degree of multicollinearity. A similar argument was made by Hoerl et al. (1986), and Swamy et al. (1978) also show that prediction using least squares is not undermined by multicollinearity.

Remember that the theoretical result of Hoerl and Kennard (1970a), described earlier in Section 12.1, is an "estimation error" result and does not pertain to prediction error. Specifically, the general form of the total mean square error in Eq. (12.3) does not directly involve prediction. Consequently, a different criterion would be needed if we were interested in prediction rather than estimation.

As in Theobald (1974) and Park (1981), we consider the *mean square error of prediction (MSEP)* given by $MSEP(\hat{Y}|\mathbf{x}) = E(\mathbf{x}'\boldsymbol{\beta}^* - \mathbf{x}'\boldsymbol{\beta})^2$ with, as in Eq. (12.2), $\boldsymbol{\beta}^*$ denoting an arbitrary estimator, and $\mathbf{x}$ representing a particular combination of the regressor values. As shown by Theobald (1974), ridge is superior to least squares in terms of MSEP if $0 < k < 2\sigma^2/\boldsymbol{\beta}'\boldsymbol{\beta}$. Therefore, we may improve upon least squares when prediction is the objective. In particular, Park (1981) shows that for points that are equidistant from the origin, a point $\mathbf{x}$ that is in the direction of the ith eigenvector could have a large value of MSEP when least squares is used, since $Var(\hat{Y}|\mathbf{x}) = \mathbf{x}'\mathbf{x}\sigma^2/\lambda_i$, where λ_i is the corresponding eigenvalue. As stated in Section 12.1, there will be at least one small eigenvalue whenever multicollinearity exists.

We should carefully distinguish between this theoretical result and results obtained using only the data points in a sample. If we were to apply the least squares solution to the 24 data points that were used in the Fearn (1983) example to estimate the coefficients, any value of k will, by definition, cause the mean square error to be larger than would be obtained with the least squares solution, since the latter minimizes the sum of squares of the prediction errors. This result strictly applies, however, only when predictions are to be made at the same combinations of regressor values that were in the sample used to obtain the parameter estimates. This will almost certainly not happen, however, as multicollinearity is a problem that will occur when the regressors are random variables, so the experimenter would not have any control over the combinations of regressor values that are in the sample. When other combinations of regressor values are subsequently used, one is essentially interpolating between the data points in the original sample, and least squares may not then provide the best predictions. Therefore, sample results and (some) theoretical arguments that favor the use of least squares for prediction in the presence of multicollinearity are not compelling.

12.5 GENERALIZED RIDGE REGRESSION

The most commonly used ridge regression technique is ordinary ridge regression, which was described in Section 12.1. An alternative is to replace $k\mathbf{I}$ by $\mathbf{K}$ in Eq. (12.1), where $\mathbf{K}$ is a diagonal matrix with diagonal elements k_i. This is termed *generalized ridge regression*. Generalized ridge regression gives the user some flexibility regarding the shrinkage of each regression coefficient, as it may be desirable to treat the coefficients differently. This form was considered by Hoerl and Kennard (1970a), who determined the optimal values of the k_i. These optimal values are functions of $\boldsymbol{\beta}$ and σ^2, however, so the user of generalized ridge regression is faced with the same type of problem as the user of ordinary ridge regression. Furthermore, as discussed by Montgomery and Peck (1992, pp. 344–353), there are problems in trying to select values of the k_i.

12.6   INFERENCES IN RIDGE REGRESSION

One of the drawbacks to ridge regression, assuming that k is a random variable, is that the usual inferences (e.g., confidence intervals) that are made when least squares is used are not applicable in ridge regression. This is not as big a drawback as it might seem, however. In particular, when ridge is used there has probably been a decision made either to use a particular subset of regressors or to use ridge regression for subset selection. Hoerl and Kennard (1970b) suggest that variables be eliminated whose coefficients are unstable in a ridge trace, in addition to eliminating variables whose coefficients approach zero in the trace. Therefore, hypothesis tests on the β_i for the purpose of variable selection are not particularly relevant in ridge regression. But the inability to make other inferences, such as a prediction interval for Y, might frustrate some ridge users.

12.7   SOME PRACTICAL CONSIDERATIONS

As when least squares is used, there is a need to look for possible outliers and influential observations. In particular, multicollinearity might be *caused* by outliers, and the removal of such outliers could obviate the use of ridge regression.

Consider the following illustrative example. Assume that there are two regressors, and values for (X_1, X_2) are (1, 2), (5, 3), (2, 4), (1, 5), (8, 7), (7, 8), (4, 4), (6, 9), (3, 10), and (26, 27). Using all 10 observations we obtain $r_{X_1 X_2} = .937$, but using only the first 9 observations produces $r_{X_1 X_2} = .487$. Thus, the extreme data point causes the high correlation between X_1 and X_2. When this situation occurs, Gunst and Mason (1985) suggest that a robust regression estimator should be considered, rather than a ridge regression estimator, with the objective of not letting the outlier strongly influence the regression coefficients.

Assume that a decimal point was misplaced in Y, X_1, and X_2. Then not only could we have outlier-induced collinearity, but R^2 might also be considerably inflated by the bad data point, perhaps resulting in a false signal that the model has considerable value. Thus, we should check to see if there are any outliers before using ridge regression. A useful five-step procedure for detecting outlier-induced collinearities is given by Gunst and Mason (1985).

Another important consideration is influential observations. Walker and Birch (1988) consider the reverse of what was illustrated at the beginning of this section: multicollinearity *causing* influential observations rather than being caused by them. They emphasize that ridge regression influence measures (analogous to those that are used with least squares) should be applied *after* the value of k has been determined. (They give a ridge version of DFFITS. See Section 2.4.2 for a discussion of DFFITS.) The existence of both influential observations and multicollinearity in a data set is illustrated by Lawrence and Marsh (1984).

12.8 ROBUST RIDGE REGRESSION?

In contrast to what was discussed in Section 12.7, some writers have suggested combining the methods of robust regression and ridge regression to obtain a robust form of ridge regression. In particular, see Montgomery and Peck (1992), Lawrence and Marsh (1984), Askin and Montgomery (1980), and Hogg (1979).

Although data sets will undoubtedly occur for which there is a need to combine the two concepts, a seemingly superior strategy would be to first determine if there are any outlier-induced collinearities, since the presence of these would obviate the use of ridge regression. For example, we might initially use a sequential approximate LTS approach (as illustrated in Section 11.5.4.2) and then determine if multicollinearity exists for the points that are used in producing the parameter estimates. That approach might not always identify outliers that are causing multicollinearity, however, especially if the data point $(y, x_1, x_2, \ldots, x_p)$ increases R^2. Therefore, it would probably be desirable at this point to additionally use a strategy such as that given by Gunst and Mason (1985), which was mentioned in Section 12.7. After outlier-induced collinearities have perhaps been identified and removed, the possible use of ridge regression would be addressed at that point.

12.9 OTHER BIASED ESTIMATORS

As indicated in Section 12.1, we may appropriately think of regression estimators in general as being biased since the correct model will almost certainly be unknown. Nevertheless, when biased estimators are discussed in the literature, the reference is to something other than the ordinary least squares estimator.

A frequently mentioned biased estimation technique is *principal component regression*, due to Massy (1965). This is more like a variable selection approach, however, as artificial variables (which are called *principal components*) are obtained by transforming the regressors. The general idea is to work with the model $\mathbf{Y} = \mathbf{Z}\boldsymbol{\alpha} + \boldsymbol{\epsilon}$ instead of the model $\mathbf{Y} = \mathbf{X}\boldsymbol{\beta} + \boldsymbol{\epsilon}$, with $\mathbf{Z} = \mathbf{XT}$, $\boldsymbol{\alpha} = \mathbf{T}\boldsymbol{\beta}$, and $\mathbf{T}$ denoting the matrix of eigenvectors corresponding to the eigenvalues of $\mathbf{X}'\mathbf{X}$. The columns of $\mathbf{Z}$ are the principal components, and a subset of the principal components is selected for use, just as a subset of the columns of $\mathbf{X}$ is used in variable selection. The selection of the subset of principal components is often made somewhat subjectively. One objective way to select the principal components is to use cross validation. Let c denote the number of principal components in the model and let $\hat{y}_{c(i)}$ denote the predicted value of y_i using the c components but not using the ith observation. The selected value of c is the one that minimizes $\sum_{i=1}^{n} (y_i - \hat{y}_{c(i)})^2$.

[See Draper and Smith (1981, p. 328) or Montgomery and Peck (1992, p. 353) for additional details.]

A related procedure is *latent root regression*, which was proposed by Webster et al. (1974) and also discussed extensively in Gunst and Mason (1980). The

essential difference between this and principal component regression is that with
the latter the response variable is ignored in selecting the principal components,
whereas multicollinearities are retained in latent root regression if they have
predictive value.

A similar technique is *partial least squares* (Wold 1975, 1984). Frank
and Friedman (1993) compared principal component regression, partial least
squares, ordinary least squares, ridge regression, and stepwise regression in a
simulation study (using average squared prediction error computed over new
observations as the criterion), and found that ridge regression came out ahead
of the other techniques. Naes et al. (1986) discuss a successful application of
partial least squares to the data given in Fearn (1983).

12.10 SOFTWARE

BMDP-4R is for ridge regression and principal components regression. The
ridge regression output can include a ridge trace in which the standardized
regression coefficients are plotted against k, in addition to plots in which the
residual sum of squares and R^2 are each plotted against k.

There are no programs for ridge regression in SAS, SPSS, SYSTAT,
S-Plus, or MINITAB. Therefore, users of these software packages would need
to write their own programs. (A ridge regression Minitab macro, RIDGE, is
described in Appendix A.) SCAN (Software for Chemometric Analysis), a
Minitab, Inc. product, does include capability for ridge regression, principal
components regression, and partial least squares.

While it is possible to use a standard least squares program to obtain the ridge
estimator for a specified value of k [see, e.g., Montgomery and Peck (1992, p.
334) or Draper and Smith (1981, p. 323)], we should seek more output (e.g., a
ridge trace) than just the ridge regression coefficients.

SUMMARY

Ridge regression has been presented as a viable alternative to ordinary least
squares. Potential users of ridge regression should not think in terms of having
to choose between a biased and an unbiased estimator, since at least one of
the $\hat{\beta}_i$ will be biased if the fitted model is not the true model, as will generally
be the case. Various methods of determining the constant k in ordinary ridge
regression have been presented. Because there is no obvious best way to make
the selection, users may wish to employ both a ridge trace and an estimator such
as the Hoerl et al. (1975) estimator or one of the suggested modifications. The
user might then determine if one or more of these estimators seem compatible
with the ridge trace.

Another possible use of ridge regression would be to apply it after the two-
stage transformation procedure given in Chapter 6 has been applied, if the appli-

cation of the Box–Tidwell procedure has produced multicollinearity among the transformed regressors.

It is important to search for outlier-induced collinearities (and bad data points, in general) before using ridge regression (or any other biased estimation technique), as discarding bad data points could obviate the use of ridge regression. Robust regression techniques may be helpful in this regard.

Some indication of the use of ridge regression in many different fields of application is apparent from the extensive list of applications papers given by Hoerl and Kennard (1981).

Ridge regression has also been applied to other areas of regression besides linear regression. See, for example, le Cessie and van Houwelingen (1992), Schaefer et al. (1984), and Schaefer (1986) for information on ridge estimators for logistic regression. Furthermore, the ridge concept has been applied to other areas of statistics besides regression with an eye toward producing improved parameter estimates in the presence of multicollinearity-type conditions.

APPENDIX

12.A Derivation of Eq. (12.3)

We will use an approach that is essentially the same as that given by Hoerl and Kennard (1970a), but which is given here with more details. It is convenient to first write $\hat{\boldsymbol{\beta}}^*_{(k)}$ in terms of $\hat{\boldsymbol{\beta}}$. We proceed as follows:

$$
\begin{aligned}
\hat{\boldsymbol{\beta}}^*_{(k)} &= (\mathbf{X}'\mathbf{X} + k\mathbf{I})^{-1}\mathbf{X}'\mathbf{Y} \\
&= (\mathbf{X}'\mathbf{X} + k\mathbf{I})^{-1}\mathbf{X}'\mathbf{X}(\mathbf{X}'\mathbf{X})^{-1}\mathbf{X}'\mathbf{Y} \\
&= [(\mathbf{X}'\mathbf{X})^{-1}(\mathbf{X}'\mathbf{X} + k\mathbf{I})]^{-1}\hat{\boldsymbol{\beta}} \\
&= [\mathbf{I} + k(\mathbf{X}'\mathbf{X})^{-1}]^{-1}\hat{\boldsymbol{\beta}} \\
&= \mathbf{Z}\hat{\boldsymbol{\beta}}
\end{aligned}
$$

It then follows that

$$
\begin{aligned}
E[(\hat{\boldsymbol{\beta}}^*_{(k)} - \boldsymbol{\beta})'(\hat{\boldsymbol{\beta}}^*_{(k)} - \boldsymbol{\beta})] &= E[(\mathbf{Z}\hat{\boldsymbol{\beta}} - \boldsymbol{\beta})'(\mathbf{Z}\hat{\boldsymbol{\beta}} - \boldsymbol{\beta})] \\
&= E[(\mathbf{Z}\hat{\boldsymbol{\beta}} - \mathbf{Z}\boldsymbol{\beta})'(\mathbf{Z}\hat{\boldsymbol{\beta}} - \mathbf{Z}\boldsymbol{\beta})] + E(\hat{\boldsymbol{\beta}}'\mathbf{Z}'\mathbf{Z}\boldsymbol{\beta} - \boldsymbol{\beta}'\mathbf{Z}\hat{\boldsymbol{\beta}} - \hat{\boldsymbol{\beta}}'\mathbf{Z}'\boldsymbol{\beta} + \hat{\boldsymbol{\beta}}\boldsymbol{\beta}) \\
&= E[(\mathbf{Z}\hat{\boldsymbol{\beta}} - \mathbf{Z}\boldsymbol{\beta})'(\mathbf{Z}\hat{\boldsymbol{\beta}} - \mathbf{Z}\boldsymbol{\beta})] + \boldsymbol{\beta}'\mathbf{Z}'\mathbf{Z}\boldsymbol{\beta} - 2\boldsymbol{\beta}'\mathbf{Z}\boldsymbol{\beta} + \boldsymbol{\beta}'\boldsymbol{\beta} \\
&= E[(\hat{\boldsymbol{\beta}} - \boldsymbol{\beta})'\mathbf{Z}'\mathbf{Z}(\hat{\boldsymbol{\beta}} - \boldsymbol{\beta})] + (\mathbf{Z}\boldsymbol{\beta} - \boldsymbol{\beta})'(\mathbf{Z}\boldsymbol{\beta} - \boldsymbol{\beta})
\end{aligned}
$$

To evaluate the expected value we recognize that we are taking the expected value of a *quadratic form*, which has the general form $\mathbf{x}'\mathbf{A}\mathbf{x}$. Here $E(\mathbf{x}) = \mathbf{0}$ since $\hat{\boldsymbol{\beta}}$ is assumed to be unbiased. Then, $E(\mathbf{x}'\mathbf{A}\mathbf{x}) = \text{tr}(\mathbf{A}\mathbf{V})$, where $\text{Var}(\mathbf{x}) = \mathbf{V}$

(see, e.g., Searle, 1971, p. 55). Thus,

$$E[(\hat{\boldsymbol{\beta}}^*_{(k)} - \boldsymbol{\beta})'(\hat{\boldsymbol{\beta}}^*_{(k)} - \boldsymbol{\beta})]$$

$$= \sigma^2 \text{tr}[\mathbf{Z}'\mathbf{Z}(\mathbf{X}'\mathbf{X})^{-1}] + \boldsymbol{\beta}'(\mathbf{Z} - \mathbf{I})'(\mathbf{Z} - \mathbf{I})\boldsymbol{\beta}$$

$$= \sigma^2 \text{tr}[(\mathbf{I} + k(\mathbf{X}'\mathbf{X})^{-1})^{-1}(\mathbf{X}'\mathbf{X} + k\mathbf{I})^{-1}] + \boldsymbol{\beta}'(\mathbf{Z} - \mathbf{I})'(\mathbf{Z} - \mathbf{I})\boldsymbol{\beta} \quad (A.1)$$

To evaluate the trace we need to break up the product inside the brackets and obtain an equivalent expression involving the traces of individual matrices. In this instance we can avoid having the trace of a product if we write the first matrix inside the brackets in an equivalent form. For two square matrices $\mathbf{A}$ and $\mathbf{B}$, $(\mathbf{I} + \mathbf{AB})^{-1} = \mathbf{I} - \mathbf{A}(\mathbf{B}^{-1} + \mathbf{A})^{-1}$. [This result can be obtained from somewhat similar results given by Rao (1973, p. 33) or Draper and Smith (1981, p. 127).) If we let $\mathbf{A} = k\mathbf{I}$ and $\mathbf{B} = (\mathbf{X}'\mathbf{X})^{-1}$, we obtain $[\mathbf{I} + k(\mathbf{X}'\mathbf{X})^{-1}]^{-1} = \mathbf{I} - k(\mathbf{X}'\mathbf{X} + k\mathbf{I})^{-1}$.

Since $[\mathbf{I} - k(\mathbf{X}'\mathbf{X} + k\mathbf{I})^{-1}](\mathbf{X}'\mathbf{X} + k\mathbf{I})^{-1} = (\mathbf{X}'\mathbf{X} + k\mathbf{I})^{-1} - k(\mathbf{X}'\mathbf{X} + k\mathbf{I})^{-2}$, we will then obtain the trace of the difference of these two matrices, and the trace of a difference of two matrices is the difference of the two traces. Since the trace of a matrix is the sum of the eigenvalues, λ_i, the trace of an inverse is the sum of the λ_i^{-1}, and the eigenvalues of $(\mathbf{X}'\mathbf{X} + k\mathbf{I})$ are $\lambda_i + k$, we thus obtain $\sum 1/(\lambda_i + k) - k \sum 1/(\lambda_i + k)^2 = \sum \lambda_i/(\lambda_i + k)^2$.

We will proceed similarly in simplifying the second term in Eq. (A.1). By using the equivalent expression for $\mathbf{Z}$ that was given above, we obtain $\boldsymbol{\beta}'(\mathbf{Z} - \mathbf{I})'(\mathbf{Z} - \mathbf{I})\boldsymbol{\beta} = k^2 \boldsymbol{\beta}'(\mathbf{X}'\mathbf{X} + k\mathbf{I})^{-2}\boldsymbol{\beta}$.

Putting these two results together gives the final result

$$E(\hat{\boldsymbol{\beta}}^*_{(k)} - \boldsymbol{\beta})'(\hat{\boldsymbol{\beta}}^*_{(k)} - \boldsymbol{\beta}) = \sigma^2 \sum \frac{\lambda_i}{(\lambda_i + k)^2} + k^2 \boldsymbol{\beta}'(\mathbf{X}'\mathbf{X} + k\mathbf{I})^{-2}\boldsymbol{\beta}$$

which is a combination of Eq. (12.3) and (12.4).

12.B   Derivation of Eq. (12.4)

This result can be easily derived by utilizing some of the results that were obtained in deriving Eq. (12.3). Using $\hat{\boldsymbol{\beta}}^*_{(k)} = \mathbf{Z}\hat{\boldsymbol{\beta}}$, it follows that $\text{Var}(\hat{\boldsymbol{\beta}}^*_{(k)}) = \text{Var}(\mathbf{Z}\hat{\boldsymbol{\beta}}) = \mathbf{Z}\text{Var}(\hat{\boldsymbol{\beta}})\mathbf{Z}' = \sigma^2\mathbf{Z}(\mathbf{X}'\mathbf{X})^{-1}\mathbf{Z}'$. Since we seek $\sum \text{Var}(\hat{\boldsymbol{\beta}}^*_{(k)})$, we thus need $\text{tr}[\mathbf{Z}(\mathbf{X}'\mathbf{X})^{-1}\mathbf{Z}']$. Since this is equivalent to $\text{tr}[(\mathbf{X}'\mathbf{X})^{-1}\mathbf{Z}'\mathbf{Z}]$, it follows from the derivation of Eq. (12.3) that $\sum \text{Var}(\hat{\boldsymbol{\beta}}^*_{(k)}) = \sigma^2 \sum \lambda_i/(\lambda_i + k)^2$. Thus, Eq. (12.3) represents the sum of the variances of the ridge estimators plus the sum of the squared bias terms, but we should remember that Eqs. (12.3) and (12.4) are based upon the assumption that k is not a random variable.

REFERENCES

Askin, R. G. and D. C. Montgomery (1980). Augmented robust estimators. *Technometrics*, **22**, 333–341.

Brown, P. J. (1977). Centering and scaling in ridge regression. *Technometrics*, **19**, 35–36.

Draper, N. R. and H. Smith (1981). *Applied Regression Analysis*, 2nd edition. New York: Wiley.

Egerton, M. F. and P. J. Laycock (1981). Some criticisms of stochastic shrinkage and ridge regression, with counterexamples. *Technometrics*, **23**, 155–159 (correction: vol. 25. p. 304).

Fearn, T. (1983). A misuse of ridge regression in the calibration of a near infrared reflectance instrument. *Applied Statistics*, **32**, 73–79.

Frank, I. E. and J. H. Friedman (1993). A statistical view of some chemometrics regression tools. *Technometrics*, **35**, 109–135 (discussion: 136–148).

Gunst, R. F. and R. L. Mason (1980). *Regression Analysis and Its Application*. New York: Dekker.

Gunst, R. F. and R. L. Mason (1985). Outlier-induced collinearities. *Technometrics*, **27**, 401–407.

Hampel, F. R., E. M. Ronchetti, P. J. Rousseeuw, and W. A. Stahel (1986). *Robust Statistics: The Approach Based on Influence Functions*. New York: Wiley.

Hoerl, A. E. (1962). Applications of ridge analysis to regression problems. *Chemical Engineering Progress*, **58**, 54–59.

Hoerl, A. E. and R. W. Kennard (1970a). Ridge regression: Biased estimation for nonorthogonal problems. *Technometrics*, **12**, 55–67.

Hoerl, A. E. and R. W. Kennard (1970b). Ridge regression: Applications to nonorthogonal problems. *Technometrics*, **12**, 69–82. (erratum: vol. 12, p. 723.)

Hoerl, A. E. and R. W. Kennard (1976). Ridge regression: Iterative estimation of the biasing parameter. *Communications in Statistics*, **A5**, 77–78.

Hoerl, A. E. and R. W. Kennard (1981). Ridge regression-1980—Advances, algorithms, and applications. *American Journal of Mathematical and Management Sciences*, **1**, 5–83.

Hoerl, A. E., R. W. Kennard, and K. F. Baldwin (1975). Ridge regression: Some simulations. *Communications in Statistics*, **A4**, 105–123.

Hoerl, A. E., R. W. Kennard, and R. W. Hoerl (1985). Practical use of ridge regression: A challenge met. *Applied Statistics*, **34**, 114–120.

Hoerl, R. W. (1985). Ridge analysis 25 years later. *The American Statistician*, **39**, 186–192.

Hoerl, R. W., J. H. Schuenemeyer, and A. E. Hoerl (1986). A simulation of biased estimation and subset selection techniques. *Technometrics*, **28**, 369–380.

Hogg, R. V. (1979). An Introduction to Robust Estimation, in *Robustness in Statistics*, R. L. Launer and G. N. Wilkinson, eds. New York: Academic Press.

Lawrence, K. D. and L. C. Marsh (1984). Robust ridge estimation methods for predicting U.S. coal mining fatalities. *Communications in Statistics*, **A13**, 139–149.

le Cessie, S. and J. C. van Houwelingen (1992). Ridge estimators in logistic regression. *Applied Statistics*, **41**, 191–201.

Massy, W. F. (1965). Principal components regression in exploratory statistical research. *Journal of the American Statistical Association*, **60,** 234–246.

Montgomery, D. C. and E. A. Peck (1992). *Introduction to Linear Regression Analysis,* 2nd edition. New York: Wiley.

Naes, T., C. Irgens, and H. Martens (1986). Comparison of linear statistical methods for calibration of NIR instruments. *Applied Statistics,* **35,** 195–206.

Park, S. H. (1981). Collinearity and optimal restrictions on regression parameters for estimating responses. *Technometrics,* **23,** 289–295.

Peele, L. C. and T. P. Ryan (1980). Comment on "A critique of some ridge regression methods" by G. Smith and F. Campbell. *Journal of the American Statistical Association,* **75,** 96–97.

Peele, L. C. and T. P. Ryan (1982). Minimax linear regression estimators with application to ridge regression. *Technometrics,* **24,** 157–159.

Picard, R. R. and K. N. Berk (1990). Data splitting. *The American Statistician,* **44,** 140–147.

Rao, C. R. (1973). *Linear Statistical Inference and its Applications.* New York: Wiley.

Schaefer, R. L., L. D. Roi, and R. A. Wolfe (1984). A ridge logistic estimator. *Communications in Statistics,* **A13,** 99–113.

Schaefer, R. L. (1986). Alternative estimators in logistic regression when the data are collinear. *Journal of Statistical Computation and Simulation,* **25,** 75–91.

Searle, S. R. (1971). *Linear Models.* New York: Wiley.

Swamy, P. A. V. B., J. S. Mehta, and J. N. Rappoport (1978). Two methods of evaluating Hoerl and Kennard's ridge regression. *Communications in Statistics,* **A7,** 1133–1155.

Thisted, R. A. (1980). Comment on "A critique of some ridge regression methods" by G. Smith and F. Campbell. *Journal of the American Statistical Association,* **75,** 81–86.

Theobald, C. M. (1974). Generalizations of mean square error applied to ridge regression. *Journal of the Royal Statistical Society, Series B,* **36,** 103–106.

Walker, E. and J. B. Birch (1988). Influence measures in ridge regression. *Technometrics,* **30,** 221-227 (correction: vol. 30, pp. 469–470).

Webster, J. T., R. F. Gunst, and R. L. Mason (1974). Latent root regression analysis. *Technometrics,* **16,** 513–522.

Wold, H. (1975). Soft Modeling by Latent Variables: The Nonlinear Iterative Partial Least Squares Approach, in *Perspectives in Probability and Statistics, Papers in Honour of M.S. Bartlett,* J. Gani, ed. London: Academic Press.

Wold, H. (1984). PLS Regression, in *Encyclopedia of Statistical Sciences,* Vol. 6, N. L. Johnson and S. Kotz, eds. New York: Wiley.

EXERCISES

12.1. Explain why ridge regression would never be used when there is only a single regressor.

12.2. Using some computer program, such as the Minitab macro RIDGE, construct the ridge trace for the last 26 observations in Table 12.1 and determine if the ridge constant $k = p\hat{\sigma}^2/\hat{\beta}'\hat{\beta}$ seems reasonable relative to the trace. If not, explain what probably causes this ridge constant to be inappropriate for this (partial) data set. Should all six of the variables be used? If some variables can be deleted, should ridge regression be applied to the subset that remains? If so, construct the ridge trace and compute the HKB ridge constant.

12.3. Assume that there are two regressors and one data point has extreme values of X_1 and X_2, with $r_{X_1X_2} = .94$. How should the data analyst proceed if a regression equation is to be used for control?

12.4. Explain why a ridge trace cannot necessarily be relied upon as a stand-alone procedure for determining k.

12.5. Explain the difference between ordinary ridge regression and generalized ridge regression.

12.6. Can the bias of a ridge regression estimator be computed? If so, then compute it for the HKB ridge estimator in Exercise 12.2. If not, then explain why not.

12.7. Critique the following statement: Since ridge regression will not always be superior to ordinary least squares in terms of mean square error, it should not be used in place of ordinary least squares.

12.8. Consider the following data:

Y:	2.4	2.8	2.9	3.2	3.3	3.1	2.6	3.7	3.8	2.9	3.7	2.5	3.0	3.5
X_1:	11	16	13	15	18	19	12	17	15	13	20	12	14	18
X_2:	35	38	34	37	40	42	33	36	38	30	40	35	36	39

Would it be reasonable to use ridge regression for this example? If so, construct the ridge trace, using appropriate software, and also compute the value of the HKB value of k. What do you suggest for the choice of k?

12.9. Would the combination of robust regression and ridge regression be necessary for the data in Exercise 12.8?

12.10. Critique the following statement: The use of ridge regression can also be motivated for orthogonal data, since the vector $\hat{\beta}$ is too long even when the data are orthogonal.

CHAPTER 13

Nonlinear Regression

In this chapter we provide an introduction to nonlinear regression, with two examples (one with real data) used to illustrate basic concepts. Recent developments are also discussed, and readers are referred to the literature for topics not covered in this chapter.

13.1 INTRODUCTION

In Chapter 6 a distinction was made between nonlinear models that are transformably linear and those that are not transformably linear. Recall that the former can be converted into a linear model by an appropriate transformation, whereas no such transformation can be found for the latter.

The model $Y = \theta_0 X^{\theta_1} \epsilon$ can be converted into the simple linear regression model $Y' = \beta_0 + \beta_1 X' + \epsilon'$ with $Y' = \ln(Y)$, $X' = \ln(X)$, $\beta_0 = \ln(\theta_0)$, $\beta_1 = \theta_1$, and $\epsilon' = \ln(\epsilon)$. If, however, the model is $Y = \theta_0 X^{\theta_1} + \epsilon$, then a linearizing transformation cannot be found. Consequently, this model is an (untransformable) *nonlinear regression model*. Such models are nonlinear in the parameters, and nonlinear estimation methods must be used to estimate the parameters.

Nonlinear regression is a subject to which an entire book could be devoted, and books on nonlinear regression include Bates and Watts (1988), Seber and Wild (1989), and Ratkowsky (1983, 1990).

13.2 LINEAR VERSUS NONLINEAR REGRESSION

There are certain complexities inherent in nonlinear regression that are not part of linear regression. In particular, with the latter an experimenter uses linear terms as a starting point, and, for a selected number of regressors, may later elect to use some nonlinear terms. What stays constant throughout the model-building process, however, is the fact that the model is linear in the parameters.

With nonlinear regression the experimenter may not have any idea as to the

416

true model but simply believes that the model is nonlinear. When there is only a single independent variable, a scatter plot can be constructed, and a plausible model determined by comparing the plot with graphs of common nonlinear models. The graphs given by Ratkowsky (1990) can be very helpful in this regard. Guidance on selecting a model from the appearance of the scatter plot can also be found in Bates and Watts (1988, p. 69).

Before looking at some simple illustrative examples, the notation for nonlinear regression will be considered. The notation differs from the linear regression notation, and the nonlinear notation also varies across different sources. We will let the general nonlinear regression model be represented as

$$Y_i = f(\mathbf{x_i}, \boldsymbol{\theta}) + \epsilon_i \qquad (13.1)$$

where f is a nonlinear function, $\mathbf{x_i}$ is a vector that contains the ith observation on each of the independent variables, and ϵ_i is the error term that has the same properties as in linear regression [i.e., $\epsilon_i \sim \text{NID}(0, \sigma_\epsilon^2)$.]

13.3 SIMPLE NONLINEAR EXAMPLE

The other differences in notation (and in methodology) will become apparent as we progress through the following example. Consider the graph in Figure 13.1. The obvious nonlinearity in the graph suggests at least three possible options: (1) fit a linear regression model with at least one nonlinear term, (2) transform Y

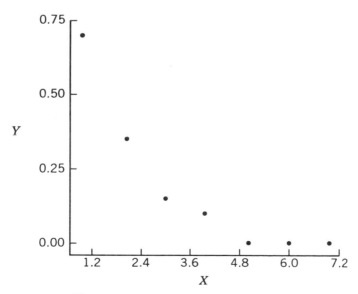

Figure 13.1 Scatter plot of Table 13.1 data.

and fit a simple linear regression model, or (3) search for a nonlinear regression model that fits the data.

Since nonlinear regression entails additional complexities, it would be preferable to first consider options 1 and 2. For the first option we might consider using a Box–Tidwell transformation approach that would search for a term of the form X^α that would account for the nonlinearity. For option 2, a Box–Cox transformation could be used. (The latter was covered in Section 6.4, and the former was discussed in Section 6.5.)

If we have only a graph to look at and no clue regarding an appropriate model, we might inadvertently fit a linear model when we should have fit a nonlinear model. In fact, this must happen very frequently, since it could be successfully argued that most true (unknown) models are, in fact, nonlinear. As George Box has stated: "All models are wrong, but some are useful." So we simply need to find useful models, either linear or nonlinear.

In this instance the model that generated the data that are given in Table 13.1 (and which were graphed in Figure 13.1), was a nonlinear model, so we will thus use the "correct" approach and search for that model. From matching Figure 13.1 with one of the curves in Ratkowsky (1990) or from knowledge of simple mathematical functions, we quickly conclude that an *exponential decay model* should provide a good fit. Thus, we are quickly led to consider a model such as

$$f(x, \boldsymbol{\theta}) = \theta_1 e^{-\theta_2 X} \tag{13.2}$$

If the error term were multiplicative rather than additive (i.e., $Y = \theta_1 e^{-\theta_2 X} \epsilon$), we could then fit the simple linear regression model

$$\log(Y) = \log(\theta_1) - \theta_2 X + \log(\epsilon) \tag{13.3}$$

In practice, we will generally not know the form of the error term, but an additive error term undoubtedly is more common than a multiplicative one.

Table 13.1   Hypothetical Data Set for Illustrating Simple Nonlinear Regression

Y	X
0.700	1
0.325	2
0.168	3
0.103	4
0.020	5
0.018	6
0.020	7

Nevertheless, we can fit the model given by Eq. (13.3) to provide "starting values" for θ_1 and θ_2. Nonlinear regression algorithms involve iteration in arriving at parameter estimates, so starting values are required. We will thus obtain a "pseudo prediction equation" of the form $\hat{Y}' = \hat{\theta}_1' + \hat{\theta}_2(-X)$ with $Y' = \log(Y)$ and $\theta_1' = \log(\theta)$. Using linear least squares, we obtain the prediction equation $\hat{Y}' = 0.165 + 0.661X$. Thus, $\hat{\theta}_2 = 0.661$ and $\hat{\theta}_1 = \exp(\hat{\theta}_1') = 1.179$.

How good are these starting values if the error term is additive rather than multiplicative? If $f(x, \boldsymbol{\theta}) = \theta_1 e^{-\theta_2 X}$ is the correct expectation function, and if the model fits the data very well, so that the model errors are quite small, the starting values should be quite good, since there will then be very little difference between $\ln(\theta_1 e^{-\theta_2 X})$ and $\ln(\theta_1 e^{-\theta_2 X} + \epsilon)$.

In fact, the fit for the linear model *is* quite good as $R^2 = .922$ for the model given by Eq. (13.3), and on the original scale $R^2_{\text{raw}} = .991$. Note that R^2_{raw} is here obtained as $R^2_{\text{raw}} = 1 - \sum(Y - \hat{Y}_{\text{raw}})^2 / \sum(Y - \overline{Y})^2$, in accordance with the recommendation of Kvålseth (1985) regarding the form of R^2 to use in nonlinear regression. Nevertheless, there are instances in which the computed value will be negative, so that the reported value will be zero, as in an application with real data described by Stromberg (1993). Consequently, there will be situations where it will be preferable to define R^2 for nonlinear regression as $R^2 = r^2_{Y\hat{Y}}$.

13.3.1 Iterative Estimation

As in linear regression, our parameter estimates will be those that minimize the residual sum of squares. That is, for our one-regressor model we wish to minimize

$$G(\boldsymbol{\theta}) = \sum_{i=1}^{n} [y_i - f(x_i, \boldsymbol{\theta})]^2$$

This minimization is performed iteratively, and one technique that can be used to accomplish this is the *Gauss–Newton method*. This entails using a Taylor series expansion about the vector of starting values, which we will here denote as $\boldsymbol{\theta}^0$, and using only the first term in the expansion.

Specifically, for our two-parameter example we have

$$f(x_i, \boldsymbol{\theta}) \approx f(x_i, \boldsymbol{\theta}^0) + d_{1i}(\theta_1 - \theta_1^0) + d_{2i}(\theta_2 - \theta_2^0)$$

with

$$d_{1i} = \left. \frac{\partial f(x_i, \boldsymbol{\theta})}{\partial \theta_1} \right|_{\boldsymbol{\theta} = \boldsymbol{\theta}^0}$$

and

$$d_{2i} = \left. \frac{\partial f(x_i, \boldsymbol{\theta})}{\partial \theta_2} \right|_{\boldsymbol{\theta} = \boldsymbol{\theta}^0}$$

Here $d_{1i} = e^{-\theta_2^0 X_i} = e^{-0.661 X_i}$ and $d_{2i} = \theta_1^0 X_i e^{-\theta_2 X_i} = 1.179 X_i e^{-0.661 X_i}$. In matrix notation we have

$$\boldsymbol{\eta}(\boldsymbol{\theta}) \approx \boldsymbol{\eta}(\boldsymbol{\theta}^0) + \mathbf{V}^0(\boldsymbol{\theta} - \boldsymbol{\theta}^0)$$

where $\boldsymbol{\eta}(\boldsymbol{\theta}) = f(\mathbf{x}, \boldsymbol{\theta})$, $\boldsymbol{\eta}(\boldsymbol{\theta}^0) = f(\mathbf{x}, \boldsymbol{\theta}^0)$, $\mathbf{V}^0$ is the $n \times 2$ matrix of partial derivatives evaluated at $\boldsymbol{\theta} = \boldsymbol{\theta}^0$, and $\mathbf{x}' = (x_1, x_2, \ldots, x_n)$.

One way to obtain the first estimates beyond the starting values is to compute $\mathbf{b}_0 = [(\mathbf{V}^0)'\mathbf{V}^0]^{-1}(\mathbf{V}^0)'[\mathbf{Y} - \boldsymbol{\eta}(\boldsymbol{\theta}^0)]$ and then solve for the new estimates $\boldsymbol{\theta}^1$ as $\mathbf{b}_0 + \boldsymbol{\theta}^0$. This process is then repeated with $\boldsymbol{\theta}^0$ replaced by $\boldsymbol{\theta}^1$ (and $\mathbf{V}^0$ by $\mathbf{V}^1$), and this produces a new set of estimates. This iterative procedure continues until convergence is achieved.

A stopping rule that signals convergence must therefore be selected. Bates and Watts (1988, p. 49) review various proposed stopping rules but point out that these rules do not guarantee convergence to the minimum value of the residual sum of squares. Rather, the use of the stopping rules can simply indicate a relative lack of progress in reducing the residual sum of squares. To overcome this deficiency, Bates and Watts (1981) proposed a *relative offset convergence criterion*, which is explained in the next section.

For the present example we obtain $\hat{\theta}_1 = 1.43$ and $\hat{\theta}_2 = 0.72$ after four iterations, at which point the criterion is satisfied. This produces $R^2 = .997$, so the fit is excellent. Such a large R^2 value might seem unrealistic, but a real data set is analyzed later in this chapter that has $R^2 = .992$.

The data were generated using the model $Y = \theta_1 e^{-\theta_2 X} + \epsilon$, with $\theta_1 = 1.5$ and $\theta_2 = 0.75$, so we can see that the parameter estimates do not differ greatly from the parameter values. In fact, when we use $\theta_1 = 1.5$ and $\theta_2 = 0.75$ as the starting values, the Gauss–Newton algorithm with the relative offset convergence criterion converges to the same parameter estimates as before. Although this might seem counterintuitive, we can also note that the sum of the squared residuals is 0.0014 using the parameter *values* and 0.0012 using the parameter *estimates*. Thus, the solution is indeed "right" in terms of providing a good fit. In general, we should, of course, not expect parameter estimates to equal parameter values, and this same type of apparent "least-squares oddity" (relative to the sum of squared residuals) could also be shown to occur in linear regression.

13.4 RELATIVE OFFSET CONVERGENCE CRITERION

As stated in Section 13.3.1, the relative offset convergence criterion of Bates and Watts (1981) was developed to overcome the potentially poor performance of existing stopping rules. The criterion is based on the use of a *QR decomposition* of the matrix **V**. That is, $\mathbf{V} = \mathbf{QR}$, where $\mathbf{Q} = [\mathbf{Q}_1|\mathbf{Q}_2]$ is an orthogonal matrix (as defined in Section 4.1.1), and

$$\mathbf{R} = \begin{bmatrix} \mathbf{R}_1 \\ \mathbf{0} \end{bmatrix}$$

where $\mathbf{R}_1$ is an upper triangular matrix. [The latter is a matrix that has elements below the main diagonal (from upper left to lower right) that are all zero.] It follows that $\mathbf{V} = \mathbf{Q}_1 \mathbf{R}_1$. We will define $\mathbf{w}_1$ and $\mathbf{w}_2$ as $\mathbf{w}_1 = \mathbf{Q}_1'\mathbf{Y}$ and $\mathbf{w}_2 = \mathbf{Q}_2'\mathbf{Y}$. Here **V** is used to denote the matrix of partial derivatives at some arbitrary stage.

The words "relative offset" refer to the extent to which the residual vector is *not* orthogonal to the estimated expectation function. Recall that in linear regression the residual vector is orthogonal to $\hat{\mathbf{Y}}$; that is, $e'\hat{\mathbf{Y}} = 0$. Similarly, $\mathbf{e}'\mathbf{X} = \mathbf{0}$, so $\mathbf{e}'\mathbf{X}\boldsymbol{\beta} = \mathbf{0}$, with $\mathbf{X}\boldsymbol{\beta}$ denoting the expectation function. But $\mathbf{X}\boldsymbol{\beta}$ is, of course, unknown and is estimated by $\mathbf{X}\hat{\boldsymbol{\beta}}$.

Similarly, in nonlinear regression the residual vector $\mathbf{e} = \mathbf{Y} - \hat{\boldsymbol{\eta}}$ should (ideally) be orthogonal to the set of estimated expectation function values, $f(\mathbf{x}, \hat{\boldsymbol{\theta}})$. (Here $\hat{\boldsymbol{\eta}}$ denotes the vector of fitted values.) Unlike linear regression, however, *exact* orthogonality cannot be realistically expected in nonlinear regression, since the estimation process is iterative rather than consisting of only a single step.

Since $\mathbf{V} = \mathbf{Q}_1 \mathbf{R}_1$ from the *QR* decomposition of **V**, and **V** corresponds to **X** in linear regression, it follows that $\mathbf{Q}_1$ should be (approximately) orthogonal to the residual vector $\mathbf{Y} - \hat{\boldsymbol{\eta}}$. Since the magnitude of the residuals at convergence will obviously depend on the data, what is needed is a unit-free measure of the extent to which the residual vector is almost orthogonal to $\mathbf{Q}_1$. With the Bates and Watts (1981) criterion one could compute

$$O = \frac{\|\mathbf{Q}_1(\mathbf{y} - \hat{\boldsymbol{\eta}}(\boldsymbol{\theta}^i))\|}{\hat{\sigma}\sqrt{p}} \tag{13.4}$$

where $\|\cdot\|$ denotes the norm of the vector that results from the indicated multiplication, with the *norm* of a vector defined as the square root of the sum of the squares of the vector elements. The vector of predicted responses at the ith iteration is denoted by $\hat{\boldsymbol{\eta}}(\boldsymbol{\theta}^i)$, p is the number of model parameters, and O defined in this way is thus a measure of the nonorthogonality of the residual vector standardized by a multiple of the estimate of σ.

Notice that the use of Eq. (13.4) requires that the residuals be computed. Naturally we would like to avoid this, and as pointed out by Bates and Watts (1988, p. 288), an equivalent expression for O is

$$O = \frac{\|\mathbf{w_1}\|}{\|\mathbf{w_2}\|} \left(\sqrt{\frac{n-p}{p}} \right)$$

with $\mathbf{w_1}$ and $\mathbf{w_2}$ directly obtainable from applying a QR decomposition to the matrix $\mathbf{V}$ augmented by the vector $\mathbf{y} - \hat{\boldsymbol{\eta}}(\boldsymbol{\theta}^i)$. See Bates and Watts (1988, p. 49) for further details regarding the relative offset criterion.

13.5 ADEQUACY OF THE ESTIMATION APPROACH

In thinking about the Gauss–Newton procedure, we realize that we are using linear regression in estimating the parameters in a nonlinear regression model. Geometrically, we are approximating the *expectation surface*, determined by the set of $f(\mathbf{x_i}, \boldsymbol{\theta})$, with a linear approximation, and the adequacy of the linear approximation will depend on the curvature in the expectation surface. Therefore, it is logical to use some measure of curvature and to determine a threshold value for that measure for identifying model–data set combinations for which the linear approximation will be inadequate.

Bates and Watts (1980) gave a *relative curvature array* approach for making this determination. That approach is somewhat involved, so the technical details are given and illustrated in the chapter appendix. The general idea is to separate the intrinsic (nonremovable) nonlinearity from the nonlinearity caused by the model parametrization. If the intrinsic nonlinearity is too large, then a linear approximation, as was used for the example in Section 13.3, should not be employed. Instead, a quadratic approximation could be used. See Hamilton et al. (1982), and Bates and Watts (1988, pp. 259–261) for details. It should be noted, however, that a quadratic approximation is much more involved than a linear approximation. In particular, a quadratic approximation involves the relative curvature array, which can be somewhat burdensome to compute when a nonlinear regression model has more than a few parameters.

It is also important to assess the nonlinearity caused by the parametrization, as such nonlinearity can undermine inferences. Bates and Watts (1988, p. 256) refer to this as *parameter effects curvature* to distinguish it from the intrinsic nonlinearity (curvature). Fortunately, parameter effects curvature can generally be removed by model reparametrization.

Bates and Watts (1988, p. 256) report a study of 67 data set–model combinations resulting from 37 distinct data sets obtained from various sources used in combination with 19 models. They conclude that 93% of the 67 combinations have acceptable intrinsic nonlinearity but found that only 10% of these had what would probably be considered acceptable parameter effects curvature.

For the example in Section 13.3 it is shown in the chapter appendix that the intrinsic curvature is quite small, and the parameter effects curvature is also acceptably small. Each of these measures is a function of both the data and the model, so for this example we might expect both of the measures to be small, since R^2 was so close to 1. But the value of R^2 has not often been recommended as a general indicator, and the question naturally arises as to what we could say if R^2 were only moderately large. (See Section 13.7 for further discussion of R^2.) We should also remember that testing the two types of nonlinearity is a test on the method of estimation rather than a test on the adequacy of the model.

13.6 COMPUTATIONAL CONSIDERATIONS

The Gauss–Newton method will often require modification. Bates and Watts (1988, p. 41) give an example for which the solution for $\mathbf{b}^0$ produces a new estimator $\mathbf{\theta}^1 = \mathbf{\theta}^0 + \mathbf{b}^0$ such that the residual sum of squares *increases* rather than decreases. The possibility of this happening was noted by Box (1960) and Hartley (1961), who recommended using $\mathbf{\theta}^1 = \mathbf{\theta}^0 + \lambda \mathbf{b}^0$, where λ would be chosen to ensure a decrease in the residual sum of squares.

A variation of this approach, in which λ is set equal to 2 if a reduction in the residual sum of squares has been achieved, and set equal to 0.5 if an increase has resulted, is discussed by Draper and Smith (1981, p. 464), although the latter cite an example in which the modified approach failed. These and other modifications are covered in detail by Seber and Wild (1989).

Just as multicollinearity is a problem in linear regression, it can also cause convergence problems in nonlinear regression. Specifically, multicollinearity among the columns in $\mathbf{V}$ can cause the Gauss–Newton method to perform erratically, thus necessitating the use of a different approach. One such approach is "Marquardt's compromise," which has also been referred to as the *Levenberg–Marquardt compromise* (Bates and Watts, 1988, p. 80) since Marquardt's (1963) work was based on the work of Levenberg (1944).

Since the objective is to overcome multicollinearity, it is not surprising that with Marquardt's compromise the Gauss–Newton increment is solved by using a generalized ridge regression approach (see Section 12.5). That is, $\mathbf{\delta}(k) = (\mathbf{V}'\mathbf{V} + k\mathbf{D})^{-1}\mathbf{V}'(\mathbf{y} - \mathbf{\eta})$, with $\mathbf{D}$ a diagonal matrix whose elements are the diagonal elements of $\mathbf{V}'\mathbf{V}$. Thus, $\mathbf{\delta}_0(k)$ would be defined analogous to $\mathbf{b}_0$, and subsequent steps would be the same as with the Gauss–Newton method. As in generalized ridge regression, the determination of k may not be easily made, however.

The word *compromise* represents the fact that the method is a compromise between the Gauss–Newton method, which would result if $k = 0$, and a "steepest descent" approach in which k is infinite. The latter is used extensively in response surface work but, as indicated by Draper and Smith (1981, p. 470), it may not work well once the neighborhood of convergence is reached. (Hereinafter we shall use $\hat{\mathbf{\theta}}$ to denote the fully iterated estimator of $\mathbf{\theta}$, and $\hat{\mathbf{V}}$ shall denote the corresponding derivative matrix.)

Obviously, one way to attempt to avoid multicollinearity, and thus the need to use Marquardt's compromise, is to use either an experimental design or to avoid overparametrization. Experimental design for nonlinear regression is discussed briefly in Section 14.9. Unfortunately, we cannot always construct designs for nonlinear regression that have certain desirable properties (e.g., orthogonality) that we seek. [See Bates and Watts (1988, p. 119).]

Since the columns of $\hat{\mathbf{V}}$ are functions of both estimated parameter values and regressor values, rather than just the latter, it is desirable to theoretically justify apparently unnecessary parameters before allowing them to cause multicollinearity. Recall that multicollinearity undermines most inferences in linear regression, and we would expect the same thing to happen in nonlinear regression. This is discussed further in Section 13.7.3.

13.7 DETERMINING MODEL ADEQUACY

If the linear approximation approach (or perhaps the quadratic approximation approach) can be deemed to be adequate, model adequacy can then be assessed. That is, the method used for estimating the model parameters must be adequate before we can determine if the model itself is adequate.

Again the issue of whether R^2 should be used must be addressed. It is not mentioned at all as a possible measure of model adequacy by Bates and Watts (1988) or Seber and Wild (1989), nor is it mentioned in the chapter on nonlinear regression in Draper and Smith (1981). Neter et al. (1989) recommend tht it should *not* be used in nonlinear regression because the regression and residual sum of squares do not necessarily add to the total sum of squares. Kvålseth (1985) indicates that R^2 can be used in nonlinear regression, however, but stresses that it must be computed as it was in Section 13.3.

Why is R^2 not mentioned more prominently in the nonlinear regression literature? First, in linear regression the number of parameters is, for an intercept model, one more than the number of regressors. In nonlinear regression, however, the number of parameters is not related to the number of regressors. This raises several questions, including how meaningful R^2 is in the comparison of models that have the same number of regressors but a different number of parameters, or the same number of parameters but a different number of regressors. Nevertheless, the general usefulness of some global measure of model adequacy would seem to override some of the shortcomings of the use of R^2 in nonlinear regression.

13.7.1 Lack-of-Fit Test

A lack-of-fit test can be used in nonlinear regression similar to the way that the test is used in linear regression. In nonlinear regression, however, it is not possible to construct an exact F-test of size α. Rather, we have to settle for an "asymptotic size α" test, which means that we need a reasonably large sample. Remember that the estimation approach utilizes a linear approximation, so distribution theory cannot be applied exactly. Hence no test or inference in nonlinear regression will have exact properties. As with the other inference procedures, the intrinsic and parameter effects curvature should be minimal for the test results to be meaningful.

The test consists of computing

$$F = \frac{n - m}{m - p} \frac{\|\overline{\mathbf{Y}} - f(\mathbf{x}, \hat{\boldsymbol{\theta}})\|^2}{\|\mathbf{Y} - \overline{\mathbf{Y}}\|^2}$$

where p is the number of parameters in the nonlinear model, n is the sample size, and m is the number of distinct combinations of regressor values (or the number of distinct values of the regressor if there is only one regressor). The vector $\overline{\mathbf{Y}}$ contains n_i repeats of each $\overline{Y}_i$, $i = 1, 2, \ldots, m$, where n_i is the number of repeats of each combination of replicated values, and $\overline{Y}_i$ is the average of the y-values for the ith combination. Similarly, $f(\mathbf{x}, \hat{\boldsymbol{\theta}})$ contains n_i repeats of each of the predicted values.

The value of F is compared against $F_{\alpha, n-m, m-p}$, recognizing that the true α-level for the test is unknown. An alternative F-test that can be used, regardless of whether the regressor values are replicated or not, is given by Neill (1988), but this is also an asymptotic α-level test.

13.7.2 Residual Plots

Standardized residuals in nonlinear regression are computed analogous to the way that they are computed in linear regression. Specifically, we compute the standardized residuals r_i as $r_i = e_i/\hat{\sigma}\sqrt{1 - \hat{v}_{ii}}$, where the $\hat{v}_{ii}$ are the diagonal elements of $\hat{\mathbf{V}}$.

As in linear regression, residual plots can be used to check the model assumptions and also to see if the model should be modified. Unlike linear regression, however, in nonlinear regression we must determine if the extent of intrinsic nonlinearity precludes the use of ordinary residuals. Cook and Tsai (1985) show that the ordinary residuals can produce misleading results when there is (considerable) intrinsic curvature and propose the use of a different type of residual [see also Seber and Wild (1989, p. 174)].

If the intrinsic curvature is small, the residual plots that can be used in nonlinear regression are the same as those that are used in linear regression. For example, the residuals would be plotted against $\hat{Y}$ and against the regressor(s) to see if the constant error variance assumption appears to be met. If so, a normal probability plot of the residuals should then be constructed to check the assumption that the error term has a normal distribution. One might question the need for meeting the normality assumption, since the inferential procedures in nonlinear regression will not be exact anyway when there is nonzero intrinsic nonlinearity and parameter effects curvature, but the asymptotic properties of the inferential procedures do depend on the normality assumption.

If there is evidence of nonnormality, the "transform both sides" (TBS) approach of Carroll and Ruppert (1984) is one possible remedy, or robust nonlinear regression might be used. Unfortunately, the latter is not well developed although some work has been done by Stromberg and Ruppert (1989), and Stromberg (1992, 1993).

If the constant error variance assumption appears not to be met, then a normal probability plot of the residuals should not be constructed until appropriate corrective action is taken, since a normal plot is for random variables that have the same variance (as was also stressed in Section 2.1.2 for linear regression).

Residuals should also be plotted against the regressors to see if the model should be modified, but a model modification might entail adding a new parameter or two rather than simply adding a nonlinear term in a regressor for which the residual plot exhibits nonrandomness, as would be done in linear regression. In general, appropriate model alteration in nonlinear regression is apt to require the experimenter to consider the physical setting in conjunction with the existing model to a greater extent than is the case in linear regression.

Partial residual plots for nonlinear regression have not been discussed in the literature, and with some reflection it becomes apparent why it is not possible to develop a nonlinear analog of partial residual plots for linear regression. Recall from Section 5.1.1 that a partial residual plot for X_k is a plot of $e + \hat{\beta}_k X_k$ against X_k. A plot of $e + \hat{\theta}_k X_k$ in nonlinear regression would be meaningless in general, however, because it is unlikely that there would be a relationship between θ_k and X_k. Consequently, we must rely upon an ordinary residual plot in determining the need for a nonlinear term in X_k.

Similarly, there is no direct nonlinear analog of an added variable plot for the same reason. Nevertheless, Cook (1987) does apply the *general idea* of an added variable plot to nonlinear regression. Specifically, a *parameter plot* was proposed in which the residuals from a nonlinear model are plotted against the residuals obtained from regressing one column of $\hat{\mathbf{V}}$ against the other columns. Such a plot is called a parameter plot since the columns of $\hat{\mathbf{V}}$ correspond to parameters rather than to regressors. As pointed out by Cook (1987), the usefulness of the plot will depend on the adequacy of the linear approximation. That is, this particular type of parameter plot can fail when there is considerable intrinsic nonlinearity. If the intrinsic nonlinearity is small, however, we can expect the plot to provide essentially the same information that is given by an added variable plot in linear regression: Outliers could be spotlighted, and the plot would also indicate how well determined is the parameter estimate. Cook (1987) also gave a modified parameter plot for use when the linear approximation is inadequate.

13.7.3 Multicollinearity Diagnostics

Residual plots might be helpful in determining if a nonlinear model needs to be enlarged, but what if a model is already too large? Bates and Watts (1988, p. 90) suggest that the parameter (approximate) correlation matrix be checked to see if there are any large correlations between parameter estimates—an indication that the model may be overparametrized. They suggest that correlations in excess of 0.99 in absolute value should be investigated. Corrective action might consist of simplifying the model or transforming the regressors and parameters. In particular, they illustrate how multicollinearity can be reduced by centering and scaling the (regressor) data.

Although examination of pairwise correlations is one way to detect multicollinearity, provided that multicollinearity is due to one or more high pairwise correlations, a nonlinear adaptation of the methods discussed in Sections 4.3.2 and 4.3.3 can also be useful. For example, Magel and Hertsgaard (1987) conclude that harmful multicollinearity exists if the condition number (see Section 3.1.1) of $\hat{\mathbf{V}}'\hat{\mathbf{V}}$ is at least 30, and two or more variance decomposition proportions in excess of 0.50 will indicate which parameter estimates are affected by the multicollinearity. This recommendation was based on a somewhat limited study using specific data matrices and specific constructed (near) dependencies. It is, however, the same conclusion that was reached by Belsley et al. (1980, p. 112) for linear regression.

13.7.4 Influence and Unusual Data Diagnostics

We should search for the possible existence of influential data and anomalous data points in nonlinear regression just as we do in linear regression. The former presents some complications not found with the latter, however.

13.7.4.1 Leverage

Leverage values can be computed as easily in nonlinear regression as they can be computed in linear regression. In nonlinear regression the "hat matrix" [called the *tangent plane hat matrix* by St. Laurent and Cook (1992)] is given by $\hat{\mathbf{H}} = \hat{\mathbf{V}}(\hat{\mathbf{V}}'\hat{\mathbf{V}})^{-1}\hat{\mathbf{V}}'$. [See also St. Laurent and Cook (1993).] We might not expect to have any large leverage values for the example in Section 13.3, since the x-values are evenly spaced, and the range is fairly small. The reader is asked to compute the leverages in Exercise 13.1. It can be shown that all of the leverages are below $3p/n$, although one is between $2p/n$ and $3p/n$.

13.7.4.2 Influence

We can encounter problems in nonlinear regression when using the simplified forms of influence statistics. For example, in linear regression the different expressions for Cook's-D (such as the two expressions given in Sections 2.4.2 and 2.4.4) are equivalent, but this is not true in nonlinear regression. In particular, it is not possible to obtain an exact expression for Cook's-D that avoids computing $\hat{\boldsymbol{\theta}}$ after deleting the ith observation.

The exact value of Cook's-D can be obtained from

$$D_i = \frac{(\hat{\boldsymbol{\theta}} - \hat{\boldsymbol{\theta}}_{(i)})\hat{\mathbf{V}}'\hat{\mathbf{V}}(\hat{\boldsymbol{\theta}} - \hat{\boldsymbol{\theta}}_{(i)})}{ps^2} \tag{13.5}$$

where p is the number of parameters. (Cook's-D may also be written in this general form in linear regression, with $\hat{\mathbf{V}}'\hat{\mathbf{V}}$ replaced by $\mathbf{X}'\mathbf{X}$, and $\boldsymbol{\theta}$ replaced by $\boldsymbol{\beta}$.) We would like to avoid having to compute $\hat{\boldsymbol{\theta}}_{(i)}$ for each observation, however. A one-step approximation is used to approximate the $\hat{\boldsymbol{\theta}}_{(i)}$ (see Cook and Weisberg, 1982, p. 182), but this approximation can sometimes be quite poor.

Consider again the data in Table 13.1. If we plot Y against X, we would expect that (x_1, y_1) would be influential because that point greatly strengthens the evidence of nonlinearity in the plot. With only seven data points the most extreme points could be highly influential for a nonlinear model. Such small data sets occur quite frequently in nonlinear regression, however. In the survey of 67 data set–model combinations reported by Bates and Watts (1988), 32 had at most seven observations, and 21 had exactly seven observations. For this data set we obtain $D_1 = 6.96$, which is much greater than the usual threshold value of 1.0 for Cook's-D statistic. Consequently, the first observation is identified as an unusual data point. This results partly from the fact that the first observation would be classified as a leverage point if we used $2p/n$ as the threshold value. The model does fit that data point very well, however.

A one-step approximation to $\hat{\boldsymbol{\theta}}_{(i)}$ is given by

$$\hat{\boldsymbol{\theta}}_{(i)}^{*} = \hat{\boldsymbol{\theta}} - \frac{(\hat{\mathbf{V}}'\hat{\mathbf{V}})^{-1}\hat{\mathbf{v}}e_i}{1 - \hat{v}_{ii}}$$

where $\hat{\mathbf{v}}_i$ is the ith column of $\hat{\mathbf{V}}$, e_i is the ith residual, and $\hat{v}_{ii}$ is the ith diagonal element of $\hat{\mathbf{V}}(\hat{\mathbf{V}}'\hat{\mathbf{V}})^{-1}\hat{\mathbf{V}}$. The use of this approximation allows D_i to be written in the same general form as was given in Section 2.4.4 for linear regression. That is,

$$D_i^1 = \left(\frac{e_i}{\hat{\sigma}\sqrt{1 - \hat{v}_{ii}}} \right)^2 \frac{\hat{v}_{ii}}{1 - \hat{v}_{ii}} \left(\frac{1}{p} \right) \tag{13.6}$$

The question naturally arises as to how good is the one-step approximation. That is, when would we expect $\hat{\boldsymbol{\theta}}_{(i)}^{*}$ to be sufficiently close to $\hat{\boldsymbol{\theta}}_{(i)}$ so that D_i^1 will be a good approximation to D_i. This is essentially an unsolved problem. Cook and Weisberg (1982, p. 188) give an example in which the one-step approximation works well but recognize the need for additional research. The need to identify conditions under which the one-step approximation will work well is also recognized by Atkinson (1985). When n is small, as is often the case with nonlinear data sets, it is not really a computational burden to compute the D_i using Eq. (13.5).

A more fundamental issue is how large values of D_i should be interpreted. In linear regression we do not want data points that are well-removed from the other points to be influential. We should expect to frequently encounter influential data points in nonlinear regression, however, as in small data sets extreme points can be very important in suggesting certain models to consider. Therefore, influence analysis in nonlinear regression should be used in conjunction with graphical analyses. [See St. Laurent and Cook (1992) for a somewhat similar discussion and additional leverage measures.]

Atkinson (1985, p. 231) presents a modification of Cook's statistic for nonlinear regression, analogous to his modification of Cook's statistic for linear regression (see Section 2.4.4). The modified statistic for nonlinear regression is

$$C_i(\boldsymbol{\theta}) = \left\{ \frac{n-p}{p} \cdot \frac{R(\hat{\boldsymbol{\theta}}_{(i)}) - R(\hat{\boldsymbol{\theta}})}{s_{(i)}^2} \right\}^{1/2}$$

where $R(\hat{\boldsymbol{\theta}})$ is the residual sum of squares using all of the observations, $R(\hat{\boldsymbol{\theta}}_{(i)})$ is the residual sum of squares that results from using all but the ith observation, and $s_{(i)}^2$ is similarly defined. Notice that the fully-iterated estimator $\hat{\boldsymbol{\theta}}_{(i)}$ is required to compute $C_i(\boldsymbol{\theta})$, however.

Just as is the case with residuals in linear regression, the question arises as to how well the residuals represent the true model errors. This issue is considered by Cook and Tsai (1985), who suggest the use of a different type of residual (projected residuals). That type of residual may also be inadequate when the intrinsic nonlinearity is large, however,

13.8 INFERENCES

Once the method of approximation (linear or quadratic) has been deemed to be adequate, and the model also seems to be adequate, it is then appropriate to consider desired inferences. We shall illustrate confidence intervals, prediction intervals, and hypothesis tests using the example from Section 13.3.

In looking at these results we must keep in mind, however, that even when a linear approximation is appropriate, any inferences that are made will only be approximate. This follows from the fact that the standard errors of the estimators are only approximate, and $\hat{Y}$ and $\hat{\theta}_i$ will be only approximately normally distributed even when the errors are exactly normally distributed.

The extent of this problem depends only on the degree of intrinsic nonlinearity, which as indicated previously, depends on the model and the regressor values but does not depend on the parametrization. This is why Ratkowsky (1990) has advocated finding and using "close-to-linear" nonlinear regression models. The advantage of using such models is that $\hat{\theta}_i$ will be approximately unbiased, normally distributed, and have minimum variance, even for relatively small sample sizes. [See also the discussion in Ratkowsky (1983, p. 185).]

Inferences can also be undermined by multicollinearity. Magel and Hertsgaard (1987) indicate that multicollinearity can increase the width of confidence intervals. The effects of multicollinearity are also discussed by Seber and Wild (1989, p. 110).

13.8.1 Confidence Intervals

Confidence intervals for the θ_i are obtained using $\hat{\theta}_i \pm ts \sqrt{c_{ii}}$, where c_{ii} is the ith diagonal element of $(\hat{\mathbf{V}}'\hat{\mathbf{V}})^{-1}$, and $t = t_{\alpha/2, n-p}$ for a $(1 - \alpha)\%$ confidence interval. Continuing with the example using the Table 13.1 data, since $c_{11} = 16.97$ and $c_{22} = 3.90$, it follows that an approximate 95% confidence interval for θ_1 is $\hat{\theta}_1 \pm ts \sqrt{c_{11}} = 1.43 \pm 2.571(0.0156)(\sqrt{16.97}) = 1.433 \pm 0.166$, so that the interval is (1.27, 1.60). A 95% confidence interval for θ_2 would be obtained in an analogous manner.

As in linear regression, multicollinearity can increase the width of a confidence interval and thus limit its worth. Thus, a "non-overparameterized" model is important.

A confidence interval for $E(Y|x_0)$ can also be produced in nonlinear regression analogous to the way that it is computed in linear regression. The computation will not be illustrated here because the confidence interval is of limited usefulness, just as it is in linear regression.

13.8.2 Prediction Interval

An approximate prediction interval for Y is produced from

$$\hat{Y} \pm t_{\alpha/2, n-p} s \sqrt{1 + \mathbf{v}_0'(\hat{\mathbf{V}}'\hat{\mathbf{V}})^{-1}\mathbf{v}_0}$$

where $\mathbf{v}_0$ is given by

$$\mathbf{v}_0 = \left. \frac{\partial f(x_0, \boldsymbol{\theta})}{\partial \boldsymbol{\theta}'} \right|_{\boldsymbol{\theta} = \hat{\boldsymbol{\theta}}}$$

Let's assume that we want to obtain a 95% prediction interval for Y when $x = 3$, which was one of the x-values in Table 13.1. The interval is obtained as $0.165 \pm 2.571(0.0156)\sqrt{1 + 0.290} = 0.1647 \pm 0.0456$. Thus, the approximate interval is (0.1190, 0.2103).

13.8.3 Hypothesis Tests

Approximate t-tests could be constructed to test $\theta_i = 0$, $i = 1, 2, \ldots, p$. We know for the current example that each of the two t-values, which are of the form $t = \theta_i/s \sqrt{c_{ii}}$, will be relatively large, because the corresponding confidence intervals did not come close to including zero. It can be shown that $t = 22.26$

for testing $H_0: \theta_1 = 0$, and $t = 23.36$ for testing $H_0: \theta_2 = 0$. Certainly we should expect such large values because we know that the fitted model is the correct model, and the R^2 value is very large.

13.9   AN APPLICATION

The Michaelis–Menten (1913) model has an expectation function given by

$$f(x, \boldsymbol{\theta}) = \frac{\theta_1 X}{\theta_2 + X} \tag{13.7}$$

The model has been used for a variety of applications, including enzymatic reaction, and has been discussed extensively in the literature. For these reasons we will apply this model to an actual data set in illustrating the material presented in this chapter. Before proceeding, we should specify the model to include the error term, but here we encounter a problem because, as reported by Ruppert et al. (1989), there does not appear to be a typical error structure.

Let's first assume that the error term is additive on a transformed scale. Specifically, let

$$Y^{-1} = \frac{1}{\theta_1} + \frac{\theta_2}{\theta_1} \left(\frac{1}{X} \right) + \epsilon \tag{13.8}$$

Notice that Eq. (13.8) is in the general form of a linear regression model, $Y' = \beta_0' + \beta_1' X' + \epsilon$, with $Y' = Y^{-1}$, $\beta_0' = \theta_1^{-1}$, $\beta_1 = \theta_2 \theta_1^{-1}$, and $X' = X^{-1}$. Least squares for simple linear regression would then be applied to this model to obtain estimates of θ_1 and θ_2.

Equation (13.8) implies that the model on the original scale is

$$Y = \frac{\theta_1 X}{X + \theta_2 + \theta_1 X \epsilon} \tag{13.9}$$

Notice that Eq. (13.9) is not in the form of Eq. (13.1) and is thus not a nonlinear regression model. Thus, although the double reciprocal transformation that was given by Lineweaver and Burk (1934) for the Michaelis–Menten model does produce a linear model, the question that must be addressed is whether the location of the error term in Eq. (13.9) makes any sense in a particular situation.

In most applications it would seem that the question would be answered in the negative (or else "indeterminable"). If so, simple regression would then be used to obtain starting values for θ_1 and θ_2 if the model is really $Y = \theta_1 X / (\theta_2 + X) + \epsilon$. [Note: As Ruppert et al. (1989) point out, there are examples where the Lineweaver–Burk transformation produces an excellent fit, although we would logically guess that the unknown raw-form model was not the one given by Eq. (13.9).]

We will *tentatively* assume that this last model with the additive error term is the appropriate model for a data set that we will now analyze. (The data set is labeled as BDRC 2 by Ruppert et al. (1989) and was obtained from the Becton Dickinson Research Center at Research Triangle Park, North Carolina.) We will note at the outset that the Becton Dickinson personnel typically analyze such data from enzyme kinetics experiments using a Michaelis–Menten model. The data are given in Table 13.2.

The postulated model in terms of the subject matter variables is

$$V_0 = \frac{V_{max}S}{K_m(I) + S} \tag{13.10}$$

where V_0 denotes the initial velocity, S is the substrate concentration, and $K_m(I)$ is a parameter that depends on I, the inhibitor concentration. It is believed that the other parameter, V_{max}, is independent of I, and we will proceed on that assumption.

Indicator variables are needed to represent the four inhibitor concentrations. Therefore, rewriting Eq. (13.10) to incorporate the indicator variables and using the notation of Eq. (13.7) produces

$$Y = \frac{\theta_1 X}{(\theta_{2(0)}\delta_0 + \theta_{2(.25)}\delta_{.25} + \theta_{2(1)}\delta_1 + \theta_{2(4)}\delta_4) + X} + \epsilon \tag{13.11}$$

where δ_0, $\delta_{.25}$, δ_1, and δ_4 are indicator variables that assume the value 1 if $I = 0$, 0.25, 1, or 4, respectively, and 0 otherwise, and $\theta_{2(0)}$, $\theta_{2(.25)}$, $\theta_{2(1)}$, and $\theta_{2(4)}$ are the corresponding parameters.

The linearized form is then

Table 13.2   Enzyme Kinetics Data (Response Variable Is Initial Velocity)

	Inhibitor Concentration			
Substrate	0	.25	1	4
25	.0195	.0134	.0092	.0055
50	.0327	.0249	.0159	.0103
100	.0540	.0428	.0298	.0205
200	.0667	.0612	.0464	.0295
400	.0729	.0706	.0582	.0452

$$Y^{-1} = \frac{1}{\theta_1} + \frac{\theta_{2(0)}}{\theta_1}\left(\frac{1}{X}\right)\delta_0 + \cdots + \frac{\theta_{2(4)}}{\theta_1}\left(\frac{1}{X}\right)\delta_4 + \epsilon'$$

$$= \beta_0 + \beta_{1(0)}\left(\frac{\delta_0}{X}\right) + \cdots + \beta_{1(4)}\left(\frac{\delta_4}{X}\right) + \epsilon'$$

Solving for the estimates of θ_1, $\theta_{2(0)}$, $\theta_{2(.25)}$, $\theta_{2(1)}$, and $\theta_{2(4)}$ produces 0.103, 106.49, 163.57, 257.90, and 443.15, respectively. These are the initial estimates that would be used in an iterative estimation approach such as the Gauss–Newton method.

Technically, before proceeding further we should check to make sure that the degree of intrinsic nonlinearity permits a linear approximation. We can use the value of R^2_{raw}, as defined in Section 6.4, as a rough indicator of the adequacy of the linear approximation and also as a measure of the adequacy of the selected model. Specifically, if R^2_{raw} is very close to 1.0, then the model is probably adequate. Similarly, if the linear approximation were not adequate, we would not likely obtain such a large value of R^2_{raw}. Furthermore, Ratkowsky (1983, p. 198) points out that rapid convergence to estimates that are considerably different from the initial estimates almost always indicates that the model behaves very much like a linear model. That is, we would expect the intrinsic nonlinearity to be small.

Here we have $R^2 = .998$ for the linearized model, and $R^2_{\text{raw}} = .981$. Therefore, the linearization works well and the model given by Eq. (13.11) also fits the data very well. The linearization works well in this case because, as can be seen from Figure 13.2, each of the four curves has virtually the same shape and transforming to Y^{-1} and X^{-1} practically straightens each curve. This accounts for the very large R^2 value on the transformed scale, and the similarity of the four curves accounts for the large R^2_{raw} value on the original scale.

Since $R^2_{\text{raw}} = .981$, we might ask why we do not simply use the estimates that resulted from the linear regression. Even though R^2_{raw} is very close to 1, we might be able to improve it slightly by using the Gauss–Newton approach, or some modification of that method. We could also see from a plot of the residuals against the predicted values that there is some evidence of heteroscedasticity, so we would hope that we could eliminate that by proceeding further. We should also remember, as has been emphasized, that if the assumptions on ϵ in Eq. (13.11) are met, then the necessary assumptions on ϵ' will not be met.

The results from the application of the Gauss–Newton algorithm to this data set are shown in Table 13.3. We can see that the initial step produced a slight improvement in the R^2_{raw} value, with four additional iterations required before convergence is declared using the Bates–Watts convergence criterion. We may note that for this example it wasn't necessary to use a modified Gauss–Newton approach, since convergence occurred before a region of the parameter space was entered that would produce an increase in the residual sum of squares.

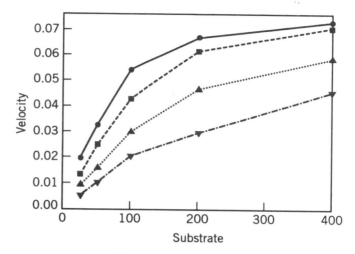

- ● = velocity at Inhibitor concentration 0
- ■ = velocity at Inhibitor concentration .25
- ▲ = velocity at Inhibitor concentration 1
- ▼ = velocity at Inhibitor concentration 4

Figure 13.2   Plot of Table 13.2 data.

Notice that we have virtually reached the final solution after the third iteration, and that this solution differs considerably from the initial estimates. Therefore, following Ratkowsky (1983, p. 198), we would expect the intrinsic nonlinearity to be small.

The plot of the standardized residuals against $\hat{Y}$ is given in Figure 13.3. We should note that the heteroscedasticity that was present in the linearized model is also apparent in the nonlinear model. Ruppert et al. (1989) state: "However, we have never encountered homoscedastic Michaelis–Menten data, and we expect nonlinear least squares to be, in general, less efficient than TBS/PX." (The latter is a modification of the transform-both-sides approach.) The TBS/PX approach

Table 13.3   Gauss–Newton Algorithm Applied to Table 13.2 Data[a]

Iteration Number	R^2_{raw}	Parameter Estimates				
		θ_1	$\theta_{2(0)}$	$\theta_{2(.25)}$	$\theta_{2(1)}$	$\theta_{2(4)}$
1	.990790	0.089	75.890	105.564	202.829	387.295
2	.991766	0.092	82.634	118.132	214.448	402.550
3	.991769	0.092	82.242	117.576	213.677	401.403
4	.991769	0.092	82.278	117.634	213.739	401.486
5	.991769	0.092	82.274	117.628	213.733	401.478

[a]Convergence is declared by the relative offset convergence criterion after the fifth iteration.

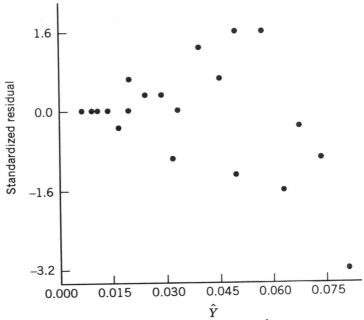

Figure 13.3 Plot of the standardized residuals against $\hat{Y}$ for Table 13.3 data.

requires the use of special software, however, so we will not pursue that here. See Ruppert et al. (1989, p. 641) for details. An important point is that the plot of the standardized residuals against $\hat{Y}$ has shown that more work is needed.

When the residuals are plotted against X, there is some evidence of quadrature, as can be seen from Figure 13.4. We might consider adding a quadratic

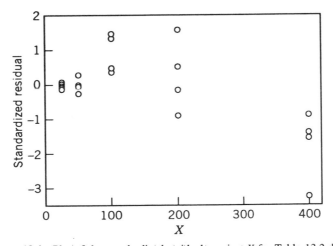

Figure 13.4 Plot of the standardized residuals against X for Table 13.2 data.

term in substrate, but Cressie et al. (1989) report that the p-value for the test for the quadratic term is .144. Consequently, we would not expect a quadratic term to greatly improve the predictive value of the model, and with $R^2 = .992$ there is very little room for improvement. At this point we know that the heteroscedasticity problem must be resolved, and we also need to compute the parameter effects curvature and intrinsic curvature after the heteroscedasticity has been removed.

13.10   ROBUST NONLINEAR REGRESSION

Relatively little research has been done on applying the methods of robust regression to nonlinear regression. The successful adaptation of these methods would seem to depend on a number of factors.

Unlike linear regression with linear terms, extreme values will often be needed to identify which nonlinear model to fit. Therefore, robust nonlinear regression is apt to be more successful when historical evidence suggests that a particular nonlinear model should be used (as frequently occurs).

Breakdown points of robust nonlinear regression estimators are discussed by Stromberg and Ruppert (1989), and Stromberg (1993) presents a two-stage estimation procedure. The latter uses a nonlinear generalization of an MM-estimator [see Rousseeuw and Leroy (1987, p. 152)] in the second stage.

As in linear regression, there is a need to determine an optimal (or at least desirable) first-stage estimator, and there is the related question of when the LMS estimator will perform poorly in nonlinear regression. Other references on robust nonlinear regression include Dutter and Huber (1981) and Lawrence and Arthur (1990).

13.11   SOFTWARE

There is (apparently) no nonlinear regression program that will directly provide output corresponding to everything that was covered in this chapter. The major statistical software packages can be used, with some work, to produce the desired output, and this is discussed in subsequent sections. Additionally, there are software packages such as PCNONLIN and NONLIN84 (described in *The American Statistician*, 1986, p. 52) that are solely for nonlinear modeling. The reader is also referred to Chapter 15 in Seber and Wild (1989) for important considerations in choosing software for nonlinear regression.

13.11.1   SAS Software

The appropriate program in NLIN is SAS/Stat, and the user can choose from five iterative procedures: modified Gauss–Newton, steepest descent, Newton, Marquardt, and DUD. [The latter is a derivative-free approach; see Ralston and

Jennrich (1978).] The number of iterations can be controlled by simply specifying a maximum number or by using the built-in convergence criterion (but see the caution in Section 13.11.1.1). A grid search can also be performed, and the parameter estimates may be bounded. Programming statements may also be used within the procedure, and programming statements must be used to determine the amount of intrinsic nonlinearity and to provide additional diagnostics and graphs. Bates and Watts (1988, p. 296) give pseudocode for the SAS/IML module NLSFIT, but their code gives only the parameter estimates, with these obtained using their convergence criterion.

13.11.1.1 Cautions

Bates and Watts (1988, p. 49) caution against the use of convergence criteria that are based on a small change in the residual sum of squares, including the one used by the NLIN program in SAS. A criterion that is similar to the relative offset convergence criterion is available in PROC MODEL, however, which can be used to fit a nonlinear regression model.

13.11.2 SPSS

SPSS has two procedures for nonlinear regression. The more general of the two is CNLR (constrained nonlinear regression), which can be used both when there are parameter value constraints and when there are no constraints. In addition to providing parameter estimates and (optionally) predicted values, derivatives, and residuals, bootstrap estimates of the standard deviations of the estimators and the correlations between them can also be obtained, in addition to bootstrap confidence intervals. (As discussed in Section 10.1.1, bootstrapping is a resampling technique for estimating standard errors of estimators. See, for example, Efron and Gong (1983) for an introduction to the bootstrap technique.) A loss function may also be specified. CNLR uses a sequential quadratic programming algorithm.

The other procedure, NLR, has fewer capabilities than CNLR. Specifically, NLR does not have a bootstrap subcommand, nor can parameter bounds or a loss function be specified. The two procedures have a different set of convergence criteria from which the user may select, but as in SAS NLIN, the Bates–Watts relative offset convergence criterion is not among them.

13.11.3 BMDP

Program 3R and Program AR are similar programs for nonlinear regression. Neither program requires the user to specify derivatives, and both programs provide the user with various options including residual plots. As with SAS and SPSS, the convergence criterion that is used is a "lack of progress" criterion. There are five other BMDP programs that handle nonlinear models.

13.11.4 Minitab

Minitab does not have a command for nonlinear regression, but a macro for computing the parameter estimates for the Michaelis–Menten model, NLIN-MICH, is described in Appendix A.

SUMMARY

True regression models are generally nonlinear, but linear regression models are often acceptable substitutes. Physical considerations and information obtained from subject matter experts will frequently suggest a specific nonlinear model, however. In the absence of such information, a nonlinear model with a single regressor might be selected using a scatter plot in combination with published curves for various nonlinear models.

Only simple nonlinear regression models have been considered in this chapter, and many of the commonly used models, such as the Michaelis–Menten model, do have only one regressor. Multiple nonlinear regression presents considerable additional complexities that are not found in multiple linear regression. With the latter the model stays linear regardless of the message received from various types of plots, whereas not being able to rely upon scatter plots in multiple nonlinear regression could make it very difficult to determine a tentative model in the absence of subject matter knowledge.

Inferences in nonlinear regression are also only approximate, and are influenced by the amount of intrinsic and parameter effects nonlinearity.

Other common uses of nonlinear regression include growth models, compartmental models, and multiresponse models, especially for kinetic modeling. Useful books on nonlinear regression topics not covered herein include Bates and Watts (1988), Seber and Wild (1989), and Ratkowsky (1983, 1990).

APPENDIX

13.A Relative Curvature Array

We will use the data in Table 13.1 and the corresponding model to illustrate the relative curvature array due to Bates and Watts (1980). The reader is referred to Bates and Watts (1980) and Chapter 7 of Bates and Watts (1988) for additional details, and to Bates et al. (1983) for particulars regarding the computation of the array.

Recall from Section 13.3.1 that the expectation function was differentiated with respect to each of the parameters in the model so as to provide the set of first derivatives for the Gauss–Newton method. To obtain the relative curvature array we additionally need to compute the second derivatives. As discussed by Bates and Watts (1988, p. 233), second derivatives with respect to each of the parameters in a linear model are zero, so it is logical to use the second derivatives of the expectation function as a measure of nonlinearity. The vectors that represent the second derivatives are called *acceleration* vectors, since they give the rate of change of the *velocity* vectors (the first derivatives) with respect to the parameters.

We will adopt the notation used by Bates and Watts (1988). For the example in Section 13.3 we have

$$\dot{v}_1 = e^{-\theta_2 X} \qquad \dot{v}_2 = -\theta_1 X e^{-\theta_2 X}$$

as the first derivatives, and

$$\ddot{v}_{11} = 0 \qquad \ddot{v}_{12} = -X e^{-\theta_2 X} \quad (= \ddot{v}_{21}) \qquad \ddot{v}_{22} = \theta_1 X^2 e^{-\theta_2 X}$$

as the second derivatives. In this example we have one redundant acceleration vector since $\ddot{v}_{12} = \ddot{v}_{21}$. Therefore, only one of these would be used. In general, the nonredundant second-derivative vectors form the matrix $\overset{\lower0.2ex\hbox{$\cdot\cdot$}}{\mathbf{W}}$ that is combined with $\dot{\mathbf{V}}$, which contains the first-derivative vectors. A *QR* decomposition (see Section 13.4) is performed on $(\dot{\mathbf{V}}, \overset{\lower0.2ex\hbox{$\cdot\cdot$}}{\mathbf{W}})$. Doing so produces

$$\mathbf{R}_1 = \begin{bmatrix} 0.56 & -1.04 & 0 & -0.73 & 1.69 \\ 0 & -0.51 & 0 & -0.35 & 2.13 \\ 0 & 0 & 0 & 0 & 0.61 \end{bmatrix}$$

The matrix $\mathbf{R}_1$ is partitioned as

$$\mathbf{R}_1 = \begin{bmatrix} \mathbf{R}_{11} & \mathbf{R}_{12} \\ 0 & \mathbf{R}_{22} \end{bmatrix}$$

where the full $\mathbf{R}$ matrix can be written as

$$\mathbf{R} = \begin{bmatrix} \mathbf{R}_{11} & \mathbf{R}_{12} \\ 0 & \mathbf{R}_{22} \\ 0 & 0 \end{bmatrix}$$

Thus, $\mathbf{R}$ contains two $\mathbf{0}$ submatrices that are not involved in the necessary calculations. Therefore, we need only be concerned with $\mathbf{R}_1$.

The number of rows in $\mathbf{R}_1$ is equal to the number of independent, nonzero vectors in $(\dot{\mathbf{V}}, \overset{\lower0.2ex\hbox{$\cdot\cdot$}}{\mathbf{W}})$. Here that number is three, since $\dot{v}_2 = \theta_1 \ddot{v}_{12}$. Also, since only one of the three second-derivative vectors, $\ddot{v}_{22}$, is independent of the first-derivative vectors, it follows that $\mathbf{R}_{22}$ must be 1×3. Furthermore, since $\mathbf{R}_{11}$ will always be a square matrix of the order given by the number of first-derivative vectors (i.e., the number of parameters), if follows that $\mathbf{R}_{12}$ must be 2×3.

We need use only R_{12} and R_{22} in determining the intrinsic nonlinearity and the (possibly removable) nonlinearity caused by the parametrization.

Since there are two parameters and the space spanned by the combined first- and second-derivative vectors is of dimension 3, it follows that the array will be $3 \times 2 \times 2$. That is, it will consist of three 2×2 matrices, with the first two matrices obtained from the rows of R_{12} and the last from the single row in R_{22}. Therefore,

$$R_{12} = \begin{bmatrix} 0 & -0.73 & 1.69 \\ 0 & -0.35 & 2.13 \end{bmatrix}$$

and

$$R_{22} = [0 \quad 0 \quad 0.61]$$

If follows [from Bates and Watts (1980, 1988, p. 236)] that

$$\ddot{A} = \begin{bmatrix} 0 & -0.73 \\ -0.73 & 1.69 \end{bmatrix} \begin{bmatrix} 0 & -0.35 \\ -0.35 & 2.13 \end{bmatrix} \begin{bmatrix} 0 & 0 \\ 0 & 0.61 \end{bmatrix}$$

The relative curvature array is then produced as $C = (R_{11}^{-1})' \ddot{A} R_{11}^{-1} s \sqrt{p}$ with $s = \hat{\sigma}$. Since SSE = 0.000122, it follows that $s^2 = 0.000122/(7 - 2) = 0.00024$, so $s = 0.01563$. With

$$R_{11}^{-1} = \begin{bmatrix} 1.798 & -3.707 \\ 0 & -1.975 \end{bmatrix}$$

we obtain

$$C = \begin{bmatrix} 0 & 0.057 \\ 0.057 & -0.090 \end{bmatrix} \begin{bmatrix} 0 & 0.028 \\ 0.028 & 0.070 \end{bmatrix} \begin{bmatrix} 0 & 0 \\ 0 & 0.053 \end{bmatrix}$$

Since the space spanned by $(\dot{V}, \ddot{W})$ exceeds the space spanned by $\dot{V}$ by only one dimension, the last face (matrix) in the C array is the only one used in computing the measure of intrinsic nonlinearity. We use the general form of Eq. (7.23) in Bates and Watts (1988, p. 254) in computing both c^I and c^θ. That equation is (slightly modified to prevent notational conflicts)

$$c^2 = \frac{1}{p(p+2)} \sum_{s} \left\{ 2 \sum_{t=1}^{p} \sum_{u=1}^{p} c_{stu}^2 + \left\{ \sum_{t=1}^{p} c_{stt} \right\}^2 \right\} \qquad (A.1)$$

with p denoting the number of parameters and c_{stu} representing the element in the tth row and uth column of the sth matrix. The index s goes from 1 to p for

c^θ and from $p+1$ to at most $p(p+3)/2$ for c^l, with the range dependent on the number of nonredundant acceleration vectors.

As stated earlier in this appendix, there is only one nonredundant second-derivative vector, so only one face in the **C** array is involved in the computation of c^l. There is only one nonzero element in that face, so the computation is simplified considerably. Specifically, we obtain

$$c^2 = \frac{1}{2(2+2)} \{2(0.053)^2 + (0.053)^2\}$$
$$= 0.001$$

so that $c^l = \sqrt{c^2} = 0.032$. (We should note that the **C** array is not unique, so different software can produce different arrays. The values of c^l and c^θ are unique, however.)

The value of $c^l\sqrt{F}$ is then computed, with $F = F(p, n - p; 0.05)$. The degree of intrinsic nonlinearity is considered unacceptable if $c^l\sqrt{F} > 0.3$. Here $F = 5.79$, so $c^l\sqrt{F} = 0.078$. The intrinsic nonlinearity is thus not excessive. Therefore, the use of a linear approximation is justified for this example. (As discussed in Section 13.5, a quadratic approximation could be used if the intrinsic nonlinearity had been unacceptably large.)

The first two faces of the **C** array are used to compute c^θ, and it can be shown using equation (A.1) that $c^\theta = 0.083$. Since $c^\theta\sqrt{F} = 0.1997$ is less than 0.3, the parameter effects nonlinearity is declared acceptable, and no reparametrization is needed.

REFERENCES

Atkinson, A. C. (1985). *Plots, Transformations, and Regression*. New York: Oxford University Press.

Bates, D. M. and D. G. Watts (1980). Relative curvature measures of nonlinearity (with discussion). *Journal of the Royal Statistical Society, Series B*, **42**, 1–25.

Bates, D. M. and D. G. Watts (1981). A relative offset orthogonality convergence criterion for nonlinear least squares. *Technometrics*, **23**, 179–183.

Bates D. M. and D. G. Watts (1988). *Nonlinear Regression and its Applications*. New York: Wiley.

Bates, D. M., D. C. Hamilton, and D. G. Watts (1983). Calculation of intrinsic and parameter–effects curvatures for nonlinear regression models. *Communications in Statistics, Series B*, **12**, 469–477.

Belsley, D. A., E. Kuh, and R. E. Welsch (1980). *Regression Diagnostics: Identifying Influential Data and Sources of Collinearity*. New York: Wiley.

Box, G. E. P. (1960). Fitting empirical data. *Annals of the New York Academy of Sciences*, **86**, 792–816.

Carroll, R. J. and D. Ruppert (1984). Power transformations when fitting theoretical models to data. *Journal of the American Statistical Association*, **79**, 321–328.

Cook, R. D. (1987). Parameter plots in nonlinear regression. *Biometrika*, **74**, 669–677.

Cook, R. D. and C.-L. Tsai (1985). Residuals in nonlinear regression. *Biometrika*, **72**, 23–29.

Cook, R. D. and S. Weisberg (1982). *Residuals and Influence in Regression*. New York: Chapman and Hall.

Draper, N. R. and H. Smith (1981). *Applied Regression Analysis*, 2nd edition. New York: Wiley.

Dutter, R. and P. J. Huber, (1981). Numerical methods for the nonlinear robust regression problem. *Journal of Statistical Computation and Simulation*, **13**, 79–114.

Efron, B. and G. Gong (1983). A leisurely look at the bootstrap, the jackknife, and cross-validation. *The American Statistician*, **37**, 36–48.

Hamilton, D. C., D. G. Watts, and D. M. Bates (1982). Accounting for intrinsic nonlinearity in nonlinear regression parameter inference regions. *Annals of Statistics*, **10**, 386–393.

Hartley, H. O. (1961). The modified Gauss-Newton method for the fitting of nonlinear regression functions by least squares. *Technometrics*, **3**, 269–280.

Kvålseth, T. O. (1985). A cautionary note about R^2. *The American Statistician*, **39**, 279–285.

Lawrence, K. D. and J. L. Arthur (1990). Robust Nonlinear Regression, in *Robust Regression: Analysis and Applications*, K. D. Lawrence and J. L. Arthur, eds. New York: Dekker, 59–86.

Levenberg, K. (1944). A method for the solution of nonlinear problems in least squares. *Quarterly of Applied Mathematics*, **2**, 164–168.

Lineweaver, H. and D. Burk (1934). The determination of enzyme dissociation constants. *Journal of the American Chemical Society*, **56**, 658–666.

Magel, R. C. and D. Hertsgaard (1987). A collinearity diagnostic for nonlinear regression. *Communications in Statistics, Series B*, **16**, 85–97.

Marquardt, D. W. (1963). An algorithm for the estimation of non-linear parameters. *Journal of the Society of Industrial and Applied Mathematics*, **11**, 431–441.

Michaelis, L. and M. L. Menten (1913). Kinetik der Invertinwirkung. *Biochemische Zeitschrift*, **49**, 333.

Neill, J. W. (1988). Testing for lack of fit in nonlinear regression. *Annals of Statistics*, **16**, 733–740.

Neter, J., W. Wasserman, and M. H. Kutner (1989). *Applied Linear Regression Models*, 2nd edition. Homewood, IL: Irwin.

Ralston, M. L. and R. L. Jennrich (1978). Dud, a derivative-free algorithm for nonlinear least squares. *Technometrics*, **20**, 7–14.

Ratkowsky, D. A. (1983). *Nonlinear Regression Modelling: A Unified Approach*. New York: Marcel Dekker.

Ratkowsky, D. A. (1990). *Handbook of Nonlinear Regression Models*. New York: Marcel Dekker.

Rousseeuw, P. J. and A. N. Leroy (1987). *Robust Regression and Outlier Detection*. New York: Wiley.

Ruppert, D., N. Cressie, and R. J. Carroll (1989). A transformation/weighting model for estimating Michaelis–Menten parameters. *Biometrics*, **45**, 637–656.

St. Laurent, R. T. and R. D. Cook (1992). Leverage and superleverage in nonlinear regression. *Journal of the American Statistical Association*, **87**, 985–990.

St. Laurent, R. T. and R. D. Cook (1993). Leverage, local influence and curvature in nonlinear regression. *Biometrika*, **80**, 99–106.

Seber, G. A. F. and C. J. Wild (1989). *Nonlinear Regression*. New York: Wiley.

Stromberg, A. J. (1992). High Breakdown Estimation in Nonlinear Regression, in L_1-*Statistical Analysis and Related Methods*, Y. Dodge, ed. Amsterdam: North-Holland, pp. 103–112.

Stromberg, A. J. (1993). Computation of high breakdown nonlinear regression parameters. *Journal of the American Statistical Association*, **88**, 237–244.

Stromberg, A. J. and D. Ruppert (1989). Breakdown in nonlinear regression. *Journal of the American Statistical Association*, **87**, 991–997.

U.S. Environmental Protection Agency (1978). A compendium of lake and reservoir data collected by the National Eutrophication Survey in eastern, north central and southeastern United States. Working paper No. 475, Corvallis Environmental Research Laboratory, Corvallis, Oregon.

EXERCISES

13.1. Compute the leverage values for the example in Section 13.3.

13.2. Can the general nonlinear regression model given by Eq. (13.1) be linearized? If so, how?

13.3. Assume that the intrinsic curvature for a data set is judged to be excessive. Could the standardized residuals still be used in residual plots?

13.4. Consider Figure 13.3. Notice that the absolute value of one of the standardized residuals exceeds 3. Does this suggest that the corresponding point should be investigated?

13.5. Critique the following statement. "Influence analysis in simple nonlinear regression is quite different from influence analysis in simple linear regression because we should expect to see influential data points in the former simply because we are fitting a curve rather than a straight line."

13.6. What computational problems are involved in influence analysis in nonlinear regression that are not present in linear regression?

13.7. Explain why a convergence criterion based only on the change in the parameter estimates might be insufficient.

13.8. Consider the following data, with X representing time and Y denoting oxygen demand.

Y	3.9	7.1	9.0	10.6	11.6	12.5	13.2	13.6	14.0	14.4	14.5	14.8	14.8	14.8	14.9
X	1	2	3	4	5	6	7	8	9	10	11	12	13	14	15

Based on physical considerations, a scientist believes that the model $Y = \theta_1[1 - \exp(-\theta_2 X)] + \epsilon$ should adequately fit the data. Use 13 and 0.28 as starting values for θ_1 and θ_2, respectively, and use some software program to obtain the parameter estimates. Then compute R_{raw}^2. Finally, plot the standardized residuals against the predicted values. Does the model appear to adequately fit the data?

13.9. Graph the data in Exercise 13.8, and then assume that there is no prior evidence to suggest a particular nonlinear model. How would you proceed to determine which, if any, nonlinear model might be suitable? Since nonlinear models are often approximated by linear models, try to obtain a good-fitting linear model by using a nonlinear term in X. (What nonlinear term is suggested by your graph?) Compare the fit of your linear model with the fit of the nonlinear model in Exercise 13.8 in terms of R^2. What does this suggest (at least for this example) about using a linear model as a substitute for an unknown nonlinear model.

13.10. Construct the $\dot{V}$ and $\ddot{W}$ matrices for the data and model in Exercise 13.8.

13.11. Construct a 95% confidence interval for θ_1 for the data in Exercise 13.8, as well as a 95% prediction interval for Y when $X = 5$.

13.12. The following data were given in U.S. Environmental Protection Agency (1978) and also discussed by Stromberg (1993). The objective was to develop a regression model that relates the mean annual total nitrogen concentration (TN) to the average influent nitrogen concentration (NIN) and the water retention time (TW).

TN	NIN	TW
2.590	5.548	0.137
3.770	4.896	2.499
1.270	1.964	0.419
1.445	3.586	1.699
3.290	3.824	0.605
0.930	3.111	0.677
1.600	3.607	0.159
1.250	3.557	1.699
3.450	2.989	0.340
1.096	18.053	2.899

TN	NIN	TW
1.745	3.773	0.082
1.060	1.253	0.425
0.890	2.094	0.444
2.755	2.726	0.225
1.515	1.758	0.241
4.770	5.011	0.099
2.220	2.455	0.644
0.590	0.913	0.266
0.530	0.890	0.351
1.910	2.468	0.027
4.010	4.168	0.030
1.745	4.810	3.400
1.965	34.319	1.499
2.555	1.531	0.351
0.770	1.481	0.082
0.720	2.239	0.518
1.730	4.204	0.471
2.860	3.463	0.036
0.760	1.727	0.721

Fit the model $TN = NIN/(1 + \alpha T W^{\beta}) + \epsilon$, as suggested by the EPA investigator, then compute R^2, using the computational formula in Section 13.3. Then compute R^2 as $r^2_{Y\hat{Y}}$. Which result is preferable, at least for this problem? Construct dot plots for NIN, TN, and TW. Do the dot plots suggest that there are any outliers? If there are any obvious outliers, delete them and recompute the regression equation. If you deleted any outliers, does your new regression equation appear to adequately fit the data? Could a better fit be obtained with a linear regression equation?

13.13. Compute the leverage values for the data in Exercise 13.12. What outliers are suggested by the leverages?

13.14. Give two examples of nonlinear regression models (different from those that were given in the chapter): one that can be transformed into a linear regression model, and one that cannot be so transformed.

CHAPTER 14

Experimental Designs for Regression

It was stated in Section 1.9 that many, if not most, regression data sets have regressors that are random variables. Nevertheless, it is frequently possible to select the values of the regressors. That is, the values of the regressors would be chosen through the use of an *experimental design*. This would permit the avoidance of multicollinearity, in particular. There are various other objectives that the user of an experimental design would seek to satisfy, and those will be discussed in this chapter.

14.1 OBJECTIVES FOR EXPERIMENTAL DESIGNS

Some seemingly desirable criteria for an experimental design follow:

1. We would want the design points to exert equal influence on the determination of the regression coefficients.
2. We would want some function (such as the sum) of the variances to be minimized.
3. The design should be able to detect the need for nonlinear terms.

[A more inclusive list can be found in Box and Draper (1975), which can also be found in Box and Draper (1986)].

14.2 EQUAL LEVERAGE POINTS

In Sections 14.2.1 and 14.2.2 we discuss methods for selecting the values of the regressors in such a way that the data points will, in terms of the x-coordinates, equally determine the regression coefficients. Of course, we have no control over the y-coordinates, so the (x, y) points will almost certainly not have equal influence. Nevertheless, by seeking equally influential x-coordinates we will at least have a chance of having points that have approximately equal influence. We shall first consider simple linear regression.

14.2.1 Simple Linear Regression

Intuitively we would expect a design to be an *equileverage design* if the design points are equidistant from the center. Clearly, for a single regressor this can happen only if we have just two values of X, and $n/2$ replications of each of the two values, for a sample of size n.

To see that this produces an equileverage design, we need only recall the expression for the leverage values given in Section 2.1.1. Specifically,

$$h_{ii} = \frac{1}{n} + \frac{(X_i - \overline{X})^2}{\sum_{j=1}^{n} (X_j - \overline{X})^2}$$

where the h_{ii} are the diagonal elements of the hat matrix (see Section 3.2.6). Clearly the h_{ii} will be equal when the X_i are equidistant from $\overline{X}$; and excluding the trivial case where all X_i are the same, this will occur only when there are two distinct values of X_i.

If the range of X is scaled to be $(-1, 1)$, and n is even, then the maximum value of h_{ii} is minimized when half of the observations are at $X = -1$, and the other half are at $X = +1$ (Seber, 1977, p. 186). This is of interest because a prediction interval for Y can be written as $\hat{Y} \pm t_{\alpha/2} s \sqrt{1 + h_{ii}}$.

It can also be seen that $\text{Var}(\hat{\beta}_1)$ is minimized with $n/2$ design points at each of the two X_i values, as becomes apparent when we recall that $\text{Var}(\hat{\beta}_1) = \sigma^2 / \sum (X - \overline{X})^2$. It could also be shown that $\text{Var}(\hat{\beta}_0)$ is independent of the choice of the X_i.

There are trade-offs that the user of experimental designs must consider, however (such as for this type of design), since this design would not permit the detection of nonlinearity. Thus, the first two objectives given in Section 14.1 would be met, but not the third objective. Consequently, an equileverage design would be inappropriate in simple linear regression, as it would be preferable to have some design points between the two extremes. If, however, we start from an equileverage design, and then make suitable modifications to facilitate the detection of possible nonlinearity, we will be able to avoid high leverage points. This is discussed further in Sections 14.2.2 and 14.4.

14.2.2 Multiple Linear Regression

We are also faced with the need to compromise when there is more than one regressor. As indicated by Dollinger and Staudte (1990), some of the commonly used experimental designs are *scalar equileverage designs*, with the authors defining a scalar (equileverage) design as one in which $\mathbf{Z}'\mathbf{Z}$ is a scalar multiple of the identity matrix, with $\mathbf{Z}$ being the matrix of centered regressors. These designs are also discussed in Staudte and Sheather (1990, pp. 218–236).

For example, commonly used designs such as the 2^k factorial designs, central composite (star) designs, and simplex designs are all scalar equileverage designs (Dollinger and Staudte, 1990). They would be modified (thus losing the equileverage property) to detect nonlinearity by the addition of center points.

It should be mentioned that we can attempt to detect nonlinearity simply by replicating a few design points so that a lack-of-fit sum of squares could be computed, but such a lack-of-fit test would indicate only that there was a need to modify the model. We need to have design points spread over the surface in order to be able to gain some understanding of the true response surface.

14.2.2.1 *Construction of Equileverage Designs—Two Regressors*

We will illustrate one method of constructing an equileverage design by altering the replicated orthogonal design that was given in Table 4.1. As indicated in previous chapters, an orthogonal design permits the regression equation to be interpreted in ways that are not possible when the data are multicollinear.

An orthogonal design will not necessarily be an equileverage design, however. Conversely, a design can be equileverage without necessarily being orthogonal. Although a scalar design has orthogonal columns, a scalar design is not an equileverage design unless the *rows* of the (centered) design matrix have equal length.

Clearly this requirement is not met by the design obtained by using the columns for X_1 and X_2 from Table 4.1, nor will the requirement be met if the columns are centered and scaled so that the columns are not only orthogonal but also have unit length. In fact, centering and scaling the regressors preserves the leverage values, as we might expect. The leverages for the Table 4.1 data are given in Table 14.1.

Recall the stack loss data set given in Table 6.1. It can be shown that the leverages for the full data set do differ considerably. Remember, though, that four of the data points were found to represent transitional states, and one of the regressors was found to be unimportant. Deleting the four points and removing the unimportant regressor still leaves a point (point 2 in the original data) that has a leverage value equal to 3.55 times p/n. As in previous chapters, p denotes the number of parameters, and $3p/n$ was mentioned in Section 2.4.5 as a possible threshold value for determining high leverage points. As mentioned in Section 6.1, point 2 signals the need for a quadratic term in X_1 and is the only point that does so. If that point were deleted, the leverages would still differ considerably, ranging from .06 to .28.

This shows how problems can be created by high leverage points, so we should seek designs that are equileverage, or at least approximately equileverage. The need to compromise somewhat on equal leverages has been discussed previously, and is discussed further in Section 14.4.

We will show how we may construct an equileverage design for X_1 and X_2. Following Dollinger and Staudte (1990) and Staudte and Sheather (1990), we will define an equileverage design as one for which $h_{ii} = p/n$, $i = 1, 2, \ldots, n$. (It will be observed in Section 14.3.2 that the leverages will all be equal to p/n when a design is *G*-optimal.)

With this restriction we have that the rows of the **X** matrix, $\mathbf{x_i}$, must satisfy

Table 14.1   Leverage Values for Table 4.1 Data

h_{ii}
0.2214
0.0929
0.0929
0.2214
0.1214
0.2214
0.0929
0.1214
0.2214
0.0929
0.0929
0.2214
0.0929
0.2214
0.1214
0.2214
0.2214
0.0929
0.0929
0.1214

$$\mathbf{x}_i'(\mathbf{X}'\mathbf{X})^{-1}\mathbf{x}_i = 3/n \tag{14.1}$$

where $\mathbf{x}_i' = (1, x_{1i}, x_{2i})$, since the left side of Eq. (14.1) is h_{ii}. Rather than work with Eq. (14.1) directly, however, we will work with the centered version of Eq. (14.1) that is given by

$$\mathbf{z}_i'(\mathbf{Z}'\mathbf{Z})^{-1}\mathbf{z}_i = 2/n \tag{14.2}$$

where $\mathbf{z}_i' = (z_{1i}, z_{2i}) = (x_{1i} - \bar{x}_1, x_{2i} - \bar{x}_2)$. Since the design is to be orthogonal, the restriction on z_{1i} is given by

$$\max_i(z_{1i}^2) \le \left(\frac{2}{n}\right) \sum_{i=1}^{n} z_{1i}^2 \tag{14.3}$$

Assume that the z_{1i}, and hence the x_{1i}, are to be equispaced (as in Table 4.1). Since the z_{1i} must sum to zero, this means that the z_{1i} will be given by $-m$, $-m+1, \ldots, m$. Using Eq. (14.3) and the fact that $\sum_{k=1}^{n} k^2 = n(n+1)(2n+1)/6$, we may obtain, as the reader is asked to show in Exercise 14.7, that m cannot exceed 2. Thus we are restricted to a "5-level design" with $n = 2m + 1$ distinct values of x_{1i}.

The values of z_{2i} (and hence those of x_{2i}) would be solved using Eq. (14.2), after a desired value of $\sum z_{2i}^2$ has been specified. Since the values of each replicated set of x_{2i} in Table 4.1 are 13, 14, 14, 17, and 17, it would be convenient to use $\sum z_{2i}^2 = \sum z_{1i}^2$. Then the values of x_{2i} in the orthogonal equileverage (OE) design will have a spread that is similar to the spread of the x_{2i} in Table 4.1 and be close to those values. The OE design is given in Table 14.2, with the values for X_2 given in juxtaposition to the values of X_2 from Table 4.1.

Although the orthogonal design given in Table 4.1 does not produce any large leverage values, the leverages do differ considerably, as they range from .09 to .22. Therefore, since the new combinations of design values shown in Table 14.2 do not differ greatly from the old combinations, we might as well go a step further and use an orthogonal equileverage design, which has leverage values all equal to .15.

Since a set of leverage values is unaffected by nonsingular transformations, we could transfer a given OE design to any desired experimental region simply

Table 14.2 Orthogonal Design of Table 1.4 Converted to an Orthogonal Equileverage Design

X_1	X_2 (orthogonal)	X_2 (orthogonal equileverage)
5	17	15.0000
6	14	16.7320
8	14	16.7320
9	17	15.0000
7	13	13.0000
5	17	15.0000
6	14	13.2679
7	13	17.0000
9	17	15.0000
8	14	13.2679
8	14	16.7320
9	17	15.0000
6	14	16.7320
5	17	15.0000
7	13	13.0000
9	17	15.0000
5	17	15.0000
6	14	13.2679
8	14	13.2679
7	13	17.0000

by using appropriate transformations of the form $X = (Z - s)/t$, for suitably chosen s and t. We shall use such transformations in conjunction with another method of constructing OE designs for two regressors. Specifically, Dollinger and Staudte (1990) and Staudte and Sheather (1990, p. 227) discuss generating the design points as

$$\left\{ \cos\left(\frac{2\pi j}{n} \right), \sin\left(\frac{2\pi j}{n} \right) \right\} \quad j = 0, 1, \ldots, n - 1 \qquad (14.4)$$

The resultant design will be orthogonal because of the relationship between the *sine* and *cosine* functions, as previously discussed in Section 8.2.1, and the rows of the design matrix will have equal lengths, since $\sin^2(\theta) + \cos^2(\theta) = 1$ for any angle θ.

In general, when the squared norm of the two columns is the same, as will be the case when the sine and cosine functions are used in this manner, then the design will be equileverage if the rows of the design have equal lengths. This would be apparent if we multiplied out the left side of Eq. (14.1). [See also Staudte and Sheather (1990, p. 225.)] This sine–cosine approach is probably the easiest way to construct an OE design for two regressors, but it has the obvious shortcoming of not allowing the user to select more than a few design values of X_1 and/or X_2. Nevertheless, the design will be useful if the experimenter is interested only in having the design values within specified ranges.

We will generate an OE design using Eq. (14.4) and then transform it to an appropriate region. Let the set of design values generated using Eq. (14.4) be represented by (Z_1^*, Z_2^*). Values for the original scale regressors X_1 and X_2 could then be produced as $X_1 = (Z_1^* - u)/w$ and $X_2 = (Z_2^* - s)/t$, for suitable choices of u, w, s, and t.

Assume that we wish to have 50–65 as the range of X_1, and 17–27 (the original range) as the range of X_2. Using simple algebra, we then obtain $u = -7.59287$, $w = 0.132198$, $s = -4.38123$, and $t = 0.199147$. This produces the OE design given in Table 14.3, given in juxtaposition to the original 17 design points that were retained in the stack loss example. Notice that the two sets of design points do not differ greatly, except at the first value of X_1, as it was that value (80) that caused the point to be a high leverage point.

One obvious shortcoming of this approach is that neither the design values for X_1 nor those for X_2 can be chosen, nor will the design values be equi-spaced. Another shortcoming is that if the design values must be integers, then obviously more rounding will be required than if we selected the values of one or more regressors. Since rounding will alter the leverages, we would expect to have a greater departure from a true equileverage design with this approach than we would have if we used an approach, such as that illustrated earlier in this section, that permits values for one or more regressors to be specified.

If the set of design points that is produced is unacceptable to an experi-

Table 14.3   Orthogonal Equileverage Design for Stack Loss Example

X_1	X_2	X_1 (OE)[a]	X_2 (OE)[a]
80	27	65.0000	22.0000
62	22	64.4892	23.8139
62	23	64.4892	20.1861
62	24	63.0258	25.3829
62	24	63.0258	18.6171
58	23	60.8073	26.4950
58	18	60.8073	17.5050
58	18	58.1335	27.0000
58	17	58.1335	17.0000
58	18	55.3655	26.8297
58	19	55.3655	17.1703
56	20	52.8770	26.0072
50	18	52.8770	17.9928
50	18	51.0042	24.6434
50	19	51.0042	19.3566
50	19	50.0000	22.9227
50	20	50.0000	21.0773

[a]OE denotes the orthogonal equileverage design values.

menter, one alternative would be to combine this approach with a modification of the method illustrated earlier in this section, which would permit the selection of design values for X_1 or X_2, provided that requirement (14.3) is satisfied. Specifically, desired values of $X_1(X_2)$ would have to be transformed to the unit sphere [i.e., transformed to $Z_1(Z_2)$], and a check would have to be made to make sure that requirement (14.3) is satisfied. If the requirement is met, the corresponding value of $Z_2(Z_1)$ would be solved for so that $Z_1^2 + Z_2^2 = 1$. The desired design points would also have to comprise an orthogonal design. But the sum of squares of the additional Z_1 values must equal the sum of squares of the additional Z_2 values, and this can be a severe constraint on the specification of additional values. Furthermore, even if this constraint is satisfied, the necessary transformation (coding) that transforms Z_1, say, back to the desired value of X_1 must also be applied to the design points that have previously been constructed, thus creating new design points.

Thus, while it is possible to add additional points to an OE design so as to have desired values for $X_1(X_2)$, doing so does pose some major problems. Consequently, if specific design values are desired, it is best to start with a method that will allow such values to be specified.

14.2.2.1.1   Inverse Projection Approach
Considerable space was devoted to a particular robust regression technique (LMS) in Chapter 11, not because it was better than the other approaches that

were presented, but rather because it is the technique that is discussed most frequently in the literature, and is also considered to be the most frequently used technique. Similarly, we will discuss in detail another technique for constructing equileverage designs because some space has been devoted to it in the literature.

Another method given by Dollinger and Staudte (1990) and Staudte and Sheather (1990) is the *inverse projection approach*. Unlike the sine–cosine approach, the inverse projection method can be used for any number of regressors. The step-by-step details of the inverse projection approach are given by Staudte and Sheather (1990, pp. 231–232). Generally, an orthogonal equileverage design is constructed on a unit sphere, and the design is then "projected" onto the desired experimental region by using the eigenvectors of a design on the raw scale. As with the sine–cosine approach, the experimenter does not have any direct control over the design values on the original scale, and the technique also requires that a (nonequileverage) design be constructed on the original scale so that the necessary eigenvectors will be available. The fact that there is no direct control over the design values means that rounding (to integer values, say) could considerably disturb the equileverage property. Thus, this approach has the same shortcomings as the sine–cosine approach, and it has the additional shortcoming that a matrix of eigenvectors must be specified.

We will consider the Gorman and Toman (1966) data in illustrating the construction of equileverage designs with the inverse projection technique. The objective of the experiment was to determine the effect of certain controllable factors on the rate of rutting of asphalt pavement. The response variable was the logarithm (base 10) of the rate of change of rut depth in inches per million wheel passes. There were six variables used in the study, but in their analysis of the data, Daniel and Wood (1980) concluded that the model should contain only three regressors: X_1, X_2, and X_6. These are the $\log_{10}$ of the viscosity of the asphalt, the percent asphalt in the surface course, and the percent voids in the surface course, respectively. We will construct a design for these three regressors rather than the original six.

Dollinger and Staudte (1990) and Staudte and Sheather (1990, pp. 218–236) discuss various methods of constructing equileverage designs. Since the particular method we will illustrate produces an even number of design points, the design will have 32 points instead of 31.

We will employ the same notation as in Dollinger and Staudte (1990) and Staudte and Sheather (1990). That is, we will construct a matrix $\mathbf{W}$ that is 32×2, with $\mathbf{W}$ representing a design that is scalar, but not equileverage. (The matrix $\mathbf{W}$ is chosen to be 32×2 rather than 32×3 because an extra column is added in the next step, as explained later in this section.) A subsequent step produces an equileverage design on the unit sphere, which is created using the inverse projection method and is represented by $\mathbf{V}$. That design is then transferred to the experimental region with centroid given by $(\overline{x_1^*}, \overline{x_2^*}, \overline{x_6^*})$. As pointed out by Dollinger and Staudte (1990), it is desirable that the points in $\mathbf{W}$ lie on different concentric circles so that the points in **V** will be spread somewhat evenly over the surface of the ellipsoid.

There is literally an infinite number of ways in which an equileverage design can be created using this approach, due to the infinite number of positions that the design points can occupy on the ellipsoidal region of interest. We will use the matrix **W** as given in Table 14.4.

Notice that four of the points lie on a circle of radius 1, eight of the points lie on a circle of radius 2, and the other four points lie on a circle with radius equal to $\sqrt{3}$.

A matrix **T** is then formed by duplicating each of the points in **W** and combining each pair with $\pm\sqrt{c_i}$, where $c_i = 3(24)/16 - \|\mathbf{w}_i'\|^2$, with 3 the number of design variables, 24 is from $\mathbf{W}'\mathbf{W} = 24\mathbf{I}$, and $\|\mathbf{w}_i'\|^2$ is the squared norm of the ith row of **W**. The 32×3 matrix **T** is given in Table 14.5. This matrix must then be scaled to the unit sphere. Since the squared norm of each row vector is 4.5, we can accomplish this by dividing each element of **T** by $\sqrt{4.5}$. This scaling produces the matrix **V** (which we will not display).

The matrix **E** is then produced as

$$\mathbf{E} = \overline{\mathbf{X}}^* + \mathbf{V}[(3/31)\mathbf{\Delta}]^{1/2}\mathbf{U}'$$

where $\overline{\mathbf{X}}^* = [\overline{x_1^*}\mathbf{1}, \overline{x_2^*}\mathbf{1}, \overline{x_6^*}\mathbf{1}]$ is a 32×3 matrix whose columns are the averages (repeated to form each column) of each of the regressor (design) variables on

Table 14.4 Orthogonal Matrix W for Inverse Projection

ROW	W_1	W_2
1	1.0000	0.0000
2	0.0000	1.0000
3	−1.0000	0.0000
4	0.0000	−1.0000
5	1.4142	1.4142
6	−1.4142	−1.4142
7	−2.0000	0.0000
8	0.0000	−2.0000
9	2.0000	0.0000
10	0.0000	2.0000
11	−1.4142	1.4142
12	1.4142	−1.4142
13	1.7320	0.0000
14	0.0000	1.7320
15	−1.7320	0.0000
16	0.0000	−1.7320

Table 14.5 T Matrix for Gorman–Toman Example

ROW	T_1	T_2	T_3
1	1.0000	0.0000	1.87083
2	1.0000	0.0000	−1.87083
3	0.0000	1.0000	1.87083
4	0.0000	1.0000	−1.87083
5	−1.0000	0.0000	1.87083
6	−1.0000	0.0000	−1.87083
7	0.0000	−1.0000	1.87083
8	0.0000	−1.0000	−1.87083
9	1.4142	1.4142	0.70711
10	1.4142	1.4142	−0.70711
11	−1.4142	−1.4142	0.70711
12	−1.4142	−1.4142	−0.70711
13	−2.0000	0.0000	0.70711
14	−2.0000	0.0000	−0.70711
15	0.0000	−2.0000	0.70711
16	0.0000	−2.0000	−0.70711
17	2.0000	0.0000	0.70711
18	2.0000	0.0000	−0.70711
19	0.0000	2.0000	0.70711
20	0.0000	2.0000	−0.70711
21	−1.4142	1.4142	0.70711
22	−1.4142	1.4142	−0.70711
23	1.4142	−1.4142	0.70711
24	1.4142	−1.4142	−0.70711
25	1.7320	0.0000	1.22474
26	1.7320	0.0000	−1.22474
27	0.0000	1.7320	1.22474
28	0.0000	1.7320	−1.22474
29	−1.7320	0.0000	1.22474
30	−1.7320	0.0000	−1.22474
31	0.0000	−1.7320	1.22474
32	0.0000	−1.7320	−1.22474

the original scale, Δ is a diagonal matrix whose elements are the eigenvalues of $X'X$, where X is the 32×3 matrix of centered regressor values, and U' is the transpose of the matrix of eigenvectors corresponding to the eigenvalues of $X'X$. (For the remainder of this chapter, X will denote the matrix of design values only; i.e., excluding the column of ones.) The matrix E is given in Table 14.6, in juxtaposition to the regressor design values given by Gorman and Torman (1966).

It is worth noting that E will not be a design that can be made orthogonal by mean centering unless X has the same property. Staudte and Sheather

Table 14.6 Original Values of Regressors Used in Model for Gorman–Toman Example and Corresponding Equileverage Design

Row	X_1	X_2	X_6	$X_1(E)$	$X_2(E)$	$X_6(E)$
1	0.44716	4.68001	4.9160	1.91365	5.15956	5.23834
2	0.14613	5.19001	4.5628	1.94357	4.60830	5.11259
3	0.14613	4.82001	5.3208	1.31237	5.27549	4.58703
4	0.51851	4.85001	4.8655	1.34229	4.72423	4.46129
5	0.23045	4.86001	3.7756	0.42329	5.21170	4.65513
6	0.46240	5.16001	4.3972	0.45320	4.66044	4.52938
7	0.56820	4.82001	4.8670	1.02457	5.09577	5.30643
8	0.23045	4.86001	4.8284	1.05449	4.54451	5.18069
9	−0.03621	4.78001	4.8654	2.43512	5.10438	4.81132
10	−0.16749	5.16001	4.0343	2.44643	4.89603	4.76380
11	0.77815	4.57001	5.4505	−0.07957	4.92397	5.00392
12	0.63347	4.61001	4.8526	−0.06826	4.71562	4.95640
13	−0.22185	5.07001	4.2574	−0.31259	5.06632	4.32441
14	0.25527	4.66001	5.1438	−0.30128	4.85797	4.27689
15	0.77815	5.42001	3.7180	0.88997	4.83446	5.62702
16	0.64345	5.01001	4.7146	0.90128	4.62611	5.57950
17	1.94448	4.97001	4.6255	2.66814	4.96203	5.49083
18	1.79239	5.01001	4.9769	2.67945	4.75368	5.44331
19	1.69897	4.96001	4.3222	1.46558	5.19389	4.18822
20	1.76343	5.20001	5.0874	1.47689	4.98554	4.14070
21	1.95424	4.80001	5.9708	0.32744	5.17813	3.98655
22	1.81954	4.98001	4.6466	0.33875	4.96977	3.93902
23	2.14613	5.35001	5.1154	2.02811	4.85023	5.82870
24	2.38021	5.04001	5.9394	2.03942	4.64187	5.78117
25	2.62325	4.80001	5.9163	2.46430	5.04528	5.43008
26	2.69897	4.83001	5.4709	2.48388	4.68440	5.34776
27	2.25527	4.66001	4.6016	1.42287	.5.24607	4.30202
28	2.43136	4.67001	5.0432	1.44246	4.88519	4.21970
29	2.23045	4.72001	5.0748	−0.11702	5.13560	4.41996
30	1.99123	5.00001	4.3336	−0.09744	4.77472	4.33764
31	1.54407	4.70001	5.7051	0.92440	4.93481	5.54802
32				0.94399	4.57393	5.46570

(1990, p. 233) suggest starting with a trial design **X** that is not necessarily equileverage, using a computer package to determine **Δ** and **U**, then constructing an equileverage design on the equileverage ellipsoid suggested by **X**.

If a design is to be constructed, we would generally want it to possess orthogonality as a minimum requirement, so the inverse projection approach essentially requires that an experimenter first construct an orthogonal design. Since the matrix **W** must also be an orthogonal matrix, the user must thus construct two orthogonal matrices (of different dimensions). Because constructing an orthogonal matrix can be somewhat of a chore in high dimensions, we would prefer to construct only one such matrix. This we can do simply by letting $X'X = I$ and thus use the eigenvalues (which are all 1) and a corresponding set of eigenvectors for the identity matrix of the appropriate size. This will ensure that E will be both equileverage *and* orthogonal.

Using this approach for the Gorman and Toman example produces the design in Table 14.7. The most obvious difference between the two designs is the difference in the X_1 values. Specifically, there were some extremely large values of the first variable before the $\log_{10}$ transformation was applied (see Daniel and Wood, 1980, Figure 6B.1). The values of X_1 for the equileverage design

Table 14.7   Orthogonal Equileverage Design for the Gorman–Toman Example

ROW	X_1	X_2	X_6
1	1.45778	4.91000	5.03051
2	0.90908	4.91000	5.03051
3	1.45778	5.05665	4.88386
4	0.90908	5.05665	4.88386
5	1.45778	4.91000	4.73721
6	0.90908	4.91000	4.73721
7	1.45778	4.76335	4.88386
8	0.90908	4.76335	4.88386
9	1.28713	5.11739	5.09125
10	1.07973	5.11739	5.09125
11	1.28713	4.70261	4.67647
12	1.07973	4.70261	4.67647
13	1.28713	4.91000	4.59057
14	1.07973	4.91000	4.59057
15	1.28713	4.61671	4.88386
16	1.07973	4.61671	4.88386
17	1.28713	4.91000	5.17715
18	1.07973	4.91000	5.17715
19	1.28713	5.20329	4.88386
20	1.07973	5.20329	4.88386
21	1.28713	5.11739	4.67647
22	1.07973	5.11739	4.67647
23	1.28713	4.70261	5.09125
24	1.07973	4.70261	5.09125
25	1.36304	4.91000	5.13785
26	1.00382	4.91000	5.13785
27	1.36304	5.16399	4.88386
28	1.00382	5.16399	4.88386
29	1.36304	4.91000	4.62987
30	1.00382	4.91000	4.62987
31	1.36304	4.65601	4.88386
32	1.00382	4.65601	4.88386

in Table 14.6 would transform back to very large values on the original scale, but the largest X_1 values in Table 14.7 would transform back to much smaller values. The largest values of X_1 in Table 14.6 are matched with some of the largest values of X_6, and this contributes to the moderately high correlation between the design values of X_1 and X_6 in Table 14.6 (.469). (This is the same correlation between X_1 and X_6 for the data, as the correlations will be the same when we use the eigenvalues and associated eigenvectors that result from using the original data matrix.)

While it may be of interest to start with a published data set and convert the **Z** matrix to one that is equileverage (and then compare the two), it is not practical to use an equileverage design that is far from being orthogonal. For example, Dollinger and Staudte (1990) and Staudte and Sheather (1990, p. 233) use the stack loss data, with all 21 design points, in illustrating the inverse projection technique. They simply modify the original design to make it equileverage, rather than constructing an orthogonal equileverage design. Consequently, the sets of values for X_1 and X_2 are closer to being collinear than they are to being orthogonal. We would not likely ever want to construct a design for which the "correlation" (loosely speaking) between the regressors is .77, as it is for their design. Not only is this fairly high, but it is not very far from what would be considered as representing multicollinearity. Recall from Section 4.3.2 that a variance inflation factor of 10 or more is often taken as a signal of multicollinearity, and in the two-regressor case this will happen when the correlation between the two regressors is at least .949.

Clearly we do not want to construct designs for which multicollinearity is even remotely approached, as this is one of the problems that we want to avoid by using a design in the first place. Consequently, an experimenter who constructs equileverage designs should not simply start with an arbitrary **X** matrix. Since orthogonality is a desirable property of an experimental design, it is preferable to construct equileverage designs that will be orthogonal (or at least near-orthogonal) for the desired experimental region, not just orthogonal on the unit sphere.

Although using $\mathbf{X'X} = \mathbf{I}$ is desirable in the design–construction process, there is still the possible problem of not having particular design values that are desired, in addition to the equileverage property being lost through rounding. Thus, methods for more than two regressors that are similar to the approach described in Section 14.2.2 may frequently be needed.

14.3 OTHER DESIRABLE PROPERTIES OF EXPERIMENTAL DESIGNS

In addition to the general criteria given in Section 14.1, there are various optimality criteria that have been judged important for experimental designs. Since more than a few of these optimality criteria have been proposed, and since a single letter has been used to represent a particular criterion, these criteria are sometimes referred to as "alphabet optimality."

14.3.1   D-optimality

A *D-optimal* design is one in which the determinant of $\mathbf{X}'\mathbf{X}$ is maximized for a given experimental region. *D*-optimality is a desirable feature since, as mentioned by Nachtstheim (1987), such a design minimizes "the volume of the confidence ellipsoid" for the regression parameters. (This refers to the construction of a confidence region for more than one regression parameter, as is sometimes done.)

In their review of *D*-optimality for regression designs, St. John and Draper (1975) indicated that *D*-optimal designs can be difficult to construct. Nachtsheim (1987) discusses software that can be used to construct designs that are *D*-optimal and/or have other optimality properties.

As discussed by Dollinger and Staudte (1990), an equileverage design that is constructed for a given ellipsoidal region will also be *D*-optimal. (Actually this is true only for a *scalar* equileverage design.) The converse is not true, however, as will be shown later in this section.

The matter of *D*-optimality is discussed in simple terms by Box and Draper (1986), who emphasize the importance of removing scale factors when comparing designs. They indicate that if we remove scale factors by making $\|\mathbf{x}_i\|^2 = n$, $i = 1, 2, \ldots, k$ (k = number of regressors), then *D*-optimality is met when the design is orthogonal. This can be easily shown for $k = 2$, since then

$$\mathbf{X}'\mathbf{X} = \begin{bmatrix} n & \sum X_1 X_2 \\ \sum X_1 X_2 & n \end{bmatrix}$$

with the data coded as $X_{ij} = (X_{ij}^* - \overline{X_i^*})/f$, with $f = \sqrt{\sum (X_{ij}^* - \overline{X_i^*})^2}$, and X_{ij}^* denoting the raw value.

Since $|\mathbf{X}'\mathbf{X}| = n^2 - r^2$, where $r = \sum X_1 X_2$, the determinant of $\mathbf{X}'\mathbf{X}$ would be maximized when $r = 0$. That is, when the design is orthogonal.

Since an orthogonal design is not necessarily an equileverage design, as was seen in Table 14.2, this disproves the claim of Dollinger and Staudte (1990) that if an n-point design in R^k (the k-dimensional space of real numbers) is *D*-optimal on some region, then it must be an equileverage design. *D*-optimality can be satisfied if the (scaled) design is a scalar design (in particular, $\mathbf{X}'\mathbf{X} = n\mathbf{I}$ with the scaling convention of Box and Draper). It is true, however, that a scalar equileverage design will be *D*-optimal since the "scalar property" is met.

In general, the equileverage designs constructed using the techniques in Dollinger and Staudte (1990) will not necessarily be orthogonal after they have been projected back to the desired experimental region, and thus will not be *D*-optimal. For example, they give an equileverage design for the stack loss data of Brownlee (1965), which when scaled so that each column of the design matrix has a squared norm equal to n, the determinant of the design matrix is 3103.5, whereas the determinant would have to be 10,648 ($= 22^3$) for the design to the *D*-optimal.

14.3.2 *G*-optimality

As noted by Snee (1985), maximizing $|\mathbf{X}'\mathbf{X}|$ (and thus producing *D*-optimality) is the most widely used design criterion, but its use is based on the assumption that the model is correct, and it may require many replications at certain design points. As explained by Snee (1985), the latter results from the fact that the maximum $|\mathbf{X}'\mathbf{X}|$ criterion requires that the number of design points be approximately equal to the number of parameters in the model. Consequently, when the desired value of *n* is much larger than *p*, replication is likely to result.

Another popular design criterion is *G-optimality*. The term was coined by Kiefer and Wolfowitz (1959) for a criterion that was apparently first proposed by Smith (1918). This is also called the *minimax criterion* because the maximum value of $\mathrm{Var}(\hat{Y}_x)$, maximized over all vectors **x** in the experimental region, is minimized over the space of candidate points for the design.

It is of interest to examine the relationship between equileverage designs and *G*-optimal designs, just as the relationship between *D*-optimal and equileverage designs was considered in Section 14.3.1. Dollinger and Staudte (1990) state that the properties of *G*-optimality and equal leverages are equivalent.

Writing $\mathrm{Var}(\hat{Y}_{x_i}) = \sigma^2 \mathbf{x}_i'(\mathbf{X}'\mathbf{X})^{-1}\mathbf{x}_i = \sigma^2 h_{ii}$, where $\mathbf{x}_i$ is one of the row vectors in **X**, we recognize that $\mathrm{Var}(\hat{Y}_{x_i})$ will be constant over the $\mathbf{x}_i$ for an equileverage design.

Dollinger and Staudte (1990) state that a (centered) scalar equileverage design with rows of length one will have leverage values of $h_{ii} = p/n$, where *p* is the number of regressors. Box and Draper (1986) indicate that (for an obviously uncentered design matrix **X**), $n\mathrm{Var}(\hat{Y}_x)/\sigma^2 = p$ with *G*-optimality, where *p* is the number of parameters. Thus, the leverages must be equal in a *G*-optimal design since $\mathrm{Var}(\hat{Y}_x)$ does not depend on *i*. Conversely, a design that is equileverage will also be *G*-optimal, since the *G*-optimality criterion will be simultaneously satisifed.

Thus, since *G*-optimal designs must also be equileverage, and equileverage designs are not necessarily *D*-optimal, it follows that *G*-optimality and *D*-optimality are not equivalent. There has been some confusion regarding *D*-optimality and *G*-optimality in the literature, and this is due largely to Kiefer and Wolfowitz (1960) having established the equivalence between *D*-optimality and *G*-optimality for measure designs. [A *measure design* satisfies the requirements $\xi(\mathbf{x}) \geq 0$, $\mathbf{x} \in \chi$, and $\int_x \xi(d\mathbf{x}) = 1$, with ξ denoting a probability measure on χ, the region of permissible values of **x**. See St. John and Draper (1975) for additional details.] This theoretical result does not apply to designs with a finite number of design points, however, as discussed by Silvey (1980, p. 23). Obviously finite designs are the ones that are of practical interest. Snee (1985) points out that *G*-efficiency is a lower bound on *D*-efficiency. See also Atwood (1969).

14.3.3 Other Optimality Criteria

Another desirable property of a regression design is one that minimizes the trace of $(\mathbf{X}'\mathbf{X})^{-1}$. Since $\sum_{i=1}^{k} \mathrm{Var}(\hat{\beta}_i) = \sigma^2 \mathrm{tr}(\mathbf{X}'\mathbf{X})^{-1}$, this type of design will minimize the sum, and hence the average, of the variances of the estimators of the regression parameters. Accordingly, this property is called *A-optimality*.

It is well known that the average of the $\text{Var}(\hat{\beta}_i)$ is minimized when the design is orthogonal and the columns have the same length (see, e.g., the proof given by Seber, 1977, p. 58). It follows that $\sum \text{Var}(\hat{\beta}_i)$ must similarly be minimized, so using the terminology of Section 14.3.1, an *A*-optimal design is a scalar design.

Related somewhat to *G*-optimality is *I*-optimality (Studden, 1977). A design with this property is one for which the average (integrated) prediction variance is minimized over the design space. Another frequently mentioned optimality criterion is *E*-optimality. A design has this property when it minimizes the maximum eigenvalue of $(\mathbf{X'X})^{-1}$.

A discussion of the optimality criteria presented to this point in the chapter can be found in Steinberg and Hunter (1984). See also Silvey (1980).

14.4   MODEL MISSPECIFICATION

In the preceding it has been tacitly assumed that all of the variables included in the design will be used in the model. We saw with the stack loss example in Chapter 6 that this will not necessarily be true, and one or more variables included in the design will frequently not be used in the eventual model. What happens to our "optimal design" when we have to delete columns from the **X** matrix, and also what happens when the true model is nonlinear?

The problem of a subset of the regressors being used when a *D*-optimal or *G*-optimal design has been constructed for the full model has been considered by a number of authors. [See St. John and Draper (1975) for a review.] Dollinger and Staudte (1990) did not present a way to "retrieve" the equileverage property when one or more columns of the design matrix are removed.

A different type of problem is presented when the true model is nonlinear. Equileverage designs do not permit a check for model misspecification. Consequently, a modified approach may be desirable. In particular, there is more emphasis in the literature on constructing designs that are only approximately *D*-optimal, *G*-optimal, and so forth, than there is in constructing exact designs [see, e.g., Cook and Nachtsheim (1982)]. Therefore, modifying an equileverage design somewhat may be desirable.

Recognizing that equileverage designs do not provide information about the interior of the experimental region since the design points lie on the surface of an ellipsoid, Dollinger and Staudte (1990) indicate that an equileverage design can be modified to include interior points and still not have any high leverage points. One obvious way to do this would be to construct different equileverage designs on the unit sphere, and then to transfer the design points to ellipsoids of different radii. (Some of the interior points might then be replicated to provide a sufficient number of degrees of freedom for a lack-of-fit test.) By specifying the different radii, the extent to which the leverages differ would thus be controlled.

Box and Draper (1959) were the first to consider in detail the issue of model misspecification. They proposed the concept of *J-optimality*, which refers to the average mean-squared error (of prediction). Welch (1983) proposed a similar criterion that was termed J^{**}-*optimality*.

14.5   RANGE OF REGRESSORS

One thing to keep in mind when we are constructing these designs is that we do not want to inadvertently construct a set of regressor values with limited range, as this might result in the regressor being declared not significant just because of the limited range. This can be easily avoided, however, simply by specifying the range of one or more regressors through the appropriate choices of the necessary constants, as was illustrated in Section 14.2.2.1.

14.6   ALGORITHMS AND SOFTWARE FOR DESIGN CONSTRUCTION

Meyer and Nachtsheim (1988) provide a summary of the available methods for constructing D-optimal designs, and Nachtsheim (1987) discusses software for constructing such designs. When the number of design points is less than 30, the branch-and-bound algorithm of Welch (1982) can be used to construct D-optimal designs when the design space is finite (i.e., when there are N candidate points and, e.g., the regressors can only assume integer values.) For continuous design spaces, Meyer and Nachtsheim (1988) proposed a simulated annealing method (which can also be used for discrete design spaces), but the approach was indicated to be computer intensive.

It is important to recognize that these algorithmic approaches require a set of candidate points, and the algorithms produce a design that is, say, D- or G-optimal *relative to the set of candidate points*. Consequently, the resultant design might maximize $\mathbf{X'X}$ over the candidate points and not be particularly close to being orthogonal. Thus, the methods of constructing G-optimal (equileverage) designs given in Sections 14.2.2.1 and 14.5 will produce designs that are G-optimal, but not unique, for the desired experimental region, whereas the algorithmic approach described by Welch (1984) will produce the unique G-optimal design for a given set of candidate points. Only when the experimenter insists upon using some subset of a set of candidate points would the latter approach seem preferable, provided that the leverages are at least approximately equal.

14.7   DESIGNS FOR POLYNOMIAL REGRESSION

Just as D-optimal designs can be constructed for linear regression with first-order terms, it is also possible to construct D-optimal designs for polynomial regression. Optimal designs for a quadratic regression model can be found, as discussed in Atkinson (1982). See also Haines (1987) and Gaffke and Kraft (1982).

14.8 DESIGNS FOR LOGISTIC REGRESSION

The construction of D-optimal designs for minimizing the area of confidence regions in logistic regression is discussed by Minkin (1987), which extends the work of Abdelbasit and Plackett (1983).

14.9 DESIGNS FOR NONLINEAR REGRESSION

In general, the principles that guide the construction of designs in linear regression are also applicable in nonlinear regression. A good summary of experimental design considerations for nonlinear regression can be found in Swain (1990), who makes the important point that when two or more models are viewed as being plausible in a particular application, a design should be chosen to permit model discrimination. Unless an experimenter has drawn a bead on one particular model by the time that the experiment is carried out, a good design will be one that facilitates model discrimination and efficient parameter estimation. Hill et al. (1968) describe a sequential procedure for selecting the first several design points to permit model discrimination, with the remaining points selected to facilitate efficient parameter estimation once a model has been selected from using the first set of points. (Note: This is a sequential procedure *for a single design*, not a sequential design approach, as the latter generally refers to the use of more than one design.)

Various aspects of experimental design for nonlinear problems are covered in the review paper of Ford et al. (1989), which also contains many references.

One problem that is encountered with nonlinear design that is not generally a problem with design for linear models is the inability to create orthogonal designs. For example, Bates and Watts (1988, p. 119) point out that orthogonality is not possible for a two-parameter Michaelis–Menten model. This occurs because the first column of the derivative matrix must have all positive entries, and the second column must have all negative entries.

See Bates and Watts (1988, pp. 121–131) and Seber and Wild (1989, pp. 250–269) for information and practical advice regarding nonlinear designs.

SUMMARY

When regressor values can be selected in regression model building, there are certain design criteria that should be considered. Orthogonality is one such criterion, and this leads to the consideration of designs that are D-optimal, with this being the most frequently used design criterion. Equal variances of the parameter estimators is another desirable property, and such designs are G-optimal, which could also be called equileverage designs. (A side benefit of an equileverage design not mentioned previously is that such a design will make the regression estimators asymptotically normal, regardless of the distribution of the error term.) It is possible to construct designs that are both D- and G-optimal; these are orthogonal equileverage designs.

Design considerations are also important in linear regression with nonlinear terms, and in nonlinear regression.

D-optimal designs require special software, whereas equileverage designs do not, but there are no guidelines available for constructing equileverage designs for any combination of n and p. When using a computer program to generate a design that satisfies an optimality criterion, it is important to recognize that "optimality" is only for the set of candidate points, and this may be quite different from global optimality.

In this chapter we have focused attention on the statistical aspects of experimental designs for regression. Important nonstatistical considerations that should be made when planning designed experiments are discussed by Cox (1958), and more recently by Coleman and Montgomery (1993), and the discussants of that paper.

REFERENCES

Abdelbasit, K. M. and R. L. Plackett (1938). Experimental design for binary data. *Journal of the American Statistical Association*, **78**, 90–98.

Atkinson, A. C. (1982). Developments in the design of experiments. *International Statistical Review*, **50**, 161–177.

Atwood, C. L. (1969). Optimal and efficient designs of experiments. *Annals of Mathematical Statistics*, **40**, 1570–1602.

Box, G. E. P. and N. R. Draper (1959). A basis for the selection of a response surface design. *Journal of the American Statistical Association*, **54**, 622–654.

Box, G. E. P. and N. R. Draper (1975). Robust designs. *Biometrika*, **62**, 347–352.

Box, G. E. P. and N. R. Draper (1986). *Empirical Model Building and Response Surfaces*. New York: Wiley.

Brownlee, K. (1965). *Statistical Theory and Methodology in Science and Engineering*, 2nd edition. New York: Wiley.

Coleman, D. E. and D. C. Montgomery (1993). A systematic approach to planning for a designed industrial experiment. *Technometrics*, **35**, 1–12 (discussion: 13–27).

Cook, R. D. and C. J. Nachtsheim (1982). Model-robust, linear-optimal designs. *Technometrics*, **24**, 49–54.

Cox, D. R. (1958). *Planning of Experiments*. New York: Wiley.

Daniel, C. and F. S. Wood (1980). *Fitting Equations to Data*, 2nd edition. New York: Wiley.

Dollinger, M. B. and R. G. Staudte (1990). The construction of equileverage designs for multiple linear regression. *Australian Journal of Statistics*, **32**, 99–118.

Ford, I., C. P. Kitsos, and D. M. Titterington (1989). Recent advances in nonlinear experimental design. *Technometrics*, **31**, 49–60.

Gaffke, N. and O. Krafft (1982). Exact D-optimum designs for quadratic regression. *Journal of the Royal Statistical Society, Series B*, **44**, 394–397.

Gorman, J. W. and R. J. Toman (1966). Selection of variables for fitting equations to data. *Technometrics*, **8**, 27–51.

Haines, L. M. (1987). The application of the annealing algorithm to the construction of exact optimal designs for linear-regression models. *Technometrics*, **29**, 439–447.

Hill, W. J., W. G. Hunter, and D. W. Wichern (1968). A joint design criterion for the dual problem of model discrimination and parameter estimation. *Technometrics*, **10**, 145–160.

Kiefer, J. and J. Wolfowitz (1959). Optimum designs in regression problems. *Annals of Mathematical Statistics*, **30**, 271–294.

Kiefer, J. and J. Wolfowitz (1960). The equivalence of two extremum problems. *Canadian Journal of Mathematics*, **12**, 363–366.

Meyer, R. K. and C. J. Nachtsheim (1988). Simulated Annealing in the Construction of Exact Optimal Designs, in *Computer Science and Statistics: Proceedings of the Symposium on the Interface*, 144–146. Alexandria, VA: American Statistical Association.

Minkin, S. (1987). Optimal designs for binary data. *Journal of the American Statistical Association*, **82**, 1098–1103.

Nachtsheim, C. J. (1987). Tools for computer-aided design of experiments. *Journal of Quality Technology*, **19**, 132–160.

Seber, G. A. F. (1977). *Linear Regression Analysis*. New York: Wiley.

Seber, G. A. F. and C. J. Wild (1989). *Nonlinear Regression*. New York: Wiley.

Silvey, S. D. (1980). *Optimal Design*. London: Chapman and Hall.

Smith, K. (1918). On the standard deviations of adjusted and interpolated values of an observed polynomial function and its constants and the guidance they give towards a proper choice of the distribution of observations. *Biometrika*, **12**, 1–85.

Snee, R. D. (1985). Computer-aided design of experiments: Some practical experiences. *Journal of Quality Technology*, **17**, 222–236.

St. John, R. C. and N. R. Draper (1975). D-optimality for regression designs: A review. *Technometrics*, **17**, 15–24.

Staudte, R. G. and S. J. Sheather (1990). *Robust Estimation and Testing*. New York: Wiley.

Steinberg, D. M. and W. G. Hunter (1984). Experimental design: Review and comment. *Technometrics*, **26**, 71–97 (discussion: 98–130).

Studden, W. J. (1977). Optimal designs for integrated variance in polynomial regression. *Statistical Decision Theory*, **2**, 411–420.

Swain, J. J. (1990). Nonlinear regression, in *Handbook of Statistical Methods for Engineers and Scientists*, H. M. Wadsworth, ed. New York: McGraw-Hill, Chapter 17.

Welch, W. J. (1982). Branch-and-bound search for experimental designs based on D-optimality and other criteria. *Technometrics*, **24**, 41–48.

Welch, W. J. (1983). A mean squared error criterion for the design of experiments. *Biometrika*, **70**, 205–213.

Welch, W. J. (1984). Computer-aided design of experiments for response estimation. *Technometrics*, **26**, 217–224.

EXERCISES

14.1. What is an easy way to determine if a design is *D*-optimal?

14.2. Construct an 18-point design for two regressors using the sine–cosine approach. It is desired that the smallest value for X_2 should be 35 and the largest should be 75. There are no restrictions on the values for X_1.

14.3. Use the inverse projection approach to construct a 24-point orthogonal equileverage design for three regressors such that $\bar{x}_1 = 8$, $\bar{x}_2 = 13$, and $\bar{x}_3 = 10$.

14.4. Obtain a 20-point orthogonal equileverage design for two regressors, using 1, 2, 3, 4, and 5 ($\pm$ of each) as the values for one of the regressors.

14.5. Modify the design given in Exercise 14.2 so that the design is orthogonal but not equileverage.

14.6. Assume that an experimental design has been used, and the leverage values have been computed. How can the leverages be used to determine if the design is *G*-optimal?

14.7. Assume that for a specified number of regressors, a desired number of design points, and certain regressor values that an experimenter wants to use, it is not possible to construct a design that is both *D*- and *G*-optimal. Which property would you prefer to have?

14.8. Assume that there are two regressors and that the design values for one of the regressors are to be equispaced. Show that there cannot be more than five such values.

14.9. Critique the following statements. "The concept of an equileverage design seems to be of dubious value. If I construct an equileverage design for a single regressor and $n > 10$, every design point will be a high leverage point since each leverage will exceed $3p/n$."

14.10. How can an orthogonal design be checked to see if it is also an equileverage design?

14.11. What is one disadvantage of an orthogonal equileverage design, and how can the design be modified to overcome that problem?

14.12. Assume that there are 30 possible design points for two regressors, and an experimenter wishes to use 20 of them in an experiment. Will it necessarily be possible to select a 20-point subset such that the design will satisfy the condition for *D*-optimality given in Section 14.3.1?

14.13. Critique the following statement. "Since true models and true effects are almost always nonlinear, we can make the effect of any regressor appear to be either linear or nonlinear simply through choice of the range of the values of that regressor in the experimental design."

14.14. Will a design that has orthogonal columns and equal row lengths necessarily be an orthogonal equileverage design? Explain.

CHAPTER 15

Applications of Regression

Various methods have been given in the preceding chapters for analyzing regression data and determining appropriate models. In this chapter we present five challenging data sets, and we will see what techniques might be applicable in meeting the objectives of the practitioners. That is, we want to see if we can apply what we have learned about regression.

We note that these data sets have not been extensively analyzed in the literature.

15.1 WATER QUALITY DATA

In an article in an engineering journal, Haith (1976) investigated the relationship between water quality and land usage. The data have also been analyzed in the statistical literature by Allen and Cady (1982), Simpson et al. (1992), and Hadi (1992).

Haith (1976) sought to determine if relationships existed between land use and water quality, and if so, which land uses most significantly impact water quality. If such relationships do exist, a model is then sought that would be useful "for both prediction and water quality management in northeastern river basins with mixtures of land uses comparable to the basins used in the study." Thus, there is a need to determine the relative importance of each land use in terms of water quality, to determine a prediction equation that is useful in predicting the water quality in other river basins in the area, and obviously to have the water quality controlled. Therefore, in terms of the typical uses of regression models, we need a model that is useful for description, estimation, prediction, and control—the four uses of regression equations.

Haith used correlation and regression analyses and cited papers that describe similar analyses of water quality data. He gave nitrogen content as one measure of water quality, with one set of land-use variables given as: (1) percentage of land area currently in agricultural use; (2) percentage of land area in forest, forest brushland, and plantations; (3) percentage of land area in residential use;

and (4) percentage of land area in commercial and manufacturing use. (Haith also broke these four land-use variables into various subcategories. We will be concerned only with the data set that has the four land-use variables.) We will use the same symbolism as Haith, so that these variables will be denoted by AG, FR, RS, and CI, respectively. We shall let NI denote total nitrogen.

The data are given in Table 15.1. The river basins constitute a sample of river basins, with the experimenter intending to apply the regression model that is developed to the population of river basins in the northeastern part of the United States, as was previously noted. Thus, we might think of these data relative to the scenario that was discussed in Section 1.2 in which a sample of students' records would be used in predicting college GPA, with the students in that example corresponding to the river basins in this example. The land-use information for each river basin was obtained from New York State's Land Use and Natural Resource Inventory.

A cursory glance at the data reveals an obvious problem. Specifically, the percentage of residential land in the Hackensack river basin is much larger than the corresponding percentage in any of the other river basins. Thus, Hackensack is a potentially influential data point, so the reader/data analyst must determine how the Hackensack river basin is to be treated. (The reader may wish to analyze these data and draw some conclusions before proceeding further in the chapter.)

Table 15.1   Water Quality Data

River Basin	Total Nitrogen[a]	Area[b]	Agriculture (%)	Forest (%)	Residential (%)	Commercial-Industrial (%)
Olean	1.10	530	26	63	1.2	0.29
Cassadaga	1.01	390	29	57	0.7	0.09
Oatka	1.90	500	54	26	1.8	0.58
Neversink	1.00	810	2	84	1.9	1.98
Hackensack	1.99	120	3	27	29.4	3.11
Wappinger	1.42	460	19	61	3.4	0.56
Fishkill	2.04	490	16	60	5.6	1.11
Honeoye	1.65	670	40	43	1.3	0.24
Susquehanna	1.01	2,590	28	62	1.1	0.15
Chenango	1.21	1,830	26	60	0.9	0.23
Tioughnioga	1.33	1,990	26	53	0.9	0.18
West Canada	0.75	1,440	15	75	0.7	0.16
East Canada	0.73	750	6	84	0.5	0.12
Saranac	0.80	1,600	3	81	0.8	0.35
Ausable	0.76	1,330	2	89	0.7	0.35
Black	0.87	2,410	6	82	0.5	0.15
Schoharie	0.80	2,380	22	70	0.9	0.22
Raquette	0.87	3,280	4	75	0.4	0.18
Oswegatchie	0.66	3,560	21	56	0.5	0.13
Cohocton	1.25	1,350	40	49	1.1	0.13

[a]Total nitrogen is given as mean concentration in milligrams per liter.
[b]Area is given in square kilometers.

Haith (1976) did not search for anomalous data, but Simpson et al. (1992) recognize that Hackensack is an extreme point in the regressor space and show that the regression coefficients with and without that data point do differ considerably. They created an urban land usage (UR) variable (equal to RS + CI), because of the high correlation between RS and CI, and reparameterized the model by creating the variable OTHER = 100 – AG – FR – UR and using a no-intercept regression model. This makes the parameters for the four land-use variables directly interpretable, since the percentage attributable to the other land uses is given by OTHER. In applying robust regression to these data, the authors use Mallows's weights (see Section 11.6), which have the effect of downweighting five data points, with the Hackensack case having a weight of virtually zero and three other points (4, 6, and 7) having weights very close to zero. (Points 4, 6, and 7 appear as extreme points in scatter plots of the regressors, both before and after deleting the Hackensack data point.)

Simpson et al. (1992) concluded that the forest land use was only marginally significant, with AG clearly significant and UR even more so, but also exhibiting nonlinearity at large values of UR. Their conclusion that FR is only marginally significant results from viewing the ratio of the parameter estimate to the standard error. But the data still exhibit multicollinearity when the Hackensack data point is excluded, as evidenced by $r_{AG, FR} = -.944$, so the t-statistics could be misleading.

We might begin our analysis by running the regression with all four possible regressors and using all 20 data points. Doing so produces the regression equation NI = 1.72 + 0.0058AG – 0.0130FR – 0.0072RS + 0.305CI, which has $R^2 = 0.709$. Although all of the t-statistics are less than two in absolute value, the numbers could be misleading if used as a measure of the importance of a variable since there is clear evidence of multicollinearity. In particular, three of the four variance inflation factors exceed 10. (We also note that the regressors have not been centered yet; this will be discussed later.)

There are four standardized residuals that exceed two in absolute value, with the Hackensack data point (point 5) having a standardized residual of –3.07. That point is also a (high) leverage point since the leverage value is 0.972, well in excess of $3p/n = 0.75$ and very close to the upper bound of 1.0 on all leverage values (for nonreplicated data points). The point also strongly influences its own fit, as $\hat{Y}_5 = 2.13$ when the point is used in the computations, and $\hat{Y}_5 = 6.85$ when the point is not used. Thus, point 5 is not only influential, but it is also very poorly predicted from the other data points.

What we might not have guessed, however, is that the Neversink river basin has a leverage value of 0.897. Close inspection of the Neversink case shows why this is true, as it has the *smallest* value for AG, the second *largest* value for FR, the fourth largest value for RS, and the second largest value for CI.

Regarding Hackensack, we should also note that this is by far the smallest of the river basins in that it is only 120 km^2, whereas 11 of the other 19 river basins have a land area well in excess of 1000 km^2. Therefore, it seems unreasonable to let the Hackensack data point be as potentially influential as the other 19 points, much less let it have more potential influence than the other points.

Although the Neversink river basin is larger than many of the other basins, its name coupled with its remoteness in the regressor space might suggest that someone was perhaps having a bit of fun, with both the name and data being fictitious. Nevertheless, we will proceed as if this were not the case.

To see the effect that each data point has on the regression coefficients, we delete the two points sequentially, starting with Hackensack. Deleting that data point has a sizable effect not only on the regression coefficients but on the summary statistics as well. Specifically, the regression equation for the remaining 19 points is NI = 1.63 + 0.002 AG − 0.013 FR + 0.181RS + 0.076 CI. The largest change occurs for the coefficients of RS and CI. This should be expected because of the extreme RS value for Hackensack and because RS and CI were highly correlated ($r = .86$) when that data point was used.

We note that R^2 has increased from a rather pedestrian .709 to a more respectable .864, and the t-statistic for RS has increased sharply to 4.08. (The other t-statistics are still less than 2 in absolute value, however.) The Neversink basin still has a large leverage value, and the model does not fit the Raquette data point very well, since the standardized residual at that point is −2.92.

Deleting the Neversink point causes R^2 to increase slightly to .882, and there are no extreme data points. The new regression equation is NI = 1.38 + 0.005AG − 0.010FR + 0.078RS + 0.675CI. Note that there is again a relatively large change in the coefficients for RS and CI, especially the latter. The large changes in the coefficient for CI as each of the two points is sequentially deleted are due in part to the instability in the relationship between RS and CI. Specifically, deleting the Hackensack data point causes $r_{RS,CI}$ to drop from .859 to .584, but it then increases to .927 when the Neversink point is additionally deleted. It is easy to see why this happens from the scatter plot of RS against CI as given in Figure 15.1.

As with the model obtained using all 20 points, none of the t-statistics is significant. This is because the multicollinearity is now much worse than it was with the original 20 points. This should not be surprising since we have removed two points that did not fit with the others.

We should also not overlook the fact that area is one of the variables in the subset that produces the smallest C_p statistic, just as it is when all 20 points are used. Therefore, although area was not of interest to Haith (1976), it should have been considered if all 20 data points were to be used without downweighting, since the subset selected by stepwise regression (FR, Area) is superior in terms of R^2 to the subset (FR, CI) that is selected by stepwise regression when Area is ignored. The R^2 values are .73 and .69, respectively. The importance of area results from the fact that the smallest basins generally have the highest nitrogen level, and the largest basins have among the lowest nitrogen levels. This might have occurred simply due to chance, but it is something that probably should have been investigated.

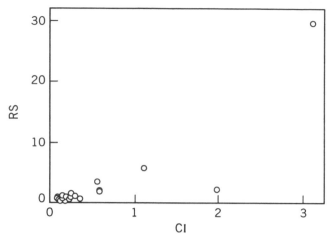

Figure 15.1   Scatter plot of RS against CI for Table 15.1 data.

Haith (1976) states

> In the present case, a suitable regression would be one which explains a signifi-
> cant portion of water quality variation with a small number of land-use variables.
> In addition, these land-use variables would be uncorrelated (independent). Such a
> regression would indicate the extent to which water quality is related to land uses
> and the relative importance of the various land uses in determining water quality.

The first objective is met with the subset (FR, CI), but the second objective is
clearly not met. That is, the land-use variables are not uncorrelated.

Therefore, our first task might be to try to determine the relative importance
of the four land-use variables, despite the high correlation between AG and
FR. There are two options that we may wish to consider: (1) ridge regression
and (2) a partitioning method given by Chevan and Sutherland (1991). We will
consider the latter first, and we will hereinafter use the data set with only the
Hackensack point deleted.

The Chevan and Sutherland method can be used in conjunction with various measures of fit, such as decomposing R^2 into additive components in the
presence of correlated regressors, and thus provide the same type of decomposition that would be obtainable directly if the regressors were orthogonal. The
authors do not prove that their method will work, however. They simply show
that their approach will decompose R^2 into individual components that will add
to R^2 for the full model, and on the basis of this they assume that they have
obtained estimates of the independent effects. With the Chevan and Sutherland
approach we first need to obtain R^2 for each possible subset model. With four
regressors there are thus 15 possible models. The necessary calculations for
each effect are shown in Table 15.2, with the calculations showing differences

Table 15.2 Calculation of Independent Effect Estimates Using Chevan and Sutherland Approach

a. AG Effect

Regressor Label Permutation	1234	1324	1423	1243	1342	1432
R^2 differences to be computed	$x_1 - x_0$	$x_1 - x_0$	$x_1 - x_0$	$x_1 - x_0$	$x_1 - x_0$	$x_1 - x_0$
	$x_{12} - x_2$	$x_{13} - x_3$	$x_{14} - x_4$	$x_{12} - x_2$	$x_{13} - x_3$	$x_{14} - x_4$
	$x_{123} - x_{23}$	$x_{132} - x_{23}$	$x_{142} - x_{42}$	$x_{124} - x_{24}$	$x_{134} - x_{34}$	$x_{143} - x_{34}$
	$x_{all} - x_{234}$	$x_{all} - x_{234}$	$x_{all} - x_{234}$	$x_{all} - x_{234}$	$x_{all} - x_{234}$	$x_{all} - x_{234}$
Values of differences	.362	.362	.362	.362	.362	.362
	.057	.292	.486	.057	.292	.486
	.001	.001	.011	.011	.278	.278
	.001	.001	.001	.001	.001	.001

Average difference = estimate of independent AG effect = .1845

b. FR Effect

Regressor Label Permutation	2134	2314	2413	2143	2341	2431
R^2 differences to be computed	$x_2 - x_0$	$x_2 - x_0$	$x_2 - x_0$	$x_2 - x_0$	$x_2 - x_0$	$x_2 - x_0$
	$x_{21} - x_1$	$x_{23} - x_3$	$x_{24} - x_4$	$x_{21} - x_1$	$x_{23} - x_3$	$x_{24} - x_4$
	$x_{213} - x_{13}$	$x_{231} - x_{13}$	$x_{241} - x_{14}$	$x_{214} - x_{14}$	$x_{234} - x_{34}$	$x_{243} - x_{34}$
	$x_{all} - x_{134}$	$x_{all} - x_{134}$	$x_{all} - x_{134}$	$x_{all} - x_{134}$	$x_{all} - x_{134}$	$x_{all} - x_{134}$
Values of differences	.508	.508	.508	.508	.508	.508
	.203	.311	.587	.203	.311	.587
	.020	.020	.112	.112	.297	.297
	.020	.020	.020	.020	.020	.020

Average difference = estimate of independent FR effect = .2695

c. RS Effect

Regressor Label Permutation	3124	3142	3214	3241	3412	3421
R^2 differences to be computed	$x_3 - x_0$	$x_3 - x_0$	$x_3 - x_0$	$x_3 - x_0$	$x_3 - x_0$	$x_3 - x_0$
	$x_{31} - x_1$	$x_{31} - x_1$	$x_{32} - x_2$	$x_{32} - x_2$	$x_{34} - x_4$	$x_{34} - x_4$
	$x_{312} - x_{12}$	$x_{314} - x_{14}$	$x_{321} - x_{21}$	$x_{324} - x_{24}$	$x_{341} - x_{41}$	$x_{342} - x_{42}$
	$x_{all} - x_{124}$	$x_{all} - x_{124}$	$x_{all} - x_{124}$	$x_{all} - x_{124}$	$x_{all} - x_{124}$	$x_{all} - x_{124}$
Values of differences	.548	.548	.548	.548	.548	.548
	.478	.478	.351	.351	.462	.462
	.295	.254	.295	.172	.254	.172
	.162	.162	.162	.162	.162	.162

Average difference = estimate of independent RS effect = .3452

Table 15.2   (*Continued*)

			d. CI Effect			
Regressor Label Permutation	4123	4132	4231	4213	4312	4321
R^2 differences to be computed	$x_4 - x_0$	$x_4 - x_0$	$x_4 - x_0$	$x_4 - x_0$	$x_4 - x_0$	$x_4 - x_0$
	$x_{41} - x_1$	$x_{41} - x_1$	$x_{42} - x_2$	$x_{42} - x_2$	$x_{43} - x_3$	$x_{43} - x_3$
	$x_{412} - x_{12}$	$x_{413} - x_{13}$	$x_{423} - x_{23}$	$x_{421} - x_{21}$	$x_{431} - x_{31}$	$x_{432} - x_{32}$
	$x_{all} - x_{123}$	$x_{all} - x_{123}$	$x_{all} - x_{123}$	$x_{all} - x_{123}$	$x_{all} - x_{123}$	$x_{all} - x_{123}$
Values of differences	.104	.104	.104	.104	.104	.104
	.228	.228	.183	.183	.018	.018
	.137	.004	.004	.137	.004	.004
	.004	.004	.004	.004	.004	.004

Average difference = estimate of independent CI effect = . 0748

of R^2 values. With 1, 2, 3, and 4 representing AG, FR, RS, and CI, respectively, we can see that when each of the variables are, in turn, listed first, there are six permutations of the other three variables. The six permutations each specify a sequence for variables being added to the model, as is illustrated by the subscripts used for the differences. Each effect is then estimated by averaging over the differences.

We observe that the "independent AG effect" is .1845, and the other estimated effects are given by FR = .2695, RS = .3452, and CI = .0748. On the basis of these numbers we would conclude that the residential land use has clearly the greatest contribution to the nitrogen level, and the effect of the commercial/industrial use is comparatively quite small.

Although it would be preferable to consider the reparameterized model used by Simpson et al. (1992), applying the Chevan and Sutherland (1991) approach to this model presents problems. Kvålseth (1985) recommends that R^2 for no-intercept models be defined as $R^2 = 1 - \sum (Y - \hat{Y})^2 / \sum (Y - \overline{Y})^2$, but this will frequently produce negative R^2 values (as was observed in Chapter 6), and that also happens with this data set. We might consider $R^2 = r^2_{Y\hat{Y}}$, but this can actually become *smaller* when a variable is added to a no-intercept model. Consequently, we are essentially forced to use the original model and thus attempt to estimate the "independent" contributions of the four land-use variables to water quality over and above the contributions of the land uses not accounted for by the model, rather than seeing what these contributions are when the other land uses are represented by the model.

Since the Chevan and Sutherland (1991) approach is somewhat controversial (see Christensen, 1992), it would be helpful if the results obtained using their approach could be corrobated using a different approach. One use of ridge regression is to make the coefficients physically meaningful, especially in regard to their sign, when multicollinearity is present. Thus, if k is wisely chosen, we would try to interpret the coefficients in the same way that we would if the regressors were orthogonal. If the regressors are centered and scaled before ridge regression is used (as is the usual case), the absolute values of the regression coefficients could be used to rank the land-use variables in terms of relative importance.

The value of the Hoerl, Kennard, and Baldwin (HKB) ridge estimator (see Section 12.2) is 0.065 for this data set, so it would be reasonable to run a ridge trace using values for k that are somewhat above and below this value. Doing so produces the ridge trace shown in Figure 15.2. Notice that the absolute values of the coefficients at the HKB value of k give the same message as was received using the Chevan and Sutherland approach. That is, RS is the primary contributor followed by FR and AG, with CI making only a small contribution.

It is important to note that this ranking is strongly dependent on the number of data points that are used. When all 20 observations are used, the ridge trace shows RS and CI essentially swapping places, and when 18 observations are used, the ridge trace suggests that the ordering should be CI, RS, AG, and FR, with the first three not differing greatly but FR making a noticeably smaller

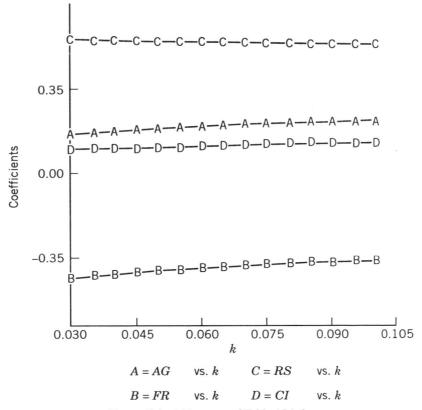

$$A = AG \quad \text{vs. } k \qquad C = RS \quad \text{vs. } k$$

$$B = FR \quad \text{vs. } k \qquad D = CI \quad \text{vs. } k$$

Figure 15.2 Ridge trace of Table 15.1 data.

contribution. (Recall the discussion of $r_{RS, CI}$ changing dramatically when the two points are sequentially deleted.) This last ranking is in agreement with the general conclusion drawn by Simpson et al. (1992). This is not surprising since they use a weight of .0662 for the Neversink river basin, in addition to virtually eliminating the Hackensack basin.

Haith (1976) also sought a model that would be useful for prediction. The desirability of using as few land uses in the model is apparent, as we would want to keep the variance of the predicted response (and the variance of the estimators) as small as possible. Thus, if we want to use the model for prediction, we should have as few land uses as possible. But when using a model for control, we would not want to exclude any land uses that account for an appreciable percentage of the total if we are trying to adjust the percentages so as to produce a desired level of nitrogen. The values of the coefficients do matter when regression is used for control, and in this instance the ridge regression coefficients might be the best choice. Thus, an experimenter might wish to use two models, one for prediction and one for control.

Haith (1976) applied stepwise regression to the full data set, and this produced the subset (FR, CI), with $R^2 = .69$. With only four potential regressors we could easily look at all possible regressions and should certainly do so. When only the Hackensack data point is deleted, both the stepwise and all subsets approaches suggest that (RS, FR) is the best subset. We would also have obtained this subset if we applied stepwise regression using the 18 data points (with $R^2 = .86$), but examination of all subsets suggests that (FR, CI) is probably the best subset, as this has $R^2 = .87$.

The selection of the model is important, because point deletion and/or point downweighting is determined by the model. Hadi (1992) also analyzes this data set, but only from the standpoint of influential observations, and shows that influence measures can disagree. Specifically, for illustrative purposes Hadi assumes that the model is (AG, CI), and for that model assumption there is considerable disagreement between influence measures. That is not the model that we are using, however, so we will not pursue that issue here.

If we use (RS, FR) as the subset, then the Hackensack data point should either be deleted or given a very small weight, since it is both extreme in the regressor space and has a standardized residual of -3.42. Deleting that point causes R^2 to increase from .64 to .86, and the coefficient of RS changes from 0.0162 to 0.195. Therefore, it seems best to use the subset (RS, FR) for our initial model and have the coefficients determined from using all but the Hackensack data point.

What about possibly transforming NI and/or one or more of the regressors? Simpson et al. (1992) did detect nonlinearity relative to the UR variable that they created, as was previously noted. We might proceed by applying the two-stage procedure that was illustrated in Chapter 6. Doing so produces results that suggest that neither NI nor RS should be transformed, but that log(FR) is a better predictor than FR. Using the log term produces $R^2 = .884$, so there is a noticeable improvement. The residual plots also do not exhibit any serious abnormalities, although the normal probability plot of the standardized residuals does show the Oswegatchie river basin to be an unusual and badly fit data point. R^2 increases from .884 to .935 when the point is deleted, but that alone is not a sufficient reason for deleting that data point. The coefficients change very little when the point is deleted, so the point is not causing any harm. We also note that the Neversink river basin, which had a large leverage value when all of the potential regressors were considered, does not show as an unusual data point for the model that is being used here. Even though multicollinearity is not a major problem, in general, when a regression model is used for prediction, here the correlation between the two regressors is only $-.202$, anyway.

It was mentioned previously that this data set has been analyzed by a few other writers. The analysis used by Haith (1976) was, additionally, critiqued by Chiang (1976) and Shapiro and Küchner (1976). The authors of each discussion pointed out that a cause-and-effect relationship cannot necessarily be inferred from a regression analysis and that residual analyses should also be performed. That point is relevant here because we are using observational data. A cause-and-effect relationship can, of course, be inferred when regression is used to analyze data from a designed experiment. Questions were also raised regarding the quality of the data. In particular, Chiang (1976) states that "another point is that magnitude and characteristics of pollutants may be altered during transport in receiving waters." Shapiro and Küchner (1976) point out that data from a land-use bank may not be kept up-to-date and mention other possible problems with the data. They also suggest that a search for possible nonlinear relationships should have been conducted. Shapiro and Küchner also pointed out that water quality data often have a nonnormal distribution. A robust regression approach would, of course, help compensate for this.

This example illustrates that both statistical and nonstatistical issues must be faced when a regression analysis is performed, and certainly these can be of equal importance. Any good data analyst would point out that Haith made a mistake in failing to identify the Hackensack data point as an anomalous and potentially influential data point, although the identification of influential points was not exactly "in vogue" in the mid-1970s. A fairer criticism would be of Haith's failure to look beyond R^2 and, in particular, not to attempt to combat the multicollinearity problem by using techniques such as ridge regression nor to check the assumptions of the model he selected.

15.2 PREDICTING LIFESPAN

Recall the discussion in Section 2.1.5 regarding the prediction of human lifespan from lifeline length in reference to the paper by Newrick et al. (1990). It was mentioned that a conclusion different from that drawn by Newrick and co-workers is reached when another data set of this type is analyzed. In a letter to the editor of the *Journal of the American Medical Association*, Wilson and Mather (1974) state that they investigated lifespan and lifeline length and found no relationship. They did not give the data that they obtained, but the data are given in Draper and Smith (1981, p. 67). That data set is given here as Table 15.3.

One point that has not been made previously is that not only can influential data points give a false signal of a significant relationship, they can also give a false signal of no relationship. We want to consider the latter possibility in analyzing the data in Table 15.3. The data in Table 15.3 are plotted in Figure 15.3. Notice the similarity between Figure 15.3 and Figure 2.15. That is, there is no apparent relationship between lifespan and lifeline length for most of the data points. But unlike Figure 2.15, there are no points in Figure 15.3 that would cause the regression line to deviate very much from being horizontal. Moreover, the only point in Figure 15.3 that is even moderately influential, the rightmost point, simply serves to make the correlation coefficient less negative than it otherwise would be. Now consider the leftmost points in Figure 2.15. Those points will pull the line down and thus cause it to deviate from being

Table 15.3   Lifespan Data

Lifeline Length	Age at Death	Lifeline Length	Age at Death
9.75	19	9.00	68
9.00	40	7.80	69
9.60	42	10.05	69
9.75	42	10.50	70
11.25	47	9.15	71
9.45	49	9.45	71
11.25	50	9.45	71
9.00	54	9.45	72
7.95	56	8.10	73
12.00	56	8.85	74
8.10	57	9.60	74
10.20	57	6.45	75
8.55	58	9.75	75
7.20	61	10.20	75
7.95	62	6.00	76
8.85	62	8.85	77
8.25	65	9.00	80
8.85	65	9.75	82
9.75	65	10.65	82
8.85	66	13.20	82
9.15	66	7.95	83
10.20	66	7.95	86
9.15	67	9.15	88
7.95	68	9.75	88
8.85	68	9.00	94

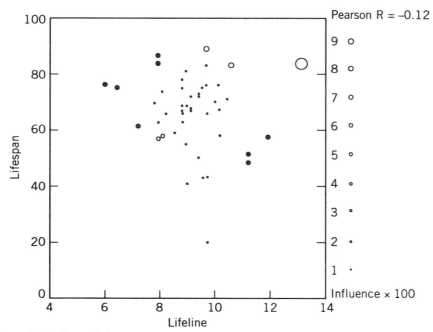

Figure 15.3 Scatter-influence plot of Table 15.3 data (open circle indicates point that increases the correlation coefficient; filled-in circle denotes point that decreases the correlation).

horizontal. Similarly, consider the near replica of Figure 2.15 in Figure 10.4. Notice that three of the four points in the lower left portion of the graph are slightly influential, and the other is moderately influential.

For the Table 15.3 data we obtain $R^2 = .015$, $t = -0.86$ (for testing $\beta_1 = 0$), and the associated p-value is .397. Thus, all signs indicate that there is no relationship. Remembering that most data sets contain errors, let's change the first data point from (19, 9.75) to (19, 15.85). The t-value is then -2.72, and the p-value is .009. Thus, there is now a "significant" relationship, in terms of these indicators, and this resulted from creating just one bad data point out of 50. (Of course, R^2 is still small, however.)

What has happened is the bad data point becomes aligned with good data (similar to the example in Section 11.5.1.2), which can be seen from Figure 15.4. Notice how influential the bad data point is, as evidenced by the very large black dot. This is quite different from Figure 2.15, where there very clearly seems to be two distinct subsets of data, with the "influential subset" having much greater scatter than the other subset.

The lesson to be learned from this is that we need to look well beyond p-values and hypothesis test results, in general. We would not be misled if we had first constructed the type of (influence) scatter plot in Figure 15.4 before looking at summary statistics. Remember the title of Section 1.3: "Graph the Data!"

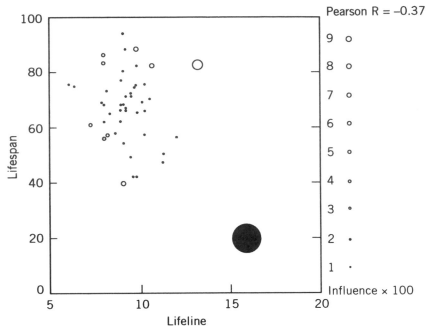

Figure 15.4 Scatter-influence plot of modified Table 15.3 data.

15.3 SCOTTISH HILL RACES DATA

These data were given by Atkinson (1986) in discussing a paper by Chatterjee and Hadi (1986). The data have also been analyzed by Staudte and Sheather (1990, p. 265) and Hadi (1992). This data set is of interest because it contains a known error, as reported by Atkinson (1988), and also because alternatives to least squares have been applied to it. The data, given in Table 15.4, are record-winning times for 35 races run in Scotland. The need to develop a regression model was apparent for the settings described in Sections 15.1 and 15.2; here we will assume that a regression model would be of value to the administrators of future races.

Applying least squares to the data set produces R^2 = .919, but there is a standardized residual of 4.57 at observation 18. We would not expect to observe such an extremely large standardized residual unless there is a problem with the data. Because of the nature of the data, we can see that this observation must be in error because the winning time for a 3-mile race will not be approximately 78 min. [Atkinson (1986) reports that 78 should have been 18, but we will not use this information in our analysis.]

Our logical first step would be to discard observation 18. After doing so, we obtain a standardized residual of 4.10 at observation 7, which might suggest that this observation is also in error. We should recognize, however, that whereas

Table 15.4 Scottish Hill Races Data

Race	Record Time (min)	Distance (mi)	Climb (ft)
Greenmantle New Year Dash	16.083	2.5	650
Carnethy 5 Hill Race	48.350	6.0	2500
Craig Dunain	33.650	6.0	900
Ben Rha	45.600	7.5	800
Ben Lomond	62.267	8.0	3070
Goatfell	73.217	8.0	2866
Bens of Jura	204.617	16.0	7500
Cairnpapple	36.367	6.0	800
Scolty	29.750	5.0	800
Traprain Law	39.750	6.0	650
Lairig Ghru	192.667	28.0	2100
Dollar	43.050	5.0	2000
Lomonds of Fife	65.000	9.5	2200
Cairn Table	44.133	6.0	500
Eildon Two	26.933	4.5	1500
Cairngorm	72.250	10.0	3000
Seven Hills of Edingburg	98.417	14.0	2200
Knock Hill	78.650	3.0	350
Black Hill	17.417	4.5	1000
Creag Beag	32.567	5.5	600
Kildoon	15.950	3.0	300
Meall Ant-Suiche	27.900	3.5	1500
Half Ben Nevis	47.650	6.0	2200
Cow Hill	17.933	2.0	900
North Berwick Law	18.683	3.0	600
Creag Dubh	26.217	4.0	2000
Burnswark	34.433	6.0	800
Largo	28.567	5.0	950
Criffel	50.500	6.5	1750
Achmony	20.950	5.0	500
Ben Nevis	85.583	10.0	4400
Knockfarrel	32.383	6.0	600
Two Breweries Fell	170.250	18.0	5200
Cockleroi	28.100	4.5	850
Moffat Chase	159.833	20.0	5000

the record time is only 12.8 min per mile, the Climb is considerably greater than in any of the other races. Moreover, the record time is 9.5 min per mile at the next highest (but considerably lower) Climb. Thus, observation 7 is not obviously in error.

Although this observation has been labeled an outlier by Hadi (1992) and Atkinson (1986), we must keep in mind that it is an outlier *relative to* the linear regression model with linear terms. It seems that the influence of Climb should be nonlinear, especially at the larger values of Climb. Therefore, if the model is to be applied to races that have a considerable climb, then we should consider either a nonlinear model or a linear model with nonlinear terms.

In the absence of any hint as to what an appropriate nonlinear model might be, we will consider the use of nonlinear terms in a linear model. The application of the two-stage transformation approach described in Chapter 6 suggests that we should use (Climb)^2 instead of Climb. Doing so produces $R^2 = .988$, which is more than halfway between the R^2 value using the linear term (.972) and an exact fit. Moreover, the standardized residual for observation 7 is now -1.84. The largest standardized residual, in absolute value, is -2.34, which occurs at the last observation, and this is the only standardized residual outside the interval $(-2, 2)$. This is about what we would expect with 34 data points and normally distributed errors. Therefore, the linear model with a linear term in Distance and a squared term in Climb appears to be more than adequate with no strong evidence that the assumptions are violated.

It is important to remember that outliers must be defined relative to a model. We should seek an appropriate model before classifying certain points as outliers and subsequently discarding or downweighting data points.

15.4   LEUKEMIA DATA

We consider a variation of a data set that was given by Feigl and Zelen (1965). The data are on 33 leukemia patients, with each observation providing information on three variables: whether the patient survived for at least 52 weeks after diagnosis (a binary response variable), whether the patient had a certain morphologic characteristic in the white blood cells (a binary *independent* variable denoted by AG), and the white blood cell count (WBC).

15.4.1   *Y* Binary

The data that were originally given by Feigl and Zelen (1965) contained the actual survival in weeks of patients who died of acute leukemia, so that the response variable was not binary. But we will first use the definition of the response variable given herein, as the data set in this form has been analyzed by Cook and Weisberg (1982), briefly discussed by Carroll and Pederson (1993), and analyzed from the standpoint of influential observations by O'Hara Hines and Carter (1993). The data, with the response variable both binary and nonbinary, are given in Table 15.5. We will then compare our results with the results that would be obtained using the actual survival times.

Since the response variable is binary, we will consider logistic regression models. Although the sample size is small, the data set is not highly unbalanced in terms of the two values of the response variable. Therefore, we will use the maximum likelihood approach for logistic regression and later comment briefly on what the results would have been if the exact logistic regression had been

Table 15.5   Leukemia Data

Y (actual)	Y (binary)	WBC	AG
65	1	2300	1
156	1	750	1
100	1	4300	1
134	1	2600	1
16	0	6000	1
108	1	10500	1
121	1	10000	1
4	0	17000	1
39	0	5400	1
143	1	7000	1
56	1	9400	1
26	0	32000	1
22	0	35000	1
5	0	52000	1
1	0	100000	1
1	0	100000	1
65	1	100000	1
56	1	4400	0
65	1	3000	0
17	0	4000	0
7	0	1500	0
16	0	9000	0
22	0	5300	0
3	0	10000	0
4	0	19000	0
2	0	27000	0
3	0	28000	0
8	0	31000	0
4	0	26000	0
3	0	21000	0
30	0	79000	0
4	0	100000	0
43	0	100000	0

used. The objective is to develop a regression model for survival using either or both of the available independent variables and perhaps to use some function of at least one of the variables.

Our first step might be to determine if there is complete or quasicomplete separation. When there is more than one regressor, as in Table 15.5, it is not necessarily easy to determine if complete or quasicomplete separation exists, and methods such as those referenced in Chapter 9 will often be needed. Nevertheless, with only two regressors we can consider the value of Y for selected pairs of regressor values in trying to determine if overlap exists. We see from Table 15.5 that overlap does exist, so we may proceed.

We might next consider a scatterplot matrix for Y and the two regressors. This is shown in Figure 15.5. The plot of Y against WBC reveals an obvious outlier on WBC. This corresponds to the patient who had survived longer than 52 weeks with a WBC of 100,000. This is a much higher WBC than for any of the other survivors, and we also see from this plot that the patient is the only one of five patients with a WBC of 100,000 who survived for at least 52 weeks. Thus, the point is outlying in both ways. Although the plot suggests that WBC may still be a significant predictor, its worth in a model will certainly be diminished if that point (17) is used in the computations.

The plot of Y against AG suggests that the latter may be a significant predictor, due to the fact that the number of zeros and ones for Y differ greatly at AG = 0. The plot of WPC against AG suggests that the two are practically unrelated, so if they are judged to be important individually they will probably be important in tandem. We will initially use point 17 in the analysis, primarily to show the effect that it has on the results.

The use of LOGMACRO (see Appendix A) or other software produces $\log[\hat{\pi}/(1 - \hat{\pi})] = -1.30734 - 0.00003\text{WBC} + 2.26107\text{AG}$, so that

$$\hat{\pi}(X) = \hat{Y} = \frac{\exp(-1.30734 - 0.00003\text{WBC} + 2.26107\text{AG})}{1 + \exp(-1.30734 - 0.00003\text{WBC} + 2.26107\text{AG})}$$

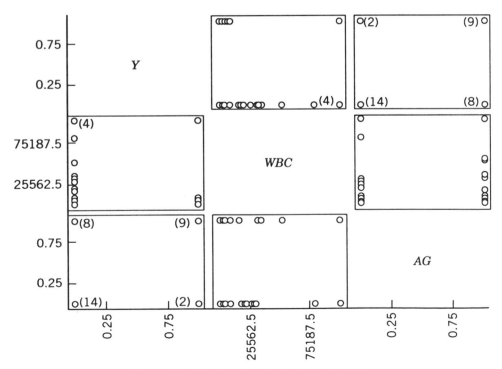

Figure 15.5   Scatterplot matrix for Table 15.5 data.

As discussed in Section 9.3, R^2 will frequently be small when the model adequately fits the data, so we will use the correct classification rate (CCR) as an indicator of the value of the model. We obtain CCR = $\frac{27}{33}$ = .82, with only 3 of 22 $y = 0$ values misclassified and 3 of 11 $y = 1$ values misclassified. Here, we are initially using a nominal cutoff of 0.5.

Although CCR = .82 would be generally acceptable, we would expect that some improvement would result from not using point 17. When point 17 is deleted, the coefficients change from -1.30734, -0.00003, and 2.26107 to 0.21192, -0.00024, and 2.55806, respectively, and the CCR changes to $\frac{28}{32}$ = .875 with two points of each type misclassified. Thus, point 17 has a considerable effect on all of the coefficients and also on the value of CCR. Furthermore, $\hat{\pi}_{17}$ is .114 when point 17 is used in the computations but less than 10^{-6} when the point is not used. The t-statistics for each of the regressor coefficients, which are actually pseudo-t (Wald) statistics, actually become smaller in absolute value, however, even though every other indicator shows that the model fits the data better without the suspicious point. (Recall the discussion in Section 9.5.1 regarding the inadvisability of relying upon the Wald statistics.)

We may observe that in this example the CCR cannot be increased by using something other than a nominal cutoff value of 0.5. We would next like to determine (1) if there are any other unusual data points, (2) if both regressors are needed in the model, and (3) if the model should be modified to include nonlinear terms.

Since we have only two potential regressors, we don't need to employ a variable selection algorithm; we can simply fit all possible subset models since there are only three of them. By performing likelihood ratio tests, as the reader is asked to do in Exercise 15.4, we may determine that both regressors are needed in the model, assuming a significance level of .05 for the tests. The results are very much affected, however, by whether or not point 17 is used.

When the point *is* used, AG is a much better single predictor than WBC, as the latter is only marginally significant ($p = .0445$). When WBC is added to the model that already contains AG, it is again only marginally significant ($p = .0338$). Here we are using the likelihood ratio p-values, *not* the Wald statistic p-values.

The numbers are much different when the point is excluded, as WBC is then a much stronger predictor than AG. (The p-value for the former when used individually is .0001.) The addition of AG to the model that contains WBC does produce a significant improvement, however. Thus, both regressors should be used in the model, in some form.

Although we were successful in identifying one outlier from a scatterplot matrix, and we saw that the outlier had a considerable effect on the analysis, we obviously cannot rely upon scatterplot matrices to show data points that would not be well fit by the model. But in searching for additional unusual data points, we face the problems discussed in Chapter 9 regarding some suggested diagnostic statistics that are somewhat flawed, and other diagnostics that are difficult to compute. There is also not a well-accepted approach for obtaining transformations in logistic regression. We might attempt to use a Box–Tidwell approach adapted to logistic regression, but as discussed by Guerrero and Johnson (1982), this will not easily detect small departures from linearity. Nevertheless, we will construct a plot of the deviance residuals defined in Eq. (9.20) in searching for unusual data points. We will use point 17 in the computations, acting as if we had not already discovered that it is an extreme data point for the sake of illustration.

The plot of the deviance residuals against $\hat{\pi}_i$ is given in Figure 15.6. There are clearly three points that stand out as not being well fit by the model: points 17, 18, and 19. We would have guessed that point 17 would appear as an extreme point in the plot, but inspection of Table 15.5 reveals why the other two points also stand out. The deviance residuals for these two points are within the interval $(-2, 2)$, however, so we might delete only point 17 and recompute the deviance residuals. Doing so causes the deviance residuals for points 18 and 19 to become considerably smaller, since the fit at those two points improves noticeably.

Since an (unsmoothed) partial residual plot is not a good plot to use for logistic regression data, as discussed in Section 9.7.5, and a smoothed partial residual plot is computationally difficult for more than one regressor, we will not use either approach in attempting to determine whether WBC should be transformed.

Feigl and Zelen (1965) mention that high WBC counts are unreliable. Since such large WBC values are of questionable predictive value, and the WBC values are also quite skewed, we might try using a log transformation on WBC, as suggested by Cook and Weisberg (1982, p. 197). Although the CCR is improved to $\frac{28}{33} = .81$ when observation 17 is included, the CCR *drops* from .875 to $\frac{26}{32}$

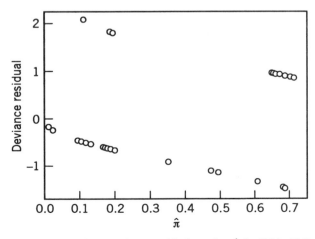

Figure 15.6 Plot of the deviance residuals against $\hat{\pi}$ for Table 15.5 data.

= .81 when that observation is excluded. Thus, a log transformation is helpful only if we use all 33 data points, and that would seem imprudent.

Since the CCR = .875 is quite high for a logistic regression model, there seems to be hardly any point in searching for a better model. In concluding our discussion of the case where Y is defined as binary, we should note that since the sample size for this data set is fairly small, using the exact logistic regression approach discussed in Section 9.2.1.2 could produce results that are somewhat different from those given here. The reader is asked to investigate this possibility in Exercise 15.4. If the exact approach had been used, there would be a memory problem with many PCs, since WBC is a continuous covariate with a wide range.

15.4.2 *Y* Continuous

Since Y was originally measured in weeks rather than being binary, we can use the data as originally given to see what is lost by dichotomizing Y. Using all of the observations we obtain $R^2 = .342$. Although this is quite low, since R^2 will typically be quite low in a logistic regression application (as was stated in Section 9.3), we should not expect R^2 to be particularly high when we use the actual values of Y. If we compare the predicted values with the observed values, after first dichotomizing both Y and $\hat{Y}$ for the purpose of determining CCR, we obtain CCR = $\frac{24}{33}$ = .73. This is less than what was obtained using Y in binary form. Therefore, it might seem as though using the actual survival times does more harm than good. This is primarily due to the fact that the relationship between Y and WBC is weaker when the actual survival times are used because of the considerable variability in the survival times for low WBC values. This can be seen by constructing the appropriate scatter plots. In particular, there appears to be no relationship between survival time and WBC for low WBC values.

It should be noted in the analysis using the actual survival times that there is some evidence of heteroscedasticity, and this issue would have to be addressed before the model could be used. We will leave it to the reader to pursue this matter.

Other published analyses of this data set include an attempt to model Y, using actual survival times, as a function of WBC when AG is positive. This would entail using the first 17 observations in Table 15.5. The model

$$Y = [\theta_1 \exp(\theta_2 X)]\epsilon \qquad (15.1)$$

with ϵ_i assumed to have a standard exponential distribution, was suggested by Fiegl and Zelen (1965) and used by Cox and Snell (1968) and Cook and Weisberg (1982, p. 180). Here $X = X' - \overline{X}'$, where X' denotes $\log_{10}(\text{WBC})$. We should note that this model is not in the general form of a nonlinear regression model, however, since the error term must be additive for the latter (and is also assumed to have a normal distribution).

Although these sources have used the model given by Eq. (15.1) directly, we can convert the nonlinear model to a simple linear regression model by taking the logarithm of each side. Doing so produces

$$Y' = \theta_1' + \theta_2 X + \epsilon' \tag{15.2}$$

where $Y' = \log(Y)$, $\theta_1' = \log(\theta_1)$, and $\epsilon' = \log(\epsilon)$. We should recognize, however, that if the normality assumption on ϵ is satisfied for the nonlinear model, then ϵ' will not have a normal distribution. Consequently, it is particularly important that the normality assumption be checked whenever we linearize a nonlinear model. Here we assume that ϵ has a standard exponential distribution, however, so $\log(\epsilon)$ may be approximately normal.

Estimating the parameters for the linear model produces $R^2 = .493$ on the log scale and $R^2 = .383$ on the original scale. There is evidence of heteroscedasticity, however, and this can be seen in the scatter plot of Y against X and in the plot of the standardized residuals against X. [We would do much better if we simply regressed Y against $\log(\text{WBC})$, as this produces $R^2 = .470$, with no evidence of either heteroscedasticity or nonnormality. We would suspect that this transformation would be helpful, since it produced an appreciable increase in the CCR when Y is in binary form.]

Remembering that observation 17 was influential when Y was in binary form, we would want to check if this observation is also influential with our linear regression model. The observation is flagged by DFFITS, and removing that observation causes R^2 to increase to .553. If we dichotomize the predicted and observed values of Y, we find that CCR $= \frac{14}{16} = .875$. Therefore, even though R^2 is not particularly high, we obtain the same value of CCR as was obtained using two regressors when Y was binary, and observation 17 was also excluded. Thus, for classification purposes the model has improved since we have the same classification rate with one less regressor.

Realistically, we should not expect to have a high R^2 value in trying to predict survival time. Certainly there are other important factors such as diet, medication, and patient attitude that should clearly affect survival, some of which cannot be easily measured. Therefore, not only should we expect to frequently encounter low R^2 values in logistic regression, but we can also encounter low R^2 values in linear regression when the model is considered to perform satisfactorily and as well as could be expected.

15.5 Dosage Response Data

Sugiyama et al. (1982) collected data to investigate hepatic dissociation and binding capacities, with the data obtained from the last author and displayed in Dixon et al. (1990, p. 401). That data set is reproduced in Table 15.6.

The objective is to see if spectrophometric response can be predicted by Rose Bengal dose. Since we have only a single regressor, we should start with

Table 15.6 Dose Response Data

Spectrophotometric Response	Concentration of Rose Bengal
12.7	.027
16.0	.044
20.4	.073
22.3	.102
26.0	.175
28.8	.257
29.6	.483
31.4	.670

an X–Y scatter plot, and that plot is given in Figure 15.7. The relationship between X and Y is clearly nonlinear, so we should consider nonlinear models. As mentioned in Section 13.2, if we have no idea as to which model(s) might be appropriate, we could try to match the curve in Figure 15.7 with one of the curves for nonlinear functions given by Ratkowsky (1990). We may also see that Figure 15.7 is very similar to Figure 2.1 in Bates and Watts (1988, p. 34). The Michaelis–Menten model was believed to be appropriate for that example, and Dixon et al. (1990, p. 401) state that data of the type given in Table 15.6 are often fit by this model.

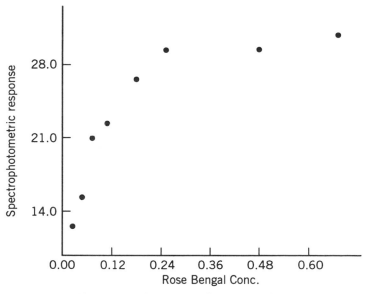

Figure 15.7 Scatter plot of Table 15.6 data.

Thus, we will use the expectation function that was given by Eq. (13.7) with the variables and parameters defined similarly to their definitions for Eq. (13.10). We will also assume that the error term is additive. Thus the model is given by

$$Y = \frac{\theta_1 X}{\theta_2 + X} + \epsilon \tag{15.3}$$

where X is the concentration of substrate (Rose Bengal in this instance), θ_1 is the maximum value of Y that is approached as $X \to \infty$, θ_2 is the value of X that produces one-half of the maximum value of Y, and Y is, of course, the response variable.

We could obviously obtain initial estimates of θ_1 and θ_2 from the sample data. That is, we could let $\hat{\theta}_1^1 = y_{max}$ and interpolate if necessary to find the value of X that corresponds to $y_{max}/2$. In their use of this data set to illustrate derivative-free nonlinear regression, Dixon et al. (1990) used a value slightly larger than y_{max} for their initial estimate of θ_1. Derivative-free methods are useful for certain nonlinear functions for which it is difficult or impossible to specify derivatives. See Seber and Wild (1989, pp. 646–649) for information on these methods. It isn't necessary to use derivative-free methods for the Michaelis–Menten model, however, since the derivatives can be easily obtained.

From Table 15.6 we can see that if we use 32 as our initial estimate of θ_2, then we won't need to interpolate to obtain our initial estimate of θ_1, since $\frac{32}{2} = 16$ is one of the y-values in the data set. Thus, we will use .044 and 32 as our initial estimates of θ_1 and θ_2, respectively. Good initial estimates should, of course, reduce the number of iterations before convergence, and in this instance convergence is achieved after three iterations with $\hat{\theta}_1 = 33.1246$ and $\hat{\theta}_2 = 0.0461$.

We also have $R^2 = .9948$, and a plot of the residuals against X does not reveal any problems, nor do the errors appear to be nonnormal. Detecting nonnormality or heteroscedasticity could be difficult with only eight data points, however. [We note that essentially the same results were given in Dixon et al. (1990, p. 402), using a derivative-free approach.]

When the model fits the data this well, there is hardly any need to consider possible model modification or to question whether the linear approximation is adequate. What about influential data points? Because of the difficulties in adapting influence statistics to nonlinear regression that were discussed in Chapter 13, and because there are only eight observations in this data set, the best approach would be simply to delete each point individually, and then obtain the parameter estimates. This would give us the $\hat{\boldsymbol{\theta}}_{(i)}$, which could be used to compute Cook's statistic as given in Eq. (13.5), or we could simply compare the parameter estimates after point deletion with those before the deletion.

Here the results of the point deletion process are quite predictable. Since θ_1 is the y-asymptote, we would expect the largest change in $\hat{\theta}_1$ to occur when the points with the largest y-values are deleted. That is what occurs here, but the change is not great, as $\hat{\theta}_1$ changes from 33.1246 to 32.8006 when point 8 is deleted and changes to 33.5403 when point 7 is deleted.

Before leaving this data set, we might wish to consider what would have happened if we had tried to linearize the model and apply linear regression. If the true model is given by Eq. (15.3), then we cannot linearize the model to obtain a linear regression model. As discussed in Section 13.9, in order to obtain the linear model given by Eq. (13.8) by taking reciprocals, the untransformed model would have to be that given by Eq. (13.9). Not only is the error term not additive, but the model is not even the Michaelis–Menten model.

Therefore, we would hope that the residuals would emit a signal that something is wrong. Unfortunately, with only eight data points the message from a residual plot may not be particularly strong, and that is what happens here. Only if we use a "connect the dots" approach do we have a signal that the variance of the error term increases as $1/X$ increases.

15.6 A STRATEGY FOR ANALYZING REGRESSION DATA

Many different tools for collecting and analyzing regression data have been presented in Chapters 1–14, and some of those techniques were used in analyzing the five data sets that were considered in this chapter. General strategies for the analysis of regression data and the selection of a model can be found in Watts (1981), Draper and Smith (1981, Chapter 8), and Ryan (1990).

It is difficult to give a simple step-by-step approach for analyzing regression data, but the following essentially summarizes the considerations that should be made.

1. Is There Prior Information to Suggest a Tentative Model?

The strategies that are used will depend considerably on whether a tentative model can be proposed based on subject-matter considerations. This can be especially important in nonlinear regression. The dialogue between a statistician or regression expert and a scientist/practitioner is especially important at this stage.

2. How Will the Data Be Obtained?

Will a designed experiment be used or will the analysis be performed on observational data?

2a. *Designed experiments.* Select a design that has desirable properties. This will generally be an orthogonal or near-orthogonal design, and it might also be an equileverage design or a modification of the latter.

2b. *Observational data.* If regression is to be applied to observational data (the usual scenario), then the limitations of the analysis should be understood (e.g., the inability to infer a cause-and-effect relationship). In addition to the type of inferences that can be drawn for observational data, the experimenter and data analyst should also recognize that the analysis of observational data can be very difficult.

3. Consider the Number of Regressions That Will Be Used

3a. *One regressor.* Frequently prior information will suggest a particular model, especially in the case of nonlinear models. In the absence of prior information, an *X–Y* scatter plot can be useful in trying to select an appropriate model. This model may be a linear model or a nonlinear model. It the scatter plot suggests a nonlinear relationship, but the data configuration does not match any of the curves that would result from any of the commonly used transformations of *X*, then a nonlinear model should be considered. Curves for various nonlinear regression models might then be compared against the data configuration. If a match is not obtained, then a nonparametric regression approach, such as using a smoother, might be employed.

3b. *Multiple regressors.* If a set of variables seems to be related to *Y*, a subset of these variables may provide a useful model. That subset may include transformations of the variables. A subset selection procedure may be used in conjunction with a transformation approach, such as the two-stage procedure described in Chapter 6, in arriving at a good subset model.

An alternative to using subset selection is to use ridge regression, or possibly use ridge regression combined with subset selection. The important point is to avoid multicollinearity, especially when the regression analysis is used primarily for parameter estimation or control.

If the true model is nonlinear, as is generally the case, determining that model or a nonlinear model that serves as a good proxy will be difficult in the case of multiple regressors and in the absence of prior information. Under such circumstances, the selected linear model will hopefully be a good proxy for the true nonlinear model.

4. Check the Model Assumptions

Frequently the model assumptions will be violated; this is especially true of the homoscedasticity assumption for nonlinear models. The assumptions of normality, homoscedasticity, and independent errors are all important, to different degrees, and should be checked whenever a regression model is formulated.

Violations of the assumptions necessitate corrective actions. A transformation of both sides of a regression model will sometimes simultaneously correct for nonnormality and heteroscedasticity. With the latter we need to see clear evidence that the heteroscedasticity is at least moderately severe before using a transformation. A transformation will not always be sufficient, and when it is not, a robust regression or nonparametric approach could be used.

5. Check for Outliers and Influential Observations

Recall the statement in Section 11.3.1 that a data set will generally contain from 1–10% gross errors. Such errors may become manifest as outliers, which might also be influential observations. We obviously want to detect outliers that result from mistakes and remove them.

As with bad data points, good data points may be influential. When there are highly influential good and/or bad data points, the use of ordinary least squares could produce poor results. The importance of diligently checking for influential observations cannot be overemphasized. Frequently, there will be data configurations, especially when the error term has a nonnormal distribution, that will mitigate against the use of least squares. In such situations the use of a robust regression technique such as LTS can produce vastly superior results.

6. Validate the Model, If Possible

The model that is formulated should perform well not only for the data that were used in determining the model, but more importantly, the model should perform well in future applications. Therefore, the model should be validated by obtaining new data, if this is possible, while being careful to avoid inadvertent (extreme) extrapolation. Even if a regression model performs well with new data, we should not necessarily expect it to be useful indefinitely, especially in the presence of changing physical conditions. Thus, regression models will often have to be modified periodically.

SUMMARY

We have analyzed five data sets; the first one in considerable detail. In analyzing the water quality data we saw the importance of analyzing a data set in detail, and we also saw the importance of supplementary nonstatistical information supplied by subject-matter specialists.

There are various far-reaching applications of regression analysis, but if, for example, regression is being used in an attempt to predict lifespan, and the results are carried by the wire services, we would certainly hope that a thorough analysis would be performed. Certainly the failure to do so could have serious ramifications.

With the Scottish hill races data we saw the importance of determining an appropriate model before determining that certain observations are outliers. Clearly we should not be surprised to find outliers in good data if we have fit a poor model.

The leukemia data showed us that we should not dichotomize a continuous random variable, and we also saw the need to eschew the use of R^2 in logistic regression (as was also seen in Chapter 9), and instead use a statistic such as the correct classification rate.

In the analysis of the dosage response data we saw how comparing a scatter plot with published curves for commonly used nonlinear models can aid in the selection of a nonlinear model. Since many nonlinear data sets are quite small, we can easily determine the possible existence of influential observations simply by deleting each data point (and also checking to make sure that there is no masking).

Combinations of regression tools will frequently have to be employed. A strategy that is frequently used in statistics is to use a parametric technique and the corresponding nonparametric technique and accept the results obtained using the nonparametric technique if the results differ considerably. This strategy could be usefully employed in regression analysis, as the results obtained from using nonparametric or robust regression should be carefully considered when they differ considerably from the results obtained using ordinary least squares.

REFERENCES

Allen, D. M. and F. B. Cady (1982). *Analyzing Experimental Data by Regression*. Belmont, CA: Lifetime Learning Publications.

Atkinson, A. C. (1986). Comment: Aspects of diagnostic regression analysis (discussion of paper by Chatterjee and Hadi). *Statistical Science*, **1**, 397–402.

Atkinson, A. C. (1988). Transformations unmasked. *Technometrics*, **30**, 311–317.

Bates, D. M. and D. G. Watts (1988). *Nonlinear Regression Analysis and its Applications*. New York: Wiley.

Carroll, R. J. and S. Pederson (1993). On robustness in the logistic regression model. *Journal of the Royal Statistical Society, Series B*, **55**, 693–706.

Chatterjee, S. and A. S. Hadi (1986). Influential observations, high-leverage points, and outliers in linear regression. *Statistical Science*, **1**, 379–416.

Chevan, A. and M. Sutherland (1991). Hierarchical partitioning. *The American Statistician*, **45**, 90–96.

Chiang, C. H. (1976). Discussion of paper by Haith. *Journal of the Environmental Engineering Division*, American Society of Civil Engineers, **102**, no. EE5, October, 1131–1132.

Christensen, R. (1992). Letter to the Editor. *The American Statistician*, **45**, 74.

Cook, R. D. and S. Weisberg (1982). *Residuals and Influence in Regression*. New York: Chapman and Hall.

Cox, D. R. and E. J. Snell (1968). A general definition of residuals. *Journal of the Royal Statistical Society, Series B*, **30**, 248–275.

Dixon, W. J., M. B. Brown, L. Engelman, and R. I. Jennrich, eds. (1990). *BMDP Statistical Software Manual*, Volume 1. Berkeley, CA: University of California Press.

Draper, N. R. and H. Smith (1981). *Applied Regression Analysis*, 2nd edition. New York: Wiley.

Feigl, P. and M. Zelen (1965). Estimation of exponential probabilities with concomitant information. *Biometrics*, **21**, 826–838.

Guerrero, V. M. and R. A. Johnson (1982). Use of the Box–Cox transformation with binary response models. *Biometrika*, **69**, 309–314.

Hadi, A. S. (1992). A new measure of overall potential influence in linear regression. *Computational Statistics and Data Analysis*, **14**, 1–27.

Haith, D. A. (1976). Land use and water quality in New York rivers. *Journal of the Environmental Engineering Division*, American Society of Civil Engineers, **102**, no. EE1, Paper 11902, February, 1–15.

Kvålseth, T. O. (1985). Cautionary note about R^2. *The American Statistician*, **39**, 279–285.

Newrick, P. G., E. Affie, and R. J. M. Corrall (1990). Relationship between longevity and lifeline: a manual study of 100 patients. *Journal of the Royal Society of Medicine*, **83**, 498–501.

O'Hara Hines, R. J. and E. M. Carter (1993). Improved added variable and partial residual plots for the detection of influential observations in generalized linear models. *Applied Statistics*, **42**, 3–20.

Ratkowsky, D. A. (1990). *Handbook of Nonlinear Regression Models*. New York: Dekker.

Ryan, T. P. (1990). Linear Regression, in *Handbook of Statistical Methods for Engineers and Scientists*, H. M. Wadsworth, ed. New York: McGraw-Hill, Chapter 13.

Seber, G. A. F. and C. F. Wild (1989). *Nonlinear Regression*. New York: Wiley.

Staudte, R. G. and S. J. Sheather (1990). *Robust Estimation and Testing*. New York: Wiley.

Shapiro, M. and J. Küchner (1976). Discussion of paper by Haith. *Journal of the Environmental Engineering Division*, American Society of Civil Engineers, **102**, no. EE6, December, 1299–1302.

Simpson, D. G., D. Ruppert, and R. J. Carroll (1992). On one-step GM estimates and stability of inferences in linear regression. *Journal of the American Statistical Association*, **87**, 439–450.

Sugiyama, Y., T. Yamada, and N. Kaplowitz (1982). Identification of hepatic Z-protein in a marine elasmobranch, Platyrhinoides triseriata. *Biochemistry Journal*, **203**, 377–381.

Watts, D. G. (1981). A task-analysis approach to designing a regression analysis course. *The American Statistician*, **35**, 77–84.

Wilson, M. E. and L. E. Mather (1974). Letter to the Editor. *Journal of the American Medical Association*, **229**, 1421–1422.

EXERCISES

15.1. **a.** Consider the data set analyzed in Section 15.1. Are there any other approaches (other than robust regression) that were presented in the preceding chapters that might be applied in analyzing the data? If so, analyze the data, using whatever method(s) seem appropriate, and compare your results with those given in Section 15.1.

b. Is the normality assumption crucial considering Haith's objectives? Would having an approximately normal error term be necessary if Haith wanted to obtain prediction intervals?

c. Consider the additional concerns (beyond nonnormality) of Shapiro and Küchner that were mentioned in the next-to-last paragraph of Section 15.1. What would you do before using any of the models that were discussed in Section 15.1? In general, what action(s) should be taken regarding data that have been used in constructing a model before the model is used?

15.2. a. Consider the lifespan data discussed in Section 15.2. Perform the regression analysis using all of the data points, and then delete those data points that have a standardized residual outside the interval $(-3, 3)$, in addition to deleting those points that have a leverage value in excess of $2p/n$. Continue doing this until there are no large standardized residuals and no large values.

What does a large leverage value mean in terms of what is being measured with this data set?

b. Consider Figure 15.2 relative to the number of points that you eventually deleted because they had large leverage values. What does this suggest about using $2p/n$ as the cutoff for leverage values?

c. Consider the "data acquisition" for these data. Would you have any concerns about the quality of these data, as Shapiro and Küchner had for the data analyzed in Section 15.1?

15.3. Consider the scenario in Section 15.3 and assume that you are the race director for a new race in Scotland.

a. As race director, why would you be interested in predicting the winning time for the race?

b. What, in particular, would you want to check before using the regression model selected in Section 15.3 for your particular race?

15.4. a. Construct the plot of the deviance residuals against $\hat{\pi}_i$ for the Table 15.5 data after first deleting point 17.

b. Construct the plot of the modified deviance residuals against $\hat{\pi}_i$ for the data in Table 15.5, using the full data set. Does the plot seem to highlight points that are not well fit by the model?

c. Consider the conclusion in Section 15.4 for the model using Y in binary form versus the conclusion when the actual values of Y were used. What would we expect to happen in general when either an independent variable or the dependent variable in a regression analysis is dichotomized?

d. If available, use the software package LogXact to obtain the exact standard errors of the estimators for the logistic regression model. Then use other available software to obtain the asymptotic standard errors, and compare the results. Is the exact approach necessary for this data set?

15.5. As was stated in Section 15.5, it is difficult to detect nonnormality or heteroscedasticity with only eight data points. Recall, however, the statement of Ruppert et al. (1989) in Section 13.9 that suggests that it is likely that we will not have homoscedastic nonlinear regression data that would be fit by the Michaelis–Menten model. What might a practitioner do to try to ensure homoscedasticity when there is a very small sample?

APPENDIX A

About the Software

The Minitab macros that were mentioned in various chapters enable the reader to apply the statistical techniques given in those chapters to data. In particular, many of these techniques are not represented in the popular statistical packages. The macros, which are a mixture of global macros and execs, are located on the accompanying diskette.

The global macros require Release 9 of Minitab or a later version. Execs will run under these versions as well as earlier versions. Macros that are global are indicated in the macro description that precedes the code, and also in the first part of the code. Other macros included are elements of the global macros or execs described here.

In addition to being of interest to Minitab users, readers familiar with programming languages such as *C* may wish to convert the Minitab code into programming languages of their choice.

The files may be installed to your hard drive by doing the following.

INSTALLING THE DISKETTE FILES

The macro files can be installed to your hard drive by using the install program on the diskette or by manually copying the files from the REGRESS directory. To use the install program:

1. Assuming you will be using drive A as the floppy drive for your diskette, at the A:> prompt type INSTALL. You may also type A:INSTALL at the C:> prompt.

2. Follow the instructions displayed by the installation program. The default choice for the installation directory is REGRESS and the default drive is C. When the files have been installed, exit from the install program and load your Minitab program.

To install the files manually:

1. Create a REGRESS directory on your hard drive.
2. Copy the files from the REGRESS directory on the diskette to the new REGRESS directory on your hard drive.

LIST OF MINITAB MACROS

Input requirements and output details for each of the 20 main macros are given at the beginning of each macro. Salient features of each macro are described below.

Chapter 2

ATKINSON and FLACFLRS were described in the Chapter 2 appendix. They produce simulation envelopes for residuals in checking for possible nonnormality. The second macro is based on a technique (due to Flack and Flores) that is essentially a robust version of the method produced by Atkinson. MATKINSO and MFLACFLR are modifications of ATKINSON and FLACFLRS, respectively, in that the former use standardized residuals rather than Studentized residuals (standardized deletion residuals).

BOXTID is a macro for producing the Box–Tidwell transformation for simple linear regression.

Chapter 6

BOXCOXA is the macro for generating the statistics that aid in the selection of the transformation parameter when the Box–Cox approach is used. The auxiliary statistics discussed in Chapter 6 and displayed in, for example, Table 6.5 are provided. BOXCOXAM is for using the two-stage Box–Cox and Box–Tidwell transformation approach with auxiliary information that was discussed in Chapter 6. These two macros can be used for simple or multiple regression.

Chapter 9

LOGISTIC is for obtaining the parameter estimates and classification results (based on a nominal cutoff of 0.5) for simple or multiple logistic regression.

Chapter 10

RLINESMO is the macro for the running line smoother when there is a single regressor. RLINSMOL is a modification of RLINESMO with an LTS adaptation. (LTS was discussed in Section 11.5.2.)

Chapter 11

LMSONEX is for performing least median of squares, using elemental point subsets, when there is a single regressor. EXACTLMS is for computing the *exact* LMS estimates in simple linear regression *with* an intercept. (The exact estimates are obtained using Chebyshev fits.) EXLMSNBO is for computing the exact LMS estimates in simple linear regression when there is no intercept.

RNDLTS1X is for approximating the exact LTS solution in simple linear regression using the RANDDIR algorithm due to David Ruppert. RNLTSMUL is the same as RNDLTS1X except that it is for multiple linear regression. LTSONEX computes the exact LTS solution for Exercise 6.7 in Chapter 6. It is not suitable for general use.

RWLS1DFB computes the first iteration of a Welsch-type bounded influence estimator for simple or multiple regression, with the weights based on *DFBETAS*. RWLS2DFB is then used for subsequent iterations. As mentioned in the Chapter 2 appendix, RWLS1DFB can also be used simply to obtain *DFBETAS*. A remark on the line in which *DFBETAS* is computed identifies its computation in RWLS1DFB.

Chapter 12

RIDGE is used for producing a ridge trace in multiple linear regression.

Chapter 13

NLINMICH is used for estimating the parameters in the Michaelis–Menten non-linear regression model.

TECHNICAL SUPPORT

If your diskette has a problem, you may obtain a new diskette by calling the Wiley technical support number at 212-850-6194.

APPENDIX B

Statistical Tables

Table B.1 Normal Distribution[a] $[P(0 \leq Z \leq z)$ where $Z \sim N(0, 1)]$

z	0.00	0.01	0.02	0.03	0.04	0.05	0.06	0.07	0.08	0.09
0.0	0.00000	0.00399	0.00798	0.01197	0.01595	0.01994	0.02392	0.02790	0.03188	0.03586
0.1	0.03983	0.04380	0.04776	0.05172	0.05567	0.05962	0.06356	0.06749	0.07142	0.07535
0.2	0.07926	0.08317	0.08706	0.09095	0.09483	0.09871	0.10257	0.10642	0.11026	0.11409
0.3	0.11791	0.12172	0.12552	0.12930	0.13307	0.13683	0.14058	0.14431	0.14803	0.15173
0.4	0.15542	0.15910	0.16276	0.16640	0.17003	0.17364	0.17724	0.18082	0.18439	0.18793
0.5	0.19146	0.19497	0.19847	0.20194	0.20540	0.20884	0.21226	0.21566	0.21904	0.22240
0.6	0.22575	0.22907	0.23237	0.23565	0.23891	0.24215	0.24537	0.24857	0.25175	0.25490
0.7	0.25804	0.26115	0.26424	0.26730	0.27035	0.27337	0.27637	0.27935	0.28230	0.28524
0.8	0.28814	0.29103	0.29389	0.29673	0.29955	0.30234	0.30511	0.30785	0.31057	0.31327
0.9	0.31594	0.31859	0.32121	0.32381	0.32639	0.32894	0.33147	0.33398	0.33646	0.33891
1.0	0.34134	0.34375	0.34614	0.34849	0.35083	0.35314	0.35543	0.35769	0.35993	0.36214
1.1	0.36433	0.36650	0.36864	0.37076	0.37286	0.37493	0.37698	0.37900	0.38100	0.38298
1.2	0.38493	0.38686	0.38877	0.39065	0.39251	0.39435	0.39617	0.39796	0.39973	0.40147
1.3	0.40320	0.40490	0.40658	0.40824	0.40988	0.41149	0.41308	0.41466	0.41621	0.41774
1.4	0.41924	0.42073	0.42220	0.42364	0.42507	0.42647	0.42785	0.42922	0.43056	0.43189
1.5	0.43319	0.43448	0.43574	0.43699	0.43822	0.43943	0.44062	0.44179	0.44295	0.44408
1.6	0.44520	0.44630	0.44738	0.44845	0.44950	0.45053	0.45154	0.45254	0.45352	0.45449
1.7	0.45543	0.45637	0.45728	0.45818	0.45907	0.45994	0.46080	0.46164	0.46246	0.46327
1.8	0.46407	0.46485	0.46562	0.46638	0.46712	0.46784	0.46856	0.46926	0.46995	0.47062
1.9	0.47128	0.47193	0.47257	0.47320	0.47381	0.47441	0.47500	0.47558	0.47615	0.47670
2.0	0.47725	0.47778	0.47831	0.47882	0.47932	0.47982	0.48030	0.48077	0.48124	0.48169
2.1	0.48214	0.48257	0.48300	0.48341	0.48382	0.48422	0.48461	0.48500	0.48537	0.48574
2.2	0.48610	0.48645	0.48679	0.48713	0.48745	0.48778	0.48809	0.48840	0.48870	0.48899
2.3	0.48928	0.48956	0.48983	0.49010	0.49036	0.49061	0.49086	0.49111	0.49134	0.49158
2.4	0.49180	0.49202	0.49224	0.49245	0.49266	0.49286	0.49305	0.49324	0.49343	0.49361
2.5	0.49379	0.49396	0.49413	0.49430	0.49446	0.49461	0.49477	0.49492	0.49506	0.49520
2.6	0.49534	0.49547	0.49560	0.49573	0.49585	0.49598	0.49609	0.49621	0.49632	0.49643
2.7	0.49653	0.49664	0.49674	0.49683	0.49693	0.49702	0.49711	0.49720	0.49728	0.49736
2.8	0.49744	0.49752	0.49760	0.49767	0.49774	0.49781	0.49788	0.49795	0.49801	0.49807
2.9	0.49813	0.49819	0.49825	0.49831	0.49836	0.49841	0.49846	0.49851	0.49857	0.49861
3.0	0.49865	0.49869	0.49874	0.49878	0.49882	0.49886	0.49889	0.49893	0.49896	0.49900
3.1	0.49903	0.49906	0.49910	0.49913	0.49916	0.49918	0.49921	0.49924	0.49926	0.49929
3.2	0.49931	0.49934	0.49936	0.49938	0.49940	0.49942	0.49944	0.49946	0.49948	0.49950
3.3	0.49952	0.49953	0.49955	0.49957	0.49958	0.49960	0.49961	0.49962	0.49964	0.49965
3.4	0.49966	0.49968	0.49969	0.49970	0.49971	0.49972	0.49973	0.49974	0.49975	0.49976
3.5	0.49977	0.49978	0.49978	0.49979	0.49980	0.49981	0.49981	0.49982	0.49983	0.49983
3.6	0.49984	0.49985	0.49985	0.49986	0.49986	0.49987	0.49987	0.49988	0.49988	0.49989
3.7	0.49989	0.49990	0.49990	0.49990	0.49991	0.49991	0.49992	0.49992	0.49992	0.49992
3.8	0.49993	0.49993	0.49993	0.49994	0.49994	0.49994	0.49994	0.49995	0.49995	0.49995
3.9	0.49995	0.49995	0.49996	0.49996	0.49996	0.49996	0.49996	0.49996	0.49997	0.49997

[a]These values were generated using MINITAB.

Table B.2 t Distribution[a]

d.f. $(\gamma)/a$	0.25	0.10	0.05	0.025	0.01	0.005	0.0025	0.001	0.0005
1	1.000	3.078	6.314	12.706	31.821	63.657	127.322	318.317	636.607
2	0.817	1.886	2.920	4.303	6.965	9.925	14.089	22.327	31.598
3	0.765	1.638	2.353	3.182	4.541	5.841	7.453	10.215	12.924
4	0.741	1.533	2.132	2.776	3.747	4.604	5.598	7.173	8.610
5	0.727	1.476	2.015	2.571	3.365	4.032	4.773	5.893	6.869
6	0.718	1.440	1.943	2.447	3.143	3.707	4.317	5.208	5.959
7	0.711	1.415	1.895	2.365	2.998	3.499	4.029	4.785	5.408
8	0.706	1.397	1.860	2.306	2.896	3.355	3.833	4.501	5.041
9	0.703	1.383	1.833	2.262	2.821	3.250	3.690	4.297	4.781
10	0.700	1.372	1.812	2.228	2.764	3.169	3.581	4.144	4.587
11	0.697	1.363	1.796	2.201	2.718	3.106	3.497	4.025	4.437
12	0.695	1.356	1.782	2.179	2.681	3.055	3.428	3.930	4.318
13	0.694	1.350	1.771	2.160	2.650	3.012	3.372	3.852	4.221
14	0.692	1.345	1.761	2.145	2.624	2.977	3.326	3.787	4.140
15	0.691	1.341	1.753	2.131	2.602	2.947	3.286	3.733	4.073
16	0.690	1.337	1.746	2.120	2.583	2.921	3.252	3.686	4.015
17	0.689	1.333	1.740	2.110	2.567	2.898	3.222	3.646	3.965
18	0.688	1.330	1.734	2.101	2.552	2.878	3.197	3.611	3.922
19	0.688	1.328	1.729	2.093	2.539	2.861	3.174	3.579	3.883
20	0.687	1.325	1.725	2.086	2.528	2.845	3.153	3.552	3.850
21	0.686	1.323	1.721	2.080	2.518	2.831	3.135	3.527	3.819
22	0.686	1.321	1.717	2.074	2.508	2.819	3.119	3.505	3.792
23	0.685	1.319	1.714	2.069	2.500	2.807	3.104	3.485	3.768
24	0.685	1.318	1.711	2.064	2.492	2.797	3.091	3.467	3.745
25	0.684	1.316	1.708	2.060	2.485	2.737	3.078	3.450	3.725
26	0.684	1.315	1.706	2.056	2.479	2.779	3.067	3.435	3.707
27	0.684	1.314	1.703	2.052	2.473	2.771	3.057	3.421	3.690
28	0.683	1.313	1.701	2.048	2.467	2.753	3.047	3.408	3.674
29	0.683	1.311	1.699	2.045	2.462	2.756	3.038	3.396	3.659
30	0.683	1.310	1.697	2.042	2.457	2.750	3.030	3.385	3.646
40	0.681	1.303	1.684	2.021	2.423	2.704	2.971	3.307	3.551
60	0.679	1.296	1.671	2.000	2.390	2.660	2.915	3.232	3.460
100	0.677	1.290	1.660	1.984	2.364	2.626	2.871	3.174	3.391
inf.	0.674	1.282	1.645	1.960	2.326	2.576	2.807	3.090	3.290

[a]These values were generated using Minitab.

503

Table B.3 F Distribution[a,b]

$.05$

$F_{\nu_1,\nu_2,.05}$

0

(a) $F_{\nu_1, \nu_2, .05}$

ν_2 \\ ν_1	1	2	3	4	5	6	7	8	9	10	11	12	13	14	15
1	161.44	199.50	215.69	224.57	230.16	233.98	236.78	238.89	240.55	241.89	242.97	243.91	244.67	245.35	245.87
2	18.51	19.00	19.16	19.25	19.30	19.33	19.35	19.37	19.39	19.40	19.40	19.41	19.41	19.42	19.43
3	10.13	9.55	9.28	9.12	9.01	8.94	8.89	8.85	8.81	8.79	8.76	8.74	8.73	8.71	8.70
4	7.71	6.94	6.59	6.39	6.26	6.16	6.09	6.04	6.00	5.96	5.94	5.91	5.89	5.87	5.86
5	6.61	5.79	5.41	5.19	5.05	4.95	4.88	4.82	4.77	4.74	4.70	4.68	4.66	4.64	4.62
6	5.99	5.14	4.76	4.53	4.39	4.28	4.21	4.15	4.10	4.06	4.03	4.00	3.98	3.96	3.94
7	5.59	4.74	4.35	4.12	3.97	3.87	3.79	3.73	3.68	3.64	3.60	3.57	3.55	3.53	3.51
8	5.32	4.46	4.07	3.84	3.69	3.58	3.50	3.44	3.39	3.35	3.31	3.28	3.26	3.24	3.22
9	5.12	4.26	3.86	3.63	3.48	3.37	3.29	3.23	3.18	3.14	3.10	3.07	3.05	3.03	3.01
10	4.96	4.10	3.71	3.48	3.33	3.22	3.14	3.07	3.02	2.98	2.94	2.91	2.89	2.86	2.85
11	4.84	3.98	3.59	3.36	3.20	3.09	3.01	2.95	2.90	2.85	2.82	2.79	2.76	2.74	2.72
12	4.75	3.89	3.49	3.26	3.11	3.00	2.91	2.85	2.80	2.75	2.72	2.69	2.66	2.64	2.62
13	4.67	3.81	3.41	3.18	3.03	2.92	2.83	2.77	2.71	2.67	2.63	2.60	2.58	2.55	2.53
14	4.60	3.74	3.34	3.11	2.96	2.85	2.76	2.70	2.65	2.60	2.57	2.53	2.51	2.48	2.46
15	4.54	3.68	3.29	3.06	2.90	2.79	2.71	2.64	2.59	2.54	2.51	2.48	2.45	2.42	2.40
16	4.49	3.63	3.24	3.01	2.85	2.74	2.66	2.59	2.54	2.49	2.46	2.42	2.40	2.37	2.35
17	4.45	3.59	3.20	2.96	2.81	2.70	2.61	2.55	2.49	2.45	2.41	2.38	2.35	2.33	2.31
18	4.41	3.55	3.16	2.93	2.77	2.66	2.58	2.51	2.46	2.41	2.37	2.34	2.31	2.29	2.27
19	4.38	3.52	3.13	2.90	2.74	2.63	2.54	2.48	2.42	2.38	2.34	2.31	2.28	2.26	2.23
20	4.35	3.49	3.10	2.87	2.71	2.60	2.51	2.45	2.39	2.35	2.31	2.28	2.25	2.22	2.20
21	4.32	3.47	3.07	2.84	2.68	2.57	2.49	2.42	2.37	2.32	2.28	2.25	2.22	2.20	2.18
22	4.30	3.44	3.05	2.82	2.66	2.55	2.46	2.40	2.34	2.30	2.26	2.23	2.20	2.17	2.15
23	4.28	3.42	3.03	2.80	2.64	2.53	2.44	2.37	2.32	2.27	2.24	2.20	2.18	2.15	2.13
24	4.26	3.40	3.01	2.78	2.62	2.51	2.42	2.36	2.30	2.25	2.22	2.18	2.15	2.13	2.11
25	4.24	3.39	2.99	2.76	2.60	2.49	2.40	2.34	2.28	2.24	2.20	2.16	2.14	2.11	2.09
26	4.23	3.37	2.98	2.74	2.59	2.47	2.39	2.32	2.27	2.22	2.18	2.15	2.12	2.09	2.07
27	4.21	3.35	2.96	2.73	2.57	2.46	2.37	2.31	2.25	2.20	2.17	2.13	2.10	2.08	2.06
28	4.20	3.34	2.95	2.71	2.56	2.45	2.36	2.29	2.24	2.19	2.15	2.12	2.09	2.06	2.04
29	4.18	3.33	2.93	2.70	2.55	2.43	2.35	2.28	2.22	2.18	2.14	2.10	2.08	2.05	2.03
30	4.17	3.32	2.92	2.69	2.53	2.42	2.33	2.27	2.21	2.16	2.13	2.09	2.06	2.04	2.01
40	4.08	3.23	2.84	2.61	2.45	2.34	2.25	2.18	2.12	2.08	2.04	2.00	1.97	1.95	1.92

(b) $F_{\nu_1,\nu_2,0.01}$

ν_2	4052.45	4999.42	5402.96	5624.03	5763.93	5858.82	5928.73	5981.06	6021.73	6055.29	6083.22	6106.00	6125.37	6142.48	6157.06
2	98.51	99.00	99.17	99.25	99.30	99.33	99.35	99.38	99.39	99.40	99.41	99.41	99.42	99.42	99.43
3	34.12	30.82	29.46	28.71	28.24	27.91	27.67	27.49	27.35	27.23	27.13	27.05	26.98	26.92	26.87
4	21.20	18.00	16.69	15.98	15.52	15.21	14.98	14.80	14.66	14.55	14.45	14.37	14.31	14.25	14.20
5	16.26	13.27	12.06	11.39	10.97	10.67	10.46	10.29	10.16	10.05	9.96	9.89	9.82	9.77	9.72
6	13.74	10.92	9.78	9.15	8.75	8.47	8.26	8.10	7.98	7.87	7.79	7.72	7.66	7.60	7.56
7	12.25	9.55	8.45	7.85	7.46	7.19	6.99	6.84	6.72	6.62	6.54	6.47	6.41	6.36	6.31
8	11.26	8.65	7.59	7.01	6.63	6.37	6.18	6.03	5.91	5.81	5.73	5.67	5.61	5.56	5.52
9	10.56	8.02	6.99	6.42	6.06	5.80	5.61	5.47	5.35	5.26	5.18	5.11	5.05	5.01	4.96
10	10.04	7.56	6.55	5.99	5.64	5.39	5.20	5.06	4.94	4.85	4.77	4.71	4.65	4.60	4.56
11	9.65	7.21	6.22	5.67	5.32	5.07	4.89	4.74	4.63	4.54	4.46	4.40	4.34	4.29	4.25
12	9.33	6.93	5.95	5.41	5.06	4.82	4.64	4.50	4.39	4.30	4.22	4.16	4.10	4.05	4.01
13	9.07	6.70	5.74	5.21	4.86	4.62	4.44	4.30	4.19	4.10	4.02	3.96	3.91	3.86	3.82
14	8.86	6.51	5.56	5.04	4.69	4.46	4.28	4.14	4.03	3.94	3.86	3.80	3.75	3.70	3.66
15	8.68	6.36	5.42	4.89	4.56	4.32	4.14	4.00	3.89	3.80	3.73	3.67	3.61	3.56	3.52
16	8.53	6.23	5.29	4.77	4.44	4.20	4.03	3.89	3.78	3.69	3.62	3.55	3.50	3.45	3.41
17	8.40	6.11	5.18	4.67	4.34	4.10	3.93	3.79	3.68	3.59	3.52	3.46	3.40	3.35	3.31
18	8.29	6.01	5.09	4.58	4.25	4.01	3.84	3.71	3.60	3.51	3.43	3.37	3.32	3.27	3.23
19	8.18	5.93	5.01	4.50	4.17	3.94	3.77	3.63	3.52	3.43	3.36	3.30	3.24	3.19	3.15
20	8.10	5.85	4.94	4.43	4.10	3.87	3.70	3.56	3.46	3.37	3.29	3.23	3.18	3.13	3.09
21	8.02	5.78	4.87	4.37	4.04	3.81	3.64	3.51	3.40	3.31	3.24	3.17	3.12	3.07	3.03
22	7.95	5.72	4.82	4.31	3.99	3.76	3.59	3.45	3.35	3.26	3.18	3.12	3.07	3.02	2.98
23	7.88	5.66	4.76	4.26	3.94	3.71	3.54	3.41	3.30	3.21	3.14	3.07	3.02	2.97	2.93
24	7.82	5.61	4.72	4.22	3.90	3.67	3.50	3.36	3.26	3.17	3.09	3.03	2.98	2.93	2.89
25	7.77	5.57	4.68	4.18	3.85	3.63	3.46	3.32	3.22	3.13	3.06	2.99	2.94	2.89	2.85
26	7.72	5.53	4.64	4.14	3.82	3.59	3.42	3.29	3.18	3.09	3.02	2.96	2.90	2.86	2.81
27	7.68	5.49	4.60	4.11	3.78	3.56	3.39	3.26	3.15	3.06	2.99	2.93	2.87	2.82	2.78
28	7.64	5.45	4.57	4.07	3.75	3.53	3.36	3.23	3.12	3.03	2.96	2.90	2.84	2.79	2.75
29	7.60	5.42	4.54	4.04	3.73	3.50	3.33	3.20	3.09	3.00	2.93	2.87	2.81	2.77	2.73
30	7.56	5.39	4.51	4.02	3.70	3.47	3.30	3.17	3.07	2.98	2.91	2.84	2.79	2.74	2.70
40	7.31	5.18	4.31	3.83	3.51	3.29	3.12	2.99	2.89	2.80	2.73	2.66	2.61	2.56	2.52

[a]These values were generated using MINITAB.
[b]ν_2 = degrees of freedom for the denominator; ν_1 = degrees of freedom for the numerator.

Answers to Selected Exercises

CHAPTER 1

1.2. $\hat{\beta}_0 = 5.37$, $\hat{\beta}_1 = 0.627$

1.7. $1 - R^2$ is the proportion of the variation in Y that is *not* explained by the regression model.

1.11. $r_{xy} = .953$

1.16. $\hat{Y} = 1.07X$

CHAPTER 2

2.2. Write the regression equation as $\hat{Y} = \overline{Y} + \hat{\beta}_1(X_0 - \overline{X})$ and then let $X_0 = \overline{X}$. Deleting a point with coordinates $(\overline{X}, \overline{Y})$ would have no effect on the parameter estimates, as follows from the discussion in Section 2.4.

2.3. Yes, the model can be transformed to a simple linear regression model by taking the log of each side of the equation.

2.6. $R^2 = .03$ *with* the three bad data points, and $R^2 = .66$ when only the good data points are used. Thus, the data should be plotted first.

CHAPTER 3

3.1. $\mathbf{x}_1'\mathbf{A}\mathbf{x}_1 = 165$. The product will always be a scalar.

3.2. The eigenvalues are 27.6015 and 0.3985.

3.8. The **X** matrix will contain a single column, which will be the regressor values. It follows that $\mathbf{X'X} = \sum X^2$ and $\mathbf{X'Y} = \sum XY$, so that $(\mathbf{X'X})^{-1}\mathbf{X'Y} = \sum XY / \sum X^2 = \hat{\beta}$, as given in Section 1.6.

CHAPTER 4

4.1. a. The regression equation is $\hat{Y} = 5.546 - 0.073\,X_1 + 1.275\,X_2$.

    c. The confidence interval is $(0.8877, 1.6629)$.

    d. No, X_1 should not be used because it makes essentially no marginal contribution when used with X_2.

4.6. a. $R^2 = .977$

    c. Yes, the t-statistics suggest that both regressors should be used in the model.

    e. The prediction interval would be of more value.

CHAPTER 5

5.3. Yes, the scatter plots are meaningful when the regressors are orthogonal.

5.5. (a) $R^2 = .52$ (b) $R^2 = .41$

5.8. CCPR plots.

CHAPTER 6

6.1. Two R^2 values computed using different scales for Y are not comparable. It is necessary to put the two sets of $\hat{Y}$ values on the same scale before the worth of the transformation of Y can be assessed.

6.6. It is necessary to check the assumptions of normality, homoscedasticity, and independence before using any statistical inference procedures. If the assumptions are not met the inferences could be badly undermined.

CHAPTER 7

7.1 The correlation coefficients might suggest what subset would be selected with a particular procedure, but we cannot rely upon the value of the correlations for selecting a subset, even when the regressors are orthogonal. In particular, the correlations tell us nothing about $\hat{\sigma}$, which is involved in all of the subset selection procedures.

7.3. 127

7.6. No, since the hypothesis tests are dependent, even having orthogonal regressors does not permit the determination of the actual significance levels.

7.10. Yes, all possible regressions should generally be performed when the number of candidate regressors is small. This does not mean, however, that this should necessarily be the only method used in selecting a subset. It would be preferable to corroborate the selection with a method, such as stepwise regression, that involves more than just the sample results.

CHAPTER 8

8.1. When the use of the Box–Tidwell transformation approach suggests that the exponent for a regressor should be between 1 and 2, the fit will frequently be virtually as good as when a linear and quadratic term are both used, and the multicollinearity that frequently results from the latter is avoided.

8.4. One or more polynomial terms should be tried.

8.6. A ninth-degree polynomial will provide an exact fit to the data. The scatter plot does not suggest that any model will provide a satisfactory fit to the data.

CHAPTER 9

9.3. $\hat{\beta}_0 = -11.5246$, $\hat{\beta}_1 = 0.5707$. CCR = .80, MCCR = .87. (The latter is obtained using any cutoff value between .23 and .34.)

9.4. The discriminant analysis approach is quite sensitive to nonnormality; logistic regression is preferred for classification.

9.6. An X–Y scatter plot.

CHAPTER 10

10.4. With linear regression there is no downweighting of points, as there is with locally weighted regression.

CHAPTER 11

11.1. For the PROGRESS approach, $\hat{\beta}_0 = 8.997$ and $\hat{\beta}_1 = -0.333$ when 9.338 is used, and $\hat{\beta}_0 = -0.34555$ and $\hat{\beta}_1 = 2.0025$ when 9.339 is used.

11.3. Several good data points could be falsely identified as outliers when .50 is used as the breakdown point. A breakdown point of .10 would usually perform better in this regard, especially when the percentage of bad data points is around 10.

11.10. A residual outlier is a point that the regression equation does not fit well when the point is used in the computations. A regression outlier is a point that does fit with the majority of the data points, and would thus be a considerable distance from the line obtained using a majority of the points.

CHAPTER 12

12.1. Ridge regression is a method for overcoming multicollinearity; the latter cannot exist when there is only a single regressor.

12.3. The extreme data point should be investigated to see if the collinearity is outlier induced. If so, the point could be downweighted or not used in the computations, with least squares perhaps being used. If the collinearity is not outlier induced, then ridge regression could be used.

12.7. Ridge regression would be preferred over least squares if we knew that it would be superior most of the time, and competitive with least squares the rest of the time.

CHAPTER 13

13.2. No, a linearized model cannot be obtained.

13.6. Influence analysis is more difficult in nonlinear regression because the computations can be much more intensive. Specifically, alternative forms of influence statistics that are easier to compute than the original forms do not exist in nonlinear regression as they do in linear regression.

13.7. It is desirable to use a criterion analogous to the minimum residual sum-of-squares criterion in linear regression. Small changes in the parameter estimates may not necessarily correspond to small changes in the relative offset convergence criterion.

CHAPTER 14

14.1. Center and scale the columns of the design matrix so that the squared norm of each column is n. If the transformed columns are orthogonal, then the design is D-optimal.

14.9. The equileverage property can be checked, for any design, by computing the leverages and seeing if they are equal.

14.10. Nonlinearity cannot be detected with an equileverage design. Doing so requires that an equileverage design be modified so that there are points inside the ellipsoidal region.

INDEX

WILEY SERIES IN PROBABILITY AND STATISTICS

ESTABLISHED BY WALTER A. SHEWHART AND SAMUEL S. WILKS
Editors
*Vic Barnett, Ralph A. Bradley, Nicholas I. Fisher, J. Stuart Hunter,
J. B. Kadane, David G. Kendall, David W. Scott, Adrian F. M. Smith,
Jozef L. Teugels, Geoffrey S. Watson*

*Now available in a lower priced paperback edition in the Wiley Classics Library.

*Now available in a lower priced paperback edition in the Wiley Classics Library.

*Now available in a lower priced paperback edition in the Wiley Classics Library.

*Now available in a lower priced paperback edition in the Wiley Classics Library.

*Now available in a lower priced paperback edition in the Wiley Classics Library.

*Now available in a lower priced paperback edition in the Wiley Classics Library.